SHIP-SHAPED OFFSHORE INSTALLATIONS

Ship-shaped offshore units are some of the more economical systems for the development of offshore oil and gas fields and are often preferred in marginal fields. These systems are especially attractive when developing oil and gas fields in deep- and ultradeep-water areas and locations remote from existing pipeline infrastructures. Recently, the ship-shaped offshore units have also been considered for application to near-shore oil and gas terminals. This book is an ideal text and reference on the technologies for designing, building, and operating ship-shaped offshore units, within inevitable space (and time) requirements. This book includes a range of topics, from the initial contracting strategy to the decommissioning and the removal of the units concerned. Coverage includes both fundamental theory and principles of the individual technologies. This book will be useful to students who are approaching the subject for the first time as well as designers working on the engineering for ship-shaped offshore installations.

Jeom Kee Paik is Professor of Ship and Offshore Structural Mechanics at Pusan National University, Korea, and is an internationally acclaimed authority on limit-state design and assessment of ships and offshore structures. Professor Paik has been chairman of the working group for development of ISO code 18072 on ships and marine technology and chairman of the International Ship and Offshore Structures Congress (ISSC) Technical Committees on Ultimate Strength, Condition Assessment of Aged Ships, and Ship Collisions and Grounding. Professor Paik is editor-in-chief of the international journal *Ships and Offshore Structures* (*SaOS*). He is a Fellow, council member, Korean branch chairman, and a publications committee member of the Royal Institution of Naval Architects (RINA), UK, and a member of the Technical and Research Steering Committee of the Society of Naval Architects and Marine Engineers (SNAME), USA. Professor Paik is the coauthor of *Ultimate Limit State Design of Steel-Plated Structures* and the author or coauthor of more than 500 publications in refereed journals, conference proceedings, research reports, and several book chapters.

Anil Kumar Thayamballi is Senior Staff Consultant and Engineering Advisor with a marine consultancy group in San Ramon, California. He is a specialist in marine structural design and life-cycle care, with 25 years of broad-ranging experience in ship-shaped structures. He has served on the American Society of Civil Engineers (ASCE) Committee for Fatigue and Fracture Reliability and on the American Petroleum Institute Resource Group RG-4 on Structural Element Behavior. He has served on the ISSC Technical Committee on Design Procedures and Philosophy and has served as its chairman. He has also served as working group chairman for the Tanker Structure Cooperative Forum and continues to be involved in the forum activities. He currently serves on the Marine Technology Committee of the SNAME in New York. Dr. Thayamballi is also a member-at-large of the Structural Stability Research Council and a member of the Royal Institution of Naval Architects. Dr. Thayamballi is the author or coauthor of more than sixty refereed technical publications and the book *Ultimate Limit State Design of Steel-Plated Structures*.

Ship-Shaped Offshore Installations

DESIGN, BUILDING, AND OPERATION

JEOM KEE PAIK

Department of Naval Architecture and Ocean Engineering
Pusan National University, Korea

ANIL KUMAR THAYAMBALLI

San Ramon, CA, USA

CAMBRIDGE UNIVERSITY PRESS
Cambridge, New York, Melbourne, Madrid, Cape Town, Singapore, São Paulo

Cambridge University Press
32 Avenue of the Americas, New York, NY 10013-2473, USA

www.cambridge.org
Information on this title: www.cambridge.org/9780521859219

First published 2007

Printed in the United States of America

A catalogue record for this publication is available from the British Library.

Library of Congress Cataloging in Publication Data

Paik, Jeom Kee.
Ship-shaped offshore installations : design, building, and operation / Jeom Kee Paik,
Anil Kumar Thayamballi.
     p.   cm.
Includes bibliographical references and index.
ISBN-13: 978-0-521-85921-9 (hardback)
ISBN-10: 0-521-85921-2 (hardback)
1. Drilling platforms.   2. Offshore structures – Design and construction.
I. Thayamballi, Anil Kumar.   II. Title.
TN871.3.P35   2006
627′.98 – dc22   2006018211

ISBN   978-0-521-85921-9 hardback

Contents

Preface   *page* xv

Acknowledgments   xix

How to Use This Book   xxi

1 Overview of Ship-Shaped Offshore Installations 1

  1.1 Historical Overview of Offshore Structure Developments   1
     1.1.1 Early History   1
     1.1.2 History from World War II to the Early 1970s   1
     1.1.3 History after the Early 1970s   2
  1.2 Process of Offshore Oil and Gas Developments   3
  1.3 System Concepts for Deep- and Ultradeep-Water Field Developments   4
     1.3.1 Semisubmersibles   5
     1.3.2 Spars   6
     1.3.3 Tension Leg Platforms   7
     1.3.4 Ship-Shaped Offshore Units   7
  1.4 A Brief History of the FPSO Installations   9
  1.5 Trading Tankers versus Ship-Shaped Offshore Units   13
  1.6 New Build versus Tanker Conversion   15
  1.7 Layout and General Arrangement of FPSOs   16
     1.7.1 Deck Area and General Arrangement   16
     1.7.2 Layout   16
     1.7.3 Relationships between Principal Dimensions   19
     1.7.4 Double-Hull Arrangements   22
     1.7.5 Tank Design and Arrangements   22
  1.8 Longitudinal Strength Characteristics of FPSO Hulls   23
  1.9 Drawings of a Hypothetical FPSO   24
  1.10 Aims and Scope of This Book   28
  References   29

2 Front-End Engineering 31

  2.1 Introduction   31
  2.2 Initial Planning and Contracting Strategies   32
  2.3 Engineering and Design   33
  2.4 Principal Aspects Driving Project and Vessel Costs   34

v

2.5   Selection of Storage, Production, and Offloading Capabilities   34
2.6   Site-Specific Metocean Data   35
2.7   Process Facility Design Parameters   36
2.8   Limit-State Design Requirements   36
2.9   Risk-Assessment Requirements   37
2.10 Project Management   38
2.11 Post-Bid Schedule and Management   39
2.12 Building Material Issues: Yield Stress   39
2.13 Building Material Issues: Fracture Toughness   41
2.14 Hull Structural Scantling Issues   42
2.15 Action-Effect Analysis Issues   43
2.16 Fatigue Design Issues   44
2.17 Hydrodynamic Impact-Pressure Action Issues: Sloshing, Slamming,
      and Green Water   45
2.18 Vessel Motion and Station-Keeping Issues   45
2.19 Topsides Design Issues   46
2.20 Mooring System Design Issues   47
2.21 Export System Design Issues   47
2.22 Corrosion Issues   48
2.23 Accommodation Design Issues   48
2.24 Construction Issues   49
2.25 Equipment Testing Issues   49
2.26 Towing Issues   50
2.27 Field Installation and Commissioning Issues   50
2.28 Inspection and Maintenance Issues   51
2.29 Regulations and Classing Issues   52
      References   53

3   **Design Principles, Criteria, and Regulations** 55
3.1   Introduction   55
3.2   Structural Design Principles   55
      3.2.1   Working Stress Design   56
      3.2.2   Limit-State Design   56
      3.2.3   Critical Buckling Strength Design   57
      3.2.4   Comparison among the Three Design Methods   60
3.3   Limit-State Criteria for Structural Design and Strength Assessment   65
3.4   Probabilistic Format versus Partial Safety Factor Format   65
      3.4.1   Probabilistic Format   65
      3.4.2   Partial Safety Factor Format   67
      3.4.3   Considerations Related to Safety Factors   68
3.5   Unified Design Requirements for Trading Tanker Hull Structures   71
3.6   Design Principles for Stability   74
3.7   Design Principles for Towing and Station-Keeping   74
3.8   Design Principles for Vessel Motions   75
3.9   Design Principles for Safety, Health, and the Environment   75
      3.9.1   Design Principles for Safety   75
      3.9.2   Design Principles for Health   77
      3.9.3   Design Principles for the Environment   77

3.10 Regulations, International Standards, and Recommended Practices   78
References   80

4 Environmental Phenomena and Application to Design 82
   4.1   Introduction   82
   4.2   Environmental Data   83
   4.3   Waves   84
         4.3.1   UKOOA FPSO Design Guidance Notes for UKCS Service   85
         4.3.2   American Petroleum Institute Recommended Practices   86
         4.3.3   Det Norske Veritas Classification Notes   86
   4.4   Winds   88
   4.5   Water Depths and Tidal Levels   91
   4.6   Currents   91
   4.7   Air and Sea Temperatures   93
   4.8   Snow and Icing   93
   4.9   Marine Growth   95
   4.10 Tank Sloshing   96
         4.10.1 Fundamentals   96
         4.10.2 Practices for Sloshing Assessment   96
         4.10.3 Measures for Sloshing Risk Mitigation   99
   4.11 Bow Slamming   99
         4.11.1 Fundamentals   99
         4.11.2 Practices for Bow-Slamming Assessment   99
         4.11.3 Measures for Bow-Slamming Risk Mitigation   100
   4.12 Green Water   100
         4.12.1 Fundamentals   100
         4.12.2 Practices for Green-Water Assessment   101
         4.12.3 Measures for Green-Water Risk Mitigation   102
   4.13 Considerations Related to the Return Period   103
   4.14 Wave Energy Spectra Expressions   105
         4.14.1 The Generalized Pierson–Moskowitz Spectrum   105
         4.14.2 The JONSWAP Spectrum   106
         4.14.3 Directional Wave Spectra   106
   4.15 Design Basis Environmental Conditions   107
   References   107

5 Serviceability Limit-State Design 111
   5.1   Introduction   111
   5.2   Design Principles and Criteria   112
   5.3   Practices for Actions and Action-Effects Analysis   113
   5.4   Elastic Deflection Limits: Under Quasistatic Actions   114
         5.4.1   Support Members   114
         5.4.2   Plates between Support Members   118
   5.5   Elastic Buckling Limits   118
         5.5.1   Elastic Plate Buckling   120
         5.5.2   Elastic Stiffener Web Buckling   123
         5.5.3   Elastic Tripping of Stiffener   125
         5.5.4   Elastic Stiffener Flange Buckling   128

5.6   Permanent Set Deflection Limits: Under Impact-Pressure Actions   128
       5.6.1   Plates between Support Members   129
       5.6.2   Longitudinally Stiffened Panels between Transverse
                Frames   131
       5.6.3   Cross-Stiffened Plate Structures   132
       5.6.4   Illustrative Examples   133
5.7   Intact Vessel Stability   134
5.8   Vessel Station-Keeping   137
5.9   Vessel Weathervaning and Heading Control   139
5.10   Vessel Motion Exceedance   140
5.11   Vibration and Noise   141
5.12   Mooring Line Vortex-Induced Resonance Oscillation   143
5.13   Corrosion Wastage   145
References   145

6   **Ultimate Limit-State Design** 148
6.1   Introduction   148
6.2   Design Principles and Criteria   148
6.3   Actions and Action-Effects Analysis   150
6.4   Structural Component Configuration   151
6.5   Ultimate Strength of Plates   153
       6.5.1   Fundamentals   153
       6.5.2   Closed-Form Expressions   154
       6.5.3   Analytical Methods   159
       6.5.4   Semianalytical Methods   164
       6.5.5   Nonlinear Finite-Element Methods   166
       6.5.6   Illustrative Examples   168
6.6   Ultimate Strength of Stiffened Plate Structures   168
       6.6.1   Fundamentals   168
       6.6.2   Closed-Form Expressions   173
       6.6.3   Analytical Methods   173
       6.6.4   Semianalytical Methods   175
       6.6.5   Nonlinear Finite-Element Methods   177
       6.6.6   Illustrative Examples   180
6.7   Ultimate Strength of Vessel Hulls   182
       6.7.1   Fundamentals   182
       6.7.2   Closed-Form Expressions   182
       6.7.3   Progressive Hull Collapse Analysis: Idealized Structural
                Unit Method   185
                6.7.3.1   Background of Idealized Structural Unit Method   185
                6.7.3.2   ISUM Structural Modeling   187
                6.7.3.3   ISUM Plate Element   188
                           (1) Nodal Forces and Nodal Displacements   189
                           (2) Strain versus Displacement Relationship 189
                           (3) Stress versus Strain Relationship   191
                           (4) Tangent Stiffness Equation   192
                           (5) Displacement (Shape) Function   193
                           (6) Failure State Considerations   194
                           (7) Post-Ultimate Strength Behavior   195
                           (8) Benchmark Study of the Plate Element   195

	6.7.3.4	ISUM Beam-Column Element	195
		(1) Nodal Forces and Nodal Displacements	197
		(2) Strain versus Displacement Relationship	197
		(3) Stress versus Strain Relationship	197
		(4) Displacement (Shape) Function	198
		(5) Tangent Stiffness Equation	198
		(6) Failure State Considerations	199
		(7) Post-Ultimate Strength Behavior	199
		(8) Benchmark Study of the Beam-Column Element	200
	6.7.3.5	Illustrative Examples	201
		(1) Test Hull Models under Vertical Bending	201
		(2) An FPSO Hull under Vertical Bending	204
		(3) A Shuttle Tanker Hull under Combined Vertical and Horizontal Bending	208
	References		215

7 Fatigue Limit-State Design 217

7.1	Introduction	217
7.2	Design Principles and Criteria	218
	7.2.1 Cyclic Stress Ranges	221
	7.2.2 S–N Curves	223
	7.2.3 Fatigue Damage Accumulation	225
7.3	Practices for Spectral-Analysis-Based FLS Design	226
7.4	Seakeeping Analysis	232
7.5	Stress Range Transfer Functions	236
7.6	Global Structural Analysis	237
7.7	Local Structural Analysis and Hot Spot Stress Calculations	238
	7.7.1 Definition of Hot Spot Stress	238
	7.7.2 Finite-Element Analysis Modeling	241
7.8	Selection of S–N Curves	243
7.9	Fatigue Damage Calculations	245
7.10	High-Cycle Fatigue versus Low-Cycle Fatigue	250
7.11	Time-Variant Fatigue Crack Propagation Models	250
	References	254

8 Accidental Limit-State Design 257

8.1	Introduction	257
8.2	Design Principles and Criteria	257
8.3	Damaged Vessel Stability: Accidental Flooding	259
8.4	Collisions	261
	8.4.1 Fundamentals	261
	8.4.2 Practices for Collision Assessment	263
	8.4.3 Nonlinear Finite-Element Modeling Techniques	265
	8.4.4 Dynamic Material Properties	271
	8.4.5 Illustrative Examples	274
8.5	Dropped Objects	277
	8.5.1 Fundamentals	277
	8.5.2 Ultimate Strength Characteristics of Dented Plates	281
	8.5.2.1 Under Axial Compressive Loads	281
	8.5.2.2 Under Edge Shear Loads	287

8.5.3   Closed-Form Expressions for Ultimate Strength of Dented
Plates   293
    8.5.3.1   Under Axial Compressive Loads   293
    8.5.3.2   Under Edge Shear Loads   296
8.6 Fire   296
   8.6.1   Fundamentals   296
   8.6.2   Practices for Fire Assessment   297
8.7 Gas Explosion   299
   8.7.1   Fundamentals   299
   8.7.2   Practices for Gas Explosion Action Analysis   301
    8.7.2.1   Prescriptive Methods   301
    8.7.2.2   Probabilistic Methods   302
   8.7.3   Practices for Gas Explosion Consequence Analysis   303
   8.7.4   Illustrative Examples   303
8.8 Progressive Collapse of Heeled Hulls with Accidental Flooding   308
8.9 Considerations for ALS Applications to Ship-Shaped Offshore Units   313
References   313

9 Topsides, Mooring, and Export Facilities Design 318

9.1 Introduction   318
9.2 Topsides Facilities   319
   9.2.1   Oil and Water Separation Facilities   319
   9.2.2   Gas Compression Facilities   323
   9.2.3   Water Injection Facilities   324
   9.2.4   Cargo Handling Systems   324
   9.2.5   Utility and Support Systems   325
   9.2.6   Safeguard Systems   326
9.3 Structural Design and Fabrication Considerations for Topsides and
Their Interfaces with the Hull   327
   9.3.1   Types of Topsides Supports   327
    9.3.1.1   Multipoint Support Columns   328
    9.3.1.2   Sliding/Flexible Support Stools   329
    9.3.1.3   Transverse Girder Supports   330
   9.3.2   Types of Topsides Flooring   330
   9.3.3   Types of Topsides Fabrication   331
    9.3.3.1   Built-In Grillage Deck   332
    9.3.3.2   Preassembled Units   332
   9.3.4   Structural Analysis of Topsides Modules and Interfaces   332
   9.3.5   Interface Management and Other Lessons Learned   333
9.4 Mooring Facilities   336
   9.4.1   Types of Moorings   336
    9.4.1.1   Spread Moorings   337
    9.4.1.2   Single-Point Moorings   338
      (1) Fixed Tower   338
      (2) CALM   338
      (3) SALM   340
      (4) ALP   340
      (5) SPAR   342
      (6) SAL   342
      (7) Turret Mooring   342

		9.4.2	Mooring System Selection for an FPSO in Deep Water	348
		9.4.3	Design Considerations for Mooring Systems	349
	9.5	Export Facilities		350
		9.5.1	Methods of Export	350
		9.5.2	Types of Shuttle Tanker Export	351
		9.5.3	Design Considerations for Export Systems	352
	References			354

10 Corrosion Assessment and Management **356**

10.1 Introduction   356
10.2 Marine Corrosion Mechanisms   357
    10.2.1 Fundamentals   357
    10.2.2 Types of Corrosion   358
        10.2.2.1 General Corrosion   359
        10.2.2.2 Pitting Corrosion   360
        10.2.2.3 Grooving   361
        10.2.2.4 Weld Metal Corrosion   361
    10.2.3 Factors Affecting Corrosion   361
10.3 Mathematical Models for Corrosion Wastage Prediction   364
    10.3.1 Overall Behavior of Corrosion   365
    10.3.2 Mechanical Models   366
        10.3.2.1 Corrosion Depth Formulations   366
        10.3.2.2 Data Collection of Corrosion Measurements   367
        10.3.2.3 Characteristics of Observed Corrosion
                 Wastage   371
        10.3.2.4 Annualized Corrosion Rates   374
    10.3.3 Phenomenological Models   379
10.4 Options for Corrosion Management   382
    10.4.1 Corrosion Margin Addition   383
    10.4.2 Coating   386
        10.4.2.1 Surface Preparation   386
        10.4.2.2 Types of Coating   387
        10.4.2.3 Selection Criteria of Coating Material   389
        10.4.2.4 Methodologies for Coating-Life Prediction   390
    10.4.3 Cathodic Protection   391
    10.4.4 Ballast Water Deoxygenation   393
    10.4.5 Chemical Inhibitors   395
References   395

11 Inspection and Maintenance **400**

11.1 Introduction   400
11.2 Types of Age-Related Deterioration   401
11.3 Methods for Damage Examination   402
    11.3.1 Corrosion Wastage Examination   402
    11.3.2 Fatigue and Other Crack Examination   404
    11.3.3 Mechanical Damage Examination   405
    11.3.4 Probability of Detection and Sizing   405
11.4 Recommended Practices for Trading Tankers   406
    11.4.1 Condition Assessment Scheme   408
    11.4.2 Enhanced Survey Programme   409

	11.4.3	Emergency Response Services	411
	11.4.4	Ship Inspection Report Programme	411
11.5	Risk-Based Inspection	411
	11.5.1	RBI Team Setup	412
	11.5.2	Component Grouping and Baselining	413
	11.5.3	Risk-Based Prioritization	413
	11.5.4	Inspection Plan Development	414
		11.5.4.1 Inspection Strategy	414
		11.5.4.2 Scope of Inspection	414
		11.5.4.3 Frequency of Inspection	415
	11.5.5	Inspection Execution	415
	11.5.6	Analysis of Inspection Results	415
	11.5.7	RBI Program Updating	416
11.6	Risk-Based Maintenance	416
	11.6.1	Time-Variant Failure Mechanisms	417
	11.6.2	Planned Maintenance	419
	11.6.3	Condition Monitoring	421
	11.6.4	Combination of Planned Maintenance and Condition Monitoring	421
	11.6.5	Failure Finding	421
11.7	Recommended Practices for Ship-Shaped Offshore Units	423
	11.7.1	Inspection Practices	423
	11.7.2	Maintenance Practices	425
11.8	Effect of Corrosion Wastage on Plate Ultimate Strength	428
11.9	Effect of Fatigue Cracking on Plate Ultimate Strength	431
11.10	Effect of Time-Variant Age-Related Deterioration on FPSO Hull Ultimate Strength Reliability: An Academic Example	433
	11.10.1 Scenario for Sea States and Operational Conditions	434
	11.10.2 Scenario for Time-Variant Corrosion Wastage	436
	11.10.3 Scenario for Time-Variant Fatigue Cracking	437
	11.10.4 Time-Variant Ultimate Hull Strength Reliability Assessment	438
	11.10.5 Considerations for Repair Strategies	439
	References	444

12 Tanker Conversion and Decommissioning 447

12.1	Introduction	447
12.2	Tanker Conversion	448
	12.2.1 Selection of Suitable Tankers	449
	12.2.2 Condition Assessment of Aged Tanker Hull Structures	450
	12.2.2.1 Inspection and Maintenance	450
	12.2.2.2 Renewal Scantlings for Tanker Conversion	452
	12.2.2.3 Repair of Defects, Dents, Pitting, Grooving, and Cracks	453
	12.2.2.4 Residual Strength Assessment	453
	12.2.3 Reusability of Existing Machinery and Equipment	453
	12.2.4 Addition of New Components	454
	12.2.5 Appraisals of Conversion Yard	456

12.3   Decommissioning   456
          12.3.1   Regulatory Framework   457
          12.3.2   Technical Feasibility Issues   458
          12.3.3   Safety and Health Issues   459
          12.3.4   Environmental Issues   459
          12.3.5   Cost Issues   460
          12.3.6   Decommissioning Practices for Ship-Shaped Offshore
                   Installations   460
       References   460

13   Risk Assessment and Management 463
       13.1   Introduction   463
       13.2   Process for Formal Safety Assessment   464
          13.2.1   System Definition   465
          13.2.2   Hazard Identification   465
          13.2.3   Risk Assessment   469
          13.2.4   Risk-Management Options   470
          13.2.5   Cost–Benefit Analysis   470
          13.2.6   Decision-Making Recommendations   471
       13.3   Qualitative Risk Assessment   472
       13.4   Quantitative Risk Assessment   475
          13.4.1   Frequency Analysis   476
          13.4.2   Consequence Analysis   479
          13.4.3   Risk Representation   480
       13.5   Risk Management during Design   482
          13.5.1   Selection of Materials   483
          13.5.2   Layout for Hazard Impact Minimization   483
          13.5.3   Limit-State Design   483
          13.5.4   Passive Safeguards for Fire and Explosion   484
          13.5.5   Accelerated Degradation Protection   484
       13.6   Risk Management during Operation   484
          13.6.1   Collisions   484
          13.6.2   Dropped Objects   485
          13.6.3   Active Safeguards for Fire and Explosion   485
          13.6.4   Inspection and Maintenance   485
       References   486

Appendix 1. Terms and Definitions 489

Appendix 2. Scale Definitions of Winds, Waves, and Swells 503
A2.1 Beaufort Wind Scale   503
A2.2 Wave Scale   503
A2.3 Swell Scale   504

Appendix 3. Probability of Sea States at Various Ocean Regions 505
A3.1 Identification of Ocean Areas Using Marsden Squares   505
A3.2 Probability of Sea States in the North Atlantic   506
A3.3 Annual Sea States in the North Atlantic   507
A3.4 Annual Sea States in the North Pacific   508

A3.5 Characteristics of 100-Year Return Period Storms at Various Ocean
       Regions   509
A3.6 Extremes of Environmental Phenomena at Various Ocean Regions   510

Appendix 4. Scaling Laws for Physical Model Testing 511

A4.1 Hydrodynamics Model Tests   511
       A4.1.1 Froude Scaling Law   511
       A4.1.2 Reynolds Scaling Law   511
       A4.1.3 Vortex-Shedding Effects   512
       A4.1.4 Surface Tension Effects   512
       A4.1.5 Compressibility Effects   513
A4.2 Structural Mechanics Model Tests   513

Appendix 5. Wind-Tunnel Test Requirements 514

Appendix 6. List of Selected Industry Standards 515

Index   531

Preface

Today, the need for development of offshore oil and gas resources in increasingly deeper waters is becoming more important because of many reasons associated with the world economy and the related energy resource development constraints and strategies.

Fixed-type offshore platforms, which have been useful for oil and gas developments in relatively shallow waters, are now much less feasible as we move further in developing oil and gas fields in deep- and ultradeep-water areas, now reaching more than 1,000m water depth. Floating-type offshore structures have to be increasingly considered to develop these deep-water areas. In addition to ship-shaped offshore units, at least three other types of floating production systems – semisubmersibles, spars, and tension leg platforms (TLP) – are also available today for that purpose. All of these types of floating systems require storage, pipeline infrastructure, and other associated field structures and systems to transport produced oil and gas to the facilities on shore, but perhaps to varying degrees.

That the use of ship-shaped offshore units remains a very attractive alternative in many cases of field development is attributable to its ability to successfully serve multiple functions, such as production, storage, and offloading, and the capability for oil or gas to be transported to shore via shuttle tankers. Ship-shaped offshore units reduce need for pipeline infrastructure and are functional on a fast-track basis.

Ship-shaped offshore units are now recognized as perhaps one of the most economical of all systems for potential developments of offshore oil and gas and are often the preferred choice in marginal fields. These systems are becoming more attractive for developing oil and gas fields in deep- and ultradeep-water areas and locations remote from the existing pipeline infrastructures. Recently, the ship-shaped offshore units have also begun to be applied to near-shore oil and gas terminals.

Although the use of ship-shaped offshore units has been in existence since the late 1970s, the complexity and size of the units have been gradually increasing, and there are still many issues related to design, building, and operation to be resolved for achieving high integrity in terms of safety, health, environment, and economics/financial expenditures.

Although ship-shaped offshore units are similar to trading tankers in structural geometry, they are different in a variety of ways. Environmental conditions are unique in each case, and structural design concepts must be tailored to a specific location. Trading tankers may avoid rough weather or alter their heading in operation, but ship-shaped offshore units must be continuously located in the same area with site-specific environments and do not have the ability to periodically dry-dock for the necessary inspection and maintenance. This is an aspect that must be reflected in some fashion in the design and long-term durability and reliability of the units concerned.

To continue further on the subject of differences from trading tankers, one should note that ship-shaped offshore units are likely to be subjected to significant environmental actions even during loading and unloading; however, trading tankers are typically loaded and unloaded at still-water condition in harbor. And, for historical reasons, the design return period of ship-shaped offshore units is typically taken as 100 years, and that of trading tankers is considered to be 20 to 25 years or so.

The application of existing procedures, criteria, and standards to the structural design of ship-shaped offshore units also requires additional thought and discussion. This can be particularly important for the many interface areas between the hull and topsides. Even for the hull part, the shipbuilding industry standards may need to be selectively upgraded to ensure the long life and onsite reliability needed. Similarly, for the topsides part, it is often not straightforward to apply the relatively more economical shipbuilding industry standards, in part perhaps because of differences in the background, experience, and culture of the operating personnel involved. In any event, the complexities of the design are enormous, and there are many interface issues (e.g., those related to the interaction between hull and topsides facilities and related consistency in design information) that need to be identified up front and addressed and managed on a continuous basis.

In such a situation, direct analyses from first principles, advanced engineering, and practices are increasingly desired so that practicing engineers and academic researchers can resolve the issues that remain, reconcile differences in standards and practices, and improve structural and other design procedures and criteria. In the never-ending quest for safe, reliable, yet economical structures and systems effectively designed and constructed, there are often demanding schedules and other constraints and challenges.

Also, many diverse international organizations in the maritime industry such as the International Maritime Organization (IMO), International Organization for Standardization (ISO), International Association of Classification Societies (IACS), and the industry in general are now increasingly applying the limit-state design approach for both trading ships and ship-shaped offshore installations, making related knowledge and training even more relevant. Another emerging and increasingly more important technology consists of risk-based approaches to design, operation, and human and environmental safety, with much of the same accompanying knowledge, training, and familiarization needs.

The intention behind writing this book is to develop a textbook and handy resource that sufficiently addresses current practices, recent advances, and emerging trends on core technologies for designing, building, and operating ship-shaped offshore units, within certain inevitable space (and time) requirements. This book covers a wide range of topics, from the initial contracting strategy to the decommissioning and even the removal of the units concerned, but not always to a depth some might have wished for. Although a large number of research papers and references and industry standards useful for specific topics in the areas do exist, we did our best to highlight selected and useful ones among them in the various chapters and appendices.

We have also tried our utmost to always refer to relevant past work, with proper acknowledgments. It is respectfully requested that any unintentional oversights in this regard be brought to our attention for correction in future editions.

We believe and hope that this book will be very useful for practicing engineers and engineers-in-training and will contribute to their increased awareness and potentially greater use of advanced and sophisticated technologies as well as existing and emerging practices. Because of its coverage of the fundamentals and principles of the individual technologies, this book will also be useful for university students who are approaching both the initial and more intensive studies of advanced engineering for ship-shaped offshore installations. With regard to the scope, emphasis, and other relevant aspects of this book, we encourage all related and pertinent feedback and suggestions for the future; these will be gratefully received.

Professor Jeom Kee Paik, Pusan National University, Korea
Dr. Anil Kumar Thayamballi, San Ramon, CA, USA

Acknowledgments

We are very pleased to acknowledge all of those individuals and organizations who helped make this book possible.

Professor N. D. P. Barltrop (Universities of Glasgow and Strathclyde, UK), editor of the excellent book in this area titled *Floating Structures: A Guide for Design and Analysis* (1998), provided invaluable suggestions to improve the initial book manuscript. Invaluable suggestions for improvements were also received from Dr. I. Lotsberg (Det Norske Veritas, Norway) for Chapter 7, "Fatigue Limit-State Design," and from Professor R. E. Melchers (University of Newcastle, Australia) for Chapter 10, "Corrosion Assessment and Management." We also wish to thank Mr. G. J. Adhia, Mr. M. R. Buetzow, and Dr. M. C. Ximenes (Chevron Shipping Company, USA) for their very useful advice, critique, and comments related in particular to Chapter 2, "Front-End Engineering," and several other chapters; and also Dr. A. Newport (Single Buoy Mooring Inc., Monaco) for his valuable comments related to mooring system designs.

Numerous individuals and organizations provided various materials in the book. Samsung Heavy Industries, Korea, provided various computer graphics and pictures of topsides modules and of FPSOs. SBM Offshore NV, The Netherlands, (via Mr. H. Peereboom) provided various illustrations and photos of mooring systems and FPSOs. These are quite helpful to better understand the layout and arrangements of FPSOs, topsides modules, and mooring systems. We also acknowledge Dr. G. Wang (American Bureau of Shipping, USA) for structural characteristics data of trading tankers and FPSOs; Dr. Z. Czujko (Nowatec AS, Norway) for numerical simulations of gas explosion actions and their consequences; Professor C. M. Rizzo (University of Genoa, Italy) for the methods of damage detection and examination; Dr. P. A. Frieze (PAFA Consulting, UK) for several useful improvements and suggestions; the UK Health and Safety Executive for illustrations regarding ship-shaped offshore design; the Society of Petroleum Engineers for illustrations regarding topsides facility design; and the American Society of Mechanical Engineers for pictures regarding FPSO construction.

Our heartfelt thanks are also due to our senior editor, Mr. Peter Gordon at Cambridge University Press, and our project manager, Ms. Katie Greczylo at Techbooks. In many ways, this book would not have been possible without their extraordinary efforts and assistance.

We also acknowledge the extensive efforts of graduate students at the Ship and Offshore Structural Mechanics Laboratory, Department of Naval Architecture and

Ocean Engineering, Pusan National University, which is a National Research Laboratory funded by the Korea Ministry of Science and Technology. In particular, special thanks are given to J. K. Seo, who developed many of the illustrations in this book.

Any opinions expressed in this book are strictly those of the authors and not those of the organizations with which the authors are associated or of the individuals and organizations who provided us invaluable assistance during this effort.

Finally, we take this opportunity to thank our wives and families for their unfailing patience and support during the writing of this second book we coauthored: Yun Hee, Myung Hoon, and Yun Jung and Nita and Neil. To them we hereby dedicate this book.

Jeom Kee Paik and Anil Kumar Thayamballi

How to Use This Book

Our intention behind writing this book is to develop a textbook and handy resource that contains current practices, recent advances, and future trends on core technologies essential for ship-shaped offshore installations. We feel that such a book, with an appropriate mixture of academic rigor and practical experience, will be a welcome addition to many bookshelves, including those of university students in key shipbuilding countries worldwide and of interested practitioners. Therefore, in this book we have attempted to cover, within a limited space, a number of pertinent topics ranging from the initial contracting strategy to the decommissioning and removal of the floating units concerned.

Chapter 1 presents an overview of ship-shaped offshore installations, including structural characteristics with general arrangement and midship section drawings of a hypothetical FPSO. Historical overview and selection strategy of various floating offshore systems (e.g., semisubmersibles, spars, tension leg platforms, and ship-shaped offshore units) to develop oil and gas offshore are also discussed.

Chapter 2 addresses the front-end engineering of ship-shaped offshore installations, including the identification and discussion of various important issues that must be resolved successfully during the design and building of such offshore units.

Chapter 3 describes principles and criteria for designing and building ship-shaped offshore units, with the emphasis on limit-state design. Some considerations for safety factor determination are given. This chapter refers to existing classification society rules, recommended practices, regulations, and international standards that will be used for designing and building ship-shaped offshore units in terms of safety, health, the environment, and economics/financial expenditures.

Chapter 4 addresses environmental phenomena and application to design, covering many types of potential environmental actions such as wind, waves, current, swell, ice, snow, temperature, and marine growth. Emerging practices for predicting impact actions arising from tank sloshing, bow slamming, and green water are presented. Some considerations for the design return period of the offshore units are addressed.

Chapter 5 presents current practices and recent advances useful for serviceability limit-state design of ship-shaped offshore units. This chapter describes the fundamentals, calculation methods, and design criteria for elastic deflection limits under quasistatic actions, elastic buckling limits, permanent set deflection limits under impact-pressure actions (arising from tank sloshing, bow slamming, and green water), intact vessel stability, watertight integrity, weathervaning (heading control), station-keeping, vessel motion exceedance, vibration and noise, mooring line vortex-induced resonance oscillations, and localized corrosion wastage.

Chapter 6 presents emerging practices and recent advances useful for ultimate limit-state design of ship-shaped offshore units. This chapter describes the fundamentals, calculation methods, and design criteria for determining the ultimate strength of plates, stiffened plate structures, entire vessel hulls, and structural systems. Closed-form expressions and progressive collapse analysis methods are presented. Illustrative examples for the ultimate strength calculations of structural components and vessel hulls are shown.

Chapter 7 presents current practices and recent advances useful for fatigue limit-state design of ship-shaped offshore units, with emphasis on the spectral-analysis-based approach. This chapter describes the fundamentals, calculation methods, and practices for fatigue limit-state design. The methods for determining hot spot stresses with finite-element modeling techniques are presented. The selection of relevant S–N curves and the calculations of fatigue damage accumulation are described. The time-variant crack propagation models that are needed for time-variant reliability assessment of aged structures with fatigue cracking are described together with illustrative examples of the calculations to be made.

Chapter 8 addresses emerging practices and recent advances useful for accidental limit-state design. This chapter describes the fundamentals, calculation methods, and practices for determining accidental actions and the consequences of damaged vessel stability due to collision, dropped objects, fire, gas explosion, progressive hull collapse due to structural damage, and accidental flooding. Closed-form expressions and numerical simulation methods are presented. Illustrative examples for analyzing the consequences of the accidental events are shown.

Chapter 9 presents an overview of the considerations and practices for designing and building topsides, cargo export, and mooring facilities. Several illustrations of FPSO systems and the structural response analyses of the interaction between vessel hull and topsides modules are presented. The importance of various interface-management issues is emphasized.

Chapter 10 presents corrosion assessment and management for ship-shaped offshore structures. Starting with pertinent corrosion mechanisms, useful mechanical and phenomenological models for predicting corrosion wastage are presented. Corrective or protective design and operational measures, such as corrosion margin addition, coating, cathodic protection, ballast water deoxygenation, and inhibitors, are described. The effect of corrosion wastage on the ultimate limit state of structural components and vessel hulls is addressed with illustrative examples of the calculations. Methods for predicting the coating durability are also presented in this chapter.

Chapter 11 presents current practices and recent advances for inspection and maintenance of ship-shaped offshore structures. Emerging practices for condition assessment of trading tankers, which may be useful for offshore units, are reviewed. Risk-based inspection and maintenance procedures are presented. The effects of age-related deterioration, such as corrosion and fatigue cracking, on the time-variant ultimate strength reliability of ship-shaped offshore units are addressed. Some considerations for repair strategies based on quantitative reliability and risk-based methodologies are provided.

Chapter 12 presents current practices for conversion and decommissioning. Although this book is focused on the core technologies for designing and building new-build units, the conversion strategies are also important because a large number of ship-shaped offshore units worldwide are tanker conversions. Today, the world community requires all of us to pay appropriate attention to the decommissioning and disposal of the used offshore units by meeting strict international and regional regulations and standards and also by proactive planning and anticipatory design. This chapter provides an overview of the current practices and the important issues related to decommissioning as well.

Chapter 13 presents emerging practices and recent advances for risk assessment and management. It is highly desirable today to apply risk-assessment methods to designing, building, and operating various types of structural systems, including ship-shaped offshore units. This chapter describes the fundamentals and salient details of selected risk-assessment methods, together with extensive references. Specific areas of the application of risk-assessment methods to the design and operation of ship-shaped offshore units are noted and discussed.

The appendices provide useful data necessary for design of ship-shaped offshore units. Important terminologies used in the book are defined. Scale definitions of wind, wave, and swell are presented. Representative data of sea states at various ocean regions, an important part of wave action predictions of ship-shaped offshore units as well as trading tankers, are provided. Selected data on annual sea-state occurrences in the North Atlantic and North Pacific are presented. Illustrative characteristics of 100-year return period storms and of extremes of environmental phenomena in various regions are provided. Scaling laws for both hydrodynamics and structural mechanics model testing are given. Wind-tunnel testing requirements are addressed. Selected industry standards, guidelines, and recommended practices useful for ship-shaped offshore installations are listed.

The methods and practices presented in this book cover all core technologies that are essential to better understand designing, building, and operating ship-shaped offshore installations in some fashion. We certainly hope that this book, with its advanced methodologies as well as emerging practices together with the list of carefully selected references, is seen and received as a handy resource and also that it meets the needs of practicing engineers and engineers-in-training to a good degree. This book should also be well suited as a textbook for university students in the fields of naval architecture and offshore civil, architectural, and mechanical engineering.

When this book is used as a textbook for undergraduate university students during a 45-hour single-semester class, the fundamentals and some current practices in all chapters should be studied. For postgraduate students, who may be approaching the topics in depth, the detailed methodologies presented in some selected chapters should be studied. For instance, those who are more likely to be interested in structural mechanics and limit-state design may begin with Chapter 1, "Overview of Ship-Shaped Offshore Installations" and focus on Chapters 5 through 8. Of course, it will also be a good idea for graduate-course students in a higher level to concentrate on and further explore any specific chapter, for example, Chapter 6, "Ultimate

Limit-State Design." Related theoretical and numerical calculations using the closed-form expressions and/or computer programs, where available, can be used.

We hope that future revisions of this book will be made more useful and even more attractive to a wide spectrum of its readers and users; therefore, pertinent feedback and suggestions are encouraged, both by the publisher and the authors, and will be fully considered for future editions.

CHAPTER 1

Overview of Ship-Shaped Offshore Installations

1.1 Historical Overview of Offshore Structure Developments

1.1.1 Early History

One of the primary necessities in the progress of civilization has been energy. Industrial advances were first stoked by coal and then by oil and gas. Today, oil and gas are essential commodities in world trade. Exploration that initially started ashore has now moved well into offshore areas, initially in shallower waters and now into deeper waters because of the increasingly reduced possibilities of new fields in shallower waters.

The quest for offshore oil began, perhaps in California, in the late 1800s and early 1900s (Graff 1981). In the beginning, the techniques and facilities used for production of oil on land were applied to an offshore field by extending the field out over the water by jetty to distances of up to 150m off the coast. By the early 1930s, oil drilling was being undertaken by derrick systems located in waters more than a mile (1.6km) offshore, although the water depth at the drill sites was still limited to less than 5m. These derrick systems were constructed using timber. Barges transported supplies and produced oil, canals were dredged, and boats pulled the barges.

As the well sites moved farther away from shore and the water depths increased, it soon became evident that there were many challenges to overcome if efficient and safe offshore operations were to be possible. The impediments to continued use of essentially land-based technologies for the drilling and production of offshore oil in such cases are primarily due to the ocean environment and its obvious effects on the structures and facilities involved. In addition to wave action, structural damage due to hurricanes, particularly in the Gulf of Mexico, is also significant.

1.1.2 History from World War II to the Early 1970s

World War II brought great advances in technologies that later could be adapted to build offshore platforms in even deeper waters and harsher environments and also operate them more safely and efficiently.

In 1946, the first steel offshore platform constructed of tubular members was built to operate in about 4.5m of water some 8km off the coast in the Gulf of Mexico (Graff 1981). The platform was 53m long by 23m wide and it stood only a few meters above the high-tide level. The derrick was supported by more than 300 steel tubular piles. Radio communications were used. The platform could withstand hurricanes with wind speeds of more than 120 knots and waves with a maximum height of about 5m.

Starting in 1947, more advanced design technology began to be used to build larger platforms in deeper waters. These platforms look almost like modern platforms called "bottom-supported platforms" or "jacket-type offshore structures" and were usually completely self-contained systems that included drilling rig and equipment. Work crews were sometimes housed on a separate platform connected to the drilling platform by a bridge. To install the platform, a number of jackets or templates fabricated onshore were carried to the site by barge, lowered into the water by a crane, and integrated by welding. The term "template" is used because the jacket legs serve as guides for the steel tubular piles. This construction method made it possible to shorten the installation period to weeks instead of months.

Furthermore, the new designs began to use tubular bracing below as well as above the water line. This feature is helpful for placing the platform into deeper water because structures without bracing below the water line can only sustain much smaller wave- and current-induced forces than those with such bracing. These types of platforms had become the norm for design and construction for many years. By 1970, the operating water depth for jacket-type offshore structures had reached more than 80m.

1.1.3 History after the Early 1970s

Until the early 1970s, ocean engineering as a discipline was primarily limited to universities, although engineers had become increasingly involved in important practical applications. But, after the impact of the first world oil shock in the early 1970s, matters began to change as the development of offshore oil moved into deeper and deeper waters at a rapid rate. The operating water depth of fixed-type offshore structures had reached more than 300m in the late 1970s and more than 550m in the late 1980s.

Somewhat different design concepts, in addition to steel jackets, started to appear in the early 1970s. The first concrete gravity platform built on land, floated to the site, and installed to the bottom appeared in 1973 in the North Sea (Randall 1997). By the middle of 1980, more than 3,500 offshore structures had been placed in the offshore waters of some 35 countries, and nearly 98 percent of them are steel structures supported by piles driven into the sea floor (McClelland and Reifel 1986). In 1977, Shell Oil Company's *Cognac* platform was installed in the Gulf of Mexico in a record water depth of about 311m.

Oil and gas reserves are, of course, found in much deeper water. For these cases, however, new design concepts, other than the traditional fixed offshore structures, are required. Thus, the 1990s began to usher in new design concepts for offshore platforms that could be placed and operated economically and reliably in increasingly deeper waters. Thus, the era of the *floating* drilling, production, storage, and offloading systems (of various types, functions, and features) began. For further historical overview of deep-water production systems, see Dunn (1994).

1.2 Process of Offshore Oil and Gas Developments

The process of developing offshore oil and gas reserves can be divided into the following major steps (e.g., Graff 1981):

- Exploration
- Exploratory drilling
- Development drilling
- Production
- Storage and offloading
- Transportation

Ships and ship-shaped offshore structures have been key to these developments. Trading tankers, which are perhaps the largest mobile waterborne structures created by humans, increasingly move oil from its sources to the refineries. Ships are involved in oil exploration, starting with the seismic surveys from specially outfitted vessels. Exploratory drilling of promising fields relies on jack-ups, semisubmersibles, and ship-shaped drilling rigs.

Fields with substantial amounts of oil may be developed – that is, the requisite number of wells drilled and subsea equipment installed – either around fully self-contained platforms or from various combinations of platforms and ships or barges for drilling, accommodations, and supplies. Production and processing equipment may be placed on platforms, or on ship- or barge-shaped structures called FPSOs (floating, production, storage, and offloading units). In addition to processing, those floating ship-shaped offshore structures serve the important functions of storage of crude oil and their offloading into shuttle tankers or even vessels of opportunity. Alternately, oil that is processed in platforms may be stored in floating ships or barge-shaped structures called FSOs (floating, storage, and offloading units), to be offloaded into shuttle tankers. Sometimes processed oil is stored directly in platforms and shipped ashore via pipelines. There are many possible alternatives to production, storage, and offloading depending on a particular development that is the most economical one under the circumstances.

Topsides facilities, in either fixed platforms or in FSOs, FPSOs, or drill ships, may by case refer to and include facilities and equipment for drilling, processing, offloading, utilities, services, safety measures (including gas leak detection), fire and gas explosion protection, accommodations, and life support. Process systems serve to separate the well stream into its components, to treat the well stream through operations such as dehydrating, and to transfer the oil.

Therefore, a process train treats the oil in various ways before the product is transferred to a shore terminal or to storage for offloading. In a typical train, the well stream is first separated into produced oil, gas, and water. The gas so obtained may be taken off for further treatment such as compression, storage, and transport; compression and reinjection; or for flaring – a practice that continues to decrease. The water is drained and disposed often by pressurizing and reinjection that in turn may serve to improve production from nearby wells. The produced oil may undergo further processing, including removal of impurities and further removal of water to obtain crude oil of the requisite specification.

The selection of an optimum processing option is an important issue, while a broad range of possibilities is considered for offshore/onshore processing (Bothamley 2004):

- From minimal offshore processing with all produced fluids sent to an onshore terminal (or terminals) for final processing to meet saleable product specifications
- To full processing offshore to make specified products on the offshore facility, with no further onshore processing required

Bothamley (2004) reviews the major processing options available for an offshore oil production facility, including comparisons, major factors, and pros and cons that can serve as a basis for evaluating processing alternatives for future projects.

This book is primarily a structural treatise; nevertheless, it is of some value to understand the processes and related systems typically involved in offshore oil and gas operations. See also Myers et al. (1969), Harris (1972), and Whitehead (1976) for additional information.

It is important to realize that there are many field-development configurations employing platforms, ships, barges, and pipelines, and for storage, processing, and transport. In shallow waters, the developed oil or gas may be transported onshore through a pipeline infrastructure. Otherwise, a storage tank is anchored next to the production platform and the developed oil or gas is transported to shore by barges or shuttle tankers.

For developing oil and gas reserves in deep and ultradeep waters reaching more than 1,000m depth, it is not straightforward to construct and maintain the pipeline infrastructure in terms of cost and technology. Employing a separate storage tank may not always be the best way. In this regard, it is now recognized that FSOs or FPSOs are, in many cases, more attractive for developing offshore oil and gas reserves in deep waters because of cost and efficiency. They house storage tanks together that can be offloaded directly, which is more efficient when the developed oil or gas can be transferred to shuttle tankers or barges.

Advances in mooring and offloading systems and in fluid swivel technology are key to the development of modern FPSOs. Carter and Foolen (1983) and Barltrop (1998) trace the evolutionary developments that advanced the FPSO concept offshore.

1.3 System Concepts for Deep- and Ultradeep-Water Field Developments

The selection of offshore field-development concepts typically involves consideration of the following (Inglis 1996; Barltrop 1998; Ronalds 2002, 2005):

- Environment, including water depth
- Production capacity
- Distance from field to shore or supporting infrastructure such as pipelines
- Required number of drilling centers and wells for each center
- Well-fluid chemistry and pressure and intervention or well-entry frequency for optimum well performance, depending on the types of offshore platforms
- Risk to personnel

Fixed-type offshore platforms that have been useful for developing oil and gas reserves in relatively shallow waters have now been much less feasible for the development of oil and gas fields in deep and ultradeep waters reaching more than 1,000m depth. In order to produce offshore oil and gas in increasingly deep waters, floating-type systems are much better candidates because the weight and cost of fixed structures exponentially increases with water depth; however, those of floating-type offshore structures increase linearly (Hamilton and Perrett 1986). Floating-type offshore structures are also useful to produce oil and gas in marginal fields, that is, for a shortened production period. They can also be designed, built, transported to the site, installed, and commissioned fairly rapidly; and removed, modified, and moved to other similar applications as needs change. Floating-type offshore structures have therefore been considered to develop deep- and ultradeep-water areas.

A floating-type offshore unit must meet the following performance requirements:

- Appropriate work area, deck load capacity, and possible storage capacity
- Acceptable stability and station-keeping during harsh environmental actions
- Sufficient strength to resist harsh environmental actions
- Durability to resist fatigue and corrosion actions
- Possible capabilities needed for both drilling and production
- Mobility when needed

Three types of floating offshore structures – semisubmersibles, spars, and tension leg platforms – have been employed for that purpose. However, all of these three types may typically require a pipeline infrastructure to transport the produced oil to the facilities on shore. The pipeline infrastructure is difficult to construct and maintain in deep and ultradeep waters. Ship-shaped offshore units with multifunctions such as production, storage, and offloading have been considered, and they have been in existence since the late 1970s. FPSOs can both process and store the produced oil or gas in their own cargo tanks until shuttle tankers offload the cargo to transport it ashore.

In Sections 1.3.1 to 1.3.4, the advantages and challenges of various floating-type offshore structures are addressed.

1.3.1 Semisubmersibles

Figure 1.1 shows a computer graphic of one type of semisubmersible. These structures have been used mainly for drilling purposes, but since the early 1980s, these have also been used as production platforms. These do not usually have any oil storage capacity. In one common concept, these structures have two submerged horizontal tubes called pontoons, which provide the main buoyancy for the platform and also act as a type of catamaran hull when in transit to or from a site at low draft. Alternatively, a ring pontoon may be used for such units meant solely for one fixed location. Typically, four to eight vertical surface-piercing columns are connected to these pontoons. The platform deck is located at the top of the columns.

Station-keeping of semisubmersibles is usually achieved by chain- or wire-mooring systems. Where moorings are not practical, dynamic positioning systems with

Figure 1.1. A computer graphic of a semisubmersible installation.

computer-controlled thrusters that respond to vessel displacements or accelerations are used. The advantages and disadvantages of semisubmersibles have been discussed by Barltrop (1998).

The advantages of semisubmersibles include the following:

- Semisubmersibles can achieve good (small) motion response and, therefore, can be more easily positioned over a well template for drilling.
- Semisubmersibles allow for a large number of flexible risers because there is no weathervaning system.

Disadvantages of semisubmersibles include the following:

- Pipeline infrastructure or other means is required to export produced oil.
- Only a limited number of (rigid) risers can be supported because of the bulk of the tensioning systems required.
- Considering that most semisubmersible production systems are converted from drilling rigs, the topsides weight capacity of a converted semisubmersible is usually limited.
- Building schedules for semisubmersibles are usually longer than those for ship-shaped offshore structures.

1.3.2 Spars

Figure 1.2 shows a computer graphic of a typical spar. In the beginning, spars were used as storage units, but spars are also now used for production. A spar usually has a vertical circular cylinder with a very large diameter, say, 15–30m, which contributes to significant reduction of heave motion of the unit by virtue of the large draught. Because of the reduced heave motion, the use of rigid risers (instead of flexible risers), which are self-buoyant, is easier.

Spars are usually moored to allow motion of all six degrees of freedom, but, alternatively, a tether-mooring system that makes it into a kind of tension leg platform

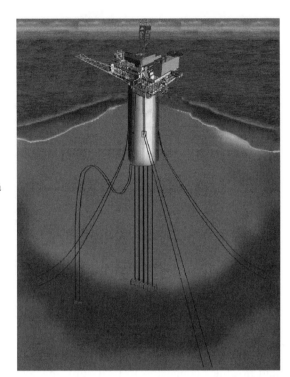

Figure 1.2. A computer graphic of a spar installation.

(Figure 1.3) with a single cylinder may be used. A production spar may or may not have oil storage and related wells at surface or subsea as shown in Figure 1.2. In general, the building cost of spars may be greater than that of ship-shaped offshore structures because of their specialized and nonmass-produced nature.

1.3.3 Tension Leg Platforms

Figure 1.3 shows a computer graphic of a tension leg platform (TLP). A TLP may have up to six vertical surface-piercing columns with a complete ring of pontoons and a number of vertical tethers. Although the motions of surge, sway, and yaw may be relatively large, the heave, roll, and pitch motions of the platform are usually well limited by the vertical tethers that can be designed so that their periods in heave, roll, and pitch are well below the significant wave periods involved.

TLPs cannot be moved from location to location. Also, TLPs are sensitive to payloads because of the tensioning effect of tethers and, as a result, they cannot usually be used as storage units. Therefore, TLPs normally need a pipeline infrastructure or FSOs plus a shuttle tanker offloading system to export the produced oil.

1.3.4 Ship-Shaped Offshore Units

A ship-shaped offshore unit may be used as a floating storage unit (FSU), an FSO unit, an FPSO, or even include drilling capabilities in some cases. Figure 1.4 shows a computer graphic of an FPSO installation with a shuttle tanker offloading system.

Figure 1.3. A computer graphic of a tension leg platform (TLP) installation.

An FPSO system stores produced oil or gas in tanks located in the hull of the vessel, and flowlines connected to risers link the subsea development wells to the FPSO system after the development wells have been drilled by other types of offshore units, such as semisubmersibles. The oil is periodically offloaded to shuttle tankers

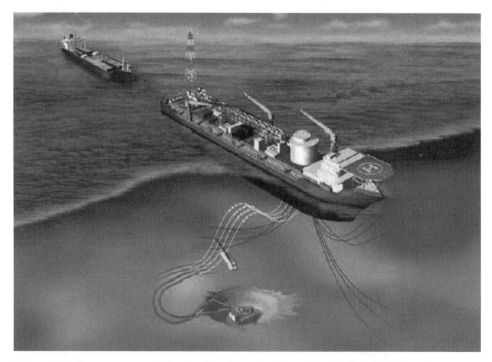

Figure 1.4. A computer graphic of a ship-shaped offshore installation (FPSO) with a shuttle tanker offloading system.

or oceangoing barges for transport to the facilities on shore. FPSO systems may also be used as the primary production facilities to develop marginal oil fields or fields in remote deep-water areas without the need of a pipeline infrastructure.

Ship-shaped offshore units have various benefits when compared to other types of floating structures in terms of ample work area, deck load, high storage capability, structural strength, shorter lead time, building/capital cost, and suitability for conversion and reusability. However, similar to other types of floating platforms, their displaced volume below the water line is comparatively large, and the response and failures of the structures under harsh environmental conditions associated with waves, winds, and currents are significant issues to consider in design and operation. Dynamic/impact-pressure actions arising from green water, sloshing, and slamming are also issues to be resolved both in design and for operation, particularly in harsh weather areas.

Careful consideration of an adequate station-keeping system and adequate design considerations of systems, such as the riser system, are necessary in order to avoid difficulties due to vessel motions. The riser system used for ship-shaped offshore units is usually flexible (rather than rigid). There are several methods of mooring the ship-shaped structures, including turret moorings, articulated towers, and soft yoke systems, which permit the unit to weathervane, that is, rotate according to the direction of external forces. Thrusters can assist the mooring system to reduce forces and motions.

In relatively benign environmental areas, FPSO systems may be spread-moored; also, rigid risers may be acceptable. However, in harsh environmental areas – for example, with revolving tropical storms such as typhoons in the South China Sea and tropical cyclones offshore of Northwestern Australia – careful consideration is required for the station-keeping with relevant mooring system designs.

FPSO systems may be either new builds or conversions from trading tankers. Challenges for their structural design are mostly related to assessment of limit states including ultimate limit states, fatigue limit states, and accidental limit states as well as serviceability limit states. The 100-year return period is usually considered for design onsite strength assessment, but tow considerations are based typically on 10-year return period environmental phenomena. For operation, relevant programs of inspection and maintenance must also be established to keep the structural integrity and reliability at an adequate level.

Useful discussions of the technical challenges and technology gaps and needs related to the use of ship-shaped offshore units to develop the offshore oil and gas in deep and ultradeep water are given, for example, by Henery and Inglis (1995), Birk and Clauss (1999), Bensimon and Devlin (2001), Lever et al. (2001), Maguire et al. (2001), Le Cotty and Selhorst (2003), and Hollister and Spokes (2004).

1.4 A Brief History of the FPSO Installations

Over the past 25 years, ship-shaped offshore units have proven to be reasonably reliable, cost-effective solutions for the development of offshore fields in deep waters worldwide. These include FPSOs or FSOs operating in harsh environmental areas and also waters of more than 1,000m depth; see FSO/FPSO performance records by Single Buoy Moorings, Inc. (http://www.singlebuoy.com) for examples.

It is hard to say with precision exactly when ship-shaped units made their appearance on the offshore oil scene. Certainly, oil storage and shuttle tanker-mooring facilities using converted trading tankers existed in the late 1960s. The first such vessels were connected by hawsers to catenary anchor leg mooring (CALM) systems. These then evolved into the now more familiar systems employing single-point mooring, where the FSO *Ifrikia* was permanently moored to a buoy via a rigid arm (rather than a hawser) in the early 1970s, with a concomitant increase in operational reliability and reduced downtime.

The first dedicated FPSO application offshore was by *Arco* in the Ardjuna field in the Java Sea offshore Indonesia in 1976 (D'Souza et al. 1994). Interestingly, this was a concrete barge with steel tanks, used to store refrigerated liquefied petroleum gas (LPG) moored to a buoy using a rigid arm system in 42.7m water depth. The first tanker-based single-point moored FPSO facility for oil is said to be the *Castellon* for Shell offshore Spain in 1976. This facility was meant to produce oil from a subsea completed well, some 65 km offshore Tarragona. It began operations in 1977, and was designed for a 10-year field life.

Compared to these early days, floating production systems have now evolved to a mature technology that potentially opens up the development of offshore oil and gas resources that would be otherwise impossible or uneconomical to tap. The technology now enables production far beyond the water-depth constraints of fixed-type offshore platforms and provides a flexible solution for developing short-lived fields with marginal reserves and fields in remote locations where installation of a fixed facility would be difficult.

Figure 1.5(a) shows a photo of an early permanently moored FSO *Ifrikia* in a side-by-side export arrangement at the Ashtart field offshore Tunisia in 1972. Figure 1.5(b) shows a photo of an early FPSO *Castellon* on the Castellon field of Shell offshore Spain in 1976. Figures 1.5(c) and 1.5(d) are photographs of a modern FPSO with an external or internal turret mooring in a tandem export arrangement, respectively.

In the early 2000s, more than ninety FPSOs were in service and more than twenty FPSOs were under construction. Some of them were newly built for site-specific environments, and others were converted from existing tankers, mostly very large crude oil carriers (VLCCs). FPSOs are now found in all offshore areas where floating production systems are used, with the notable exception of the Gulf of Mexico thus far. The largest presence of FPSOs appears to be in the North Sea and off of Africa. They range in size from 50,000-barrel tankers with capability to process 10,000–15,000 barrels per day to VLCC size units, which can process more than 200,000 barrels per day and store 2 million barrels.

Although a majority of FPSOs have so far been installed in relatively benign environmental areas such as Southeast Asia, West Africa, and Offshore Brazil near the Equator, the FPSO applications for oil and gas exploration in deeper marginal waters and harsh environmental areas, for example, with tropical cyclones and storms, are challenging. For instance, the effect of hurricanes on the station-keeping capability of a mooring system and the structural failures is a major concern of regulatory bodies as operators consider the FPSO installations for deep-water developments in the Gulf of Mexico. A mooring system failure of an FPSO can lead to collisions with adjacent offshore installations causing major oil spills.

a

b

Figure 1.5. (a) An early permanently moored FSO *Ifrikia* in a side-by-side export arrangement, at the Ashtart field offshore Tunisia in 1972 (courtesy of SBM Offshore NV). (b) An early FPSO *Castellon*, at the Castellon field offshore Spain in 1976 (courtesy of SBM Offshore NV).

Figure 1.5. (*cont.*) (c) A photograph of a modern FPSO with an external turret mooring in a tandem export arrangement (courtesy of SBM Offshore NV).

Figure 1.5. (*cont.*) (d) A photograph of a modern FPSO with an internal turret mooring in a tandem export arrangement (courtesy of SBM Offshore NV).

1.5 Trading Tankers versus Ship-Shaped Offshore Units

Although the hull structural arrangement of a ship-shaped offshore unit used for the offshore oil and gas development is similar to that of a trading tanker, it is important to realize that large differences exist between them in a variety of items, as indicated in Table 1.1.

Table 1.1. *Differences between trading tankers versus ship-shaped offshore units in terms of strength and fatigue design*

Trading tankers	Ship-shaped offshore units
Design condition: North Atlantic wave environment	Design condition: Site- and tow-route-specific environments
20- to 25-year return period	100-year return period
Predominantly wave actions	Currents as well as wind and wave actions
Limited number of loading/offloading cycles; loading occurs in sheltered situations	More frequent loading/offloading cycles; loading occurs with relatively more environmental effects present
Limited number of loading conditions	More numbers and variety of loading conditions
At open sea for about 70 percent of the time	Offshore for 100 percent of the time
Weather in any direction; rough weather avoidance possible	Highly directional weather and weathervaning; rough weather avoidance not possible once on site
Regular dry-docking every 5 years	Continuous operation usually without dry-docking
Without topsides	With topsides and associated interaction effects between hull and topsides

A key difference between trading tankers and ship-shaped offshore units is in the consideration of design environmental conditions. For the design of trading tankers, the North Atlantic wave environment is typically considered as the design premise for an unrestricted vessel to make worldwide travel possible. However, the design loads of ship-shaped offshore units will be based on the environmental phenomena specific to their operational sites, their transport to field before installation and mooring, and the commencement of operations as the case may be. Appendix 3 presents sample wave-scatter diagrams for the North Atlantic and some selected site environmental data.

For historical reasons, the return period of waves for the hull girder strength design of ship-shaped offshore units is typically taken as 100 years, but that of trading tankers for the same purpose is considered to be 20–25 years or so.

Winds and currents, as well as waves among other factors, may induce significant actions and action effects on offshore structures, whereas waves are often the primary source of environmental actions on trading ships at sea.

Trading tankers are typically loaded and unloaded at still-water conditions in harbor, but ship-shaped offshore units are subjected to significant environmental loads even during loading and unloading. The number of loading/offloading cycles on ship-shaped offshore units is greater than that on trading tankers. Ship-shaped offshore units are typically offshore for 100 percent of the time of their design life, but trading tankers are on the open sea for approximately 70 percent of the time. Certainly, the fatigue failure characteristics of ship-shaped offshore structures may differ somewhat in comparison to trading tankers, for example, in the need to consider low-cycle fatigue related to loading and offloading. This can be important because large still-water forces and still-water moments can be created in ship-shaped offshore units because loading patterns may be very different from those of trading tankers, and because loading/unloading cycles are much more frequent.

In terms of operating conditions, trading tankers normally operate in either laden or ballast condition, but ship-shaped offshore units will be in varying states of loading and unloading. These characteristics of loading and unloading in turn imply the possibility of large draft variations among the fully loaded, the minimally loaded, and ballast conditions, compared to trading tankers. It follows that strength considerations must then address a number of loading conditions at varying drafts and a number of environmental conditions with different return periods.

Trading tankers may avoid rough weather or alter their heading in operation by "weather routing" (Olsen et al. 2006), but ship-shaped offshore units must be continuously located in the same area with site-specific environments.

Trading tankers are regularly dry-docked in 5-year intervals, but ship-shaped offshore units will generally not be dry-docked (and, in any event, are preferred not to be dry-docked) during the entire production period in the field, possibly more than 10 years to even 20 years. This means that repairs in a dry dock are not economically realistic in many cases, primarily because of the potential production interruptions that must be dealt with. Also, welding or flame cutting that is common for traditional repairs of trading tankers in a dry dock may not be used for the repair work of offshore structures in situ for reasons of high fire and explosion risk.

Unlike trading tankers, ship-shaped offshore units have topsides, a turret, flare towers, riser porches, and a drill tower, which are items of a large mass, a high center of gravity, and a large windage area, which affect vessel motions and responses to environmental phenomena. Undesirable motion characteristics leading to green water, sloshing, slamming, mechanical downtime on equipment, and crew discomfort are then very specific design considerations. For a turret-moored ship-shaped offshore unit, the vessel may head into the weather and other differences can arise. For instance, in comparative terms, the hull girder strength for FPSOs meant for turret-moored operation in the North Sea must be significantly greater than that of trading tankers in unrestricted service. However, in some areas, such as West Africa, the wave environment can be considerably benign, and this can be an advantage in terms of the strength required, whether or not it is turret-moored.

In any event, it is important to realize that the design considerations for ship-shaped offshore units may be more complex than those for trading tankers. This is not necessarily because ship design is any less complicated in principle, but rather because of the relative importance of site-specific conditions offshore and the need to consider many aspects in their design explicitly and specifically, unlike a trading ship wherein many of the same considerations may be made implicitly by well-established rules and procedures.

1.6 New Build versus Tanker Conversion

Advantages and disadvantages do exist and need to be evaluated when deciding between a new-build option versus a tanker-conversion option. The advantages of a new-build option (Parker 1999) include the following:

- Design and fatigue lives for a field can be achieved easier.
- Technical, commercial, and environmental risks can be more easily contained.
- A system can be more easily designed to survive harsh environments.
- Resale and residual values can be maximized.
- Reusability opportunities can be improved.

On the other hand, the advantages of a tanker-conversion option include the following:

- Capital costs can be reduced.
- Design and construction schedule can be faster and less extensive.
- Construction facility availability is increased.
- Overall project supervision requirements can be less.

The best option for a particular situation needs to be chosen taking their advantages and disadvantages into account (see Section 2.2 of Chapter 2). One of the key drivers for selection of either the new-build or conversion option may be the field life that corresponds to the economic depletion duration of the reservoir. When the design life for continuous operation on site is more than 20 years, a new build will invariably be desirable. For marginal fields, the design life may often be 5, 10, or 15 years, and a conversion option may be more economical.

Building cost of new-build FPSOs may, of course, vary depending on the many aspects including the capacity of production and storage. For instance, an FPSO operating in a marginal field may cost 60 million US$ for a converted tanker with topsides plant installed, but an FPSO newly built for a large field may cost more than 100–200 million US$ or more depending on the size of the hull and size and complexity of the topsides installations.

The building project of an FPSO can be divided into different work packages, for example, those related to hull, topsides, their integration, and, of course, project management. The related contracts are sometimes awarded separately or as a whole as described in Chapter 2 of this book. Although most parts of this book deal with the technologies for design and construction of new-build systems, Chapter 12 presents further considerations for a tanker-conversion option.

1.7 Layout and General Arrangement of FPSOs

1.7.1 Deck Area and General Arrangement

The deck area required in an FPSO depends on the process plant size, footprint, and complexity. Preferences such as single-level construction with "pancake"-type process plant design will affect the deck area required to accommodate the process plant.

The production plant (process) capacity is usually project-specific and is dependent on field economics. Its optimal selection can be an involved matter because the overall field development costs will vary not only with the plant capacity but also with the complexity such as gas handling, water injection, crude characteristics, flow assurance, and chemical treatments required.

Many other factors also affect deck area and the general arrangement, including:

- Hull form
- Turret location and size
- Accommodation location and size
- Ballast capacity and distribution
- Double-side or double-bottom requirements
- Escape, evacuation, and rescue arrangements
- Offloading arrangements
- Margins for future process upgrading and expansion

1.7.2 Layout

The layout of an FPSO can be divided into the following:

- Main deck
- Topsides deck
- Mooring system
- Accommodation
- Machinery room
- Cargo and ballast tanks
- Offloading area

Figure 1.6 illustrates an example of the overall field layout involving both oil and gas. Figure 1.7 shows typical layouts of an FPSO topsides facility. The accommodation

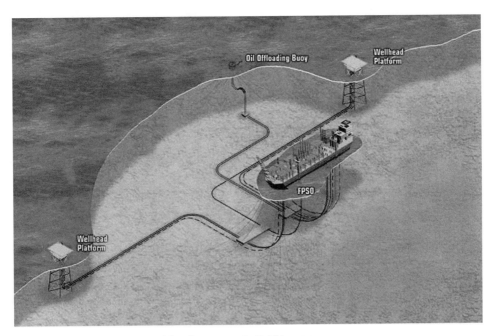

Figure 1.6. A computer graphic of an example field layout.

is located in the bow area as shown in Figure 1.7(a) or in the stern area as shown in Figure 1.7(b). The vessel layout is, in principle, configured so that the separation between the accommodation (including the principal evacuation systems) and the major hydrocarbon hazards should be maximized. The accommodation and the turret are separated as far away as possible when a turret-mooring system is adopted with risers and mooring facilities located at the bow; see Figure 1.7(b). This configuration is beneficial also because turret motions can be minimized while weathervaning capacity can be maximized. Also, when the accommodation with a helideck is located at the stern in a conversion, the proximity so achieved, to the engine room that contains many of the major vessel systems including utility systems, can be an advantage.

As shown in Figure 1.7(b), a turret is often located as far forward as possible, although the accommodation with the helideck must usually be sited aft, with the process modules and power generation in the cargo length and flare tower in the forward area. One aim of placing the turret location as far forward as possible is to make active heading control by thrusters easier. However, in a tanker conversion with an internal turret, how far forward it can be placed also depends on its size and the number of risers that must be served. When a larger turret is required, it may be sited in the section of 0.2L–0.35L from the forward end (L = vessel length) and the accommodation may be sited forward of the turret. This allows the cargo region length aft of the turret to be maximized.

The topsides facilities are located above the main deck in between the turret and the accommodation. The main deck needs to be strong enough for the support columns of topsides modules and also have space for the required piping for cargo loading/offloading, inerting and venting, and hatches for access to the tanks. It will also contain the main cranes, perhaps two, one on the port side and one on the

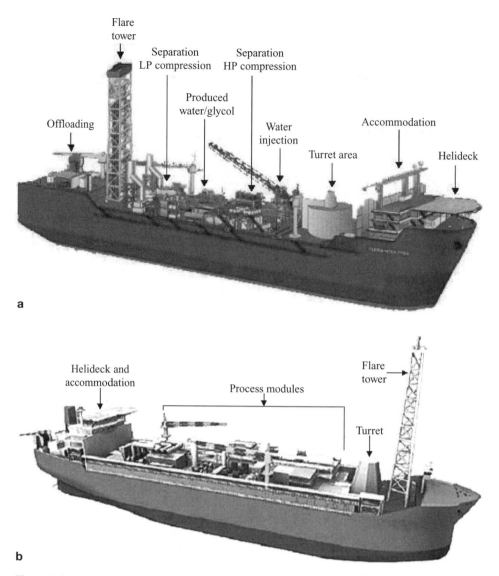

Figure 1.7. Sample topsides layout of a ship-shaped offshore unit FPSO: (a) with an accommodation forward and an internal turret; (b) with an accomodation aft and an internal turret.

starboard side. The oil-metering skid for fiscal metering during offloading may be located on the main deck, usually in front of the accommodation. Shielded escape routes may run on both the port and the starboard sides of the main deck from the bow to the front of the accommodation (in the case of Figure 1.7(b)). Stairways and ladders may be used for intermediate access from the elevated process deck onto the escape routes.

The topsides modules may be divided into the process area and the utility area. The process area includes spaces for hydrocarbon-containing equipment, flare tower, compression equipment, and separation equipment. The utility area includes spaces for utility equipment and power-generation equipment.

Table 1.2(a). *Average principal dimensions of FPSOs and trading tankers (MacGregor and Smith 1994)*

Vessel type	L/B	B/D	T/D	B/T
New-build FPSO/FSU				
North Sea	6.0	1.9	0.65	2.9
Worldwide	5.2	1.8	0.67	2.7
Tanker conversion (worldwide)	6.6	2.0	0.76	2.6
Trading tanker				
50,000–70,000 dwt	6.3	2.5	—	—
70,000–100,000 dwt	5.6	3.0	—	—
100,000–200,000 dwt	5.6	2.8	—	—

Note: L = length between perpendiculars; B = breadth; D = depth; T = freeboard at light draught.

Considering the layout of topsides facilities, one of the most important aspects is to maximize the safety of the personnel on board. In this regard, it is important to minimize potential hazards from process equipment. For example, the amount of piping must be minimized, and piping must be adequately protected from hazards such as dropped objects and the dynamic flexing of the vessel's hull. The process area must be located as far as possible from the accommodation. The utility area may be located in between the process area and the accommodation. Additional considerations related to the design of topsides facilities are presented in Chapter 9.

1.7.3 Relationships between Principal Dimensions

The vessel's dimensional relationships depend on and also affect storage, stability, motion characteristics, mooring, station-keeping, and, of course, the environmental actions that the vessel is subjected to; for a good exposition, see Parker (1999).

The interrelationships between vessel principal dimensions affect various features of the design. For instance, an increase in a vessel's length for a given storage capacity will generally increase mooring forces and hazardous zone extents, all in turn affecting the cost. Generally, the lowest building cost for an FPSO can be expected for the lowest L/B ratio (L = vessel length, B = vessel breadth).

The depth must be maximized with respect to the length and breadth. The breadth must also be maximized for deck area design. Although the draft is the smallest dimensional parameter, it should also be maximized for overall storage efficiency and should be deep enough to avoid excessive bottom slamming. The block coefficient of the vessel should be maximized for the storage capacity and for the construction efficiency.

MacGregor and Smith (1994) made some investigations in principal dimensions of FPSOs/FSUs or trading tankers that were in service from 1986 to 1990, as indicated in Table 1.2(a). Deluca and Belfore (1999) investigated the length (L) to breadth (B) ratio of converted and newly built FPSOs, which were in operation in various geographical locations in 1999, as indicated in Table 1.2b.

Wang (2003) compared the principal dimensions of 35 new-build FPSOs together with 140 single-hull trading tankers and 46 double-hull trading tankers, all of which were in service in the early 2000s. The length of vessels studied by Wang (2003) ranges between 170m and 400m. The majority of the single-hull tankers that Wang

Table 1.2(b). *Average principal dimensions of FPSOs in operation at different sites (Deluca and Belfore 1999)*

Site	Length/Breadth	
	New-build FPSO	Tanker conversion
Angola	—	4.1
Australia	6.0	6.2
Brazil	6.8	8.8
China	6.8	6.2
Congo	4.7	6.8
UK	5.2	6.2

studied were built in the 1970s and 1980s, although some were built in the 1960s, and most of the double-hull tankers were built in the 1990s.

It was found that the breadth characteristics of FPSOs are similar to those of trading tankers except for a few cases where the length of the FPSOs is relatively small. The ratio of length to breadth was in the range of 4–7 for both FPSOs and trading tankers. Figures 1.8 and 1.9 show the relationship between the length and the depth of FPSOs, shuttle tankers, and trading tankers. It is observed that the depth of FPSOs tends to be slightly larger than that of trading tankers. In particular, most new-build FPSOs are deeper than single-hull trading tankers.

This may be because the depth of FPSOs with double sides may have been increased to meet ballast water requirements. A couple of exceptions to this rule are seen in Figure 1.9, where the depth of the new-build FPSOs is smaller than that of trading tankers. This may be because these particular FPSOs were built for operation in shallow water. Another interesting observation from the figures is that a new-build FPSO having a storage capacity equivalent to an ultralarge crude oil carrier (ULCC) more than 400m long has not appeared yet.

Figures 1.10 and 1.11 show the relationships between breadth (B) and depth (D) or between freeboard (T) and depth (D) of FPSOs and shuttle tankers, respectively. These are a summary of the data studied by OPL (2001, 2002) for FPSOs and shuttle tankers operating up to the year 2000. It is seen in Figure 1.10 that the B/D

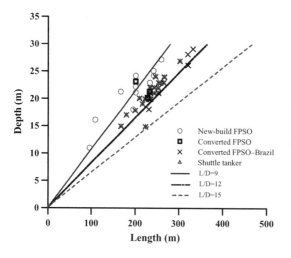

Figure 1.8. Relationship between length (L) and depth (D) of FPSOs and shuttle tankers.

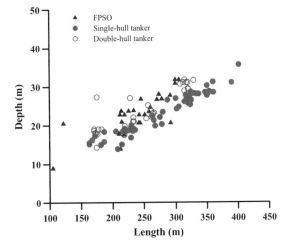

Figure 1.9. Relationship between length and depth of FPSOs and trading tankers (courtesy of Wang 2005/ American Bureau of Shipping).

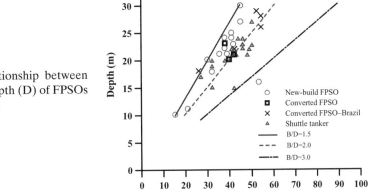

Figure 1.10. Relationship between breadth (B) and depth (D) of FPSOs and shuttle tankers.

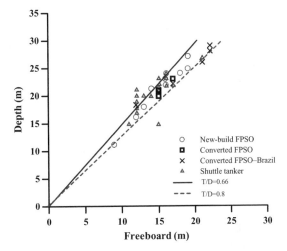

Figure 1.11. Relationship between freeboard (T) and depth (D) of FPSOs and shuttle tankers.

ratio of FPSOs is in the range of 1.5–3.0, which is similar to that of shuttle tankers. The freeboard requirement can be important for green-water loading considerations. Although the T/D ratio of trading tankers is typically in the range of 0.7–0.8, it is seen in Figure 1.11 that the T/D ratio of FPSOs is in the range of 0.6–0.8; see also Table 1.2a.

1.7.4 Double-Hull Arrangements

An aspect that needs to be considered for design and operation of an FPSO is the protection of cargo tanks from damage caused by collisions with shuttle tankers, particularly when a side-by-side configuration is to be used for exporting cargo. Supply boats or passing vessels will also be sources of collisions. A new-build FPSO hull usually has double sides, but the bottom is not necessarily double-skinned. A single-skin tanker conversion may attach sponsons that are equivalent in function to double sides.

Double bottoms are generally not required because damage from hull grounding is unlikely. However, if the FPSO is a disconnectable type and may need to leave the station from time to time under its own power, a double bottom may be required by regulation. Also, if the FPSO is located in a very shallow location with some chance of contact with the sea bottom, a double bottom may be necessary. For heavy oil and, in particular, in cold climates, a double bottom may be required to reduce the heating load. Double bottoms with complex bottom shell stiffeners may be difficult to strip and clean when they are used as cargo tanks; thus, double-bottom tanks either usually remain empty or are used for water ballast.

1.7.5 Tank Design and Arrangements

The vessel hull in a new-build option will have several cargo tanks placed centrally and several ballast water wing tanks arranged on both the port and the starboard sides (see Figure 1.16). The number of cargo and ballast water tanks is determined by the production capacity and whether a shuttle tanker will be moored to offload the produced oil. The areas for mooring and offloading contain the hose storage, handling reel, and mooring hawser.

In an FPSO, various tanks such as cargo oil tanks, ballast tanks, slop tanks, portable water tanks, fresh water tanks, diesel oil tanks, methanol tanks, and hydraulic oil storage tanks need to be incorporated. The following issues need to be considered for the tank design in the hull:

- Number, location, and size of cargo and ballast tanks
- Location and size of tanks required for special services, such as methanol tanks slop tanks, chemical tanks, reception tanks and off-spec oil tank(s)
- Pumping arrangement for tanks
- Tank strength, corrosion protection, and access

Each of these issues needs to be considered fully for the hull design and tank layout. The number of tanks is an important consideration from cost and operational

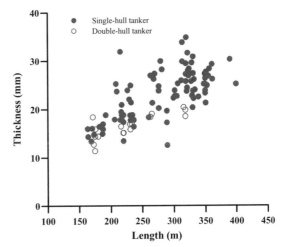

Figure 1.12. Relationship between bottom-plate thickness and length of trading tankers (courtesy of Wang 2005/American Bureau of Shipping).

viewpoints because adding more tanks usually means higher costs. Some factors that need to be considered are:

- Number of cargo production grades
- Export parcel size and production rate
- Hull stresses for the various loading cases, particularly for those related to on-site maintenance and repair
- Required flexibility for operations, inspections, and maintenance with special consideration for hot-work isolation

1.8 Longitudinal Strength Characteristics of FPSO Hulls

Similar to trading tankers, the longitudinal hull strength is a key design aspect of FPSOs. Figures 1.12 and 1.13 show the thickness variations of bottom and side shell plates for single- and double-hull trading tankers as a function of vessel length. It

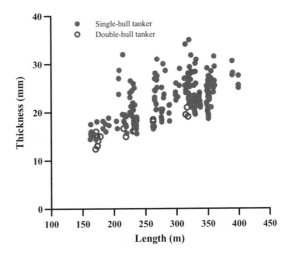

Figure 1.13. Relationship between side shell-plate thickness and length of trading tankers (courtesy of Wang 2005/American Bureau of Shipping).

Table 1.3. *The ratio of FPSO wave-induced bending moments at different sites to trading tanker wave-induced bending moments calculated for the North Atlantic (Wang 2003)*

North Sea	Gulf of Mexico	Offshore Brazil	West Africa
1.1–1.7	0.8–1.1	0.5–0.7	0.3–0.7

is surmised from the figures why the older single- or double-hull trading tankers, in many cases, retain the potential for successful conversions to FPSOs, particularly for benign wave environments. For new-build FPSOs, however, the hull cross-sectional properties will be determined for site-specific requirements to a greater extent.

As may be expected, the wave-induced vertical bending moments for FPSOs designed for site-specific environments are different from those of trading tankers typically designed for unrestricted services worldwide and depend on the installation site. Also, it appears that the ratio of sagging to hogging wave-induced vertical-bending moments may be different (higher) for new-build FPSOs than for converted tankers and may range from 1.0 to 1.33, presumably because of hull-form differences between new-build FPSOs and converted tankers (HSE 2003).

Based on seakeeping analyses, Wang (2003) obtained useful insights relating the wave-induced bending moments of new-build FPSOs at different sites to those for trading tankers, with results as indicated in Table 1.3. It is apparent that the wave-induced bending moments of FPSOs in harsh environments can be larger than those of trading tankers. However, the wave-induced bending moments of FPSOs in benign environments are much smaller than those of trading tankers because the North Atlantic wave environment is commonly used for the design of trading tankers to make the worldwide travel possible.

1.9 Drawings of a Hypothetical FPSO

This section presents sample drawings of a hypothetical FPSO hull that will be used for the illustrative examples of ultimate limit-state assessment described in Chapter 6. The principal dimensions and important features of this vessel are:

- Length overall (L) = 317m
- Breadth (B) = 58m
- Depth (D) = 32m
- Cargo capacity at 98 percent full = 2,300,000 barrels
- Slop tank capacity at 98 percent full = 90,000 barrels
- Total capacity at 98 percent full = 2,411,000 tons
- Estimated deadweight = 339,500 tons
- Estimated full-load draft = 23.5m
- Estimated light-ship draft = 4.6m
- Estimated process deck weight = 31,000 tons
- Estimated total production riser weight = 3,200 tons
- Mooring loads = 1,900 tons
- Block coefficient = 0.87

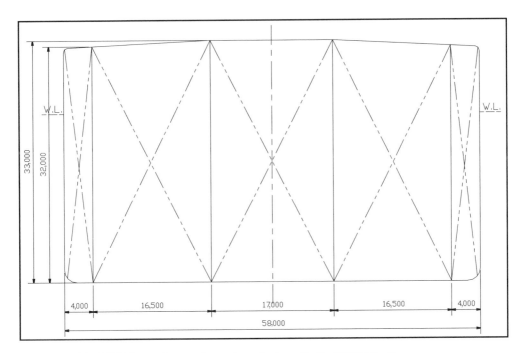

Figure 1.14. Midship section configuration of a hypothetical FPSO hull.

The vessel has double sides and its bottom is single-skinned, as shown in Figures 1.14 and 1.15. Figure 1.16 shows the general arrangement of the hull. Figure 1.17 shows the midship section drawing of the hull. Mild steel with the yield stress of 235N/mm^2 is used for most parts of the hull, except for deck and bottom

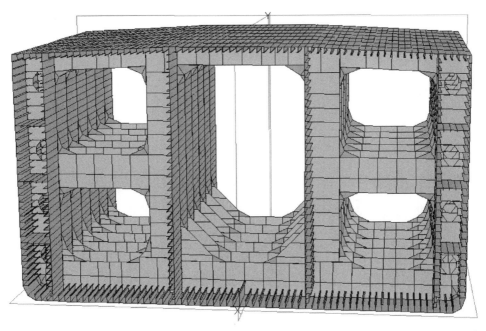

Figure 1.15. Three-dimensional midship configuration of a hypothetical FPSO hull, developed by MAESTRO modeler (MAESTRO 2006).

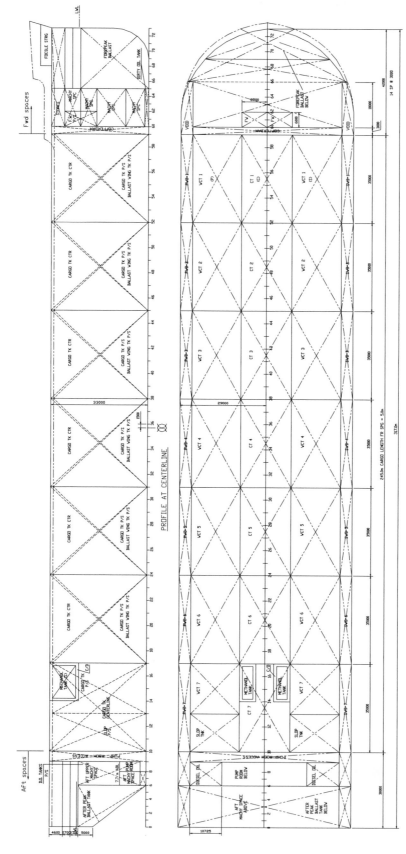

Figure 1.16. General arrangement drawing of a hypothetical FPSO.

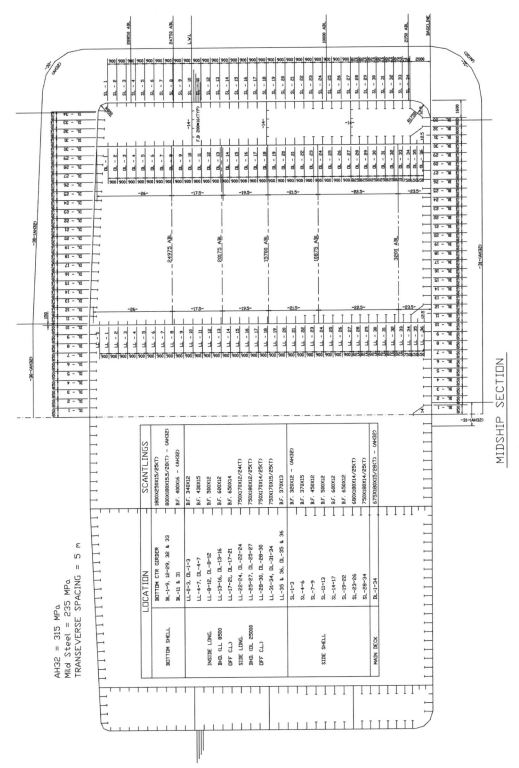

Figure 1.17. Midship section drawing of a hypothetical FPSO.

areas, where high-tensile steel with the yield stress of 315 N/mm^2 is used. Sections 2.12 and 2.13 in Chapter 2 address some considerations for building materials. The block coefficient of an FPSO is typically in the range of 0.85 to 0.90.

1.10 Aims and Scope of This Book

The trend of offshore oil and gas development has been shifting from fields in shallow and medium waters to fields in deeper waters. Ship-shaped offshore units such as FPSOs have been recognized as one of the most reliable, economical solutions to develop marginal offshore oil and gas reserves in deep-water areas.

An FPSO installation consists of the following major parts:

- Vessel (hull)
- Topsides (processing system, accommodation, machinery space, helideck)
- Mooring system
- Export system (offloading, shuttle tanker)
- Subsea systems and flowlines

Considering the entire unit as a whole, a ship-shaped offshore installation typically needs to be designed to satisfy multiple requirements, such as the following:

- Design life: vessel to typically remain on site for its entire design life
- High uptime for the production storage and offloading facilities: targets of 90–95 percent of uptime are common in design
- House and support the required crew and staff safely by meeting habitability requirements
- Receive and process well fluids, store quality crude oil to specifications, and offload the same to shuttle tankers periodically
- Provide for lifting, treating, exporting and/or reinjecting, or otherwise disposing of associated gas and produced water for pressure maintenance
- Operate such that the facility's impact on the environment conforms to high standards

Although ship-shaped offshore units have been in existence since the late 1970s, their complexity and size have been gradually increasing, and there are still a number of problem areas related to designing, building, and operating these units that must be resolved for achieving the high integrity in terms of safety, health, the environment, and economics/financial expenditures.

This book introduces and describes the technical fundamentals and engineering practices for designing, building, and operating the ship-shaped offshore units with the focus on FPSOs as the primary example. Emerging practices, recent advances, and future trends on core technologies essential for ship-shaped offshore installations are addressed with particular emphasis on structure expertise. This book covers a wide range of the subjects from design to decommissioning. It is our intention that this book will be a handy source from which the reader should be able to obtain an extensive and systematic insight into the functioning of ship-shaped offshore units in both an academic and a practical sense.

REFERENCES

ABS (2004). *Guide for building and classing floating production installations*. American Bureau of Shipping, Houston, April.

Barltrop, N. D. P. (1998). *Floating structures: A guide for design and analysis*. The Centre for Marine and Petroleum Technology (CMPT). Herefordshire, England: Oilfield Publications Ltd.

Bensimon, L. F., and Devlin, P. V. (2001). *Technology gaps and preferred architectures for deepwater FPSOs*. Offshore Technology Conference, OTC 13088, Houston, May.

Birk, L., and Clauss, G. F. (1999). *Efficient development of innovative offshore structures*. Offshore Technology Conference, OTC 10774, Houston, May.

Bothamley, M. (2004). *Offshore processing options for oil platforms*. SPE Annual Technical Conference, Society of Petroleum Engineers, SPE 90325, Houston, September.

Carter, J. H. T., and Foolen, J. (1983). *Evolutionary developments advancing the floating production storage and offloading concept*. SPE Exploration and Production Environmental Conference, SPE 11808, Society of Petroleum Engineers, April.

DeLuca, M., and Belfore, L. (1999). *Economics still support FPSO fleet expansion*. Offshore, pp. 54–56.

D'Souza, R. B., Delepine, Y. M., and Cordy, A. R. (1994). *An approach to the design and selection of a cost effective floating production storage and offloading system*. Offshore Technology Conference, OTC 7443, Houston, May.

Dunn, F. P. (1994). *Deepwater production: 1950–2000*. Offshore Technology Conference, OTC 7627, Houston, May.

Graff, W. J. (1981). *Introduction to offshore structures – Design, fabrication, installation*. Houston: Gulf Publishing Company.

Hamilton, J., and Perrett, G. R. (1986). "Deep water tension leg platform designs." *Proceedings of International Symposium on Developments in Deep Waters*. The Royal Institution of Naval Architects, London, October.

Harris, L. M. (1972). *An introduction to deepwater floating drilling operations*. Tulsa, OK: Petroleum Publishing Company.

Henery, D., and Inglis, R. B. (1995). *Prospects and challenges for the FPSO*. Offshore Technology Conference, OTC 7695, Houston, May.

Hollister, H. D., and Spokes, J. J. (2004). *The Agbami project: A world class deepwater development*. Offshore Technology Conference, OTC 16987, Houston, May.

HSE (2003). *Margins of safety in FPSO hull strength*. (Research Report 083), Health and Safety Executive, UK.

IACS (2003). *Requirements concerning strength of ships – Longitudinal strength standard*. International Association of Classification Societies, London.

IACS (2005). *Common structural rules for double hull oil tankers*. International Association of Classification Societies, London, December.

Inglis, R. B. (1996). *Production facilities selection for deep water oil and gas field development*. Institution of Engineers and Shipbuilders in Scotland (IESIS) (Paper No. 1559).

Le Cotty, A., and Selhorst, M. (2003). *New build generic large FPSO*. Offshore Technology Conference, OTC 15311, Houston, May.

Lever, G. V., Dunsmore, B., and Kean, J. R. (2001). *Terra Nova development: Challenges and lessons learned*. Offshore Technology Conference, OTC 13025, Houston, May.

MacGregor, J. R., and Smith, S. N. (1994). "Some techno-economic considerations in the design of North Sea production monohulls." *Proceedings of Offshore '94 Conference*, Paper No. 5, pp. 53–71.

MAESTRO (2006). Version 8.7.2. Proteus Engineering (http://www.proteusengineering.com), Stevensville, MD.

Maguire, M. J., Ewida, A. A., and Leonard, C. M. (2001). *Terra Nova FPSO: Certification and technical integrity challenges*. Offshore Technology Conference, OTC 13023, Houston, May.

McClelland, B., and Reifel, M. D. (1986). *Planning and design of fixed offshore platforms*. New York: Van Nostrand Reinhold Company.

Myers, J. J., Holm, C. H., and McAlister, R. F. (1969). *Handbook of ocean and underwater engineering*. New York: McGraw Hill.

Olsen, A. S., Schrφter, C., and Jensen, J. J. (2006). "Wave height distribution observed by ships in the North Atlantic." *Ships and Offshore Structures*, 1(1): 1–12.

OPL (2001). *Mobile production systems of the world*. Herefordshire, England: Oilfield Publications Ltd.

OPL (2002). *Floating storage units and shuttle tankers of the world*. Herefordshire, England: Oilfield Publications Ltd.

Parker, G. (1999). *The FPSO design and construction guidance manual*. Houston: Reserve Technology Institute.

Randall, R. E. (1997). *Elements of ocean engineering*. Jersey City, NJ: The Society of Naval Architects and Marine Engineers.

Ronalds, B. F. (2002). *Deepwater facility selection*. Offshore Technology Conference, OTC 14259, Houston, May.

Ronalds, B. F. (2005). "Applicability ranges for offshore oil and gas production facilities." *Marine Structures*, 18: 251–263.

Wang, G. (2003). *Experience based data for FPSO's structural design*. Offshore Technology Conference, OTC 15068, Houston, May.

Wang, G. (2005). Private communication. American Bureau of Shipping, Houston, September.

Whitehead, H. (1976). *An A–Z of offshore oil and gas*. Houston: Gulf Publishing Company.

CHAPTER 2

Front-End Engineering

2.1 Introduction

A new-build FPSO project may take 3–4 years to complete. For a successful project, a variety of steps must be carefully considered and implemented, including front-end engineering, design development, performance specifications, detailed specifications, vetting and selection of candidate yards and contractors, contract award, detailed engineering, construction, precommissioning (dock trials), sea trials, delivery, onsite commissioning, and contract acceptance.

Ship-shaped offshore units for developing offshore oil and gas fields in deep waters have been in existence in fair numbers since the early 1970s. In recent years, the complexity and size of ship-shaped offshore units have been increasing gradually; therefore, it is only natural that the issues related to designing, building, and operating these systems arise or may need revisiting in specific circumstances for the purposes of achieving a high level of the system integrity.

Moreover, in contrast to trading tankers and other types of offshore platforms, internationally agreed standards for ship-shaped offshore units are in a state of progress. In Chapter 3, we discuss how the classification societies (ABS 2004; BV 2004; DNV 2000, 2002; LR 1999) and institutions (API 2001) have, in recent years, developed guidance and rules specifically for floating, production, storage, and offloading systems (FPSOs), but there, for example, is not the unified approach to the rules that now exist for trading tankers (IACS 2005).

The requirements for designing and building ship-shaped offshore units are, in principle, different from the requirements of trading tankers because of the high level of onsite reliability necessary for long periods without the possibility of dry-docking. Further, FPSOs are very complex facilities that require coordinated efforts from all parties including the owners, shipyards, topsides integration contractors, hull engineering contractors, classification societies, and operators. Careful consideration and indepth engineering practices are required to properly design, construct, and commission any FPSO project. For instance, very detailed and comprehensive structural design specifications are essential.

In this chapter, we discuss engineering practices for ship-shaped offshore units and focus on FPSO systems. Selected key decisions and issues for designing, constructing, and commissioning FPSOs are presented. We emphasize front-end engineering and discuss overall project considerations but to a much lesser extent. For a greater exposition of project planning, see Parker (1999).

2.2 Initial Planning and Contracting Strategies

As we discussed in Section 1.6 of Chapter 1, it is important to decide if the FPSO will be either newly built or converted from a trading tanker, and owned or leased. These decisions are made considering many aspects, such as (Adhia et al. 2004a, 2004b):

- Economics
- Field life and how the asset is amortized over the field life
- Residual value of the used system
- Opportunities for redeployment

For the purpose of production, owners will usually consider the new-build or tanker-conversion option rather than leasing because drilling rigs are typically leased for the intermittent periods of drilling time during production. However, FPSOs can of course be leased for various periods of times, both long and short. The choice in this regard is primarily one of economics unique to a project. It is actually very rare, for many reasons, including justification for funds and host-country needs and regulations, that a company will design and build an FPSO for use in multiple fields over its design life.

In the initial planning, it is important to establish relevant contracting plans that could largely determine the success of the project. FPSOs today are increasingly constructed with shipyard involvement so that the related construction contracts are administered in a culture and atmosphere that is similar to typical shipbuilding. For cost and schedule reasons, the FPSO hull is usually built to ordinary shipbuilding practice, with specific enhancements where needed. Topsides processing facilities, however, are designed and fabricated by a separate contractor using offshore practice more akin to fixed platforms than to ships and are integrated onto the FPSO hull by the shipyard.

The design and building processes of FPSOs are unique and more complex compared to trading tankers. An FPSO project involves many elements, such as engineering of the hull, topsides, mooring system, integration of the topsides onto the hull, towing to site, installation on site, and commissioning. Compared to tankers, which are built in shipyards, there are several major interfaces to be managed, such as between the topsides facilities design and the hull design and between the multiple contractors involved. Usually, for financial and risk-spreading reasons, some sort of a consortium is formed involving owners, joint venturers, and operators; all have a say and can affect one or more of the work elements required in an FPSO contract. The success of such a contract depends on three major factors:

- Good engineering capability
- Good fabrication capability, including quality control systems
- Good project management, including cost and schedule control

Not all shipyards are necessarily specialized in all areas of required types of expertise including design, custom engineering, and project management. Some important aspects for success then include good front-end engineering; comprehensive technical specifications; clear scope of work; clear identification and management of all interfaces; effective and accurate communications among owners, fabrication yards and classification societies; an adequate and detailed construction plan; adequate site teams for construction supervision; adequate health, safety, and environment (HSE) and quality-control systems; standardization of equipment and materials; appropriate planning for procurement and supply of long-lead-time items; and avoidance of changes after contract. Useful guidelines and strategies for establishing good contracting plans are described by Parker (1999) and Adhia et al. (2004a).

2.3 Engineering and Design

Past FPSO projects indicate that many problems associated with costs and schedules can be traced to inadequate project definition and requirements, which ultimately necessitated related changes during the project execution phase (Adhia et al. 2004a). Therefore, front-end engineering and design (FEED), involving substantial engineering capabilities and taking account of lessons learned from FPSO projects, is necessary for any new project; and this needs to be carried out to the necessary extent before the development of specifications, the invitation to tender package, and, in general, the bidding phase.

After award of contract, and before fabrication starts, the relevant parts of detailed engineering must, of course, be completed. Preliminary safety studies, such as fire and explosion analyses and gas dispersion analyses for process facilities, may significantly influence the layout and design of the FPSO. These studies are carried out as part of the FEED, although detailed and specific studies of that nature will necessarily be part of the detailed design phase. Engineering, whether FEED or detailed, will deal with many of the same aspects to different degrees of sophistication; aspects considered may include the following:

- Vessel principal particulars and general arrangement
- Hull stability and strength analyses
- Vessel motion analysis
- Mooring system and station-keeping analyses
- Riser system analysis
- Turret system analysis and design
- Process plant layout and determination of support loads
- Operational and safety philosophies and plan development
- Risk assessment and management planning

When developing necessary specifications for an FPSO, adequate consideration of operational factors is particularly important to achieve the required high levels of onsite reliability and reduced downtime. In the case of the structure and safety systems, the vessel owner may often opt for classification, implying that in the design and construction phase the vessel will need to meet owner requirements as well as classification society rules, and various offshore industry standards. Although the classification rules do provide certain minimum requirements for structural integrity, these rules, in general, do not address the FPSO operational aspects that are equally important to the owner.

The owner requirements will then invariably involve functional and performance features including detailed prescriptive requirements for items that are not

Table 2.1. *A sample breakdown of an FPSO project cost*

Item	Cost division
Engineering and management	10%
Vessel hull and systems	40–50%
Process topsides	20–30%
Moorings and installation	4–5%
FPSO installation and commissioning	2–3%

adequately or clearly covered by the classification society rules. Similar enhancements may also be necessary regarding structural aspects, for reasons of the required onsite reliability. In addition, the owner will also be involved in the design review through the plan approval process and in the monitoring of the construction quality in order to ensure that all the requirements are met.

2.4 Principal Aspects Driving Project and Vessel Costs

The parameters driving the cost of FPSO projects and vessels (Parker 1999) can include the following:

- Field production profile over time
- Water depth, vessel size, and capacities
- Operational requirements for uptime and reliability
- Site and tow conditions and needs
- Process facility deck-space requirements
- Subsea design and manifold arrangements
- Support functions such as power generation and utilities
- Design life and the related structural integrity management philosophy
- Classification, verification, and regulatory compliance
- Safety in design

It is difficult to come by published cost data in the literature, but Table 2.1 shows a sample breakdown of the costs in an FPSO project. Additional examples of cost data can be found in Kennedy (1993) and Parker (1999). In general, such data must be interpreted with great care. The cost of a new-build FPSO is proportional to the hull size. The size, in turn, depends on storage production and offloading capacities required. Construction "friendliness" affects fabrication costs. Cost specifications and relative cost proportions vary from project to project and case to case; whether a new build or a conversion, design and operation philosophies and management priorities all affect the cost of an FPSO project.

2.5 Selection of Storage, Production, and Offloading Capabilities

The factors affecting the storage capacity include the following:

- Rate of production
- Size of export cargo parcels
- Number of different grades of production or export fluids
- Offloading system efficiency and other characteristics
- Buffer storage capacity requirements

The simplest way to select the storage capacity is to determine the most frequent large export parcel size and then add spare capacity to deal with, say, a 5-day production amount at the largest production rate to allow for additional storage due to events such as export shuttle tanker arrival delay and awaiting the favorable weather conditions for offloading. Additional flexibility is occasionally possible by reducing the production rate during unwanted events.

The optimization of storage capacity for a vessel can be performed by a cost–benefit analysis that takes into account hull size, export system capacity, production life cycles, and related costs. If more than one grade of production is planned, each must be segregated with its associated piping arrangements.

2.6 Site-Specific Metocean Data

In Chapter 4, we describe the meteorological and oceanographic (metocean) data for the operational site required for the development of an FPSO design. Wind, waves, and current data for new fields must often be obtained by measurements, hindcasting, or from comparable situations. For anchoring, piling, and subsea construction design activities, bathymetric and geophysical data also need to be developed.

Based on the site-specific metocean data, various design parameters must be determined, generally in terms of 1-, 10-, 50-, and 100-year return periods, and stated in a design basis. Relevant information in the design-basis document includes the following:

- Wind in terms of extreme speed and direction, vertical profile, gust speeds, and spectra
- Waves in terms of joint probability of significant wave height and period; extreme wave crest elevation; extreme wave height, direction, and range of associated period; cumulative frequency distribution of individual wave heights and steepness; and wave spectra and direction spreading
- Water depth in terms of water depth below mean sea level and extreme still-water-level variations
- Currents in terms of extreme current speed and direction, variations through the water depth, current speed for fatigue design, joint probability of wave, current occurrence, and extremities
- Temperatures in terms of extreme air temperatures (maximum and minimum) and extreme sea temperatures (maximum and minimum)
- Snow and ice accretion in terms of maximum snow thickness, and maximum ice thickness, densities of snow and ice
- Marine growth in terms of type of growth, permitted thickness, and terminal thickness profile

The determination of these parameters is very important for establishing the different environmental conditions for the different operational and extreme responses such as for mooring forces, hull-bending moments, green-water loading, bow slamming, and steep-wave impacts. It is also important to note that FPSOs behave in a much more complicated way than, say, fixed offshore platforms, and that a much more detailed understanding of the environment is needed because of the much greater importance of wave period and joint probabilities of waves, current, and wind that affect the responses to waves.

2.7 Process Facility Design Parameters

The design requirements for the process facilities can be established based on the following, also to be stated in the design-basis document:

- Maximum oil, gas, and water production
- Well fluid characteristics
- Water and gas injection rates and pressures
- Produced oil storage temperature

The process facility design will also depend on minimal or full offshore processing. The minimal offshore processing option sends all fluids produced to an onshore terminal (or terminals) for final processing, while full offshore processing makes saleable products on the offshore facility. For a useful review of various offshore/onshore processing options, see Bothamley (2004).

2.8 Limit-State Design Requirements

The structure must be designed with a high level of structural integrity during its design service life so that it should achieve uninterrupted and safe operation on site. Repairs on site as well as by dry-docking can be largely impractical considering their high costs and difficulties with hot work; this is in marked contrast to trading tankers that can be dry-docked every 5 years.

A factor complicating an FPSO design is that long-term service data related to certain design aspects may be lacking. For example, even for benign environments off Africa, conditions of high year-round temperatures and humidity may be present but data are not available. Related data on coating durability and corrosion wastage may be limited.

An important aspect that should be considered in establishing the requirements of FPSO structural integrity is that FPSOs are different from trading tankers in many ways, as we described in Section 1.5 of Chapter 1. The following features are some of the significant differences (Adhia et al. 2004a, 2004b):

- Environmental actions of FPSOs are different as compared to trading tankers, for example, long-crested waves versus short-crested; waves highly directional weather versus multidirectional.
- FPSOs operate at constantly changing draft with frequent offloading cycles.
- Loading patterns of FPSOs can be quite different from those of trading tankers.
- Production systems of FPSOs can be subjected to differing levels of motions and are impacted by the hull flexing and deformation of the main deck.

For the design of an FPSO structure, the following analyses need to be performed:

- Vessel motion analysis considering interface among hull, topsides, and mooring systems and taking account of omnidirectional and noncollinear features associated with wind, waves, and currents together with weathervaning

- Action-effect analysis at global and local levels
- Serviceability limit-state assessment
- Ultimate limit-state assessment
- Fatigue limit-state assessment
- Accidental limit-state assessment

Vessel motion analysis will define the site-specific design actions on the offshore unit involving hull structures, topsides, moorings, and risers, and the action-effect analysis will define the corresponding action effects (e.g., working stresses, deformation); see Section 2.15.

Serviceability limit-state assessment needs to be undertaken to check exceedance of criteria governing normal functional or operational use of the offshore unit. As we describe in Chapter 5, various factors and their limits must be considered. In particular, structural damage due to impact-pressure actions is a primary concern in terms of (a) bow for impact from steep waves; (b) forebody for slamming; (c) internal structures for sloshing; and (d) deck structures and topsides for green-water loading.

Ultimate limit-state assessment must deal with buckling and collapse of individual structural components and the hull to assure the adequacy of structural safety at local and global levels. Chapter 6 presents the ultimate limit-state assessment in detail.

During a fatigue limit-state assessment, parts that are typically found to need the most careful analysis and design include the following: (a) bottom, deck, and side shell details; (b) internal structures subjected to stress ranges from loading/unloading cycles; (c) hull openings; (d) mooring turret and connections to hull; (e) process plant and pipe run seatings to hull; and (f) interface structures such as module support stools, and flare tower base. Chapter 7 describes emerging practices of fatigue limit-state assessment.

Accidental limit-state assessment needs to be performed to check situations of accidental or abnormal events such as unintended flooding and subsequent progressive hull collapse or loss of stability or survival buoyancy; collisions; impacts due to dropped objects; fire and heat; and gas explosion and blast. Chapter 8 describes current practices and recent advances of accidental limit-state assessment.

2.9 Risk-Assessment Requirements

Qualitative and quantitative risk assessments, described in Chapter 13, are required to consider all potential hazards. Major hazardous and risk scenarios that are considered include the following:

- Unintentional release of flammable or explosive materials
- Hydrocarbon fires and explosions
- Extreme weather and structural failure effects
- Dropped objects
- Collisions
- Helicopter accidents
- Smoke and gas ingress into safe refuge
- Loss of mooring and station-keeping integrity
- Green-water risk (to person as well as to structure)

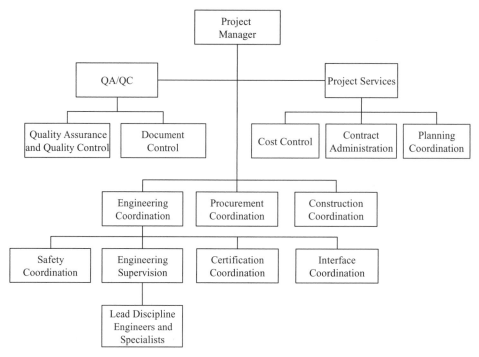

Figure 2.1. A sample project-management organization.

Risk assessment using qualitative and quantitative approaches needs to be performed to determine the overall characteristics and performance specifications for the following systems and purposes: (a) accommodations and temporary safe refuge; (b) passive and active fire protection systems; (c) gas explosion and blast protection; (d) escape, evacuation, and rescue; (e) fire and gas detection systems; (f) emergency shutdown systems; (g) emergency power generation system; and (h) relief, blow-down, and flare systems.

Chapter 11 describes risk-based technologies that are also used to establish cost–benefit schemes for inspection and remedial actions for aged hull structures. Ideally, risk considerations should also be used during the design stage to select steel grades, fatigue lives, and design schemes to ensure that the designed structure is safe enough.

2.10 Project Management

Project management is of great importance to the success of FPSO projects; the same should effectively control and manage any restraints or constrictions relating to the projects, which may be created by other concurrent projects, production and fabrication limitations within shipyards, and a number of other factors such as material procurement and cost control.

Figure 2.1 shows a sample chart of the project-management organization. Project management encompasses various aspects relating to project engineering, procurement and construction, and the related planning, contracting, monitoring, and cost control. Each key function is to be supported by an adequately staffed team with the right resources and expertise.

2.11 Post-Bid Schedule and Management

One of the most important elements for the success of any FPSO project is effective monitoring and follow-up action in order to keep the planned schedule of construction and delivery. The planned schedule itself is a function of various factors including capabilities of the building yard and its contractors.

Hence, before any final commercial commitment is given to any shipyard, potential construction facilities must be evaluated in terms of physical facilities (e.g., steelwork prefabrication, dry-docks); management systems (e.g., project-management system, quality assurance and quality control procurement, and preoutfitting experience); discipline and trade resources (e.g., engineering manning levels; steelwork and outfitting trade levels, hook-up, and commissioning resources); and corporate considerations (e.g., previous offshore sector experience, fiscal stability).

The construction schedule for a typical FPSO project can be broadly broken down into four quartiles; namely, those related to (a) engineering and procurement; (b) prefabrication and preoutfitting; (c) vessel erection, outfitting, and process installation; and (d) final outfitting, hook-up, commissioning, and completion. Table 2.2 shows a sample schedule for an FPSO construction during a 24-month period.

Figure 2.2 shows the so-called 'S' progress curve for FPSO projects from engineering and procurement to commissioning, following Parker (1999). In Figure 2.2, an example of schedule slippage is illustrated, where the slippage begins in the first quartile regarding engineering and procurement. The recovery of a delayed schedule during the second quartile can be difficult because the steelwork prefabrication, preoutfitting, and pipework production activities can be significantly affected by delayed engineering and procurement.

It is indeed one of the primary functions of project management to anticipate and avoid schedule slippages altogether to the greatest extent possible. Recovery from delays can be accomplished in many ways, but usually at the cost of additional resources because a greater work volume than planned must be made up in a shorter period of time than originally planned. For a good discussion on project management, see Parker (1999).

2.12 Building Material Issues: Yield Stress

The selection of building materials is an important design consideration for the structural design of FPSOs. Vessel hull steel is usually selected according to classification society rules (ABS 2004; BV 2004; DNV 2000; 2002; LR 1999) or recommended practices (API 2001); see also Chapter 3 of this book.

To achieve high levels of buckling and fatigue performance, it is usually recommended to limit the amount of high-tensile steel (HTS) used in the FPSO to as small a proportion as possible. Mild steel of the rule minimum yield strength of 235 N/mm^2 will then be used for as much of the hull structure as possible. HTS would only be allowed for longitudinal strength members at limited extents of the bottom and deck areas within the cargo tank region (e.g., hull, cargo tanks, slop tanks, and ballast tanks). HTS may also be allowed in case of plate thickness and scantlings of mild steel exceeding 30mm to avoid heavy welding and for ease of construction.

Table 2.2. *A sample schedule for an FPSO project (post-bid).*

Schedule in month

	Activity	1	2	3	4	5	6	7	8	9	10	11	12	13	14	15	16	17	18	19	20	21	22	23	24
Vessel construction	Basic vessel engineering															↑									
	Detail vessel engineering																			↑					
	Vessel procurement																						↑		
	Vessel fabrication																↑								
	Vessel pre-outfitting																	↑							
	Vessel berth erection																				↑				
	Vessel outfitting on board																							↑	
	Vessel testing and trials																								↑
Turret construction	Turret engineering											↑													
	Turret procurement													↑											
	Turret fabrication																		↑						
	Turret outfitting																								
	Turret testing																								
Topsides construction	Topsides engineering										↑														
	Topsides procurement											↑													
	Topsides fabrication																	↑							
	Topsides installation																			↑					
	Topsides hook-up and commissioning																							↑	

40

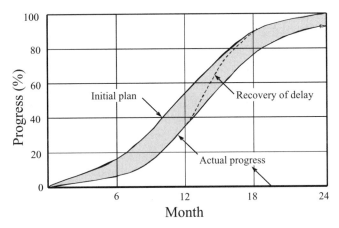

Figure 2.2. A sample progress curve for an FPSO project – post-bid, following Parker (1999).

The use of only HTS not exceeding the minimum yield strength of 315 N/mm^2 is recommended and allowed by some companies when mild steel cannot be used. The reason for this requirement is mainly due to the long-life requirement on site without dry-docking and the relatively greater strength impact of corrosion on thinner HTS plates. It is of interest that there has been limited research so far to establish corrosion rates and minimum thickness requirements considering FPSOs that need to be on site for many years and potentially subject to specific site characteristics with potential year-round high temperatures and humidity (Paik et al. 2003). For a detailed description of corrosion assessment and management, see Chapter 10 of this book.

2.13 Building Material Issues: Fracture Toughness

Structural fractures associated with cracks are often classified into three modes: brittle fracture, ductile fracture, and rupture (Paik and Thayamballi 2003). When the strain at fracture of material is very small, it is called brittle fracture. In steel structures made of ductile material with adequately high fracture toughness, however, the fracture strain can be comparatively large. When the material is broken by necking associated with large plastic flow, it is called rupture. As a failure mode, ductile fracture is an intermediate phenomenon between brittle fracture and rupture. Ductile behavior is a very desirable phenomenon in structural design.

Although fatigue cracking is still relatively common in trading ships, brittle fracture is generally not (Sumpter and Kent 2003). However, it is also worth noting that brittle fractures still occasionally occur, for example, in the bulk carrier Lake Carling as described by Drouin (2006). The risk associated with brittle fracture, although highly dependent on temperature and strain rate, can be significant.

Many common ship-grade steels do not require specific toughness control by testing, even though subject to a material qualification test such as the Charpy V notch impact test. The presumption in those cases is that for certain applications and ranges of thickness, the chemistry and steel-making methods by themselves are adequate, and no toughness testing is required. Owners, designers, and operators need to be aware that this presumption is not necessarily fail-proof.

Because mild steel is desired in FPSOs, a high proportion of steel therein can potentially be grade A, for which there are no classification society requirements for fracture toughness testing. Grade A steel could, under certain circumstances, pose a concern to structural safety, albeit in relatively few cases historically. A similar situation exists for grade B steels under a certain thickness limit. The initiation toughness of such steels is generally adequate, but it is also variable; hence the risk of brittle fracture (Drouin 2006). It is therefore recommended that the steel and welding used for the FPSO structure be demonstrated to meet adequate Charpy V notch impact-energy requirements applicable for the intended service. Related Charpy test absorbed-energy requirements are recommended to be included in the specifications for all hull steel.

2.14 Hull Structural Scantling Issues

Trading tanker design rules have been developed over many years based on a semiempirical approach, and it is only in more recent years that *first-principles approaches*, with considerations of dynamic load components, have come into being, and are indeed now being unified for tankers (IACS 2005). However, there is not a generally universally accepted set of rules for FPSO designs yet, although various recognized classification societies have developed their own rules and guidelines (ABS 2004; BV 2004; DNV 2000, 2002; LR 1999).

FPSO hull structures for locations such as the North Sea, which are subject to an environment more severe or harsher than that of trading tankers in worldwide service, are ideally new builds. In practice, some FPSO hull structures are sited in environments more benign than that of trading takers in unrestricted service, but are usually designed to be at least as strong as trading tankers. Their design then is at least suitable in terms of strength for the so-called unrestricted oceangoing service based on the classification society rules.

Due to the vast and good experience of trading tanker structures designed by classification society rules, such an approach is usually recommended wherever possible. The classification society approach also provides some flexibility in case the vessel is relocated to another site or needs to be taken to a shipyard for major modifications or repairs. In the case of benign environments, classification societies may allow a limited amount of reduction in hull girder section modulus for site-specific operation. Considering the related lack of service experience behind such recommendations, it may not be prudent to accept such reduced scantlings.

In principle, assuming that service-proven first-principles approaches to structural design are available, an FPSO structure for a benign environment can be designed solely for the site-specific service demands and the transit route environmental conditions, with motions and acceleration forces associated with the extreme condition based on a 100-year return period for site-specific conditions, 10 years for the tow, and perhaps 1 year for onsite inspection and maintenance conditions. Although many components of such direct analysis procedures now exist and are indeed used for checking designs, the related risk is still difficult to evaluate correctly because relevant service experience is generally lacking.

The fatigue performance of an FPSO can be designed to be equivalent to that of trading tankers, but there is an important difference to bear in mind: the fatigue design procedures and targets used in trading tanker designs are more appropriate for a structure that can be dry-docked every 5 years for thorough inspection and repairs than for an FPSO that may be meant to stay on station, ideally without repairs, for the life of a field.

In addition, the side shell structure in the fender areas should be designed for absorbing the energy impact of supply boats and of export vessels in case of side-by-side offloading operations. Slamming, sloshing, and green-water loads, as well as accidental actions, are to be considered. Lay-down areas may need additional margins of thickness to allow for coating damage and possible abrasion in these areas.

2.15 Action-Effect Analysis Issues

For the requisite structural reliability and integrity, and also to appropriately consider circumstances unique to an FPSO, rigorous structural analysis or action-effect analysis must usually be performed using direct load analysis and refined finite-element models of the whole FPSO structure including topsides for unrestricted, towing, and onsite conditions.

However, the classification society rules for minimum requirements of trading tankers may not be allowed to be reduced as a result of these analyses, at least to any significant extent even in conversions. This is important to ensure the required reliability with reduced probability of repairs and associated economic consequences. Comprehensive stress-range analysis is required for the fatigue limit-state assessment of a whole FPSO hull, topsides, topsides supporting structures, and other structural details. The results are also used to identify critical locations prone to fatigue cracking that require special attention during construction (i.e., enhanced quality assurance and quality control) and also for inspections during operations.

Advanced structural analyses include dynamic load analysis (Liu et al. 1992; ABS 2004), spectral fatigue analysis, vibration analysis, and sloshing analysis. In some cases, the effect of expected fabrication deviation may need to be included explicitly in the strength evaluation for fatigue limit states as well as ultimate limit states. Analysis for loading and off-take low-cycle fatigue effects may also be necessary.

Use of thick sections and associated design assessment with tertiary stresses is not common in ordinary ship designs, and finite-element analysis is usually based on primary and secondary membrane behavior, with classification society rules-based allowable stress reaching as high as 90 percent of yield stress in some cases (or even yield, with a so-called net structure approach). Although such a high allowable stress-based approach may be justifiable for unrestricted service of trading tanker hulls through experience alone, their unquestioned use in the case of FPSOs is more problematic. If the FPSO has thick sections, however, the effect of tertiary bending stresses should be taken into account in the analysis and design where necessary (Paik and Thayamballi 2003).

The towing condition should be considered carefully because towing, and associated actions such as slamming, should be a primary strength issue in structural analysis and design. Short-term conditions with tow-line failure and a resulting direct beam sea case should be considered in structural analysis. Also, hindcast data, including storms and cyclones if relevant, could be more widely used for structural analysis instead of using global weather statistics type of visual data gathered mainly from trading merchant ships for tow load determination, because the latter type of data can often have a significant degree of rough weather avoidance implicit in them (Olsen et al. 2006); whereas a towed FPSO may have relatively smaller opportunity to avoid rough weather depending on the circumstances, such as speed of tow, storm development, and distance to safe harbor.

2.16 Fatigue Design Issues

The fatigue safety factors should be based not only on the required high degree of structural reliability but also on abilities to easily and safely inspect, maintain, and repair the structure. Fatigue safety factors to be applied may be used to appropriately take into account the degree of consequence of failure of structural details.

Generally, the classification society guidelines for FPSOs do specify safety factors, which are based on the ability to inspect and on the criticality of the structural components with respect to consequence of failure. However, the classification societies often associate consequence of failure primarily with structural strength, safety, and pollution, whereas in addition to these, the owners and/or operators need to consider consequences related to operability and economic aspects such as the impact on production (financial impact from loss of production or downtime) and the difficulties in making repairs on site.

For trading tankers, classification societies typically require a fatigue safety factor of 1 in unrestricted service, presuming regular inspections and dry-docking, implying that all structural components are accessible for inspection. For FPSO hulls, some classification societies only note that safety factors greater than 1 need to be considered for critical or noninspectable locations, without specifying a required increase. For example, DNV (2006) requires a safety factor of 2 for submerged parts of the outer shell and internal structure directly welded to it, and a safety factor of 1 for most of the ordinary, inspectable locations (i.e., the majority of the hull). For fatigue limit-state design involving fatigue safety factors, see Chapter 7 of this book.

Implicit structural reliability levels in the classification society rules for trading tankers are lower than those for permanently moored offshore structures. This may be due to the fact that ship rules were developed originally from a semi-empirical approach as opposed to a rational-based approach, higher structural redundancy is inherent in ship hull structures, and ship hulls are relatively easy to inspect in a dry-dock as compared to other offshore structures on site.

For new-build FPSOs, therefore, more stringent requirements on fatigue safety factors than those inherent in ship rules must be specified because FPSOs are intended to operate for prolonged periods with a high degree of reliability. In practice, for the transit and site conditions, fatigue safety factors varying from 2 to 4 for the hull are recommended depending on location characteristics (i.e., inspectability, repairability, maintainability, redundancy, and consequences of failure of structural components). For topsides structures, the offshore standards with the fatigue safety factor of 2 are often used (API 2001). For the unrestricted oceangoing condition checks, a fatigue safety factor of 1 or 2 may be used for the hull, depending on the location and weather conditions.

2.17 Hydrodynamic Impact-Pressure Action Issues: Sloshing, Slamming, and Green Water

For FPSO designs, structural damage due to impact-pressure actions arising from sloshing, slamming, and green water can be a significant issue. The profile of impact-pressure actions in terms of pressure versus time history must be identified and the consequences of the actions need to be analyzed. Risk assessment may also be needed to develop appropriate measures for risk mitigation.

A detailed description of this issue is given in Sections 4.10–4.12 of Chapter 4. For recent studies on green-water and wave-impact phenomena on ships and similar offshore units, see Barltrop and Xu (2004), Fyfe and Ballard (2003, 2004), Guedes Soares (2004), Hodgson and Barltrop (2004), Kleefsman and Veldman (2004), Voogt and Buchner (2004), and HSE (2005).

2.18 Vessel Motion and Station-Keeping Issues

FPSOs may be subjected to large motions due to several reasons, including the occurrence of beam sea conditions, which can occur not only in particular cases for an FPSO moored with a spread-mooring system but also in case of line failure or unusual weather conditions. Conceivably, it can occur even for a weathervaning system when heavy currents are moving in different directions than the wind and the waves.

In addition, a high center of gravity due to topsides equipment can potentially lead to larger vessel motions of an FPSO than a trading tanker. Large motions can result in oil–water and gas–liquid separation process problems, slamming, and green-water occurrence. Motion and hydrodynamic analysis using advanced analysis methods may be considered to predict and keep such phenomena within limits. The roll motion is strongly governed by nonlinear viscous damping. Accurate estimation of roll damping and related validation with experimental results is important to ensure that structural and functional or operational requirements of the FPSO are adequately met.

For operational safety and efficiency, a vessel's motion must be minimized or optimized by considering overall behavior and interaction of the complete vessel, including its moorings, risers, and offloading systems at the design stage. To avoid any special issues related to the process facilities, motions are commonly limited to pitch within 10 degrees double amplitude (±5 degrees) or similar and roll within 20 degrees double amplitude (±10 degrees) or similar.

Station-keeping or heading-control systems may be required to minimize a vessel's motion. The required degree of a heading-control system should account for the weathervaning capacity of the vessel considering wind, waves, and current actions together with the turret location. The following are some of the parameters that must be considered for the design of heading-control systems: (a) offloading system uptime; (b) heading control for survival; (c) reduction in peak line loading; (d) heading control for human, process, and riser comfort; (e) line failure conditions; and (f) vessel and helicopter transfer operations.

The cost-effective method to increase the roll damping and reduce the roll motions is to increase the length and size of the bilge keels. Strength and fatigue performance of bilge keel structural details should be checked in this case.

2.19 Topsides Design Issues

The topsides facilities and specific equipment will need to be commensurate with what the production needs, as we describe in Chapter 9. Considerations need to be given for marinization, motions, specifications, layout, and integration with the hull. Some of the standard topsides equipment used for processing on an FPSO may have been developed for fixed-type offshore platforms or onshore applications.

Both for hull and topsides, the equipment vendor must be selected carefully. Standardization, spare parts availability, maintenance, and equipment alliances available to the owner should be considered when selecting the equipment vendors. These requirements should be considered against any standard equipment vendor preferences and alliances that the FPSO contractor or shipyard may have. Relying on the FPSO contractor's approved or preferred equipment vendor list may in some cases result in initial cost and schedule savings, but this should be balanced against potential future operating and maintenance concerns (Adhia et al. 2004a, 2004b).

Special considerations need to be given for placing equipment on a vessel in motion. This may include using special baffling or packing in a vessel to absorb the motion; structures and foundations designed for vessel accelerations; providing access to the equipment; handling of the equipment; and taking drainage issues into account.

In addition, gas accumulation, especially between the hull main deck and topsides, and venting of hull systems need to be considered specifically. The vessel's predicted motions need to be considered in the facility equipment design and layout in order to design for or around the various operational constraints from sea conditions. Equipment sensitive to motions of the vessel may be located near or at the midship and center line where the motions and accelerations are the least. Horizontal instruments containing liquids should be orientated longitudinally to minimize the sloshing effect.

Typically, topsides facilities design will follow offshore standards and practices, and the marine facilities design will follow the shipyard or marine standards and practices, as described in Chapter 3. It is important to consider maximizing commonality and standardization in equipment and components for ease of future operations and maintenance. For example, instrumentation units for equipment may be different between marine and topsides equipment; but, if appropriate, a common instrumentation unit methodology may be developed and used.

The interaction of the topsides to the hull is an important issue, as discussed in Chapter 9. In addition to the deformation of hull and topsides, transferring the topsides facility loads to the hull structure efficiently must be taken into account. Because the vessel is subject to continuous motions and loading changes, the effects of fatigue on the topsides facilities are an important consideration.

The overall aim in design should be to obtain an efficient, cost-effective, and complete FPSO system, incorporating topsides and marine components to optimal extents. Chapter 9 further discusses topsides design and building technology with an the emphasis on structures.

2.20 Mooring System Design Issues

It is important to appropriately select the relevant mooring system. Typical mooring arrangements used by many FPSOs in recent years are as follows:

- Spread mooring
- External turret mooring
- Internal turret mooring
- Disconnectable turret mooring

Some illustrations of various types of mooring systems noted here are presented in Chapter 9. The chosen mooring system will vary from one project to another, and many factors go into selecting the appropriate mooring arrangement. Regarding design, one consideration is the risk of collision, which includes factors such as offtake frequency, environmental conditions, and export tanker size.

A spread-mooring system is usually suitable for reasonably deep – say, deeper than about 150m – to ultradeep waters. These can be used only in moderately benign environments. There is little practical limit on the number of risers. With a spread-mooring system, it is usual practice to provide a catenary anchor leg mooring (CALM) buoy for export operations. In recent years, FPSOs in benign areas such as West Africa and Indonesia have quite often used a spread-mooring system.

External turret-mooring systems are suitable for reasonably shallow – say, about 30m deep – to deep waters. These can be used in moderate to severe environments. For a severe environment, protection of risers from wave damage has to be considered, and this could be a limiting factor. The typical limit on the number of risers is about twenty. With an external turret, it may be possible to use tandem as well as side-by-side offloading.

Internal turret-mooring systems are suitable for reasonably deep – say, deeper than 150m – to ultradeep waters. These can also be used in moderate to severe environments. A typical limit on the number of risers is about 100. With an internal turret, it may also be possible to use tandem as well as side-by-side offloading. The integration of turret into hull has to be carefully designed to ensure strength of hull structure. Also, an internal turret could result in some reduction in available deck area for the topside footprint; this is a consideration that is relatively more important to conversions than new builds.

2.21 Export System Design Issues

Offloading system design has some challenging issues that include system layout and capacity. Typical layout arrangements used for oil-export operations are tandem offloading, side-by-side offloading, and a CALM buoy located at a distance from the FPSO, as described in Chapter 9. Various factors need to be considered in selecting the export arrangement; and all three types have been used successfully in past projects. A backup arrangement is also highly desirable in case of failure of the primary export means.

An offloading system should be able to safely unload the oil, liquefied petroleum gas (LPG), or liquefied natural gas (LNG) from the FPSO concerned to the export tanker, and it should also be able to accurately measure the quantity and quality of the export product. The product also has to be exported at sufficient rates to avoid incurring demurrage on the export vessels.

For instance, it may be considered that a million-barrel oil parcel needs to be offloaded in less than 36 hours from the arrival of the export tanker at the FPSO to its departure. The loading rate selected for a design should also allow for connection time, slow-down during start and finish (topping-up) operations, paperwork, and disconnection times. This means that offloading capacity should allow for full parcel offloading within 24–26 hours with remaining times for all other activities such as connection, start, topping-up, and paperwork. For a larger parcel size, the industry may use longer periods, for example, 72 hours for a 2-million-barrel parcel size.

2.22 Corrosion Issues

As an FPSO structure ages, it deteriorates by corrosion and by fatigue. Corrosion management by adequate measures such as corrosion margin addition and coatings, especially for the inaccessible areas of the hull and tanks, is important to ensure integrity and survival of the FPSO without the need to dry-dock during the long design life in the field, as we describe in Chapter 10. In selecting and using corrosion-protection measures, relevant environmental regulations must also be considered; for example, the use of tributyl-tin-based antifouling paints is not permitted.

Most marine and topsides systems of an FPSO hull may be outfitted after the hull leaves dry-dock. This outfitting work is typically carried out at a wharf where the hull is berthed. It is not unusual for the outfitting work to take 1–2 years.

In such cases, one concern is that some corrosion might occur; another is that antifouling paint will start to leach out as soon as the paint comes in contact with water. This antifouling paint will have a finite life in water and, therefore, any reduction at the outfitting stage will reduce its life in the field accordingly. In such cases, a dry-docking for restoration of hull coatings will need to be considered. For additional discussion on this subject, see Ximenes et al. (1997) and Adhia et al. (2004a, 2004b). A detailed description of corrosion management and control is given in Chapter 10.

2.23 Accommodation Design Issues

The foremost consideration that needs to be given for an accommodation facility design is its size. The size and number of beds in the accommodation facility is dependent on the number of people necessary to operate and maintain the vessel, along with the necessary support crew for food services and so on. Additionally, spare beds may need to be allocated for installation and commissioning as well as future maintenance or construction periods. Adequate life-saving and emergency-escape equipment will need to be available for the full complement.

FPSO accommodation design should use flat-plate-type construction with stiffeners instead of corrugated-plate-type construction; this makes for easier coating of the flat-plate surface during a long life in the field. This can also more efficiently provide the necessary strength to withstand blast pressure, where required.

Consideration should be given to the prevailing wind and its direction(s) when locating the accommodations, which should be upwind of the processing facilities, from the helipad location, and helicopter approach route. Additionally, the relative location of the accommodation facilities to the process facilities should be considered for fire, gas explosion, and blast protection requirements.

2.24 Construction Issues

To ensure high integrity and uninterrupted long life without dry-docking, the selected areas of the hull and interface structure must be fabricated with higher standards than those used for trading tankers (Ximenes et al. 1997; Adhia et al. 2004a, 2004b). Selective applications can be considerably more cost effective than across the board enhancements in standards. Therefore, for important parts of structure construction, the following types of enhancements to typical shipbuilding standards are recommended:

- Reduced limits for acceptable misalignment of cruciform joints and verification by measurements and statistical analysis of actual fabrication deviation
- Enhanced quality assurance and quality control of selected critical joints, including additional nondestructive testing (NDT), and fit-up inspections to achieve good fit-up and sound welding
- Enhanced statistical control of fabrication deviations and more stringent fabrication tolerances for selected higher-strength steel structural components, and also for LNG or LPG tank areas, when applicable

Critical areas for application of enhanced standards may be selected based on the closeness of the predicted stresses and fatigue lives to the corresponding allowable measures, and on generic experience with the type of structural details involved. It is also recommended that all critical areas be identified in the design drawings and the production drawings; such information can be used for inspection purposes during service to ensure structural integrity.

2.25 Equipment Testing Issues

Before the FPSO leaves the fabrication site for commissioning, it is necessary to test systems and equipment as much as possible. Adequate and complete testing minimizes risks that require potentially higher costs associated with correcting deficiencies offshore. In general, every effort must be made to make sure that the FPSO is precommissioned and tested as far as practicable prior to its departure from the fabrication yard.

Consideration should then be given to what extent testing and commissioning must be completed prior to the vessel leaving the construction site. This should be clearly understood and communicated to all parties and should be spelled out in the contract and allowed for in the project schedule because there may be towing and offshore installation commitments that force departure from the yard prior to complete testing.

The offshore industry has adopted a very structured checklist approach to testing and commissioning that normally applies to topsides facilities. The process usually includes phased testing, which consists of three phases: (a) prior to energizing a system; (b) tests performed as the systems are energized; and (c) run testing. Test results in each phase should be well documented and signoffs are required at the various phases.

It is interesting to note that testing in the shipbuilding practice is slightly different. Although all of the same tests are usually performed for trading ships by shipyards, signoffs are normally not done until the run tests are complete. Although both methods can work, it is necessary, by contract and specifications, to ensure that a consistent methodology for the testing, commissioning, and acceptance is established and used.

2.26 Towing Issues

Ship-shaped offshore installation projects require transoceanic tows because the structures are often built thousands of kilometers from the sites of operation. A typical tow method is to use several oceangoing tugs to move the ship-shaped offshore unit or a barge-mounted structure at relatively slow speed. Towing to the site may take several months in some cases; for example, a tow of a barge-mounted structure from the Far East to the Gulf of Mexico typically takes 3 months.

The issues for towing arise because the structure to be towed can be subjected to extreme waves, wind, and currents during tow. Appropriate tow design procedures and criteria involving metocean route statistics in terms of wave heights, wind speeds and current velocities must then be developed and used. It is important to use high-quality metocean data from reliable sources to make tow response predictions and for strength and fatigue assessments related to tow. Hindcasting technology has been used successfully for the purpose, but a good tow contractor typically uses multiple sources of data and methods. Useful technology for tow simulation of ship-shaped offshore structures together with their tow criteria can be found in Lacey et al. (2003).

Towing the FPSO to site is an expensive operation in terms of towing costs and insurance. A successful tow is critical to the overall success of the FPSO because there is a significant probability of matters going wrong during a tow.

Past experience indicates that single or multitug tows have been used successfully depending on the circumstances. The size and number of tugs is a function of the FPSO size and shape, the tow environment, and the distances involved. It is important to ensure that adequate redundancy exists in the number of tugs to be used. In particular, the number of tugs used in the high-risk areas of the tow route may need to be considered specifically. As the number of tugs increases, so do the costs. However, with greater tug redundancy, contingency plans are more easily developed and implemented in the event that there is a problem with any of the essential tugs.

Unmanned tows have, of course, been accomplished successfully. The advantages and disadvantages of manned tows or unmanned tows need to be evaluated. Manned tows are more expensive than unmanned tows. This aspect needs to be balanced by the benefit of potentially performing some commissioning and start-up activities prior to reaching the field. It has also been argued that a manned tow inherently makes for "better" preservation and maintenance than an unmanned tow.

2.27 Field Installation and Commissioning Issues

Field installation, commissioning, and start-up require good coordination with the shore base. It is important to arrange for relevant importation and customs clearance issues, logistics for men and supplies, scope of commissioning, and handover of operations. An undefined or poorly understood customs clearance procedure can delay the arrival and have a domino effect on associated activities. Therefore, it is necessary to use well-established clearing agents who have developed relationships with customs and government authorities to assist with the import, arrival, and clearance issues.

Logistics are important upon arrival of the FPSO to the site. There are logistical requirements for manning the vessel for the commissioning crew, catering, and bunkering. The sequence and the extent of these logistical operations must be developed in an adequate way. The manning-up plan must take a practical approach for consideration of available accommodations at site on arrival and the length of time necessary to commission and make ready the available accommodations. Transportation constraints and availability must be considered for the manning-up plan.

Installation activities and operations need to be carefully planned to ensure the integrity of the FPSO because the activities and operations are usually sensitive to the weather. Close monitoring of weather forecasts is necessary, and proper contingencies should be planned in case of bad weather.

Supplies necessary for commissioning and start-up and the local availability need to be planned for, including but not limited to fuel, potable water, hydro- and leak-testing mediums, and inerting medium. The full commissioning scope required after the FPSO arrives at the site needs to be understood and planned in detail well in advance. The scope of commissioning will have an impact on how fast the start-up can occur.

2.28 Inspection and Maintenance Issues

Application of a good inspection and maintenance scheme is important to maintain the safety and integrity of an FPSO installation over its long life without dry-docking. Keeping the FPSO on site for long periods without dry-docking requires in-water surveys in lieu of dry-docking and other inspections and maintenance. In-water surveys require proper marking by weld bead and long-lasting nonleaching paint of the hull and tank boundaries to allow diver inspection of the hull condition. Also, adequate means of access for general inspection of tanks need to be built in.

The following design features and considerations are also pertinent to ease the inspection, maintenance, and relatively uninterrupted operation offshore:

- Sea chests and sea valves capable of inspection without dry-docking
- Adequate means for plugging and redundancy
- Adequate and safe isolation to allow simultaneous operations with inspection and maintenance
- Long-life equipment to minimize maintenance
- Adequate access and means of handling equipment
- Appropriate material selection for valves, pipes, and equipment to ensure long life and trouble-free service

In addition, many philosophical issues, such as pump room versus in-tank pumps, equipment in hull versus topsides, and gas turbine power plant versus steam turbine or diesel-engine power plant, need to be considered and resolved. In general, because many possibilities exist in association with repair and maintenance, all such issues should be evaluated carefully to make the appropriate selection particular to the circumstances. Chapter 11 presents advanced methodologies for inspection and maintenance.

2.29 Regulations and Classing Issues

In contrast to trading ships, the regulatory framework and standards for FPSOs are more varied. All FPSOs are required to meet some regulatory regime. The details differ depending on local regulations and also the owner's philosophy; that is, the owner may opt to comply with additional nonmandatory codes and regulations as a matter of good practice. In the larger context, important basic decisions related to regulations and classing of the FPSO tend to involve the following aspects (Adhia et al. 2004a, 2004b):

- What regulatory rules to apply?
- To class the FPSO or not?
- To certify the FPSO or not?
- To have an independent verification by a third party or not?
- To flag the FPSO or not?

Decision as to classing of the FPSO is dependent on local regulatory requirements and the owner's philosophy. If classification is required, then the extent and the details of classification need to be decided, and there are many choices. For example, some owners may opt not to class topsides equipment, and some may class topsides facilities using a risk-based approach rather than opting for traditional, prescriptive classification.

Similarly, the decision to verify and certify parts of the FPSO design and related activities is again dependent on local regulatory requirements and the owner's philosophy. Some owners, and also some jurisdictions, may also require risk-based "safety cases" to be made by the owner or operator.

Flagging an FPSO is usually not mandatory. However, there are many commercial and legal reasons why an FPSO may need to be registered with a flag nation. The decision to flag the FPSO is dependent on the owner's philosophy and other commercial and legal requirements. Historically, most FPSOs are flagged; however, not all are.

Similarly, diverse options exist regarding standards for structural design. Most owners usually apply offshore industry standards to the topsides equipment and marine industry standards to the hull. However, this is easier said than done because many interface areas need to be defined as to which standards they fall under. There are also many details to be developed further and appropriately specified and considered.

Even for part of topsides, it is not straightforward to apply offshore industry standards because some impact of FPSO motions must be considered. Similarly, for the hull part, because the FPSO cannot be dry-docked regularly, marine standards have to be upgraded selectively to ensure long life without the need for dry-docking. There is an excellent discussion on this issue in more detail in Costa et al. (2003).

REFERENCES

ABS (2004). *Guide for building and classing floating production installations.* American Bureau of Shipping, Houston, April.

Adhia, G. J., Pellegrino, S., and Ximenes, M. O. (2004a). "Practical considerations in the design and construction of FPSOs." *Proceedings of OMAE FPSO 2004, OMAE Specialty Symposium on FPSO Integrity, OMAE FPSO'04-0090,* August 30–September, 2, Houston.

Adhia, G., Pelleguino, S., Ximenes, M. C., Awashima, Y., Kakimoto, M., and Ando, T. (2004b). *Owner and shipyard perspective on new-build FPSO contracting scheme, standards and lessons.* Offshore Technology Conference, OTC 16706, Houston, May.

API (2001). *Recommended practice for planning, designing, and constructing floating production systems.* (Recommended Practices, 2FPS), American Petroleum Institute, March.

Bannister, A. C., and Stacey, A. (1999). "Literature review of the fracture properties of grade A ship plate." *Proceedings of OMAE Conference,* St. John, Newfoundland, July.

Barltrop, N. D. P., and Xu, L. (2004). "Research on bow impact loading in Glasgow." *Proceedings of OMAE Specialty Symposium on Integrity of Floating Production, Storage and Offloading (FPSO) Systems, OMAE-FPSO'04–0063,* Houston, August 30–September 2.

Bothamley, M. (2004). *Offshore processing options for oil platforms.* SPE Annual Technical Conference, Society of Petroleum Engineers, SPE 90325, Houston, September.

BV (2004). *Hull structure of production storage and offloading surface units.* (Rule Note No. 497), Bureau Veritas, Paris, October.

Costa, R., Manuel, L. P., Isaac, R., des Deserts, L., Harris, R. J. S., and de Bonnafos, O. (2003). *A generic FPSO hull for Angolan waters.* DOT, Marseilles, November.

DNV (2000). *Structural design of offshore ships.* (Offshore Standards, OS-C102). Det Norske Veritas, Oslo, October.

DNV (2002). *Rules for classification of floating production and storage units.* (Offshore Service Specifications, OSS-102). Det Norske Veritas, Oslo, April.

DNV (2006). *Fatigue methodology for offshore ships.* (Recommended Practices, DNV-RP-C206,) Det Norske Veritas, Oslo.

Drouin, P. (2006). "Brittle fracture in ships – a lingering problem." *Ships and Offshore Structures,* 1(3): 229–233.

Fyfe, A. J., and Ballard, E. J. (2003). "Prediction of green water events on FPSO vessels." *Proceedings of OMAE 2003, OMAE 2003-37452,* Cancun, Mexico, June 8–13.

Fyfe, A. J., and Ballard, E. J. (2004). "A design evaluation methodology for green water and bow impact type problems." *Proceedings of OMAE Specialty Symposium on Integrity of Floating Production, Storage and Offloading (FPSO) Systems, OMAE-FPSO'04-0065,* Houston, August 30–September 2.

Guedes Soares, C. (2004). "Probabilistic models of wave parameters for the assessment of green water on FPSOs." *Proceedings of OMAE Specialty Symposium on Integrity of Floating Production, Storage, and Offloading (FPSO) Systems, OMAE–FPSO'04-0061,* Houston, August 30–September 2.

Hodgson, T., and Barltrop, N. D. P. (2004). "Structural response of bow type structures to impact by steep fronted waves and resulting structural design." *Proceedings of OMAE Specialty Symposium on Integrity of Floating Production, Storage and Offloading (FPSO) Systems, OMAE–FPSO'04-0064,* Houston, August 30–September 2.

HSE (2005). *Wave slap loading on FPSO bows.* (Research Report 324), Health and Safety Executive, UK.

IACS (2005). *Common structural rules for double hull oil tankers.* International Association of Classification Societies, December.

Kennedy, J. R. (1993). *Cost trends in international petroleum activities.* SPE Conference, Society of Petroleum Engineers, SPE 26410, Houston, October.

Kleefsman, K. M., and Veldman, A. E. P. (2004). "An improved volume-of-fluid (IVOF) method for wave impact type problems." *Proceedings of OMAE Specialty Symposium on Integrity of Floating Production, Storage, and Offloading (FPSO) Systems, OMAE–FPSO'04-0066,* Houston, August 30–September 2.

Lacey, P., Hee, D., Chen, H., and Cardone, V. (2003). "Tow simulation." *SNAME Transactions*, 111: 79–96.

Liu, D., Spencer, J., Itoh, T., Kawachi, S., and Shiegmatsu, K. (1992). "Dynamic load approach in tanker design." *SNAME Transactions*, 100: 143–172.

LR (1999). *Rules and regulations for the classification of a floating offshore installation at a fixed location*. Lloyd's Register, London.

Olsen, A. S., Schroter, C., and Jensen, J. J. (2006). "Wave height distribution observed by ships in the North Atlantic." *Ships and Offshore Structures*, 1(1): 1–12.

Paik, J. K., Lee, J. M., Hwang, J. S., and Park, Y. I. (2003). "A time-dependent corrosion wastage model for the structures of single- and double-hull tankers and FSOs and FPSOs." *Marine Technology*, 40(3): 201–217.

Paik, J. K., and Thayamballi, A. K. (2003). *Ultimate limit state design of steel-plated structures*. Chichester, UK: John Wiley & Sons.

Parker, G. (1999). *The FPSO design and construction guidance manual*. Reserve Technology Institute, Houston.

Sumpter, J. D. G., and Kent, J. S. (2004). "Prediction of ship brittle fracture casualty rates by a probabilistic method." *Marine Structures*, 17: 575–589.

Voogt, A., and Buchner, B. (2004). "Wave impacts excitation on ship-type offshore structures in steep fronted waves." *Proceedings of OMAE Specialty Symposium on Integrity of Floating Production, Storage, and Offloading (FPSO) Systems, OMAE–FPSO'04-0062*, Houston, August 30–September 2.

Ximenes, M. C., Adhia, G., and Abe, A. (1997). "Design and construction of a floating storage and offloading vessel *Escravos* LPG FSO." *SNAME Transactions*, 105: 455–489.

CHAPTER 3

Design Principles, Criteria, and Regulations

3.1 Introduction

Although substantial efforts are now being directed by the maritime industry toward the application of limit-state design approaches, the shipbuilding industry has traditionally used classification society rules for design of trading ships. On the other hand, the offshore industry has more extensively applied first-principles methods based on limit states. It may be said that the design approach for moored ship-shaped offshore structures, such as FPSOs, often takes a form that is a fusion of the two industry approaches.

In a ship-shaped offshore installation, the structures of the vessel are of primary importance because they serve to house and support the systems and equipment needed for the overall success of the enterprise. The ability to correctly and consistently provide the necessary safety margins while meeting the twin requirements of structural safety and economy is key to the design of successful structures. This is where design principles, procedures, and criteria play an important part. Needless to say, successful structures during their life cycle also need to adequately meet the various requirements and regulations on health, safety, and the environment.

This chapter presents principles and criteria for design and strength assessment of ship-shaped offshore structures with a focus on the limit-state approach. The importance of safety, health, and the environment is emphasized. The regulatory framework and international standards pertinent to design and operation are addressed. For additional information, see Barltrop (1998).

This book is largely about structural design, and the related principles and their details of implementation are discussed in this chapter and throughout the book. At this stage, two very important principles that we would like to state upfront are as follows: first, that structural safety depends on various factors of design, construction, operation, inspection, and maintenance, and not design alone; and second, in today's world, social accountability considerations, including but not limited to various aspects of health, safety, and the environment, are also an ever-present aspect to be aware of and to address. Engineering projects do not occur in a vacuum.

3.2 Structural Design Principles

While in service, ship-shaped offshore structures, like other structures such as ships and other types of offshore platforms, are subjected to various types of actions and

Table 3.1. *Definition of the high-tensile-strength steel factor (IACS 2005)*

σ_Y	235	265	315	340	355	390
K	1.0	0.93	0.78	0.74	0.72	0.70

Note: σ_Y = Minimum yield strength of material in N/mm^2; K = high-tensile-strength steel factor.

action effects arising from service requirements that may range from the routine to the extreme or accidental.

In design, a structure must withstand such demands or applied actions throughout its expected lifetime. Three types of structural design approaches are considered: working stress design (WSD), critical buckling strength design (CBSD), and limit-state design (LSD). LSD is also often termed LRFD (load and resistance factor design).

3.2.1 Working Stress Design

In WSD, the design is undertaken so that the working stresses resulting from the design actions would not exceed a given level that may be called the *allowable stress*. Successful similar past experience is usually employed to determine the allowable stress. In the marine context, the value of the allowable stress is typically specified as some fraction of the mechanical properties of materials; for example, yield stress or ultimate tensile stress.

For example, the allowable stress σ_a of trading ship hulls under the vertical bending moment, as specified by recognized classification societies, is given by

$$\sigma_a = \frac{175}{K}(\text{N/mm}^2), \qquad (3.1)$$

where K = high-tensile-strength steel factor, as defined in Table 3.1.

3.2.2 Limit-State Design

In contrast to WSD, LSD is based on the explicit consideration of limit states that aim to define the various conditions under which the structure may cease to fulfill its intended function. For these conditions, the applicable load-carrying capacity is calculated and used in design or strength assessment as a limit for the related structural behavior.

It is now well recognized that the limit-state approach is a more rational procedure than the traditional working stress approach for design and strength assessment of various types of structures including ships, offshore structures, aerospace structures, bridges, and other land-based structures.

Limit states are classified into four categories: serviceability limit states (SLS), ultimate limit states (ULS), fatigue limit states (FLS), and accidental limit states (ALS) (Paik and Thayamballi 2003). SLS represents the exceedance of criteria governing normal functional or operational use. ULS represents the failure of the structure and/or its components usually when subjected to the maximum or near maximum values of actions or action effects. FLS represents damage accumulation (usually fatigue cracking damage) under repetitive actions, often considered on a component-by-component basis. ALS represents situations of accidental or abnormal events.

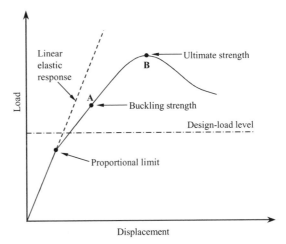

Figure 3.1. Structural design considerations regarding the ultimate limit state for a global system structure under extreme actions.

Figure 3.1 illustrates some considerations of the progressive nonlinear behavior for a global system structure under extreme actions. Even after the structural components buckle at point A, the system structure may be able to sustain further loading until the ultimate strength represented by point B is reached. However, as long as the strength level associated with point B remains unknown, as it is with the traditional WSD, it is not possible to determine the true safety margin or otherwise quantitatively use such information for design purposes.

As depicted in Figure 3.1, the true safety margin of a particular structure can be evaluated by a comparison of its ultimate strength with the extreme applied actions or action effects. This means that it is essential to determine accurately the limit-state load-carrying capacity, as well as the applied actions or action effects; this is necessary to determine the true safety margin of the structure and, therefore, to enable the most efficient design. Also, to the extent the progression of local failures and their interacting effects can be predicted in a unified manner during an ultimate limit-state calculation, a good understanding of the robustness of the structure can be obtained.

3.2.3 Critical Buckling Strength Design

A limit-state-like approach termed critical buckling strength-based design (CBSD) has typically been applied in the maritime industry for scantlings and structural design related to buckling. In CBSD, the design criterion is based on the so-called "critical" buckling strength that is usually obtained by a simple plasticity correction of the elastic buckling strength. The most typical formula for the plasticity correction is the so-called Johnson–Ostenfeld equation, given by

$$\sigma_{cr} = \begin{cases} \sigma_E & \text{for } \sigma_E < p_r \sigma_F \\ \sigma_F \left(1 - \dfrac{\sigma_F}{4\sigma_E}\right) & \text{for } \sigma_E \geq p_r \sigma_F, \end{cases} \tag{3.2}$$

where σ_E = elastic buckling stress; σ_{cr} = critical (elastic/plastic) buckling stress; σ_F = reference yield stress; $\sigma_F = \sigma_Y$ for compressive stress; $\sigma_F = \tau_Y = \sigma_Y/\sqrt{3}$

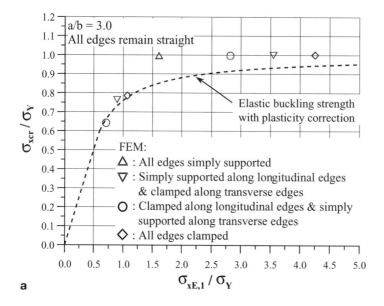

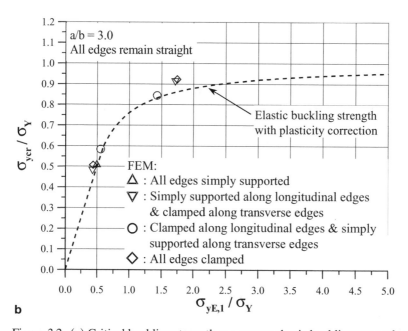

Figure 3.2. (a) Critical buckling strength σ_{xcr} versus elastic buckling strength $\sigma_{xE,1}$ for a rectangular plate under longitudinal axial compression. (b) Critical buckling strength σ_{ycr} versus elastic buckling strength $\sigma_{xE,1}$ for a rectangular plate under transverse axial compression. (Note that FEM results are ultimate strength obtained by elastic/plastic large deflection analysis.)

for shear stress; and σ_Y = material yield stress. For the plate-stiffener combinations when the yield stress of plating is different from that of stiffeners, σ_Y may be taken as the equivalent yield stress; that is, $\sigma_Y = \sigma_{Yeq} \cdot p_r$ is a coefficient accounting for the plasticity sensitivity, which is typically taken as $p_r = 0.5 \sim 0.6$.

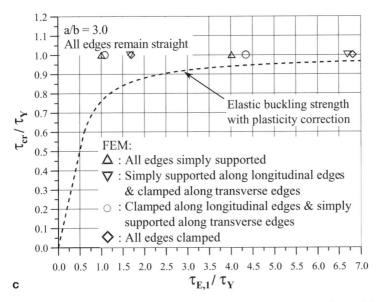

Figure 3.2. (*cont.*) (c) Critical buckling strength τ_{cr} versus elastic buckling strength $\tau_{E,1}$ for a rectangular plate under edge shear. (Note that FEM results are ultimate strength obtained by elastic/plastic large deflection analysis.)

In using Eq. (3.2), the sign of the compressive stress is taken as positive. Because the critical buckling strength does not necessarily mean that the actual ultimate limit state has been reached, the CBSD approach can be termed a pseudo-LSD approach.

Figures 3.2(a)–(c) plot Eq. (3.2) for a rectangular plate with the plate length, a, to breadth, b, ratio of 3 under longitudinal axial compression, transverse axial compression or edge shear, respectively, under varying plate-edge conditions. The ultimate strengths calculated by elastic/plastic large deflection finite-element method (FEM) are also plotted in the figures for a comparison. It is evident from Figure 3.2 that Eq. (3.2) can be useful for predicting a limit-state-like critical strength, although strictly speaking, the strength prediction considers perfect structural components, that is, without cutouts, structural damage, or defects.

However, one should be cautioned that Eq. (3.2) can give optimistic evaluations of ultimate strength for imperfect structural components, for example, with cutouts or initial imperfections when axial compressive loads are predominant (Paik and Thayamballi 2003). Figures 3.3–3.5 show examples for rectangular plates with a cutout and under axial compressive loads or edge shear, indicating that the critical buckling strengths determined from Eq. (3.2) can be greater than the ultimate strengths obtained by nonlinear finite-element methods, depending on the plate thickness, aspect ratio, and opening (cutout) size, where an *average* level of initial deflection w_{opl} was considered for the plate ultimate strength analyses although the plate critical buckling strength calculations were made without the effect of initial deflections. It turns out that for relatively thick plates with cutouts, the CBSD approach may produce unsafe design results. For a further description with closed form strength formulae of perforated plates, see Paik and Thayamballi (2003).

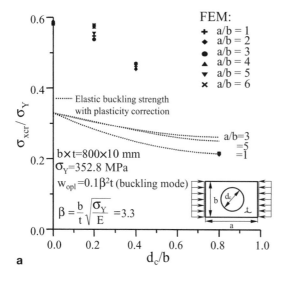

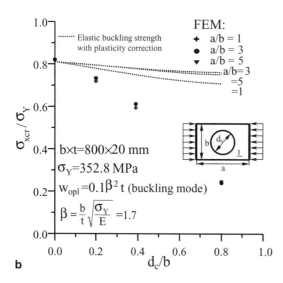

Figure 3.3. (a) Critical buckling strength versus ultimate strength for simply supported rectangular plates with a cutout and under longitudinal axial compression; t = 10mm. (b) Critical buckling strength versus ultimate strength for simply supported rectangular plates with a cutout and under longitudinal axial compression; t = 20mm. (Note that FEM results are ultimate strength obtained by elastic/plastic large deflection analysis.)

3.2.4 Comparison among the Three Design Methods

To see the characteristics of LSD application, a simple design example for a steel rectangular plate surrounded by support members (e.g., stiffeners or frames) under uniaxial compressive loads in the plate length direction is now used to compare the WSD, CBSD, and ULS design methods noted in the previous sections. The plate length and breadth are taken as a × b = 3,200 × 800mm, and the plate thickness t will now be determined for a required performance level. The yield stress of the plate material σ_Y is considered to be 352.8 N/mm^2 and Young modulus E is 205,800 N/mm^2. The Poisson ratio, v, is 0.3.

The structural design criterion, described in Eq. (3.10), is defined so that the design capacity or strength should not be smaller than the design demand or actions (or action effects), together with some margin of safety considering

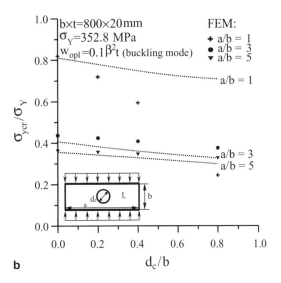

Figure 3.4. (a) Critical buckling strength versus ultimate strength for simply supported rectangular plates with a cutout and under transverse axial compression; $t = 10$mm. (b) Critical buckling strength versus ultimate strength for simply supported rectangular plates with a cutout and under transverse axial compression; $t = 20$mm. (Note that FEM results are ultimate strength obtained by elastic/plastic large deflection analysis.)

the uncertainties associated with the capacity and the demand. Table 3.2 summarizes the corresponding demand and capacity used for each design method. For convenience of the present design-method comparisons, we assume that the effects of the uncertainties associated with demand and capacity can be neglected. In this case, the design demand for all three design methods corresponds to $2{,}205/t$ N/mm^2.

With the design load taken to be $P_x = 1.764$MN, the working stress σ_x of the plate can be calculated as follows:

$$\sigma_x = \frac{\text{design load}}{\text{plate sectional area}} = \frac{1{,}764{,}000}{800 \times t} = \frac{2{,}205}{t} \text{ N/mm}^2. \tag{3.3}$$

The design capacity for WSD is equivalent to the allowable stress σ_a, which is normally defined based on past experience – say, by 50 percent of yield stress or

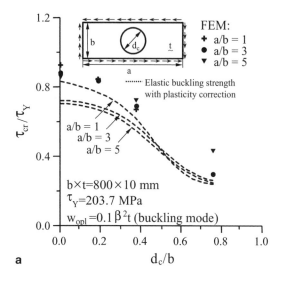

a

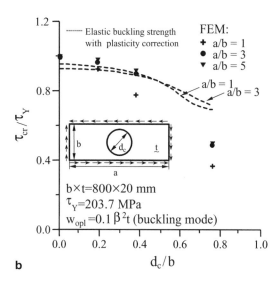

b

Figure 3.5. (a) Critical buckling strength versus ultimate strength for simply supported rectangular plates with a cutout and under edge shear; t = 10mm. (b) Critical buckling strength versus ultimate strength for simply supported rectangular plates with a cutout and under edge shear; t = 20mm. (Note that FEM results are ultimate strength obtained by elastic/plastic large deflection analysis.)

$\sigma_a = 0.5\sigma_Y$. In this case, the minimum required plate thickness will be determined from the WSD approach as follows:

$$\frac{2,205}{t} < 0.5 \times 352.8 \quad \text{or} \quad t > 12.5\text{mm}. \tag{3.4}$$

In addition, if $\sigma_a = 0.8\sigma_Y$, the minimum required plate thickness is given by WSD as follows:

$$t > 7.8\text{mm}. \tag{3.5}$$

When the CBSD approach is applied, the elastic buckling stress σ_E of the plate under uniaxial compressive loads, assuming the plate to be simply supported at all (four) edges, is obtained as follows:

$$\sigma_E = k\frac{\pi^2 E}{12(1-\nu^2)}\left(\frac{t}{b}\right)^2 = 1.1625t^2\,\text{N/mm}^2, \tag{3.6}$$

Table 3.2. *Demand and capacity used for typical design methods*

Design method	WSD	CBSD	ULSD
Demand	Design working stress (action effect)	Design working stress (action effect) or action (load)	Design working stress (action effect) or action (load)
Capacity	Allowable stress	Design critical buckling strength	Design ultimate strength

Note: WSD = working stress design; CBSD = critical buckling strength-based design; ULSD = ultimate limit-state design. "Design" implies that associated uncertainties are taken into account; see Sections 3.3 and 3.4.

where k = elastic buckling coefficient, which in this case is taken as k = 4.

A thick plate that has a high value of computed elastic buckling strength will not buckle in the elastic regime, but will buckle with a certain degree of plasticity. To account for this behavior, the maritime industry often uses the Johnson–Ostenfeld formula, Eq. (3.2), for a plasticity correction.

For CBSD, therefore, the corresponding capacity is equivalent to σ_{cr}, and thus the minimum required plate thickness is obtained from the criterion given by $\sigma_{cr} > 2205/t$. By solving the resulting second-order equation with regard to t, the minimum plate thickness to be structurally safe is determined as follows:

$$t > 12.3\text{mm}. \tag{3.7}$$

Therefore, it can be surmised that the allowable stress level in the WSD approach must be 50.8 percent (= (2,205/12.3)/352.8) of the yield stress to get the same design result as the CBSD approach.

In the ULS design, the capacity of the plate is the ultimate strength σ_u, which can be calculated by closed-form expressions or by more refined analysis methods such as the nonlinear finite-element method. For instance, the plate ultimate strength may be predicted using the following formula based on the well-known effective width concept (Paik and Thayamballi 2003):

$$\sigma_u = \sigma_Y \frac{b_e}{b}, \tag{3.8a}$$

where b_e = plate effective width at the ULS, which may be given for a simply supported rectangular plate as follows:

$$\frac{b_e}{b} = \begin{cases} 1.0, & \text{for } \beta < 1 \\ \dfrac{2}{\beta} - \dfrac{1}{\beta^2}, & \text{for } \beta \geq 1, \end{cases} \tag{3.8b}$$

with $\beta = \frac{b}{t}\sqrt{\frac{\sigma_Y}{E}}$ = plate slenderness ratio.

For ULS design, the ultimate strength (stress) must be greater than the working stress so that $\sigma_u > 2,205/t$ should be satisfied. The minimum required plate thickness is then determined as a solution of a third-order equation with regard to t as follows:

$$t > 11.1\text{mm}, \tag{3.9}$$

which turns out to be 89.5 percent of the CBSD result.

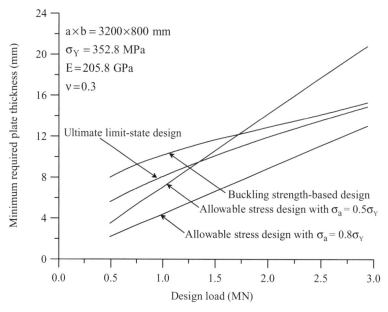

Figure 3.6. The minimum required plate thickness versus design load as obtained by the three design methods.

Figure 3.6 shows the minimum plate thickness against the design load, as obtained by the three design methods: WSD, CBSD, and ULSD. For the cases considered, it may be observed from Figure 3.6 that in the WSD method with a higher level of allowable stress, the designed minimum plate thickness becomes smaller.

For the allowable stress of $0.5\sigma_Y$, the WSD method may produce unsafe plates under small loads, and it may give plates that are too thick under large loads. For the plate-design problem presented, to be more accurate, a lower level of the allowable stress should be used for small design loads, but a higher level of the allowable stress may be adopted for the larger design loads. In reality, however, it is not always an easy task to define the relevant level of the allowable stress with such logic because design procedures need to cover a wide range of the design loads in general.

Similarly, the comparisons of this example indicate that the minimum required plate thickness as obtained by the CBSD approach is generally larger than that by the ULS design approach; and that the safety margin by the CBSD approach may be larger for smaller design loads, and the safety margin becomes smaller for larger design loads. Because we assume that the loads and strength are deterministic, we can presumably do better than that if we are striving to achieve an efficient design.

From this example, it is apparent that the safety measure calculations by the traditional allowable WSD method may not correlate well with those by the ULS-based method. The WSD method appears to evaluate optimistically the structural capacity in some cases but pessimistically in the other cases, potentially leading to inconsistent levels of safety in a design sense. This shows the primary disadvantage of the traditional allowable WSD procedures. The ULS design procedure can avoid such a problem because it can better determine the true safety margin, and therefore potentially lead to a more economically designed structure.

3.3 Limit-State Criteria for Structural Design and Strength Assessment

In structural design, designers are required to assure that the structure has an adequate margin of safety against applied actions or demands. The safety margin is necessary to account for the effects of various uncertainties due to natural variability; inaccuracy in procedures used for the evaluation; control of actions or action effects (e.g., stress, deformation); similar uncertainties in structural resistance (capacity) and its determination; and variations introduced by construction procedures.

The "demand" is analogous to action or load, and the "capacity" is analogous to the strength necessary to resist that action, both consistently measured (e.g., as stress; deformation; resistive or applied load; moment; energy, either lost or absorbed; and so on). To be safe, therefore, the following criterion must be satisfied:

$$G = C_d - D_d > 0, \tag{3.10}$$

where $G =$ a performance function; C_d, $D_d =$ "design capacity" and "design demand," respectively, taking into account the effects of the associated uncertainties. The term "design" in Eq. (3.10) does not necessarily apply only to structural design itself, but it can also relate to any performance function where associated uncertainties are involved and need to be addressed in a relevant manner.

3.4 Probabilistic Format versus Partial Safety Factor Format

It is noted that D_d and C_d in Eq. (3.10) are functions of the basic variables that characterize actions or action effects, material properties, geometric parameters, and structural failure modes and their consequences. For design assessment, two types of formats may be used: a probabilistic assessment format and a partial safety factor assessment format.

3.4.1 Probabilistic Format

Consider that the limit-state function G in Eq. (3.10) can be rewritten as a function of the basic variables, $x_1, x_2, \ldots, x_i, \ldots, x_n$, as follows:

$$G(x_1, x_2, \ldots, x_i, \ldots, x_n) = 0. \tag{3.11}$$

When $G > 0$, the structure is in the desired state. When $G \leq 0$, the structure is in the undesired state.

The basic variables are random in nature, and they always have uncertainties. As a result, the obtained characteristic values of D_d and C_d have some errors as well. Furthermore, the modeling functions of D_d and C_d are also uncertain due to lack of knowledge or simplification in developing the models.

Based on the first-order approximation, Eq. (3.11) can be written by the Taylor series expansion as follows:

$$G \cong G(\mu_{x1}, \mu_{x2}, \ldots, \mu_{xi}, \ldots, \mu_{xn}) + \sum_{i=1}^{n} \left(\frac{\partial G}{\partial x_i}\right)_{\bar{x}} (x_i - \mu_{xi}), \tag{3.12}$$

where μ_{xi} = mean value of the variable x_i; $\bar{x}$ = mean values of the basic variables = $(\mu_{x1}, \mu_{x2}, \ldots, \mu_{xi}, \ldots, \mu_{xn})$; $(\partial G/\partial x_i)_{\bar{x}}$ = partial differentiation of G with respect to x_i at $x_i = \mu_{xi}$.

The mean value of the function G is then given by

$$\mu_G = G(\mu_{x1}, \mu_{x2}, \ldots, \mu_{xi}, \ldots, \mu_{xn}), \quad (3.13)$$

where μ_G = mean value of the function G.

The standard deviation of the function G is calculated by

$$\sigma_G = \left[\sum_{i=1}^{n} \left(\frac{\partial G}{\partial x_i} \right)_{\bar{x}}^2 \sigma_{xi}^2 + 2 \sum_{i>j} \left(\frac{\partial G}{\partial x_i} \right)_{\bar{x}} \left(\frac{\partial G}{\partial x_j} \right)_{\bar{x}} \text{covar}(x_i, x_j) \right]^{1/2}, \quad (3.14)$$

where σ_G = standard deviation of G; σ_{x_i} = standard deviation of the variable x_i; $\text{covar}(x_i, x_j) = E[(x_i - \mu_{xi})(x_j - \mu_{xj})]$ = covariation of x_i and x_j; $E[\ ]$ = mean value of $[\ ]$.

When the basic variables $x_1, x_2, \ldots, x_i, \ldots, x_n$ considered are independent of each other, $\text{covar}(x_i, x_j) = 0$. In this case, Eq. (3.14) can be simplified to

$$\sigma_G = \left[\sum_{i=1}^{n} \left(\frac{\partial G}{\partial x_i} \right)_{\bar{x}}^2 \sigma_{xi}^2 \right]^{1/2}. \quad (3.15)$$

The reliability index for this case can be defined by using the first-order second-moment method (FOSM) as follows (Benjamin and Cornell 1970; EDRH 2005):

$$\beta = \frac{\mu_G}{\sigma_G}, \quad (3.16)$$

where β = reliability index.

For a function G of two parameters, x_1 and x_2, which are considered to be statistically independent, with mean μ_G, and standard deviation σ_G, the reliability index β can be obtained in closed form as follows:

$$\mu_G = \mu_{x1} - \mu_{x2}, \quad (3.17a)$$

$$\sigma_G = \sqrt{(\sigma_{x1})^2 + (\sigma_{x2})^2}, \quad (3.17b)$$

$$\beta = \frac{\mu_{x1} - \mu_{x2}}{\sqrt{(\sigma_{x1})^2 + (\sigma_{x2})^2}} = \frac{\mu_{x1}/\mu_{x2} - 1}{\sqrt{(\mu_{x1}/\mu_{x2})^2 (\eta_{x1})^2 + (\eta_{x2})^2}}, \quad (3.17c)$$

where μ_{x1}, μ_{x2} = mean values of x_1 or x_2; σ_{x1}, σ_{x2} = standard deviations of x_1 or x_2; η_{x1}, η_{x2} = coefficient of variation (i.e., standard deviation/mean value) of x_1 or x_2.

From Eq. (3.10), $x_1 = C_d$ and $x_2 = D_d$ can be applied to Eq. (3.17) for the calculation of reliability index as long as C_d and D_d are considered statistically independent. This safety index has certain undesirable properties, and can, of course, be improved on by more refined methods such as the first-order reliability method (FORM) and the second-order reliability method (SORM). Further, the safety index can be related to the probability of limit-state exceedance, that is, "failure," but these are subjects for books on structural reliability; see Madsen et al. (1986) and Mansour (1989). The reliability engineering handbook (EDRH 2005), which contains a number of more recent advances and developments, is also a good resource.

3.4.2 Partial Safety Factor Format

The design capacity C_d in Eq. (3.10) can be expressed by considering the associated uncertainties as follows:

$$C_d = \frac{C_k}{\gamma_C}, \tag{3.18}$$

where C_k = characteristic value of load-carrying capacity; $\gamma_C = \gamma_p\gamma_m$ = a partial safety factor associated with capacity; γ_p = a partial safety factor taking into account the uncertainties due to material properties; γ_m = a partial safety factor taking into account the uncertainties on the capacity of the structure, such as modeling methodology, construction quality, and structural degradation.

The design demand D_d, on the other hand, is expressible for multiple actions as follows:

$$D_d = \gamma_o \sum_{i=1}^{n} D_{ki}(F_{ki}, \gamma_{fi}), \tag{3.19}$$

where $D_{ki}(F_{ki}, \gamma_{fi})$ = characteristic value of demand for action type i at its worst situation, calculated from the characteristic value of action F_{ki} and magnified by the corresponding partial safety factor γ_{fi}, to take into account the uncertainties related to actions in the safety check; γ_o = a partial safety factor that takes into account the degree of seriousness of the particular limit state in regard to safety and serviceability, accounting for economical and social consequences as well as any special circumstances (e.g., the criticality of the mission of the system or interaction of the limit state considered with the others).

For the application of a single type of action, Eq. (3.19) can be simplified to

$$D_d = \gamma_D D_k, \tag{3.20}$$

where D_k = characteristic value of demand; $\gamma_D = \gamma_o\gamma_f$ = a partial safety factor associated with demand; γ_o, γ_f = as defined in Eq. (3.19).

The measure of structural adequacy can then be determined as follows:

$$\eta = \frac{C_d}{D_d} = \frac{1}{\gamma_C\gamma_D}\frac{C_k}{D_k}, \tag{3.21}$$

where η = measure of structural adequacy; $\gamma_C\gamma_D$ = combined safety factor. To be safe, η must be greater than 1.0 by a sufficient margin.

Similar to Eq. (3.16), Eq. (3.21) can be used for the strength assessment of damaged structures as well as intact or undamaged structures. For damaged structures, C_d may be taken as equivalent to the residual load-carrying capacity, although it would typically be the reserve load-carrying capacity in the case of intact structures.

In a safety check, D_k in Eq. (3.21) is usually defined as a characteristic value of the demand calculation model, and γ_D is consistently defined by taking into account the possibility of unfavorable deviations of the action values from the representative values and the uncertainties in the model of action effects (or actions).

Similarly, C_k in Eq. (3.21) is defined as a characteristic value of the capacity or strength in the limit-state equation. γ_C is consistently defined by taking into account (a) the possibility of unfavorable deviations of material properties from the specified values; (b) the possibility of unfavorable deviations of geometric parameters from the specified values, including the severity (importance) of variations; (c) the tolerance specifications; (d) the control of the deviations; (e) the cumulative effect of a simultaneous occurrence of several geometric deviations; and (f) the model uncertainties calculated as deviations from measurements or benchmark calculations; and are, perhaps, even adjusted for the possibility of unfavorable consequences of progressive collapse.

3.4.3 Considerations Related to Safety Factors

The adequacy of safety factors for the overall structure and the structural components or details are of obvious importance. Component failure is the starting point, so to speak; the overall structure safety factor can be based on the reserve strength beyond first nominal component failure. This section addresses safety factor considerations for both trading ships and offshore structures. It serves to illustrate the calculation methodology and environmental load return period dependency of safety factors. It also provides some food for thought regarding design for what are called "abnormal" waves. The following thought-provoking discussion is courtesy of Frieze and Paik (2004).

The partial safety factors may depend on the design situation and on the types of limit states. Also, it is important to realize that the partial safety factors can be different for different levels of refinement of calculation methodologies of demands (actions) and capacities (strengths). This means that any safety factors specified by classification societies or structural codes are, in principle, applicable only if the calculation methodologies of demands and capacities that they recommend are also employed.

As the basis for the design of offshore structures, the industry today uses a 100-year return period environmental event, but a 20–25-year return period environmental event is applied for trading ship designs. For a typical jacket structure, the mean reserve of strength to collapse beyond the nominal 100-year return period storm loading is perhaps about 1.85 (ISO CD 19902). For an API RP 2A-WSD component design (API 1993), ignoring components influenced by buckling effects, the safety factor against first yield under extreme environmental load–dominated conditions is about 1.25. This is based on an allowable design stress of 60 percent of yield stress that is increased by one third because the loading is considered to be dominated by extreme environmental conditions.

Thus, the reserve ratio beyond first yield with respect to the mean collapse strength is $1.85/1.25 = 1.48$. However, because the present comparison is based on nominal stress values, the reserve ratio of 1.48 needs to be reduced to account for two biases. First, the bias of yield stress can be taken as 1.12, corresponding to a coefficient of variation of 4 percent (Frieze 1992), where a nominal value coincides with the 0.1 percent fractile. Second, the bias between the mean strength of beam-column members, which normally dictate initial failure in offshore structures, and the design formulation value can be taken together as 1.10 (Frieze et al. 1997). Thus, the measure of redundancy in offshore structures based on nominal values is $1.48/(1.12 \times 1.10) =$ 1.20, which leads to a factor of $1.20 \times 1.25 = 1.50$ as the safety factor between nominal overall collapse of an offshore structure and its load.

According to the recommended practices for trading ship designs based on the application of beam theory for the hull girder (e.g., IACS 2005), stresses in mild steel plating under combined still-water conditions and wave-induced bending moment conditions are limited to 175 N/mm^2, as previously indicated in Eq. (3.1). In passing, it is of interest that allowable stress by some class society guidelines can be as high as 90 percent of yield stress when coarse-mesh finite-element analysis is used. The yield stress for such steel has a guaranteed minimum yield stress of 235 N/mm^2, so the safety factor is $235/175 = 1.34$. This relates to the onset of (nominal) yield. For the midship cross section, again ignoring buckling effects, a reserve of some 11 percent exists beyond first yield to allow the hull girder to develop its plastic moment capacity. This suggests that the overall safety factor for trading ships with respect to the 20-year return period storm approximates $1.34 \times 1.11 = 1.49$.

These overall safety factors – that is, 1.50 for offshore platforms versus 1.49 for trading ships – may seem similar, but they do not account for the difference in the return period of the considered storm. If trading ships were assessed for the 100-year return period storm rather than the 20-year return period value as at present, based on the simple assessment performed herein using the DEn wave height equations (DEn 1990a), their present day overall safety factor would be $1.49/1.12 = 1.33$. However, if offshore structures were assessed against the 20-year return period event, their current overall safety factor would be $1.50 \times 1.26 = 1.89$.

On this simple assessment, offshore structures seem relatively safe compared with trading ships. Despite this, however, the offshore industry viewed these safety margins as inadequate because, when the LSD alternative for offshore structure design was introduced, these margins were increased. For this LSD, the partial safety factor for structural component capacity varied according to the degree of uncertainty reflected in the test data associated with each component. The main factors varied from 1.05 (actually, the inverse of 0.95) for components dictated by tension and bending, to 1.18 (1/0.85) for components dictated by compression, and to 1.25 (1/0.8) for components dictated by hydrostatic pressure. The partial safety factor for environmental loads was 1.35 so that the minimum component factor varied from $1.35 \times 1.05 = 1.42$ to $1.35 \times 1.25 = 1.69$. The corresponding overall safety factors varied from $1.42 \times 1.20 = 1.70$ to $1.69 \times 1.20 = 2.03$ when assessed to the storm of the 100-year return period and varied from $1.70 \times 1.26 = 2.14$ to $2.03 \times 1.26 = 2.56$ when assessed to the storm of the 20-year return period.

The results of these calculations are summarized in Table 3.3, from which it can be clearly seen that there are important differences between the safety factors applicable to offshore structures compared with those applicable to trading ship structures. Indeed, except for the 100-year return period storm assessed to WSD, the first component safety factors for offshore structures are all larger than the overall safety factors for trading ship structures.

Table 3.3. *Comparison of primary component and overall failure safety factors for trading ships and offshore structures*

Storm return period	Ship structures	Offshore structures	
		WSD	LSD
First component failure			
20 years	1.34	1.58	1.79–2.13
100 years	1.19	1.25	1.42–1.69
Overall failure			
20 years	1.49	1.89	2.14–2.56
100 years	1.33	1.50	1.70–2.03

Note: WSD = working-stress design; LSD = limit-state design.

Some explanation and justification is required for this difference in safety level and, by implication, structural reliability between two structural forms that can, in principle, be exposed to the same environmental conditions. Trading ships can of course reroute when faced with severe weather conditions. This is a normal and perhaps prudent course of action open to ships' masters as acknowledged in the findings of the *Re-Opened Formal Investigation into the M. V. Derbyshire* (HMSO 2000), which stated that "Vessels of the size and design of the Derbyshire were at the time assumed to be quite capable of withstanding such conditions, even if they had to reduce speed or be hove to." The offshore industry does not so fundamentally rely on crew intervention to be part of its structural safety-management process, and, for that matter, it is unclear that any other industry does.

The ignorance of coexisting lateral loading may be a contributing factor to the idea that the classing of trading ships implicitly assumes that the design relies on crew intervention in the likelihood of encounters with severe storms. The corollary is perhaps that vessels may not survive storms without crew intervention. The MaxWave project (Faulkner 2002) found, admittedly for a relatively fast moving vessel (20 knots), midship bending moments well in excess of classification society requirements. Paik and Faulkner (2003) and Guedes Soares et al. (2003) suggest that to deal at least with abnormal waves, the present 20-year return period requirement for trading ships is inadequate. It is conjecture, but it would perhaps be surprising, if the offshore platform failures in the Gulf of Mexico in hurricane conditions were not due in part to abnormal waves, bearing in mind that hurricane conditions are likely to be a good source of such waves.

On the basis of offshore experience, it would appear that an overall safety factor of 1.89 is required if a platform is designed to a 20-year return period storm or 1.50 if designed to a 100-year return period storm to ensure that, without human intervention, a design is realized that the structure suffers only minor damage when subjected to storm approaching the design event. Considering that ship-shaped offshore structures are designed for a 100-year return period event, it is important to consider the implications of this design requirement on the safety factors derived previously. Simply, this would have the effect of increasing the trading ship structure overall safety factors by 12.4 percent (based on linear interpolations).

Thus, the corresponding first-component safety factors would be $1.124 \times 1.34 = 1.51$ when assessed on the basis of a 20-year return period storm or $1.124 \times 1.19 = 1.34$

when assessed on the basis of a 100-year return period storm. The corresponding overall safety factors would be $1.124 \times 1.49 = 1.67$ and $1.124 \times 1.33 = 1.49$ when assessed on the basis of 20-year and 100-year return period storms, respectively.

Using the 100-year values as the basis for comparison, the first-component safety factor of 1.34 exceeds that applicable to a WSD approach to offshore structures, and the overall safety factor of 1.49 is almost identical to that applicable to a WSD design, which is redundant for offshore structures. It would be prudent to adopt the 100-year return period storm as the basis for trading ship assessment until the effects on ship safety of weather routing, reduced speed, and other crew intervention measures are extensively quantified.

This proposed increase can be seen as achieving a number of objectives. We have discussed some of these objectives, but another objective relates to the issue of abnormal waves. When modeling a 3-hour storm, a wave height of 1.86 times the significant wave height is widely used to identify an appropriate design-wave height for that storm. Abnormal waves have traditionally been defined as those with a ratio of wave height to significant wave height greater than 2.0 (Wolfram et al. 2001), although this is somewhat arbitrary because each wave causes a fundamentally different response.

However, this same work also appears to demonstrate that on a probability distribution basis, a wave height greater than 2.3 times the significant wave height is required before the wave in question follows another distribution. Using the traditional definition, to design for an abnormal wave requires that the wave height to be considered be increased beyond that presently adopted by $2.0/1.86 = 1.08$. Wolfram et al. (2001) suggest a wave height increase of $2.3/1.86 = 1.24$. This hypothetical increase proposed above 1.24 can therefore be interpreted as an increase in the design requirement up to a level where it deals (perhaps only in part) with abnormal waves.

3.5 Unified Design Requirements for Trading Tanker Hull Structures

For trading tanker designs, the North Atlantic wave environment is typically adopted as the design premise for an unrestricted service vessel, although a reduced wave climate due to worldwide trade may have been applied for fatigue design purposes by some classification societies in the past. Also, these requirements varied to an extent among various classification societies, including, for example, details of the wave scatter diagram said to represent the North Atlantic service. However, as of April 2006, the various classification society rules have been unified into the Common Structural Rules (CSR) for tankers (IACS 2005).

For the design of ship-shaped offshore units in a benign environment, the trading tanker design requirements are often considered. Therefore, this section introduces some of these unified design requirements and concepts for trading tanker hull structures. For more details, see IACS (2005) and/or the websites http://www.iacs.org.uk or http://www.jtprules.com.

Although shipyards and tanker vessel owners have good reasons for wanting such a change, there are several arguments one may make against it as well. Among these are that such an effort could hamper real technological progress in the future, both because the change process could itself be considerably more difficult and because incentives for future research in related areas might be diminished. Also, it may remove important elements of engineering judgment, creativity, and competition from the scene. To be fair, it should also be pointed out that a major driver for the change to unified structural rules for tankers is, apparently, to remove the steel weight-based competition.

The minimum requirements for moment of inertia and section modulus of trading tanker hulls are given on a net ship basis (i.e., before corrosion margins are added), as follows (IACS 2005):

$$I_{v-min} = 2.7C_{wv}L^3B\,(C_b + 0.7)\,cm^4, \tag{3.22a}$$

$$Z_{v-min} = 0.9KC_{wv}L^2B\,(C_b + 0.7)\,cm^3, \tag{3.22b}$$

where I_{v-min} = minimum moment of inertia; Z_{v-min} = minimum section modulus; L = vessel length in meters; B = vessel breadth in meters; C_b = block coefficient; K = high-tensile-strength factor as indicated in Table 3.1; C_{wv} = coefficient, which may be given as a function of vessel length in meters as follows:

$$C_{wv} = \begin{cases} 10.75 - \left(\frac{300-L}{100}\right)^{1.5} & \text{for } 90m \le L \le 300m \\ 10.75 & \text{for } 300 < L \le 350m \\ 10.75 - \left(\frac{L-350}{150}\right)^{1.5} & \text{for } 350m < L \le 500m. \end{cases} \tag{3.23}$$

A detailed distribution of the still-water moment along a ship's length can be calculated by a double integration of the difference between the weight force and the buoyancy force, using classical beam theory. The sectional shear force $F(x_1)$ at location $x = x_1$ in the ship-length direction is estimated by the integral of the load curve that represents the difference between weight and buoyancy curves:

$$F(x_1) = \int_0^{x_1} f(x)dx, \tag{3.24}$$

where $f(x) = b(x) - w(x)$ = net load per unit length in still water; $b(x)$ = buoyancy per unit length; and $w(x)$ = weight per unit length.

The bending moment $M(x_1)$ at location $x = x_1$ is estimated as the integral of the shear curve indicated in Eq. (3.24), as follows:

$$M(x_1) = \int_0^{x_1} F(x)dx. \tag{3.25}$$

In the IACS common structural rule requirements applicable to trading tanker designs, the design still-water vertical bending moment is calculated in the above conventional manner based on the loading manual, addressing the various loading conditions in the design specifications and also those required by statutory considerations. Approximate formulae of the minimum still-water vertical bending moments of trading tankers at sea are given as follows [for symbols used in the following equation, refer to Eq. (3.22) unless specified]:

$$M_{sw} = \begin{cases} +0.01f_{sw}C_{wv}L^2B\,(11.97 - 1.9C_b)\,(kNm) & \text{for hogging} \\ -0.0775f_{sw}C_{wv}L^2B\,(C_b + 0.7)\,(kNm) & \text{for sagging,} \end{cases} \tag{3.26}$$

where f_{sw} = coefficient that is taken as $f_{sw} = 1.0$ at 0.4L amidship or $f_{sw} = 0.15$ at 0.1L from aft-perpendicular or fore-perpendicular.

The minimum still-water bending moments for trading tankers at harbor are allowed to be 25 percent larger than those at sea:

$$M_{sw}^* = 1.25 M_{sw}, \qquad (3.27)$$

where M_{sw}^* = minimum still-water bending moment of a trading tanker at harbor; M_{sw} = as defined in Eq. (3.26).

The minimum wave-induced vertical bending moment M_w applicable to trading tankers is given for wave conditions occurring once in 25-year under North Atlantic wave conditions (in comparison to the old 20-year return period criterion, although the wave-load formula itself is apparently essentially unchanged). With the hogging moment taken as positive and the sagging moment taken as negative, the applicable formulae are as follows (for symbols used in the following equation, refer to Eqs. (3.22) and (3.23) unless specified):

$$M_w = \begin{cases} + f_{prob} 0.19 f_{wv-v} C_{wv} L^2 B C_b \, (kNm) & \text{for hogging} \\ - f_{prob} 0.11 f_{wv-v} C_{wv} L^2 B (C_b + 0.7) \, (kNm) & \text{for sagging,} \end{cases} \qquad (3.28)$$

where f_{prob} = coefficient that is taken as $f_{prob} = 1.0$ for scantlings and strength assessment; and f_{wv-v} = coefficient that is taken as $f_{wv-v} = 1.0$ in between 0.4L and 0.65L from aft-perpendicular.

In addition to the conventional elastic section modulus checks using an allowable bending stress per Eq. (3.1), the new rules also introduce an ultimate strength check for the hull girder. The following equation is used in this new case for checking that the total bending moment M_t does not exceed the ultimate hull girder bending moment:

$$M_t = \gamma_{sw} M_{sw} + \gamma_w M_w \leq \frac{M_u}{\gamma_r}, \qquad (3.29)$$

where $\gamma_{sw}, \gamma_w, \gamma_r$ = partial safety factors associated with the still-water bending moment, wave-induced bending moment, and ultimate hull girder moment, respectively; M_u = ultimate hull girder bending moment. The partial safety factors are apparently also to account for the nonsimultaneous occurrence of extreme still-water loads and wave-induced loads. For ultimate hull girder strength calculations in sagging condition, the IACS common structural rules suggest using $\gamma_{sw} = 1.0$, $\gamma_w = 1.3$, and $\gamma_r = 1.1$.

One other aspect of the IACS common structural rules for tankers should also be stated here, if only for completeness: the hull girder wave-induced shear force used in shear checks for the hull girder, both by simplified means and in the three hold finite-element analysis-based structural assessment procedures used in the new rules, corresponds not to a one in 25-year return period value but to a much more frequent return period. This apparently is for historical reasons. When considering application of tanker procedures to offshore structures, however, one obviously needs to be well aware of such differences and also account for these differences as necessary.

3.6 Design Principles for Stability

While in service, the floating production systems must be able to remain upright even in extreme environmental conditions. In this regard, the design for stability is a key part of the floating production system design. Relevant calculation and control procedures must be applied during operation so as to keep the vessel stable. Even under certain damage scenarios, including unintended flooding, the vessel must be designed so that it remains afloat and sufficiently upright, with an amount of reserve of stability.

Interestingly, in most cases, direct assessments of stability from first principles, with still-water conditions, wind, and waves accounted for, are still not possible today. We must rely on time-honored experience-based procedures and trading tanker criteria to design for stabilizing in floating production system design. Semisubmersible mobile offshore drilling units are an exception in that their stability requirements have been revised recently on the basis of research and model tests.

Stability criteria for trading ships are specified by International Maritime Organization (IMO) conventions. Such statutory intact and damage stability requirements will need to be met during all operating inspections and maintenance conditions on site and when towing to the field. Following trading tanker practices, a trim and stability booklet covering the various loading conditions of interest will be prepared by the shipyard and provided on board. One or more "loading computers" capable of calculating and displaying the same information (e.g., drafts, trim, stability, and longitudinal strength parameters) will also be provided on board.

3.7 Design Principles for Towing and Station-Keeping

Prior to installation on site, the floating offshore structure usually must be towed from the construction yard to the site. Considering a typical tow, which uses several oceangoing tugs to move the structure at relatively slow speed, the tow may take several months, and the structure to be towed can be subjected to extreme waves, wind, and currents during tow. In this regard, in addition to its structural integrity during tow, its hull form, directional stability, maneuvering, and keeping capabilities are of interest. These aspects are usually addressed through appropriate analyses and model tests.

On site, the floating production systems are subjected to steady and unsteady actions that may cause some large movements from the original target location, but for dynamic positioning, thrusters, tethers, moorings, or a combination of these are used to limit such movements.

The station-keeping of the vessel is a critical part of the floating production system design because the station-keeping capability will govern the integrity of various subsystems, such as risers and gangway bridges. Therefore, design with refined analyses must be performed for moorings, dynamic positioning systems, and tethers to predict and limit the vessel movements as necessary.

One usually prefers that the mooring system not require any tension adjustments for the various conditions on site. Where this is not possible, means are provided to allow line tensioning during initial installation, and retensioning as required over the design life.

The mooring system design needs to be such that vessel excursions at surface are limited to acceptable levels; adequate clearances can be maintained between mooring lines, hull, and risers under all loading and environmental conditions on site. A case involving failure of a mooring line is also checked in the design phase. The design process for mooring systems usually involves an extensive series of model tests, primarily to calibrate the computer tools subsequently used for mooring system design and related studies.

3.8 Design Principles for Vessel Motions

Unlike fixed structures, floating structures are compliant to varying degrees. Again, it is undesirable to allow large motions of the vessel in terms of dynamic action characteristics because to do so would increase extreme environmental loading and increasing the vessel movements, which may make the subsystems nonoperable.

In design, the influence of motions must be minimized particularly for operations such as drilling. Therefore, for a drill ship, pitch and heave motions may be minimized by selecting a relevant ratio of water-plane area given the vessel displacement (Barltrop 1998). Also, by designing so that the natural period of the hull structure is sufficiently removed from the wave-period range, the vessel motions may be reduced.

Vessel motions will affect tank sloshing loads that are ever present in offshore operations. The structure involved usually needs to be designed so that sloshing-related restrictions on tank filling levels are eliminated completely.

3.9 Design Principles for Safety, Health, and the Environment

In the design of floating production systems, important considerations must be given to the following:

- Safety and welfare of the personnel and systems during construction, installation, commissioning, and operation
- Minimization of potential damage on the environment around the systems during their lifetime and during/after their decommissioning and removal

To meet these requirements, related regulations and company policies must be adhered to rigorously through the application of appropriate advanced technologies for design, operation, and risk assessment. Therefore, usually it is required that all key members of the project-management and engineering teams who design and build the systems must be familiar with the related safety principles, regulations, and codes at the earliest stage of the project engineering.

3.9.1 Design Principles for Safety

The design for safety encompasses at least the following:

- Layout, including separation and containment of hazardous zones and equipment
- Escape, temporary refuge, evacuation, and rescue including access, and muster zones

- Personnel protection and life-saving equipment
- Fire protection and fire and gas detection
- Emergency shutdown and depressurization for topsides systems
- Considering and reducing the risk of major collisions

Technical specifications for a project must then appropriately address the consideration of these various important factors. Regulatory guidance associated with design for safety usually also covers the same areas; see, for example, HSE (1996) for design and construction; HSE (1992) for safe operation, decommissioning, and removal; and HSE (1995a) in the case of fire, gas explosion, emergency response, and personnel evacuation. Note that the particular references provided as examples are not necessarily the latest; visit http://www.hse.gov.uk for the latest ones.

Once the principal dimensions and the types of the vessel are determined with particular features including deck modules, turret, and living quarters (accommodation), a qualitative risk assessment (or concept-safety case study) needs to be undertaken, and a quantitative risk assessment performed at a later stage to identify significant hazards and potential major accident scenarios that can affect the safety of the systems and/or the crew.

The aim of the qualitative risk assessment is to identify any fundamental deficiencies in the initial design of the selected concept and also to identify particular issues or areas that must be emphasized during the various design phases to prevent the occurrence of hazardous events. Such early-stage risk assessment also makes any design changes easier before detailed design begins. As necessary, the qualitative risk assessment may be repeated until the necessary safety improvements are achieved.

As we discuss in Chapter 13, for any risk assessment, some types of risk-analysis techniques are required although the level of sophistication will differ depending on whether it is an early-stage evaluation or one undertaken at a later stage. Also, there are specific types and formats of risk assessment that regulatory bodies might expect. One example is the UK "safety case regulations" (e.g., HSE 1996 and successors) that specify, among other matters, that an operator of a facility should identify the major hazards, evaluate the risks involved, and demonstrate that appropriate actions have been taken to reduce the risks to as low as reasonably practicable (ALARP) levels, and, in general, that the design of an installation be based on current good engineering practices.

Although the UK safety case regulations do not now require prescriptive and mandatory compliance with specific codes, standards, and guidelines, many other jurisdictions take a more traditional approach where such codes and guidelines may be applied more strictly.

Nevertheless, most classification societies with prescriptive rules now recognize that a rigorous risk-based assessment and demonstration of equivalent safety must be an acceptable alternative to complying with their prescriptive requirements, either in part or in whole. Certainly, it is recommended that one should perform a rigorous risk-based assessment and demonstration of safety cases, no matter what prescriptive classification rules or regulations might be applied. Chapter 13 presents details of useful risk-assessment and management technologies for such a purpose.

3.9.2 Design Principles for Health

Occupational-health-related factors that must be addressed by appropriate specifications and procedures in a floating offshore installation may include the following:

- Avoidance of carcinogenic materials, heavy metals, and asbestos
- Provision of emergency medical care and related medical facilities
- Human-factor engineering, including workplace layout, access, ergonomics, and noise
- Food and water quality and hygiene requirements
- Workplace security requirements

Examples of regulations that cover such aspects include HSE (1996) for the working environment; HSE (1995b) for the health and welfare of crew members; and HSE (1994) for the control of substances hazardous to health. (Again, these are study examples and not necessarily the latest applicable in the jurisdictions concerned; visit http://www.hse.gov.uk for the most recent ones.)

It is important to realize that the health and welfare of the workforce employed on the floating production systems can be significantly affected by features of the design as well as operations and management. Therefore, the regulations typically cover various matters that can have an influence on the working environment and subsequently the health and welfare of crew members, in addition to matters pertaining to physical or material safety.

For instance, the use of materials hazardous to health must be avoided. Some regulations on control of substances hazardous to health (e.g., HSE 1994) provide practice and guidance for the correct selection and specification of materials that are not hazardous in the day-to-day life of the workforce and/or crew members involved.

3.9.3 Design Principles for the Environment

Design related to environmental friendliness today must consider at least the following aspects:

- Minimization, control, and treatment of emissions to air; for example, of volatile organic compounds (VOC) such as SOx and NOx
- Minimization, control, and treatment of emissions to water (e.g., oil in produced water, safety of bunkering areas, biocides) including those used to treat seawater, and disposal of waste and sewage
- Avoidance of small oil spills and containment and response to oil spills
- Avoidance of ozone-depleting substances such as halogen

All current international and national regulations including MARPOL 73/78 (IMO 1978) aim to require operators to demonstrate that their current and future activities in relation to the operation of ocean vessels and systems will have neither short-term nor long-term hazardous effects to the surrounding environment. Jurisdictions may mandate these for floating offshore structures to various extents, and companies on their own may mandate these regulations for their particular facilities and operations.

Emissions and discharges from offshore installations include products of flaring, exhaust gases from prime movers, oily water discharges from slop tanks or produced water that cannot be reinjected, and sewage. Flaring or venting of excess gas is now prohibited in many jurisdictions. Produced gas may be exported, if economically advantageous. Potential spillage of offshore oil due to collisions, impacts, fire, and gas explosions is of great significance in terms of environmental impact.

In general, the design of ocean systems must be performed by using the best available technologies to prevent environmental pollution. In developing a new project for the floating production system installation, a well-thought-out and comprehensive strategy for environmental risk management will help provide the designer a clear direction and also identify required means and measures for preventing or mitigating the environmental damage. The strategy will also be helpful to control, monitor, and report the environmental impact while in operation.

3.10 Regulations, International Standards, and Recommended Practices

For the development of floating production systems, significant guidance in terms of classification society rules, international standards, local laws, and international regulations exists. The classification society rules and international design standards are primarily concerned with the structural design of a vessel's hull and marine systems, and the legislation includes a wide range of issues including safety, health, environmental protection, pollution prevention, and pollution control.

It is important to realize that compliance with existing prescriptive codes, standards, and guidelines may not be sufficient to design, construct, and safely operate production systems. Rather, these guidelines must be appropriately interpreted and supplemented for the particular structure, facility, and circumstances involved; also, currently available advanced engineering practices, which are based on concepts such as risk or formal safety assessment must be used concurrently as necessary (Lassagne et al. 2001).

We emphasize that formal safety assessment and other risk-based techniques make possible a greatly proactive approach to safety because such methodologies are used to identify and evaluate risk areas and then implement cost-effective risk-mitigation and containment measures such as basic design changes, monitoring systems, safety equipment, procedural controls, and training.

The use of classification society rules (e.g., ABS 2004; BV 2004; DNV 2000a, 2000b, 2002; LR 1999) is in fact not mandatory for the design of floating production systems, but most owners select the rules of a specific classification society to build their vessel, and it usually remains classed during tow and even while in service. This is for many reasons: insurance, mortgage, marketing purposes, and company policy (Millar and White 2000).

Appendix 6 presents a list of selected industry standards, regulations, and recommended practices for designing, building, and operating ship-shaped offshore units, specified by recognized classification societies and other institutions. These publications are well written and are of considerable educational value.

The following are the Internet Website addresses of recognized classification societies and international organizations, in alphabetical order, that have provided international or national design codes or professional guidance notes that are often useful for design and construction of floating production systems:

ABS (American Bureau of Shipping): http://www.eagle.org

AISC (American Institute of Steel Construction): http://www.aisc.org

ANSI (American National Standards Institute): http://www.ansi.org

API (American Petroleum Institute): http://www.api.org

ASTM (American Society for Testing and Materials): http://www.astm.org

AWS (American Welding Society): http://www.aws.org

BSI (British Standards Institute): http://www.bsi-global.com

BV (Bureau Veritas): http://www.bureauveritas.com

CEN (European Committee for Standardization): http://www.cenorm.be

DNV (Det Norske Veritas): http://www.dnv.com

DTI (Department of Trade and Industry, UK): http://www.og.dti.gov.uk

HSE (Health and Safety Executive, UK): http://www.hse.gov.uk

IACS (International Association of Classification Societies): http://www.iacs.org.uk

ICS (International Chamber of Shipping): http://www.marisec.org/ics

IEC (International Electrotechnical Commission): http://www.iec.org

IEEE (Institute of Electrical and Electronics Engineers): http://www.ieee.org

IMO (International Maritime Organization): http://www.imo.org

INTERTANKO (Association of Independent Tanker Owners): http://www.intertanko.com

ISA (Instrumentation, Systems, and Automation Society): http://www.isa.org

ISO (International Organization for Standardization): http://www.iso.org

LR (Lloyd's Register): http://www.lr.org

NACE (National Association of Corrosion Engineers, USA): http://www.nace.org

NEMA (National Electrical Manufacturers Association, USA): http://www.nema.org

NFPA (National Fire Protection Association, USA): http://www.nfpa.org

NIOSH (National Institute for Occupational Safety and Health, USA): http://www.cdc.gov/niosh

NORSOK (Standardization Organizations in Norway): http://www.nts.no/norsok

NPD (Norwegian Petroleum Directorate): http://www.npd.no

OCIMF (Oil Companies International Marine Forum): http://www.ocimf.com

SSPC (Society for Protective Coatings): http://www.sspc.org

UKOOA (Offshore Operators Association, UK): http://www.ukooa.co.uk

In an FPSO project, regardless of the use of classification society rules and other guidelines, the owner or operator must prepare and implement clear design philosophies and use the design basis and functional requirements for the vessel hull and subsystems through a comprehensive front-end engineering process, as described in Chapter 2. Adequate considerations of safety, health, and environmental factors must also be as integral a part of any front-end engineering as these considerations would be during detailed design, operation, and decommissioning.

REFERENCES

ABS (2004). *Guide for building and classing floating production installations.* American Bureau of Shipping, Houston, April.

API (1993). *Recommended practice for planning, designing, and constructing fixed offshore platforms – Working stress design.* (RP 2A-WSD), American Petroleum Institute.

API (2001). *Recommended practice for planning, designing, and constructing floating production systems.* (Recommended Practices, 2FPS), American Petroleum Institute, March.

Barltrop, N. D. P. (1998). *Floating structures: A guide for design and analysis.* The Centre for Marine and Petroleum Technology (CMPT). Herefordshire, England: Oilfield Publications Ltd.

Benjamin, J. R., and Cornell, C. A. (1970). *Probability, statistics, and decision for civil engineers.* New York: McGraw-Hill.

BV (2004). *Hull structure of production storage and offloading surface units.* (Rule Notes, No. 497), Bureau Veritas, Paris, October.

DEn (1990a). *Offshore installations: Guidance on design, construction, and certification.* Department of Energy, UK.

DEn (1990b). *The public enquiry into the Piper Alpha Disaster.* (Cullen report), Department of Energy, UK.

DNV (2000a). *Design of offshore steel structures, general (LRFD method).* (Offshore Standards, DNV-OS-C101), Det Norske Veritas, Oslo, October.

DNV (2000b). *Structural design of offshore ships.* (Offshore Standards, OS-C102), Det Norske Veritas, Oslo, October.

DNV (2002). *Rules for classification of floating production and storage units.* (Offshore Service Specifications, OSS-102), Det Norske Veritas, Oslo, April.

EDRH (2005). *Engineering design reliability handbook.* Edited by E. Nikolaidis, D. M. Ghiocel, and S. Singhal. New York: CRC Press.

Faulkner, D. (2002). "Shipping safety – A matter of concern." *Ingenia, Royal Academy of Engineering, London* (August/September), 13: 13–20.

Frieze, P. A. (1992). "Structural reliability analysis in offshore safety assessment." *Proceedings of International Conference on Offshore Safety,* The Institute of Marine Engineers (Paper 9), London, May 20–21.

Frieze, P. A., Hsu, T. M., Loh, J. T., and Lotsberg, I. (1997). "Background to draft ISO provisions on intact and damaged members." *Proceedings of Behaviour of Offshore Structures (BOSS'97)* (Edited by J. Vugts), Vol. 3: 111–126.

Frieze, P. A., and Paik, J. K. (2004). "General requirements for limit state assessment of ship structures." *SNAME Transactions,* 112: 368–384.

Guedes Soares, C., Fonseca, N., and Pascoal, R. (2003). "An approach for the structural design of ships and offshore platforms in abnormal waves." *Proceedings of MaxWave Final Meeting,* Geneva, WP 6, October.

HMSO (2000). *Re-opened formal investigation – Loss of the M. V. Derbyshire.* (Executive Summary to Report), HMSO (Her Majesty Stationary Office), UK, November.

HSE (1992). *The offshore installations (safety case) regulations.* (SI 1992/2885), Health and Safety Executive, UK.

HSE (1994). *The control of substances hazardous to health regulations.* (SI 1994/3246), Health and Safety Executive, UK.

HSE (1995a). *The offshore installations (prevention of fire and explosion and emergency response) regulations.* (SI 1995/743), Health and Safety Executive, UK.

HSE (1995b). *The offshore installation and pipeline works (management and administration) regulations.* (SI 1995/738), Health and Safety Executive, UK.

HSE (1996). *The offshore installations and wells (design and construction, etc.) regulations.* (SI 1996/913), Health and Safety Executive, UK.

IACS (2005). *Common structural rules for double hull oil tankers.* International Association of Classification Societies, London, December.

IMO (1978). *MARPOL 73/78: International convention for the prevention of pollution from ships.* International Maritime Organization, London.

ISO CD 19902 (2001). *Petroleum and natural gas industries – Fixed steel offshore structures.* (Committee Draft), International Organization for Standardization, Geneva, June.

Lassagne, M. G., Pang, D. X., and Vieira, R. (2001). *Prescriptive and risk-based approaches to regulation: The case of FPSOs in deepwater Gulf of Mexico.* Offshore Technology Conference, OTC 12950, Houston, April 30–May 3.

LR (1999). *Rules and regulations for the classification of a floating offshore installation at a fixed location.* Lloyd's Register, London.

Madsen, H. O., Krenk, S., and Lind, N. C. (1986). *Methods of structural safety.* Englewood Cliffs, NJ: Prentice Hall.

Mansour, A. E. (1989). *An introduction to structural reliability theory.* Ship Structure Committee, SSC-351, Washington, DC.

Millar, J. L., and White, R. J. (2000). *The structural integrity of FPSOs and FSUs – A regulator's view.* Offshore Technology Conference, OTC 12145, Houston, May.

NPD (2000). *Orientation concerning the arrangement of regulatory supervision relating to safety and the working environment in the petroleum activities.* Norwegian Petroleum Directorate, Norway.

Paik, J. K., and Faulkner, D. (2003). "Reassessment of the M. V. Derbyshire sinking with the focus on hull-girder collapse." *Marine Technology,* 40(4): 258–269.

Paik, J. K., and Thayamballi, A. K. (2003). *Ultimate limit state design of steel-plated structures.* Chichester, UK: John Wiley & Sons.

Wolfram, J., Linfoot, B., and Stansell, P. (2001). "Long- and short-term extreme wave statistics in the North Sea: 1994–1998." *Proceedings of a Symposium on Rogue Waves,* edited by M. Olagnon and G. Athanassoulis. Germany: Springer-Verlag.

CHAPTER 4

Environmental Phenomena and Application to Design

4.1 Introduction

Actions arising from environmental phenomena on a ship-shaped offshore unit are different from those on a trading tanker. The nature of the offshore structures and their operation are such that winds, currents, and waves, among other factors, may induce significant actions and action effects on structures. Whereas waves are often the primary source of environmental actions on trading ships at sea, considerations related to specialized operations such as berthing are somewhat different. In the case of offshore structures, a good knowledge of the environmental conditions in the areas where the structures will be installed is necessary in order to design for and assure the required high-operational uptimes. Such information is also important for specialized weather-sensitive operations such as installation on site, the berthing of supply boats, and the design of mooring and station-keeping.

This chapter presents environmental phenomena and discusses selected engineering practices helpful for the determination and treatment of environmental conditions for ship-shaped offshore units, considering design, transport, installation, and operations. Primary environmental phenomena that induce significant actions and action effects on offshore structures are presented. Although winds are typically regarded as a more elementary source of actions than waves because waves are caused by winds, this chapter starts its discussion with waves first. This is perhaps appropriate only because waves are a major source of actions on the particular types of offshore structures with which we are concerned.

In this chapter we use, for illustration purposes, various references to and data from publications by the UK Health and Safety Executive (HSE) and its predecessor, the Department of Energy (DEn). Many of these references originated in connection with UK HSE's certification regime SI 289 Offshore Installations: Construction and Survey Regulations (1974). Note that SI 289 has now been replaced by the verification regime of the Offshore Installations (Safety Case) Regulations 19962. This change in regulatory philosophy has also meant that documents such as UK HSE's *Offshore Installations: Guidance on Design, Construction, and Certification* (1990 and later amendments) now stands withdrawn, although the documents are quite helpful for pedantic purposes and are used as such in this book. See UK HSE (http://www.hse.gov.uk) for the latest information.

Table 4.1. *Metocean design parameters for offshore installations (HSE 2001a)*

Parameter	Required information	Influential parameters
Wind	Extreme wind speed and direction; vertical profile; gust speeds and wind spectra; persistence	Averaging time; height above sea level
Wave and swell	Extreme wave crest elevation; extreme wave height, direction, and range of associated periods; cumulative frequency distribution of individual wave heights; joint probability of significant wave height and period; persistence of sea states; wave spectra and directional spreading	Water depth; current; length of measurement period
Water depth and sea level variations	Depth below mean sea level; extreme still-water-level variations	Long-term changes in water depth; tide and storm surge
Current	Extreme current speed and direction; variation through the water depth; fatigue design current speed	Tidal and other currents; averaging time
Temperature	Extreme air temperatures (maximum and minimum); extreme sea temperatures (maximum and minimum)	Depth below sea surface
Rain and squall	Intensity in cm/hour for given return periods	Averaging time
Snow and ice	Maximum thickness of snow; maximum thickness of ice; densities of snow and ice	Geographical area; season; part of the structure
Marine growth	Type of growth; permitted thickness; terminal thickness profile	Water depth; growth rate

4.2 Environmental Data

Reliable data on various environmental phenomena are necessary for the design and operation of the hull, topsides, moorings, and risers of ship-shaped offshore units. Table 4.1 indicates many of the relevant metocean (i.e., meteorological oceanographic) parameters that may be required.

The required data can be obtained in many different ways and from many sources; some sources are used more often than others. The environmental data may be obtained, for example, from in-situ instrumental measurements, "remotely sensed" measurements from satellites, visual observations (from ships and platforms), and through ocean environmental energy transport numerical modeling or "hindcasting." Extensive measurements of environmental data suitable for design exist for the UK and Norwegian sectors of the North Sea; for example, see Lonseth and Kvitrud (1997) and HSE (2001a). In several other offshore areas where such data may not exist, specific measurements usually will be required. FPSO projects and companies undertake such measurements routinely.

Useful data in other cases may also be obtained from sources such as the following (a more extensive list can be found in Barltrop [1998]):

BMT Fluid Mechanics Ltd., UK

British Antarctic Survey, UK

British Oceanographic Data Centre, UK

Canada Hydrographic Service, Canada

Det Norske Meteorologiske Institutt, Norway

Norwegian Polar Institute, Norway

U. S. National Oceanographic Data Center, USA

Internal or external consultants gather together and make recommendations on the data to be used for a project considering the site, season, and other factors. This reflects the often specialized nature of the effort involved. Generally, it is worth noting that the more particular and more extensive the data, the better. The greater the environmental uncertainties, the larger the number of safety factors that need to be applied to achieve a target level of structural adequacy and reliability.

4.3 Waves

The wave parameters used for offshore designs include heights, periods, and directions with associated probabilities and persistence times. It is important to realize that the waves inducing the most severe response in the global system structure may be different from those resulting in the maximum response in structural components and also that the response of ship-shaped offshore structures is wave-period dependent. It is noted also that more frequent waves rather than extreme waves will govern fatigue life, although their magnitude may be smaller.

Wave-induced maximum actions and action effects may be applied for design by using any one of a few approaches – for example, extreme-amplitude design waves, extreme-response design waves, or the more fundamental wave-energy spectra-based methods.

An extreme-amplitude design wave may be calculated for a specified return period, usually 100 years for strength design of long-term deployment, as we describe in Section 4.13. General methodologies for estimating the parameters of design waves in deep-water conditions for ship-shaped offshore units are available in standard references (e.g., Faltinsen 1990; Barltrop 1998). Ochi (1978) presents wave information that is useful for predicting responses of ships and offshore structures in a seaway and discusses specific application methods for design consideration. For normal operations, a 10-year return period environment may be specified, which may be reduced to a 1-year return period considering inspection and repair conditions.

In floating production systems, some maximum actions may often develop from a wave or group of waves with a lower amplitude than a wave with a higher amplitude because of the potential sensitivities of the wave actions to the wave frequencies involved. Also, several different design-wave combinations from various directions and frequencies with crests and troughs at various locations need to be considered for the different types of responses (e.g., maximum roll, maximum vertical hull girder bending moment). See Liu et al. (1992) for tanker-related examples.

In Sections 4.3.1–4.3.3, selected recommended practices pertaining to wave actions, which may be used for reference or for initial guidance in the absence of site-specific wave data, are presented. We should caution that the information described in these sections is updated periodically; thus, the latest versions of the original documents cited must be used for FPSO designs.

4.3.1 UKOOA FPSO Design Guidance Notes for UKCS Service

The UK Offshore Operators Association (UKOOA) FPSO design guidelines (UKOOA 2002) describe generic wave data and their application for the UK Continental Shelf (UKCS) waters and its adjacent areas. The UKOOA guidelines indicate that the wave information must cover a range of available combinations of wave height and period to determine the most severe loading conditions acting on FPSOs.

The significant wave heights are said to be 14m for the central North Sea, 16m for the northern North Sea, and 18m for the West of Shetlands, based on the 100-year return period sea state. The 10,000-year significant wave height based on average zero up-crossing analysis is approximately 25 percent greater than the corresponding 100-year values; also, the corresponding wave period is about 5 percent greater than the 100-year wave period.

It is noted that the mean wave zero up-crossing period lies in the following range:

$$3.2H_s^{0.5} < T_z < 3.6H_s^{0.5}, \tag{4.1}$$

where H_s = significant wave height in meters; and T_z = mean zero up-crossing wave period in seconds.

Ignoring the small risk (1 percent or so) that waves with $H_{max} = 2.5H_s$ may exist, it is considered that the expected (most probable) maximum wave height can usually be estimated by

$$H_{max} = 1.86H_s, \tag{4.2}$$

where H_{max} = most probable maximum wave height; and H_s = as defined in Eq. (4.1).

Also, the wave periods associated with maximum wave height are considered to lie in the range of $1.05T_z$–$1.4T_z$. For the design of FPSOs, wave influences related to all values within the range of periods must be accounted for. The following may need to be addressed:

- First-order motions of the vessel at wave frequency (heave, surge, sway, roll, pitch, yaw)
- Low-frequency motions particularly for surge and sway near the natural frequency of the vessel and of the mooring system
- Steady or mean drift forces

The UKOOA guidelines suggest that one predicts the wave-induced actions from model testing or motion analysis using diffraction theory. The two-dimensional strip theory may, however, be used as an approximation for initial design. In addition, for bow-wave impact and green-water effects, the response to short steep waves needs to be considered. For this purpose, see NORSOK N003 (1999) and HSE (2005a).

It is noted that hull-fatigue calculations must account for the distribution of wave encounters in number for all possible wave periods. A scatter diagram of significant wave height versus mean zero up-crossing periods for the specific location can be used to obtain the fatigue-design-wave data referred to.

4.3.2 American Petroleum Institute Recommended Practices

The American Petroleum Institute (API) Recommended Practices (API 1993a, 1993b) provide information on environmental parameters for U.S. (United States) waters where offshore structures are installed. These include more than twenty areas such as the Gulf of Mexico, the West and East Coasts of the United States, and the coasts of Alaska. More detailed information is presented for the Gulf of Mexico, including information on the variation of location and water depth of design-wave height and the directionality of waves associated with hurricanes (i.e., tropical cyclones in the Gulf of Mexico).

Although both API RP 2A-WSD (API 1993a) and API RP 2A-LRFD (API 1993b) deal with fixed offshore structures, and API RP 2FPS (API 2001) deals with floating production systems, these documents are useful for obtaining indicative information on the environmental parameters for U.S. waters.

In the API-recommended practices for floating production system installations, the main reference parameter for design is the 100-year return maximum individual wave height. Two sets of environmental criteria are considered: (a) the 100-year return period waves with associated winds and currents, and (b) the 100-year return period wind with associated waves and currents. The most severe directional combination of waves, winds, and currents should be specified consistent with the environmental conditions to be experienced at the operational field. In some cases, it is likely that extremes of waves and winds may approach a specific operational field from different directions so that a weathervaning floating production system may be exposed to higher actions than where waves and winds act in the same direction.

The API-recommended practices emphasize that accurate measured and/or hindcast data must be used to determine the design-wave actions. The relationship between wave height and wave period is important, particularly for the prediction of surge and sway amplitudes and mean drift forces that are affected significantly by the wave period.

It is indicated that with the same wave height, swell-induced wave periods are approximately 40 percent higher than wind-driven wave periods in the regions considered. For any specific return period, the ratio of maximum wave height to significant wave height is said to lie in the range of 1.7–1.9, although the UKOOA design guidelines indicate that the ratio is 1.86.

4.3.3 Det Norske Veritas Classification Notes

Det Norske Veritas (DNV) Classification Notes 30.5 (DNV 1991 and later amendments) presents information on environmental parameters including waves, winds, currents, snow, ice, and temperature for various geographical areas worldwide that are classified into two groups: harsh and benign.

In determining design-wave actions, the wave characteristics may be divided into two types, regular waves and irregular waves. It is indicated that it is sufficient to regard the waves as regular waves when the corresponding wave periods are in the following range:

$$\sqrt{6.5H} < T < \sqrt{11H}, \tag{4.3}$$

where T = wave period in seconds; and H = crest to trough wave height in meters.

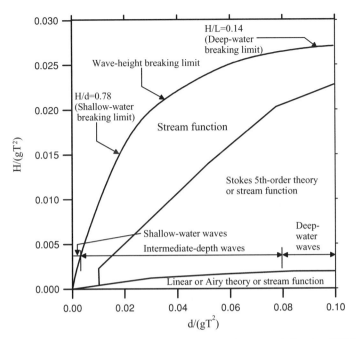

Figure 4.1. Regular wave theory selection diagram, following HSE (1989b, 2001b; courtesy of HSE) [$H/(gT^2)$ = dimensionless wave steepness; $d/(gT^2)$ = dimensionless relative water depth; H = wave height (crest to trough); d = mean water depth; T = wave period; L = wave length (distance between crests); and g = acceleration of gravity].

The regular wave characteristics can then be described by relevant analytical or numerical theories that may be classified depending on the ratio of water depth to wave length, as follows:

- Solitary wave theory for $d/L \leq 0.1$
- Stokes 5th-order wave theory for $0.1 \leq d/L \leq 0.3$
- Linear wave theory (e.g., Airy) for $0.3 < d/L$

where d = water depth; and L = wave length. Similar classifications for water depth and wave height are also made by HSE (1989b, 2001b). Figure 4.1 illustrates a regular wave theory selection diagram, following HSE (1989b, 2001b).

On the other hand, short-term irregular sea states can be described by a wave spectrum in terms of the power spectral density function of the vertical sea surface displacement. Although various wave spectra expressions are given in Section 4.14, the Pierson–Moskowitz spectrum can be applied for open seas, and the JONSWAP spectrum can be applied for fetch-limited growing seas. Short-crested waves (wave spreading) may be applied as relevant. In these cases, the peak period typically lies in the following range:

$$\sqrt{14.5H_s} < T_p < \sqrt{25H_s}, \tag{4.4}$$

where T_p = period at peak frequency in seconds; and H_s = significant wave height in meters.

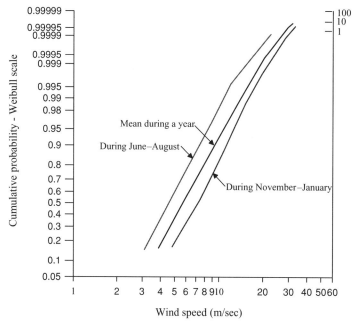

Figure 4.2. Cumulative distribution function for hourly mean wind speed at 10m above mean sea level for the northern North Sea, following Faltinsen (1990).

It is interesting to note that the highest crest elevation is approximately equal to the significant wave height (H_s) for an irregular short-term stationary sea state in visual observations, and the highest individual crest to trough wave height is approximately equal to $1.8H_s$. The DNV Classification Notes 30.5 also provides much useful information for estimating the long-term wave statistics considering geographical location and storm duration.

4.4 Winds

Wind is a primary metocean parameter that is important to the design of offshore units, for example, during normal operations. The structure must withstand the forces exerted by the wind, and this depends not only on the structural characteristics such as windage area but also on the speed and direction of the wind.

For design, extreme wind speeds for specified return periods must be obtained and are specified with averaging times ranging from 3 seconds (i.e., an extreme gust value) to 24 hours, for example. The speeds are usually estimated at a standard height of 10m above mean sea level, with corrections to more specific values at other heights.

In addition, the spectra of fluctuating wind gusts are necessary because wind gusts can excite resonant oscillations of offshore structures (Faltinsen 1990). For example, slow-drift horizontal motions of moored structures can be caused by wind gust. Also, wind can lead to phenomena such as vortex shedding, together with associated vibrations in some instances, including flare tower.

Figure 4.2 shows the cumulative distribution function for hourly mean wind speed at 10m above mean sea level for wind data from the northern North Sea (Faltinsen

Table 4.2. *Relationship between 50-year return period wind speed and extreme wind speeds at other return periods (HSE 1989a)*

N (years)	2	5	10	20	50	100	200	500	1000
V_N/V_{50}	0.75	0.83	0.88	0.93	1.00	1.05	1.11	1.17	1.23

Notes: (1) N = return period; (2) V_N = N year return speed; (3) V_{50} = 50-year return speed; and (4) these values were obtained from $V_N = 0.71(1 + 0.106 \ln N)V_{50}$.

1990). Figure 4.2 shows that the extreme wind speed with the 100-year return period is about 41m/s. Table 4.2 shows the relationship between the extreme 50-year return period wind speed used for design in some cases, and the extreme wind speeds at other return periods. Table 4.3 indicates 100-year return period design wind speeds for UK waters.

In the absence of specific wind data, the UKOOA FPSO design guidelines applicable in UK waters recommend certain design wind speeds depending on the areas involved; see Table 4.4. Wind speeds for the 10,000-year return period are approximately 16 percent greater than the speed for the 100-year return period indicated in Table 4.4. The UKOOA guidelines suggest the use of NORSOK Standard N003 (NORSOK 1999) formulations to describe the wind-speed variation with the height above sea level.

In determining the design-wind and design-wave actions, it is necessary to know the information on the variation of winds with height above sea level, the direction that the wind blows, and the joint probability between waves and winds. In this regard, Table 4.4 indicates a simplified picture representing the relationship among significant wave height, wave period, and wind speed for open seas in the North Atlantic and North Pacific (Lee et al. 1985).

The wind force on each part of the FPSO may be estimated from the following:

$$F = 0.0625\,AV^2 C_s, \tag{4.5}$$

where F = wind surface force in kgf; A = projected area in m^2; V = wind speed in m/s; and C_s = shape coefficient as defined in Table 4.5.

In API RP 2FP1 (API 1991), two methods are suggested to evaluate the wind effects: (a) as a constant applied value where the wind speed is taken as the extreme 1-minute mean wind speed; or (b) as a fluctuating force based on the extreme 1-hour average velocity together with a time-variant component calculated from a suitable wind-gust spectrum. Formulae are also provided to estimate the wind forces.

The DNV Classification Notes 30.5 (DNV 1991) suggests taking the reference averaging period of wind as 10 minutes and the reference height as 10m above sea level. The average wind speed and its profile with height may then be estimated using a closed-form formula.

Table 4.3. *Illustrative 100-year return period design wind speeds for UK waters (UKOOA 2002)*

Wind speed	Central North Sea	Northern North Sea	West of Shetlands
1-hour average	37 m/s	38 m/s	40 m/s
10-min. average	40 m/s	41 m/s	43 m/s

Table 4.4. *Annual sea-state occurrences in the North Atlantic and North Pacific (Lee et al. 1985)*

Sea state no.	Significant wave height (m) Range	Mean	Sustained wind speed (knots)[a] Range	Mean	North Atlantic Percentage probability of sea state	Modal wave period(s) Range[b]	Most probable[c]	North Pacific Percentage probability of sea state	Modal wave period(s) Range[b]	Most probable[c]
0–1	0–0.1	0.05	0–6	3	0.70	–	–	1.30	–	–
2	0.1–0.5	0.3	7–10	8.5	6.80	3.3–12.8	7.5	6.40	5.1–14.9	6.3
3	0.5–1.25	0.88	11–16	13.5	23.70	5.0–14.8	7.5	15.50	5.3–16.1	7.5
4	1.25–2.5	1.88	17–21	19	27.80	6.1–15.2	8.8	31.60	6.1–17.2	8.8
5	2.5–4	3.25	22–27	24.5	20.64	8.3–15.5	9.7	20.94	7.7–17.8	9.7
6	4–6	5	28–47	37.5	13.15	9.8–16.2	12.4	15.03	10.0–18.7	12.4
7	6–9	7.5	48–55	51.5	6.05	11.8–18.5	15.0	7.00	11.7–19.8	15.0
8	9–14	11.5	56–63	59.5	1.11	14.2–18.6	16.4	1.56	14.5–21.5	16.4
>8	>14	>14	>63	>63	0.05	18.0–23.7	20.0	0.07	16.4–22.5	20.0

Notes: [a] Ambient wind sustained at 19.5m above surface to generate fully developed seas. To convert to other altitudes, apply $V = V_0(H/19.5)^{1/7}$, where V, H = sustained wind speed (knots) or altitude (m), respectively, and V_0 = sustained wind speed for $H = 19.5$m; [b] Minimum is 5 percentile and maximum is 95 percentile for periods given wave height range; [c] Obtained on the basis of periods associated with central frequencies included in the U.S. Navy hindcast climatology.

Table 4.5. *Shape coefficient for estimation of wind force*

Shape	Sphere	Cylinder	Large flat surface[a]	Support members[b]	Isolated shapes[c]	Clustered deck houses
C_s	0.40	0.50	1.00	1.30	1.50	1.10

Notes: [a] Hull, deck house, smooth underdeck areas; [b] Exposed beams or girders under deck; [c] Cranes, booms, etc.

As an alternative to the approaches only by model tests or by statistical-methods-based model tests, the computational fluid dynamics (CFD) technique is also available to determine wind loads on offshore structures (Aage et al. 1997).

4.5 Water Depths and Tidal Levels

The overall depth of water at any location can be characterized by the mean depth and its variations from mean sea level. The mean water depth is defined as the vertical distance between the sea bed and an appropriate near-surface datum. The variations of water depth are primarily due to tides and storm surges. The tide-related variations are usually regular and predictable in terms of the highest astronomical tide and the lowest astronomical tide.

Meteorologically generated storm surges, however, are irregular in nature. The effects of tides can be superimposed on the effect of storm surges to estimate the total mean water levels; these could in some cases be above the highest astronomical tidal level or below the lowest astronomical tidal level.

4.6 Currents

Currents, together with waves and swells, can affect the orientation of the offshore structure and, therefore, directly and indirectly affect both short-term and long-term loads imposed on the structure and its mooring system. Currents can increase the hull drag forces over and above the values due to the wave system alone. Currents also ultimately affect the station-keeping of the offshore unit and the performance of its thrusters (where used).

The nature of currents is very complex, depending on the local conditions. A number of current types may be relevant – for example, oceanic currents, eddy currents, thermal currents, wind-driven currents, tidal currents, surge currents, and inertial currents (Barltrop 1998). The common ones are usually astronomical tide and storm-surge related. But this is by no means a certainty in any specific case or region, and if at all possible, specific onsite measurements need to be made before locating an offshore unit at any given site.

An offshore production vessel hull and its mooring system are affected first by surface currents. In the design of risers, one needs to appropriately consider currents at lower sea levels. Generally, the major open-ocean currents below the surface can be more predictable and subject to less change, although for currents closer to the ocean surface, the effects of wind will mean greater variability than in the open sea. As always, there are many exceptions to these generalizations; for instance, in the Indian Ocean and China Seas, current directions can change significantly seasonally and even reverse during monsoons.

Table 4.6. *Generic surface current data for UK waters (UKOOA 2002)*

Location	Central North Sea	Northern North Sea	West of Shetlands
100-year return period surface current speed	1.03 m/s	0.99 m/s	2.00 m/s
1-year return period surface current speed	0.88 m/s	0.89 m/s	1.64 m/s
Current direction	N/S	N/S	NE/SW

The current data must be obtained from the measurements made at or close to the operation field for at least a year or longer to build up an accurate picture of the current characteristics including speed and direction. Generic surface current data from the UKOOA FPSO design guidelines for UK waters are provided in Table 4.6. The UKOOA guidelines also suggest that the current forces on FPSOs should be calculated using the method presented by OCIMF (1994).

In the absence of detailed field measurements for currents, the DNV Classification Notes 30.5 suggests that one may be able to describe the current profile as the sum of tidal current and wind-generated current profiles. In the classification notes cited, the tidal current is postulated to be subject to 1/7 power exponential decay over the water column and wind-generated current is said to decay linearly from 1.5 percent of hourly mean wind speed (i.e., which gives the current speed in m/s) at still-water level to zero at 50m depth. Other types of currents must also, of course, be considered in describing the currents if relevant.

Lonseth and Kvitrud (1997) have presented data from current measurements for the Northern Norwegian Continental Shelf. Figure 4.3 shows sample profiles of extreme current velocity for the 100-year return period at three locations in the

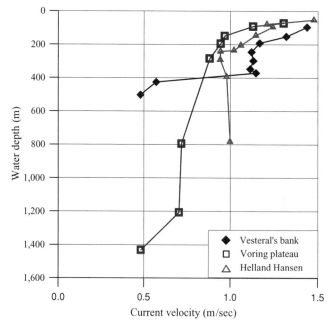

Figure 4.3. Sample 100-year return period extreme current velocity profiles at three locations in the Norwegian Sea.

Norwegian Sea, from the work of Lonseth and Kvitrud (1997). The current velocity is a function of water depth; even in deep water, the current velocity can be more than 1.0 m/s in some areas of the world. This underscores the need for site-specific measurements as the preferred approach in many cases.

4.7 Air and Sea Temperatures

For temperatures, the concept of probable extremes defined as the values probably never exceeded may be used instead of statistical extreme values with a specified return period. It is interesting to note that the probable extremes at the sea surface are sometimes more severe than the corresponding 50-year return period temperatures, and also that extremes of sea surface temperature occur less frequently than air temperature extremes (HSE 1989a, 2001a).

The information on sea temperatures is important for fracture toughness design, in many cases, although air temperature information is of interest for applications where a structure and various onboard systems respond to air temperature changes. Year-round temperatures and humidity are also of interest to the design of heating and cooling systems. In some geographical areas, such as West Africa, the year-round temperatures and humidity may remain uniformly high, a factor that must be considered in the selection of corrosion margins and corrosion protection schemes.

4.8 Snow and Icing

Depending on the areas of operation, the extent to which snow and ice may accumulate on various parts of the offshore units may need to be estimated. Associated risk mitigation measures include the provision of adequate strength and stability and local heating. Physical deicing and snow removal procedures also need to be specified based on the maximum permitted accumulation specifications for the offshore unit.

In many cases, snow accumulations may be more likely than icing, particularly on windward-facing nonhorizontal parts of the unit. Snow, if it remains, can freeze into ice and, therefore, will need to be removed before that happens by blowing it dry or by other means.

Icing takes place when the temperature of the water is around 6°C or less and the air temperature is below 0°C. As a result of water breaking or spraying over the deck, ice can form on the deck and topsides. Stability issues can then occur when the metacentric height is reduced due to icing. Temperature levels in the polar area can be below –35°C; thus, icing is always an issue when an offshore unit operates in the arctic area. Icebound regions, including the arctic, are thought to contain significant hydrocarbon reserves.

For structures operating in icebound areas, impacts due to growlers, bergy bits, and even icebergs need to be considered in design as they can cause structural damage; an example of denting due to ice is shown in Figure 4.4. The design of structures strengthened to withstand the effects of snow and ice is a specialized subject that can be of interest to ship-shaped offshore units and shuttle tankers operating in areas with seasonal or year-round ice.

Important logistics considerations for operations, such as those related to bringing in spares and supplies needed, can also arise in such areas because non-icebound

Table 4.7. *Mean ice thickness in various cold seas*

Sea	Mean ice thickness (mm)
The Kara Sea (arctic)	1,800
The Sea of Okhotsk (east Siberia/Sakhalin)	1,400
Barents Sea (arctic)	1,200
The White Sea	800
The Black Sea (The Sea of Azov)	700
The Caspian Sea	700
The Baltic (Gulf of Finland/Gulf of Bothnia)	400–800

time can often be short. As would be expected, the ice thickness and related pressures will be an important factor in the design of vessels such as icebreakers. As an example, Table 4.7 summarizes the ice thickness in various cold seas. Plastic design concepts and, more recently, nonlinear finite-element analyses are useful tools in the structural design for ice (Wang and Wiernicki 2004; Wang et al. 2005a, 2005b; Wang and Liu 2006). Similar acceptance criteria to Eq. (4.7) can be applied.

We will now present a brief historical review on the study of ice loads and strength predictions. Research for arctic structures, related to ice mechanics predictions of the maximum design ice loads, must be considered at both global and local levels. Bruen et al. (1982) review the methods of selecting local design ice pressures for arctic offshore structures including the use of the Hertzian contact theory, plasticity theory, field observations, and empirical methods. A wide scatter is found in local ice

Figure 4.4. Plastic deformation on side shell structure of a trading ship due to ice loads (Kujala 1994).

pressures, leading to the use of probabilistic criteria such as ice pressure exceedence curves. The importance of scaling law is discussed particularly for brittle ice behavior.

Vivatrat and Slomski (1983) present a probabilistic procedure for determining the maximum loads and pressures on a fixed offshore platform due to multiyear ice floes (icebergs) during winter loading events. Mechanical models for estimating the flow contact width, peak indentation pressure, pressure versus displacement, and interaction between the multiyear floe and the surrounding first-year ice are described. Hnatiuk (1984) reviews the offshore production activities in Canada's harsh arctic offshore areas; the experience and lessons learned can also be pertinent for similar applications in other icebound areas.

Kreider et al. (1985) present a probabilistic approach to develop ice load criteria for offshore structures operating in the Beaufort Sea. Williford and Winkler (1987) present the design experience for a self-propelled turret-moored icebreaker drilling unit operating in an ice environment; the unit was designed for station-keeping in 1m ice and 10m unconsolidated pressure ridges. Operational experience for a drilling unit in the Beaufort Sea is presented by Hinkel et al. (1988). Operational concerns described include wellhead protection from ice scour. Truskov (1999) provides useful ice conditions and metocean and seismic data offshore northeastern Sakhalin Island.

For polar trading ship designs, substantial efforts are now directed toward the implementation of more sophisticated methodologies into design standards. The IACS polar ship rules (IACS 2001) deal explicitly with the load-carrying capacity of the structures under ice loads in the plastic regime. The Finnish–Swedish ice class rules (FMA 2002) are commonly used for design of ice-strengthened vessels operating in the Baltic Sea. As an alternative to the Finnish–Swedish ice class rules, the Finnish Maritime Administration (FMA) has now also published guidelines for the application of first-principles methods for the structural design against ice loads (FMA 2003a, 2003b, 2004). Classification societies have also provided the guidelines of first-principles approaches applying nonlinear finite-element methods (ABS 2004a, 2005).

The *Terra Nova* FPSO operating in Newfoundland, Canada, is perhaps one of the first of offshore floating units designed for ice-infested, relatively harsh environments. To avoid the threat of icebergs, the FPSO was designed so that it could quickly disconnect from its mooring and proceed under its own power. Doyle and Leitch (2000) describe the development of the *Terra Nova* FPSO hull from design through construction to delivery to meet the requirements of operating in the harsh Canadian environment of the Grand Banks. Maguire et al. (2001) present a description of measures undertaken and implemented to ensure the fitness of the *Terra Nova* FPSO within the context of a complex regulatory climate by minimizing the related risk. An overview of the *Terra Nova* development and related prominent technical challenges are also presented.

4.9 Marine Growth

Floating offshore structures are likely to become fouled with marine growth, much like a ship or any other marine structure. In the case of offshore units such as drill ships, this may increase resistance and powering when underway. Its removal is quite simple once dry-docked.

On site, however, the removal of marine growth by cleaning prior to underwater structural inspections can be expensive. Typically in such cases, to keep the situation controllable, marine growth will be removed periodically; that is, when they reach certain predefined growth levels. In the early life of the offshore installation on site, the occurrence of marine growth can be reduced or avoided by the coating system used, including an antifouling component. Admittedly, most antifouling paint is more effective in a moving object rather than a stationary object.

For design purposes, there usually is a marine growth profile (thickness and roughness as a function of water depth) specified as part of the metocean data in a design-basis document.

4.10 Tank Sloshing

4.10.1 Fundamentals

The accelerations arising from the motions of a ship in a seaway can produce sloshing actions on the structures of partially filled tanks. Motions of liquid cargo in oil tanks may often produce significant sloshing actions, and the affected structure must be engineered to withstand them. This is of particular concern in tanker conversions because it is not always the case that trading tankers were designed for partially filled cargo tanks, unlike their ballast tanks. Cargo tanks of moored ship-shaped offshore structures are continuously loaded and unloaded and, therefore, sloshing in the tanks may not be avoidable.

Resonance between the natural sloshing period of the tank with liquids and the roll or pitch periods of the structure is of concern. The recent trend toward adopting large tanks, which serves to reduce the number of tanks, does not help in this regard because the result may be larger tanks with longer natural periods (with a more attractive construction cost, without doubt). Such trends may also complicate design for maintenance and hot work on site where, generally, the larger the number of tanks, the better.

4.10.2 Practices for Sloshing Assessment

Like other sources of impact-pressure actions such as bow slamming or green water (described in this chapter) or explosion (described in Chapter 8), sloshing can also result in impact-pressure actions and subsequent structural damage. Increased pressures of a nonimpact nature are also possible. The work to resolve the impact-pressure issue can be classified into two parts: the hydrodynamics-related study and the structural mechanics–related study. The aim of the hydrodynamics-related study is to identify the impact-pressure profile in terms of pressure versus time history, and the structural mechanics study is aimed at calculating the dynamic structural response, including damage due to the applied impact-pressure actions.

Sloshing considerations in classification society rules and procedures today are fairly well advanced; for example, reference is made to LR (2004) and IACS (2005). If structural efficiency is the consideration, more refined approaches involving impact-pressure parameters such as peak pressure and impact duration must be applied to analyze the wave-impact problem, including structural damage (Paik et al. 2004).

This is because in current classification society rules, the structural design criteria against impact-pressure actions are typically based on a quasistatic equivalence

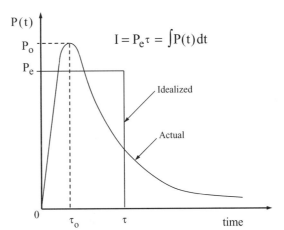

Figure 4.5. An impact-pressure action profile and its idealization.

$$I = P_e \tau = \int P(t)\,dt$$

concept that defines an equivalent quasistatic pressure situation in place of the real impact-pressure situation. We recognize that this approach does not necessarily reflect the impact-pressure characteristics relevantly. The structural damage by this concept may be underestimated in some cases and overestimated in other cases, indicating that the concept is not consistent in terms of strength assessment. However, by appropriate calibration in comparison to cases of damage versus no damage, it appears that workable design procedures can be attained as well.

For practical design purposes, the problem of impact-pressure actions in terms of structural behavior can be idealized within three domains of behavior depending on the ratio of the duration of impact actions to the natural period of the structure, as follows (NORSOK 1999):

- Quasistatic domain when $3 \leq t/T$
- Dynamic/impact domain when $0.3 \leq t/T < 3$
- Impulsive domain when $t/T < 0.3$

where t = duration of impact actions; and T = natural period of the structure.

The impact-pressure action arising from green water, bow slamming, or sloshing is generally characterized by four parameters: (1) rise time until the peak pressure, (2) peak pressure, (3) pressure decay type beyond the peak pressure, and (4) pressure duration time, as illustrated in Figure 4.5. The peak pressure value often approaches some 2–3 times the collapse pressure loads of structural components under quasistatic actions. But the rise time is very short, a few milliseconds or less. The duration (persistence) time of impact pressure is often in the range of 10–50 milliseconds. It is important to realize that, unless anticipated and designed for, the structural damage due to impact-pressure actions can be significant even though the duration time is very short as long as the associated impulse itself is large enough (Paik et al. 2004).

When the rise and duration times of impact pressure are very short, however, it is possible that the impact-pressure response can be approximated to an impulsive type of action that is characterized by only two parameters: equivalent peak pressure P_e and duration time τ, as long as the corresponding impulse is identical (Paik et al. 2004). In this case, it can then be approximated that the impact-pressure actions arising from sloshing, slamming, or green water can be characterized by P_e and τ, as shown in Figure 4.5. The two parameters may be defined so that the actual and idealized impulses of the impact-pressure action are equal:

$$I = P_e\tau = \int P(t)\,dt, \tag{4.6}$$

where I = impulse of the impact-pressure action; t = time; P_e = effective peak pressure; and τ = duration time of P_e. Simple formulae to calculate sloshing impact-pressure distribution for trading tankers that may be useful for ship-shaped offshore units are given by IACS (2005), applying the quasistatic equivalence concept.

Taking P_e as the same as P_o (peak pressure value) can be unduly pessimistic for obvious reasons; thus, P_e is often obtained by multiplying a relevant knock-down factor to P_o. Once the impulse I and the effective peak pressure value P_e are defined, the duration time τ can then be determined from Eq. (4.6). In predicting structural damage due to impact-pressure actions, P_o and τ will be dealt with as parameters of influence.

An acceptance criterion to be safe against impact-pressure actions can be based on the serviceability limit state in terms of the permanent set deflection of ship-shaped offshore structure panels, as follows:

$$w_p \leq w_{pa}, \quad \text{or } \eta_1 = \frac{w_{pa}}{w_p} \geq 1, \tag{4.7a}$$

where w_p = factored permanent set deflection; w_{pa} = allowable (factored) target value of permanent set deflection, which may be taken as a few times the plate thickness; and η_1 = measure of structural adequacy related to the permanent set deflection.

The acceptance criterion should also be considered for the ultimate limit state in terms of maximum pressure loads or associated impulse capacity, as follows:

$$P_d \leq P_u, \quad \text{or } \eta_2 = \frac{P_u}{P_d} \geq 1, \tag{4.7b}$$

$$I_d \leq I_u, \quad \text{or } \eta_3 = \frac{I_u}{I_d} \geq 1, \tag{4.7c}$$

where P_d, I_d = design (factored) peak pressure or impulse at the design duration time, respectively; P_u, I_u = factored maximum impact pressure or impulse capacity at the corresponding duration time until structural failure (e.g., buckling, fracture) takes place, respectively; and η_2, η_3 = measures of structural adequacy related to impact pressure or impulse capacity, respectively.

As indicated in Eq. (4.6), the impulse can be calculated by integrating the area below the impact pressure versus time history. The maximum impact pressure and impulse capacity can be obtained by dynamic nonlinear structural behavior analyses using numerical methods such as those presented in Section 8.4.3 in Chapter 8. It is important to realize that the dynamic nonlinear structural behavior may depend significantly on dynamic material properties (e.g., strain-rate sensitivity, viscoelasticity, damping) and, therefore, the effects of dynamic material properties should be taken into account in the dynamic structural capacity analyses.

Another important issue is damage accumulation. In reality, impact-pressure actions may be applied repeatedly; thus, the resulting structural damage can be accumulated, causing fatigue cracking and fracture. In this regard, the hydro-dynamics-related study should also identify the relevant information in terms of short-term or long-term time histories of impact-pressure actions for low-cycle fatigue and fracture analysis.

4.10.3 Measures for Sloshing Risk Mitigation

Sloshing risk mitigation measures for ship-shaped offshore structures can be similar to those for trading ships. Obviously, the size of the tank or compartments exposed to sloshing impact should ideally be decreased by minimizing bulkhead spacing or by fitting partial sloshing bulkheads. This could move the tank natural period away from the range of hull resonance. As an alternative or in conjunction, the tank boundary scantlings should be adequately increased to withstand the sloshing impact. This is normally a workable alternative because not all of the structure in a tank is dispro-portionately and adversely affected by sloshing; that is, the increases required are usually localized to certain parts of the tank (and its support structure, as relevant).

4.11 Bow Slamming

4.11.1 Fundamentals

Bow structures are likely subjected to impact-pressure actions arising from what is termed "bow flare slamming," when the vessel bow encounters the waves. Bow slamming and wave-slap impact has been known to cause structural damage (e.g., buckling, tripping) in the forecastle plating, bow flare plate, and stiffeners. Depend-ing on the hull form, the wave environment, and several other factors including forward speed and heading, bow slamming may need to be investigated for ship-shaped offshore structures in transit or during operation. At a fixed relatively benign location, bow impact-pressure actions may be less serious than those for normal trad-ing tankers. However, bow slamming may be of interest for weathervaning vessels in harsh environments with the bow pitching downward in certain cases, particularly when the waves approach with heading angles within 15–30 degrees off the bow.

  Data on bow-slamming pressures on ship-shaped offshore structures and tools to analyze such conditions are not so mature. There is relatively little information or tools available for the direct first-principles identification of the levels of bow-slamming pressures that might occur along the side of a vessel for seas from moderate to large heading angles. Susceptibility to bow-wave impact increases with harsh envi-ronments and certain bow shapes, for example, round ones (UKOOA 2002; Barltrop and Xu 2004; HSE 2005a). Section 2.17 in Chapter 2 lists some recent studies on bow-slamming impacts.

4.11.2 Practices for Bow-Slamming Assessment

In principle, for direct analysis, the same discussion described in Section 4.10.2 using an impact-pressure profile characterization is applicable to a bow-slamming problem. A first-principles design approach may be used to calculate the wave crest velocity and stagnation pressure with wave height and period known; from this, we can obtain the impact-pressure distribution on local plate panels or even larger areas (HSE 2000b, 2005a).

However, the bow-wave impact analysis for ship-shaped offshore units is more commonly based on the approaches for normal trading ships by modifying them for parameter differences, for example, forward speed. In this regard, simple formula approaches to calculate bow-wave impact-pressure distribution of trading tankers taking into account bow shape and position on bow from waterline and freeboard deck are given by IACS (2005), which may be applicable to ship-shaped offshore units with bow shapes typical of merchant ships. For other shapes and situations – for example, knuckled flat-plate bows – direct calculation methods of FPSO bow slamming (Wang et al. 2002; Barltrop and Xu 2004; HSE 2005a) usually need to be applied.

4.11.3 Measures for Bow-Slamming Risk Mitigation

It needs to be recognized that bow-wave impact is significantly affected by the shape of the bow. Full rounded bow shapes with raised forecastles and bulwarks can be difficult to protect due to high bow impact-pressure levels. This is in contrast to measures for mitigation of green water (refer to Section 4.12.3) because forecastles and bulwarks must be raised to avoid freeboard exceedance at bow. Although the bow shape of a ship-shaped offshore structure converted from a trading tanker will not usually be changed, that of a new-build vessel can be better optimized in terms of mitigating bow-slamming impact and green water by appropriate analyses, and also related sensitivity and trade-off studies.

For instance, two types of bow shape may be considered: full rounded (or semi-ellipsoidal) and sharp bow with or without a rounded extremity. The rounded bow provides maximum buoyancy for minimum steel weight due to its low surface and maximizes the buoyancy particularly at bow that may then better rise in wave crests. Also, it may assist the natural weathervaning capabilities of the vessel. However, a rounded bow has a larger flat surface area that can be more vulnerable to damage due to wave impact.

On the other hand, a sharper bow shape can minimize the wave-impact pressures, although in oblique seas with noncollinear conditions of winds, waves, and currents, the impact can still be relatively severe. Bows with a complex shape, of course, will cost more to construct. They also may result in comparatively less tank space and deck area forward. Integration of structures such as a forward-mounted turret can also become complicated in such cases.

4.12 Green Water

4.12.1 Fundamentals

Green water can be considered to consist of unbroken waves overtopping the bow, side, or stern structures of ship-shaped offshore units; its occurrence depends on various factors including the relative motion between the offshore unit and the waves, the speed, the freeboard, and the harshness of the environment. The occurrence of green water implies that the available freeboard is exceeded. The green-water problem on ship-shaped offshore structures can be an important design issue under harsh environmental conditions because green water can cause damage to deck houses, deck-mounted equipment (e.g., switch room compartments), watertight doors, walkway ladders, and cable trays (HSE 1997, 2000b).

Morris et al. (2000) reported that from 1995, over a 5-year period, seventeen greenwater incidents occurred on twelve FPSOs in UK waters of the North Sea, with more than one incident in some installations. Such experience has also been noted in Ersdal and Kvitrud (2000).

Although green-water occurrence may not cause a direct threat to integrity of the vessel hull girder, it may make the vessel more vulnerable to unintended flooding during accidental events and, therefore, constitute a threat to the workforce; also, any green-water damage requires repairs and perhaps production downtime.

For a turret-moored offshore unit, the bow of the unit may always be exposed to the waves because of the weathervaning feature and, depending on several factors, the wave heights may exceed the freeboard. Green water along the sides can also occur due to wind, waves, and currents. In ship-shaped offshore structures without a poop deckhouse, green water has also been observed at the stern (HSE 2001a) in some cases. Section 2.17 in Chapter 2 lists some recent studies on green-water impact.

4.12.2 Practices for Green-Water Assessment

Prediction of freeboard exceedance at various locations around the deck is highly dependent on relative orientation of the hull to approaching waves. It is not always straightforward to predict quasistatic and dynamic components of vessel heading relative to incoming waves and, in general, this requires nonlinear ship motion or similar calculations. Model tests can also be performed.

For the green-water problem of moored ship-shaped offshore structures, the same approach previously discussed for sloshing problems in Section 4.10.2 can, in principle, be applied for green-water impact-profile characterization and damage prediction, although the characteristics of green-water impact profiles will be different from those of sloshing or bow-slamming impact.

Traditionally in tankers, the deck structure is designed for about 2.44m quasistatic head of water to provide some strength for green-water effects. This load criterion has its basis in experience. And, in tankers in heavy weather and under laden conditions, the occurrence of green water is common and does not usually appear to affect the transportation mission of the vessel involved; a possible exception appears to be the so-called abnormal, freak, or rogue waves.

The same cannot be said with certainty for a floating offshore unit whose basic mission is likely to be more affected by green water. Because of this and also the fact that design conditions for green-water occurrences are difficult to predict, a relatively conservative view may need to be taken for green-water design of ship-shaped offshore units, to the extent possible.

When green-water events by freeboard exceedance occur at bow, head seas may cause the most severe action effects, in particular for the deck housings, the fixings, and the equipment located along the middle of the deck (HSE 2005b). The application of head-sea data may provide practical results somewhat on the conservative side for bow equipment design because the waves at the bow may approach from various heading angles of $+/-$ 30 degrees off the bow.

Determination of design conditions needed to evaluate green-water occurrence along the vessel side or at the stern of a ship-shaped offshore structure is usually more difficult than that at the bow. A more complex process is usually involved in this case because the level of freeboard exceedance along the side and the subsequent flow of water across or along the deck is strongly influenced by relative direction of the incoming wind-driven sea and any static heel and the roll motions of the vessel.

It is noted that the roll motion of the FPSO hull may be caused by a totally different mechanism than local waves; for example, it may be due to swells approaching from a beam-on direction with a frequency close to the natural frequency of the vessel in roll. In some cases, it may also be affected by the wind. In addition, it is reportedly true that green-water events along the vessel side together with severe roll motions can occasionally occur due to unusual yaw motion of the vessel and/or a breakdown of the heading-control system, including thrusters.

Simplified methods are available to evaluate green-water events from the vessel side as well as at bow or stern of the vessel, but a number of problem areas remain to be resolved. Model testing may usually need to be performed for a particular design (HSE 2000a) in the end. The results of the joint industry project (JIP) performed by Maritime Research Institute Netherlands (MARIN) that led to a computer design tool called "GreenLab," and also much relevant work by researchers at MARIN, can also be referred to for useful ways to analyze green-water behavior and response of ship-shaped offshore structures (Buchner 1995, 1996, 1999, 2002; Buchner et al. 2000).

In the GreenLab method, an initial step is to predict freeboard exceedance. The relative motions between the waves and the vessel form the basis of green-water analysis, with green water defined to flow over the deck or bulwark if the water heights exceed the freeboard available. Linear diffraction analysis with nonlinear corrections based on the JIP model test results are applied for this purpose.

Once the maximum freeboard exceedance is known, other aspects such as water heights and impact-pressure loads are calculated based on certain relationships obtained from model tests. More details for the basis of GreenLab analysis can be found in HSE (2001c) and Buchner (2002).

Other relevant literature related to the analysis and design of green-water impact on decks and topsides of ship-shaped offshore structures under harsh environment also exist. For example, Wang et al. (2001) studied green-water impact on decks and topsides for the *Terra Nova* FPSO project (Doyle and Leitch 2000; Maguire et al. 2001). This work identifies the wave period (spectral peak period), vessel heading, and vessel draft or freeboard as the three most important parameters that affect the green-water occurrence in that case. Also, Buchner et al. (2000) evaluated green-water impact problems, among other issues, in the context of future FPSO installations in the Gulf of Mexico.

4.12.3 Measures for Green-Water Risk Mitigation

The following design considerations are relevant for green-water-related risk reduction:

- Design the deck structures and equipment to withstand green-water-impact loads.
- Increase the local freeboard along the vessel length to reduce the green-water occurrences coming onto the deck.

- Optimize the shapes of deck structures, including flare and camber, to minimize green-water ingress and impact loading.
- Arrange physical or operational measures to protect the structure from green-water occurrences.

Physical and operational measures to protect against green-water occurrence may include the following (HSE 2001c):

- Bow and side protection structures, such as higher bulwarks
- Raising the poop deck or bulwark aft
- Raising equipment and piping to reduce the loading
- Appropriate protection of process, deck equipment, cable trays, hydrant, and evacuation equipment
- Operation with stern trim when bow green water is of concern
- Heading changes to reduce wave incidence angles and side green water
- Provision of safe access from green-water zones

As noted in Section 4.12.2, classification society rules for trading tankers commonly use a 2.44m head to account for green-water effects; this practice has found its way into the FPSO guidelines from some classification societies as well. In addition, in certain cases, the class guidelines may permit a further reduction in head (ABS 2004b).

In practice, FPSO specifications have often required local strength design of the deck structure to resist green-water heads of 2.5–5m, depending on the site and towing conditions as necessary; see Adhia et al. (2004). Certain environmentally harsh areas and circumstances of operation may justify a higher design sea-water head for local strength.

4.13 Considerations Related to the Return Period

For design environmental conditions of both ship-shaped offshore units and trading ships, it is important to define relevant return periods (Frieze and Paik 2004). Larger trading ships on unrestricted service are normally classed for a service life of 20–25 years. Today, the offshore industry uses a 100-year return period environmental event as the basis for the strength design of its structures. This, however, was not determined by any rational assessment of the likelihood of failure during events to which offshore structures were exposed or quantification of societal expectations with regard to loss of life or environmental pollution. This arose, as do many engineering solutions, as the rather pragmatic consequence of certain damages to jacket platforms in the Gulf of Mexico during hurricane events.

At first (early 1960s), a design-wave height was selected on the basis of a 25-year return period. With several jacket platform damages, a progressive increase to a wave height corresponding to a 50-year return period improved the situation but not completely. The 100-year return period was then selected and, to date, no jacket or floating platforms in the Gulf of Mexico, or elsewhere, designed to this criterion have been lost or suffered major damage due to severe environmental actions; however, this is now being debated in light of the recent experience with hurricanes such as *Katrina* in 2005.

It is interesting to recount the probability of encountering a storm of R-year return period. If the exposure duration of the structure (service life) in a region or site is

Y years, then the probability P of encountering an R-year return period storm can be estimated as follows (Lacey et al. 2003):

$$P = 1 - (1 - 1/R)^Y, \text{ for the return period concept} \qquad (4.8)$$

$$P = 1 - \exp(-Y/R), \text{ for a Poisson process,} \qquad (4.9)$$

where R = return period in year; Y = exposure duration in the region in year. If the service life in a region is Y = 25 years for 100-year return period storm – that is, R = 100 – the probability P of encountering the storm becomes

$$P = 1 - (1 - 1/100)^{25} = 0.222, \text{ for the return period concept} \qquad (4.10a)$$

$$P = 1 - \exp(-25/100) = 0.221, \text{ for a Poisson process.} \qquad (4.10b)$$

It may seem that a probability of more than 22 percent is high, but this does not necessarily mean that the design is unsafe. In fact, this notional probability of occurrence is one factor that enters into the risk level that is implicit in the structural design procedures involved.

Certainly, an upward change in return period usually results in an increase in loads. For example, wave actions on offshore jacket structures are dominated by *drag loading* as distinct from *inertia loading*. In simple terms, drag loading increases with wave height raised to the power 2. In real terms, the "power" may tend to be greater than 2 in part because of complex platform-framing patterns. The North Sea guidance on wave heights (DEn 1990) shows that as the return period of waves doubles, the wave height increases by some 5 percent and, therefore, the drag loading by about 10 percent. The increase of drag loading due to the increase of the return period from 25 to 100 years is then found to be 22 percent.

Although these increases are probably underestimates, it is evident that the increase in loading can be relatively significant when moving from a 25- to a 100-year return period event. On the other hand, the increase from a similar 50- to a 100-year return period event is lower, perhaps about 10 percent. Experience to date has so far shown that this was sufficient to move from a design event in which major hurricane damage generally occurred to one in which major damage did not generally occur. In view of recent experience with hurricane damage on the U.S. Gulf Coast, the physical damage that occurs may, in some cases, be very significant in terms of lost production and has an impact on various subsequent societal events and circumstances if only because of the supply disruptions that may arise and persist for a time. One observes that risk perception varies with the perceiver and that the levels of acceptable risk can change for various reasons.

We should perhaps also note that the 100-year return period criterion is not necessarily universal for strength design, even in the offshore industry. A 10,000-year return period has also been suggested by some in the context of ultimate strength-based design of offshore structures. Some investigators have also proposed a longer return period of 1,000–10,000 years for the assessment of green-water and bow-impact loads, depending on the environmental conditions of FPSOs (Barltrop and Xu 2004, HSE 2005a).

4.14 Wave Energy Spectra Expressions

Wave energy spectra are the basis for the analysis of actions and action effects due to waves. These represent the distribution of the sea-height variance as a function of frequency in a given sea state; the wave height is proportional to wave energy transported (HSE 2001b). Wave spectra are useful for various purposes, including determining design waves; determining the relative importance of waves with different frequencies in exciting the response of the global system or structural component; and also for obtaining the stress range response spectrum in fatigue calculations using the spectral method.

Because waves are caused by winds, the distribution of wave height in the immediate vicinity of the wind field will have a direct correlation to the local wind field. A part of these waves, however, can under certain conditions travel far distances as *swells* and they should be superposed on or affect the local wind-driven waves at a distant location being considered. For floating offshore structures including FPSOs at a certain site, these swell waves can also be important to design, in addition to the waves generated by local wind systems. Swells can have different wave energy spreading and directionality characteristics when compared to local waves. Swell waves generally have a longer period than locally generated waves, which can travel farther distances without decay when compared to short-period waves.

For a specific site, the wave spectrum consists of both locally generated sea and swell components, sometimes more than one of each kind. This is another reason why, for site-specific design, data obtained by relatively long-term measurements at that site should preferably be used to establish the wave spectra for design. In Sections 4.14.1–4.14.3, some generalized spectral forms useful for offshore design use are discussed; these and other forms are in use.

4.14.1 The Generalized Pierson–Moskowitz Spectrum

An early and still very useful function to describe wave spectrum was developed by Pierson and Moskowitz (1964). A generalized form of the Pierson–Moskowitz spectral function is given as follows:

$$S(f) = Af^{-5} \exp[-Bf^{-4}], \tag{4.11}$$

where f = wave frequency; $S(f)$ = distribution of sea surface variance in m^2/Hz; and A, B = variables to be determined for the prevailing sea state.

Replacing A and B in terms of the sea-state parameters, H_s and T_z, respectively, the distribution of sea-surface variance may be approximated to

$$S(f) \approx 0.080 H_s^2 T_z (T_z f)^{-5} \exp[-0.318(T_z f)^{-4}], \tag{4.12}$$

where H_s = significant wave height of the sea state in meters; and T_z = mean zero up-crossing period of the sea state in seconds.

The peak value of the wave frequency corresponding to the maximum value of $S(f)$ can be obtained from

$$f_p = (0.8B)^{1/4}, \tag{4.13}$$

where f_p = peak value of wave frequency.

4.14.2 The JONSWAP Spectrum

The Joint North Sea Wave Project (JONSWAP) spectrum was derived from environmental data measured in the North Sea off Denmark, originally for describing fetch-limited growing seas in the absence of swell (Hasselmann et al. 1976). It is now commonly used for this and a few other cases by appropriate adjustments to its parameters. The spectrum is, in this case, given by [for symbols, unless specified below, see Eq. (4.11)]

$$S(f) = Af^{-5}\gamma^q \exp[-Bf^{-4}], \tag{4.14}$$

where A, B = variables to be determined for the prevailing sea state, but these are not the same as in Eq. (4.11); γ = variable peak enhancement parameter for a particular region of interest (e.g., this variable has a mean value of 3.3 and varies by more than $\pm$ 50 percent in the North Sea); $q = \exp\left[-\frac{(f-f_p)^2}{2c^2f_p^2}\right]$; f_p = peak value of wave frequency corresponding to the maximum value of S(f); and c = constant for a particular region (e.g., c = 0.07 for $f \leq f_p$ and c = 0.09 for $f > f_p$ in the North Sea).

The relationship between H_s and T_z is in this case given by

$$S(f) \approx 0.0749 H_s^2 T_z (T_z f)^{-5} 3.3^q \exp[-0.4567 (T_z f)^{-4}], \tag{4.15}$$

where $q = \exp\left[-\frac{(1.286T_z f-1)^2}{2c^2}\right]$; c = constant for a particular region.

The period at the peak frequency f_p can be given by

$$T_p = \frac{1}{f_p}, \tag{4.16}$$

where T_p = period at the peak frequency, which can become $T_p \approx 1.286 T_z$ for $\gamma = 3.3$.

4.14.3 Directional Wave Spectra

The wave spectrum can be modified to consider the wave direction as follows [for symbols, unless specified below, see Eq. (4.11)]:

$$S(f) = \int_{-\pi}^{+\pi} S(f, \theta) d\theta, \tag{4.17}$$

where θ = direction from which the wave component is traveling.

In Eq. (4.17), S(f, θ) is often split into S(f) and G(f, θ), as follows:

$$S(f, \theta) = S(f) G(f, \theta), \tag{4.18}$$

where $\int_{-\pi}^{+\pi} G(f, \theta)d\theta = 1$.

The spreading function G(f, θ) in Eq. (4.18) is expressible as follows:

$$G(f, \theta) = N \cos^{2s}[(\theta - \theta_m)/2], \tag{4.19}$$

where θ_m = dominant direction as a function of f; s = spreading factor as a function of f; and N = normalizing constant ensuring that G(f, θ) integrates to 1.0, given by

$$N = \frac{1}{2\sqrt{\pi}} \frac{\Gamma(s+1)}{\Gamma(s+0.5)}, \tag{4.20}$$

where Γ = gamma function. For a simple evaluation, s = 10 is often used when s is considered to be independent of wave frequency. In this case, N = 0.903. In cases where the wave energy is more narrowly spread about the predominant direction, such as for swells, a "cosine to the power 4" spreading function is often used; that is, s = 2. Then, of course, for s = 1, we obtain the typically used cosine-squared spreading function.

In addition to the wave spectral function, the expected number of wave encounters that the structure is likely to experience during its service life as a function of wave amplitude must also be estimated for fatigue assessment.

4.15 Design Basis Environmental Conditions

In summary, the design environmental conditions of moored ship-shaped offshore structures must be established to determine the most severe actions during the entire service life. Relevant considerations were discussed previously in Section 4.13. The conditions for this purpose may include, for example, the following:

- 100-year return period winds and waves associated 10-year return period current
- 100-year return period currents with associated 10-year return period winds and waves

For design, both collinear and noncollinear directions of winds, waves, and currents must usually be taken into account together with their angular separation; see, for example, UKOOA (2002). Winds, waves, and currents must ideally be based on site-specific metocean data for offshore installations. Operational data for winds and waves will also be required to shuttle tanker loading (FPSO offloading) operations and to analyze and account for the downtime associated with various offshore operations in general.

For each offshore project, there is a metocean design-basis document that specifies all related environmental data for design, together with a commentary. Such a document typically includes the applicable metocean design data such as (a) the extreme wind, wave, and current cases; (b) operating criteria for winds, waves, and currents; and (c) design data for tide levels, rain and squalls, water temperatures, salinity and density, air temperatures, air pressure and humidity, water chemistry, and marine growth.

REFERENCES

Aage, C., Hvid, S. L., Hughes, P. H., and Leer-Andersen, M. (1997). "Wind loads on ships and offshore structures estimated by CFD." *Proceedings of the 8th International Conference on the Behaviour of Offshore Structures* (BOSS'97), Delft, The Netherlands.

ABS (2004a). *Guidance notes on nonlinear finite-element analysis of side structures subject to ice loads.* American Bureau of Shipping, Houston.

ABS (2004b). *Guide for building and classing floating production, storage, and offloading systems.* American Bureau of Shipping, Houston.

ABS (2005). *Guidance notes on ice class.* American Bureau of Shipping, Houston.

Adhia, G., Pelleguino, S., Ximenes, M. C., Awashima, Y., Kakimoto, M., and Ando, T. (2004). *Owner and shipyard perspective on new build FPSO contracting scheme, standards, and lessons.* Offshore Technology Conference, OTC 16706, Houston, May.

API (1991). *Recommended practice for design, analysis and maintenance for mooring for floating production systems.* (API RP 2FP1), American Petroleum Institute.

API (1993a). *Recommended practice for planning, designing, and constructing fixed offshore platforms – Working stress design.* (API RP 2A-WSD), American Petroleum Institute.

API (1993b). *Recommended practice for planning, designing, and constructing fixed offshore platforms – Load and resistance factor design.* (API RP 2A-LRFD), American Petroleum Institute.

Barltrop, N. D. P. (1998). *Floating structures: A guide for design and analysis.* The Centre for Marine and Petroleum Technology (CMPT), Herefordshire, England: Oilfield Publications Ltd.

Barltrop, N. D. P., and Xu, L. (2004). *Research on bow impact loading in Glasgow. Proceedings of OMAE Specialty Symposium on Integrity of Floating Production, Storage and Offloading (FPSO) Systems, OMAE–FPSO'04-0063,* Houston, August 30–September 2.

Bruen, F. J., Byrd, R. C., Vivatrat, V., and Watt, B. J. (1982). *Selection of local design ice pressures for arctic systems.* Offshore Technology Conference, OTC 4334, Houston, May.

Buchner, B. (1995). *The impact of green water on FPSO design.* Offshore Technology Conference, OTC 7698, Houston, May.

Buchner, B. (1996). *The influence of the bow shape of FPSOs on drift forces and green water.* Offshore Technology Conference, OTC 8073, Houston, May.

Buchner, B. (1999). "Green water from the side of a weathervaning FPSO." *Proceedings of OMAE'99, OFT 4022.* St. Johns, Newfoundland, July.

Buchner (2002). *Green water on ship-type offshore structures.* Ph.D. Thesis, Delft University of Technology, Delft, The Netherlands.

Buchner, B., Voogt, A. J., Duggal, A. S., and Heyl, C. N. (2000). *Green water evaluation for FPSOs in the Gulf of Mexico.* Offshore Technology Conference, OTC 14192, Houston, May.

DEn (1990). *Offshore installations: Guidance on design, construction, and certification.* The Department of Energy, UK.

DNV (1991). *Environmental conditions and environmental loads.* (Classification Notes, No. 30.5), Det Norske Veritas, Oslo.

DNV (2000). *Structural design of offshore ships.* (Offshore Standards, DNV-OS-C102), Det Norske Veritas, Oslo.

DNV (2002). *Structural design of offshore units (WSD method).* (Offshore Standards, DNV-OS-C201), Det Norske Veritas, Oslo.

Doyle, T., and Leitch, J. (2000). *Terra Nova FPSO design and construction.* Offshore Technology Conference, OTC 11920, Houston, May.

D'Souza, R. (1999). *Major technical and regulatory issues for monohull floating production systems in the Gulf of Mexico.* Offshore Technology Conference, OTC 10704, Houston, May.

Ersdal, G., and Kvitrud, A. (2000). "Green water on Norwegian production ships." *Proceedings of ISOPE Conference,* International Society of Offshore and Polar Engineers, Seattle, May.

Faltinsen, O. M. (1990). *Sea loads on ships and offshore structures.* Cambridge, UK: Cambridge University Press.

FMA (2002). *Finnish-Swedish ice class rules.* (FMA Bulletin No. 13/1.10), Finnish Maritime Administration, Finland.

FMA (2003a). *Tentative guidelines for application of direct calculation methods for longitudinally framed hull structure.* Finnish Maritime Administration, Finland, June.

FMA (2003b). *Background for the tentative guidelines for application of direct calculation methods for longitudinally framed hull structure.* Finnish Maritime Administration and Helsinki University of Technology, Finland, June.

FMA (2004). *Guidelines for the application of the Finnish–Swedish ice class rules.* Finnish Maritime Administration, Finland.

Frieze, P. A., and Paik, J. K. (2004). "General requirements for limit state assessment of ship structures." *SNAME Transactions,* 112: 368–384.

Hasselmann, K., Ross, B., Muller, P., and Sell, W. (1976). "A parametric wave prediction model." *Journal of Physical Oceanography*, 6: 200–228.

Hinkel, R. M., Thibodeau, S. L., and Hippman, A. (1988). *Experience with drillship operations in the U.S. Beaufort Sea.* Offshore Technology Conference, OTC 5685, Houston, May.

Hnatiuk, J. (1984). *Current and future offshore activities in Canada.* Offshore Technology Conference, OTC 4707, Houston, May.

HMSO (2000). *Re-opened formal investigation – Loss of the M. V. Derbyshire.* (Executive Summary to Report), HMSO (Her Majesty Stationary Office), London, November.

HSE (1989a). *Metocean parameters – Parameters other than waves – Supporting document to "Offshore installations: Guidance on design, construction, and certification – Environmental consideration."* (Offshore Technology Report, OTH 1989/299), Health and Safety Executive, UK.

HSE (1989b). *Wave parameters – Supporting document to "Offshore installations: Guidance on design, construction and certification – Environmental consideration."* (Offshore Technology Report, OTH 1989/300), Health and Safety Executive, UK.

HSE (1997). *Green seas damage on FPSOs, and FSUs.* (Offshore Technology Report, OTH 1997/486), Health and Safety Executive, UK.

HSE (2000a). *Review of model testing requirements for FPSO's.* (Offshore Technology Report, OTO 2000/123), Health and Safety Executive, UK.

HSE (2000b). *Review of green water and wave-slam design and specification requirements for FPSO/FSUs.* (Offshore Technology Report, OTO 2000/004), Health and Safety Executive, UK.

HSE (2001a). *Environmental considerations.* (Offshore Technology Report, OTC 2001/010), Health and Safety Executive, UK.

HSE (2001b). *Loads.* (Offshore Technology Report, OTC 2001/013), Health and Safety Executive, UK.

HSE (2001c). *Analysis of green water susceptibility of FPSO/FSU's on the UKCS.* (Offshore Technology Report, OTO 2001/005), Health and Safety Executive, UK.

HSE (2005a). *Wave slap loading on FPSO bows.* (Research Report, No. 324), Health and Safety Executive, UK.

HSE (2005b). *Findings of expert panel engaged to conduct a scooping study on survival design of floating production storage and offloading vessels against extreme metocean condition.* (Research Report, No. 357), Health and Safety Executive, UK.

IACS (2001). *Unified requirements for polar ships.* International Association of Classification Societies, London, November.

IACS (2005). *Common structural rules for double hull oil tankers.* International Association of Classification Societies, London, December.

Kreider, J. R., Zahn, P. B., and Chabot, L. G. (1985). *Probabilistic design criteria for Beaufort Sea structures: Combining limited driving force and limit-stress predictions.* Offshore Technology Conference, OTC 5052, Houston, May.

Kujala, P. (1994). *On the statistics of ice-loads on ship hull in the Baltic.* Finnish Academy of Technology, Finland.

Lacey, P., Hee, D., Chen, H., and Cardone, V. (2003). "Tow simulation." *SNAME Transactions*, 111: 79–96.

Lee, W. T., Bales, W. L., and Stowby, S. E. (1985). *Standardized wind and wave environments for North Pacific Ocean areas.* (Research Report, R/SPD-0919–02), David W. Taylor Naval Ship Research and Development Center, Washington, DC.

Liu, D., Spencer, J., Itoh, T., Kawachi, S., and Shiegmatsu, K. (1992). "Dynamic load approach in tanker design." *SNAME Transactions*, 100: 143–172.

Lonseth, L., and Kvitrud, A. (1997). *Twenty years of metocean data collection on the Northern Norwegian Continental Shelf.* Offshore Technology Conference, OTC 8270, Houston, May.

LR (2004). *ShipRight structural design – Sloshing loads and scantlings assessment.* Lloyd's Register, London, May.

MacMillan, A. (2001). *Effective FPSO/FSO hull structural design*. Offshore Technology Conference, OTC 13212, Houston, May.

Maguire, M. J., Ewida, A. A., and Leonard, C. M. (2001). *Terra Nova FPSO: Certification and technical integrity challenges*. Offshore Technology Conference, OTC 13023, Houston, May.

Morris, W. D. M., Buchner, B., and Millar, J. (2000). "Green water susceptibility of North Sea FPSO/FSUs." *IBC Conference on Floating Production Systems (FPS)*, London, December.

NORSOK N003 (1999). *Actions and action effects*. Norwegian Standards, Norway.

NORSOK N004 (1999). *Design of steel structures*. Norwegian Standards, Norway.

Ochi, M. K. (1978). "Wave statistics for the design of ships and ocean structures." *SNAME Transactions*, 86: 47–76.

OCIMF (1994). *Prediction of current loads on VLCCs*. Oil Companies International Marine Forum, UK.

Paik, J. K., Lee, J. M., Shin, Y. S., and Wang, G. (2004). "Design principles and criteria for ship structures under impact pressure actions arising from sloshing, slamming and green seas." *SNAME Transactions*, 112: 292–313.

Pierson, W. J., and Moskowitz, L. M. (1964). "A proposed spectral form for fully developed wind, seas, etc." *Journal of Geophysical Research*, 69: 5181–5190.

Regg, J. B. (1999). *Floating production storage and offloading systems in the Gulf of Mexico OCS: a regulatory perspective*. Offshore Technology Conference, OTC 10701, Houston, May.

Truskov, P. A. (1999). *Metocean, ice, and seismic conditions offshore Northeastern Sakhalin island*. Offshore Technology Conference, OTC 10816, Houston, May.

UKOOA (2002). *FPSO design guidance notes for UKCS service*. Offshore Operators Association, UK.

Vivatrat, V., and Slomski, S. (1983). *A probabilistic basis for selecting design ice pressures and ice loads for arctic offshore structures*. Offshore Technology Conference, OTC 4457, Houston, May.

Wang, G., Basu, R., Chavda, D., and Liu, S. (2005a). "Rationalizing the design of ice strengthened side structures." *Proceedings of International Congress of International Maritime Association of the Mediterranean (IMAM 2005)*, Lisbon, Portugal, September 26–30.

Wang, G., Basu, R., Chavda, D., Liu, S., Lee, M., Suh, Y., and Han, Y. (2005b). "Rationalization of design of side structure of ice-strengthened tankers." *International Journal of Offshore and Polar Engineering*, 15(3): 210–214.

Wang, M., Leitch, J., and Bai, Y. (2001). *Analysis and design consideration of green water impact on decks and topsides of FPSO*. Offshore Technology Conference, OTC 13208, Houston, May.

Wang, G., and Liu, S. (2006). "Recent advances in structural design of ice strengthened vessels." *Proceedings of International Conference and Exhibition on Performance of Ships and Structures in Ice (ICETACH 2006)*, Banff, Alberta, Canada, July 16–19.

Wang, G., and Wiernicki, C. (2004). *Using nonlinear finite element method to design ship structures for ice loads*. World Maritime Technology Conference (WMTC), Washington, DC, September 29–October 1.

Wang, G., Tang, S., and Shin, Y. (2002). "Direct calculation approach and design criteria for wave slamming of an FPSO bow." *International Journal of Offshore and Polar Engineering*, 12: 297–304.

Williford, F. B., and Winkler, R. S. (1987). *SF-500 turret-moored icebreaker drillship*. Offshore Technology Conference, OTC 5357, Houston, May.

Wolfram, J., Linfoot, B., and Stansell, P. (2001). "Long- and short-term extreme wave statistics in the North Sea: 1994–1998." *Proceedings of Rogue Waves*. Edited by M. Olagnon and G. Athanassoulis. Germany: Springer-Verlag.

CHAPTER 5

Serviceability Limit-State Design

5.1 Introduction

The performance of a structure and its components is described using limit-state functions that separate desired states from undesired states. The physical effects of exceedance of a limit state may be either reversible or irreversible. For the reversible case, removal of the cause of the exceedance allows the structure to return to a desired state. For the irreversible case, the same is not true and certain consequences, such as damage, may occur depending on the nature of the limit state. The consequences may, in turn, be either recoverable or unrecoverable from the deformed state. For example, if the damage is limited, say, in the form of a localized permanent set in a case where the same is not desired, the condition may be repairable, for example, by replacing the affected parts.

As discussed in Chapter 3, limit states are usually classified into four types:

(1) Serviceability limit states (SLS) that represent criteria governing normal functional or operational use.
(2) Ultimate limit states (ULS) that represent the failure of the structure and its components usually when subjected to extreme values of actions or action effects.
(3) Fatigue limit states (FLS) that represent damage accumulation (leading to cracking when certain limits are exceeded) under repetitive actions.
(4) Accidental limit states (ALS) that represent situations of accidental or abnormal events.

In limit-state assessment, such various limit states are considered against different target safety levels; the target to be attained for any particular type of limit state is a function of the consequences and ease of recovery from that state.

This chapter presents SLS design principles and criteria together with related practices for ship-shaped offshore structures. Various types of SLS criteria are addressed, relating to elastic deflection limits under quasistatic actions; elastic buckling limits; permanent set deflection limits under impact-pressure actions arising from tank sloshing, bow slamming, and green water; intact vessel stability; watertight integrity; weathervaning (heading control); station-keeping; vessel motion exceedance; vibration and noise; vortex-shedding-induced vibrations and oscillations; and localized corrosion wastage.

Admittedly, some of these considerations are not what other experts may normally define to be in the realm of SLS; however, to the extent they can adversely affect operations, increase downtime, and potentially reduce revenue, we think it is appropriate to include them under SLS considerations. The economic viability of the offshore installation is inextricably linked to its performance in the various serviceability limit states.

5.2 Design Principles and Criteria

SLS for ship-shaped offshore structures addresses the following:

- Unacceptable deformations that affect the efficient use of structural or non-structural components or the functioning of equipment affected by them
- Local damage (including corrosion, small dents, and limited permanent set) that reduces the durability of the structure or affects the efficiency of structural or nonstructural components
- Intact vessel stability and watertight integrity
- Vessel station-keeping
- Vessel weathervaning or heading control
- Vessel motions (or excursions) that exceed the limitations of equipment, mooring systems, risers, and so on
- Vibration or noise that can injure or adversely affect the habitability of the unit and the performance of personnel or affect the proper functioning of equipment (especially if resonance occurs)
- Deformations that may spoil the aesthetic appearance of the structure

The divisions are one of convenience in that the limit-state behaviors can be interlinked. For example, excessive deformation of a structure may also be accompanied by excessive vibration or noise as well as buckling. The acceptable SLS limits will be defined by the operator of a structure, the primary aim being efficient and economical in-service performance, usually together with a planned program of maintenance and upkeep for the unit. The SLS criterion is expressed as follows:

$$\delta_{max} < \delta_a, \tag{5.1a}$$

where δ_{max} = factored maximum value of the serviceability parameter in terms of actions effects (e.g., displacement, stress); and δ_a = factored serviceability limit value of the consistent parameter.

Although the SLS criterion in Eq. (5.1a) is expressed in terms of action effects, it may be sometimes cast in terms of actions (e.g., forces, load-carrying capacity) and given in the following form:

$$F_{max} < F_a, \tag{5.1b}$$

where F_{max} = factored maximum applied actions (loads); and F_a = factored load-carrying capacity.

A *factored value* indicates that an appropriate factor of safety associated with uncertainties is multiplied for loads or divided for strength. The acceptable limits necessarily depend on the type, mission, and arrangement of the structure. Further, in defining such limits even for structural behavior, other experts, such as machinery and equipment designers, will also need to be consulted.

5.3 Practices for Actions and Action-Effects Analysis

For SLS design of ship-shaped offshore units, in-service actions in terms of pressures or forces must be determined by vessel motion analysis based on site-specific environmental data (e.g., waves, wind, currents) together with operational conditions (e.g., loading, offloading), as previously described in Chapter 4.

For limit-state design and assessment, it is essential to analyze the action effects of individual structural components, particularly in terms of working stresses. Methodologies similar to those used for trading tankers can also be applied to the action-effects analysis of ship-shaped offshore units. The finite-element method (FEM) is typically employed for such purposes.

Regarding structural behavior, the following five levels are often approximately considered:

- Global structure (or hull girder)
- Cargo hold (or hull module)
- Grillage
- Frame and girder
- Local structure and details

For each load case, the resulting load effects are combined appropriately, using correlation factors relevant for the load case. The response at each level may provide the boundary conditions for the next lower-level analysis. The structural behavior being addressed may take the following forms:

- Static or dynamic
- Deterministic or probabilistic
- Linear or nonlinear

The analysis at each structure level may need to include a dynamic structural analysis, depending on whether that level of structure is subjected to any significant dynamic loads, that is, loads for which the shortest component period is the same order of magnitude or shorter than the longest natural period of that level of structure. At the hull girder and cargo hold levels, a wave-excited dynamic analysis is usually not required for structures such as FPSOs, but a calculation of hull girder natural frequency is almost always necessary. It is interesting to note that a dynamic analysis may be required at hull girder and/or cargo hold levels for relatively flexible trading ships, including some container vessels or naval ships that are susceptible to springing.

At the principal member and local structure levels, a vibration analysis may be required if there are some significant and unavoidable sources of excitations (e.g., machinery). In many cases, however, the preferred approach throughout the industry is to calculate the natural frequencies and to design the structure so as to avoid resonance.

When the characteristics of actions are certain, the deterministic analysis can be adopted, but the probabilistic analysis is usually required to characterize the uncertainties and irregularities associated with environmental and/or operational actions. For the practical purposes of limit-state design, the probabilistic character-istics of individual action-effect variables are identified separately, and then they are combined for limit-state assessment of the overall system structure together with the probabilistic characteristics of structural capacities.

Environmental actions due to waves, wind, and currents can be complex, including the dynamic, probabilistic, and nonlinear characteristics in nature. For simplicity, a linear analysis is often used under several simplifying assumptions. For example: (a) the irregular wave surface of the ocean can be represented as the linear sum of a large number of individual regular waves of different heights and frequencies; (b) the hydrodynamic forces on a vessel hull can be obtained using strip-theory simplifications that, for certain parameters, consider each transverse section of the vessel separately and then combine the results linearly for the overall vessel; and (c) the wave force acting on each section may be assumed to be linearly proportional to the difference between the local wave height and the vessel's still-water-plane level.

The accuracy of the first two assumptions is usually satisfactory; however, the third is valid for vessels that are approximately wall-sided in the water-plane region. If this is not so, or if there is any other source of nonlinearity, an appropriate nonlinear method of action-effect analysis should be employed.

For a more detailed consideration of action-effect analysis for ship-shaped struc-tures, see Paik and Hughes (2006).

5.4 Elastic Deflection Limits: Under Quasistatic Actions

The hull of ship-shaped offshore structures may be subjected to significant lon-gitudinal and vertical elastic distortions due to static loads, static load variations, and the dynamic effects of wind and waves. Under normal service conditions, the maximum deflection of structural components must not exceed certain acceptable limits per Eq. (5.1b) for certain applications. The related load effects need to be accounted for in the equipment support structure that is likely to be affected. Total maximum deflections in specific cases – for example, for crane supports – may be specified by equipment vendors. This section presents useful analytical formulae of the maximum deflections for main types of structural components under quasistatic actions.

We note that in classification society rules, it is not common to specify relative deflection limits for most major structural members within the hull except in special cases – for example, at crane supports and in the vicinity of certain types of equipment. This is simply because some or many of such limits may be considered implicitly by other prescriptive aspects of the rules. In designing a structure purely by first-principles-based procedures, however, one would need to define and consider those explicitly.

5.4.1 Support Members

In calculating the deflections for support members in a stiffened plate structure, illustrated in Figure 5.1, the attached plating must be considered with the support member, which is often called a plate-stiffener (beam) combination. Typical types of

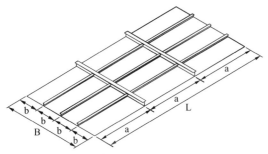

Figure 5.1. A continuous stiffened plate structure.

plate-beam combination models, consisting of a stiffener and its attached effective plating, are used for this purpose, as illustrated in Figure 5.2.

The cross-sectional shape of the plate-beam combination model will then, in general, be that of a nonsymmetric I-beam type. Some important properties of the plate-beam combination sections with (full or effective) width of attached plating are given in Table 5.1. The effective width or breadth of the attached plating may need to be used to reflect the structural ineffectiveness due to applied actions once certain applied end strains are exceeded, or for approximating complex behavior such as shear flow using beam approximations. For an elaborate description of effective width or breadth of attached plating, see Paik and Thayamballi (2003).

The span of the plate-beam combination will normally be measured between stronger support members or structures. Therefore, the actual end conditions for the plate-beam combination will be affected by the joining methods and rigidities of support members in the orthogonal direction.

Support members such as stiffeners, frames, or girders are likely subjected to bending, axial loads, or these combinations, as shown in Figure 5.3. When a one-dimensional member is subjected to bending arising from distributed lateral load q, concentrated lateral load Q, or direct bending M_A or M_B, it is called a "beam." However, the one-dimensional member under axial compressive loads is called a "column," but it is termed a "rod" when axial tension is predominantly applied. When both bending and axial compressive loads are simultaneously applied, it is called a "beam-column."

Typically, limiting values of vertical deflections for beams in steel structures vary between L/200 and L/500 in land-based structures, where L is the span of the beam measured between supports. For cantilever beams, L may be taken as twice the projecting length of the cantilever. Limits are stricter for nonstationary equipment

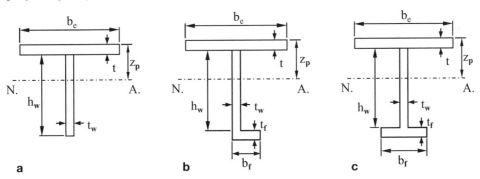

Figure 5.2. Typical types of plate-beam combinations made up of a stiffener and its attached effective plating (b_e = effective plate width between support members): (a) flat bar; (b) angle bar; and (c) tee bar.

Table 5.1. *Properties of a plate-stiffener combination section with a given width of plating; stated for an unsymmetric I-beam configuration*

Property	Expression
Cross-sectional area	$A = A_p + A_w + A_f$, $A_e = A_{pe} + A_w + A_f$ where $A_p = bt$, $A_{pe} = b_e t$, $A_w = h_w t_w$, $A_f = b_f t_f$
Equivalent yield strength over the cross section	$\sigma_{Yeq} = \dfrac{A_p \sigma_{Yp} + A_w \sigma_{Yw} + A_f \sigma_{Yf}}{A}$
Distance from outer surface of attached plating to elastic horizontal neutral axis	$z_o = \dfrac{0.5bt^2 + A_w(t + 0.5h_w) + A_f(t + h_w + 0.5t_f)}{A}$ $z_p = \dfrac{0.5b_e t^2 + A_w(t + 0.5h_w) + A_f(t + h_w + 0.5t_f)}{A_e}$
Moment of inertia	$I = \dfrac{bt^3}{12} + A_p\left(z_o - \dfrac{t}{2}\right)^2 + \dfrac{h_w^3 t_w}{12} + A_w\left(z_o - t - \dfrac{h_w}{2}\right)^2$ $\quad + \dfrac{b_f t_f^3}{12} + A_f\left(t + h_w + \dfrac{t_f}{2} - z_o\right)^2$ $I_e = \dfrac{b_e t^3}{12} + A_{pe}\left(z_p - \dfrac{t}{2}\right)^2 + \dfrac{h_w^3 t_w}{12} + A_w\left(z_p - t - \dfrac{h_w}{2}\right)^2$ $\quad + \dfrac{b_f t_f^3}{12} + A_f\left(t + h_w + \dfrac{t_f}{2} - z_p\right)^2$
Radius of gyration	$r = \sqrt{\dfrac{I}{A}}$, $r_e = \sqrt{\dfrac{I_e}{A}}$
Column slenderness ratio	$\lambda = \dfrac{L}{\pi r}\sqrt{\dfrac{\sigma_{Yeq}}{E}}$, $\lambda_e = \dfrac{L}{\pi r_e}\sqrt{\dfrac{\sigma_{Yeq}}{E}}$
Plate slenderness ratio	$\beta = \dfrac{b}{t}\sqrt{\dfrac{\sigma_{Yp}}{E}}$

Note: The subscript e represents the effective cross section, L = length of the plate-beam combination; b = full breadth of attached plating; b_e = effective width or breadth of attached plating; σ_{Yp} = yield stress of plating; σ_{Yw} = yield stress of stiffener web; and σ_{Yf} = yield stress of stiffener flange.

compared to more stationary equipment, and limits as built are usually stricter than limits in operation. For structures supporting shipboard pedestal cranes, L/300 may be an example value of a deflection limit. For other examples, see Eurocode (1992), which is specified for land-based structures.

Here, the maximum deflection formulae of beams with varying selected end conditions and loading conditions are given for a case where the depth of beam web is relatively small so that the effect of shear can be neglected.

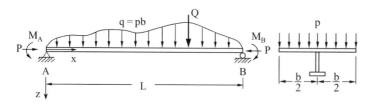

Figure 5.3. Typical load applications on a plate-beam combination.

Both ends simply supported, under uniformly distributed lateral loading:

$$\delta_{max} = \frac{5qL^4}{384EI},$$ (5.2)

where δ_{max} = maximum deflection; L = beam span; E = elastic modulus; I = moment of inertia of the beam section with regard to the neutral axis; q = uniform lateral line load ($=$ pb) as defined in Figure 5.3; b = plate breadth between support members (stiffeners); and p = uniform lateral-pressure load on the attached plating.

One end fixed and the other end simply supported, under uniformly distributed lateral loading:

$$\delta_{max} = \frac{qx^2}{48EI}(3L^2 - 5Lx + 2x^2),$$ (5.3)

where $x = \frac{15-\sqrt{33}}{16}L$.

Both ends fixed, under uniformly distributed lateral loading:

$$\delta_{max} = \frac{qx^2}{24EI}(L^2 - 2Lx + x^2),$$ (5.4)

where $x = \frac{L}{2}$.

Strong support members such as deep girders will have a large web depth, typically with the web depth greater than 20 percent of the beam span (length). In this case, the effect of shear cannot be neglected. For a beam with both ends simply supported and subjected to a concentrated lateral loading at its midspan, the maximum deflection, taking into account the shear effect, can be estimated from the following:

$$\delta_{max} = \frac{PL^3}{48EI}\left(1 + \frac{12\alpha EI}{GAL^2}\right),$$ (5.5)

where $G = \frac{E}{2(1+\nu)}$; ν = Poisson ratio; A = cross-sectional area; and P = concentrated lateral load applied at midspan; α = ratio of shear stress at neutral axis to average shear stress over the web, which may be taken as $\alpha = 1.5$ for the rectangular cross section.

For a rectangular cross-sectional steel web beam, that is, without flange, Eq. (5.5) becomes

$$\delta_{max} \approx \frac{PL^3}{48EI}\left(1 + 3.9\frac{h^2}{L^2}\right),$$ (5.6)

where h = depth of beam web.

For columns or beam-columns, significant lateral deflections may take place after the inception of buckling when axial compressive loads are predominant. Theoretically, the magnitude of lateral deflections in this case becomes infinite in the context of small deflection elastic theory suddenly after buckling, which is typical of an instability problem solved by linear theory. Note that the term "sudden" is relative to what went on before in terms of deformation; note also that with initial deformations present, some mathematical transformations need to be applied to the ongoing deflection records in order to identify the onset of bifurcation buckling. This phenomenon is also related to the consideration of ultimate limit states to the structures, which is discussed further in Chapter 6; see also Paik and Thayamballi (2003).

Table 5.2. *Coefficient α in Eq. (5.7)*

a/b	1.0	1.1	1.2	1.3	1.4	1.5	1.6	1.7	1.8	1.9	2.0	3.0	4.0	5.0	∞
α_s	406	485	564	638	705	772	830	883	931	974	1013	1223	1282	1297	1302
α_f	192	251	319	388	460	531	603	668	732	790	844	1168	—	—	1302

Note: a = plate length; b = plate breadth; and α_s, α_f = coefficient α for a plate all edges simply supported or fixed, respectively.

Various design formulations for lateral deflections of beams, columns, and beam-columns can be found in standard textbooks; for example, Timoshenko and Gere (1961) and Chen and Atsuta (1977).

5.4.2 Plating between Support Members

The maximum lateral deflection formulae are now presented for a steel plate under uniformly distributed lateral pressure loads derived using small deflection plate theory, and the following nomenclature: the plate length a; the plate breadth b($\leq$ a); the plate thickness t; the plate material elastic modulus E; and Poisson ratio v.

$$\delta_{max} = \alpha \frac{pb^4}{D} \times 10^{-5}, \tag{5.7}$$

where δ_{max} = maximum elastic lateral deflection; p = uniformly distributed lateral pressure in quasistatic condition; $D = \frac{Et^3}{12(1-v^2)}$ = plate bending rigidity; and α = coefficient according to the plate aspect ratio and boundary condition, which is indicated in Table 5.2 for steel with $v = 0.3$.

Other useful solutions for plate deflection under various types of loading and edge conditions may be found in, for example, Timoshenko and Woinowsky-Krieger (1981) and Szilard (2004). As the plate deflection limits for SLS design, a few times the plate thickness or less are usually considered. For the structural design of plating to various levels of efficiency, including consideration of a permanent set, see Hughes (1983) and Hughes and Caldwell (1991).

5.5 Elastic Buckling Limits

Steel plate elements in ship-shaped offshore structures are likely to be subjected to lateral pressure loads and/or buckling loads (e.g., axial compressive loads). For the lateral pressure loading, the plate deflects in a stable manner, but under buckling loads, the plate deflection increases "suddenly," although again it does not physically become infinite.

For compressive loading, elastic buckling control-based criteria are often employed for SLS design, in some cases to prevent such occurrence and in other cases to control elastic buckling to a known degree depending on the limits specified. For example, preventing elastic buckling can be the case in the topsides deck structures of floating offshore units. In the main load-carrying members of the hull, however, the approach by controlling elastic buckling is used, although elastic buckling is typically not allowed, following marine practices.

In principle, elastic plate buckling and related deflections must be prevented if these effects are likely or known to be detrimental. Because a plate may have some

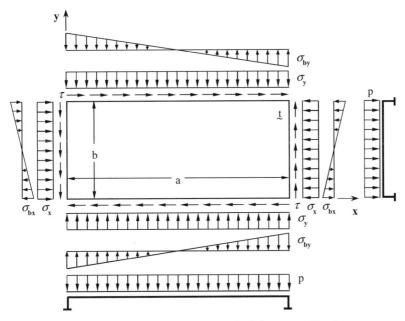

Figure 5.4. A rectangular plate under a total of six types of load components.

reserve strength beyond elastic buckling until its ultimate strength is reached, allowing elastic buckling in a controlled manner can in some cases lead to a more efficient structure. In Chapter 6, the use of ultimate strength-based design methodologies for controlling elastic buckling is presented and discussed further. An extensive description on buckling and ultimate strength of steel-plated structures is presented in Paik and Thayamballi (2003).

This section presents closed-form expressions of elastic buckling strength of plates and support members. In the elastic buckling check under a single stress component using Eq. (5.1a), δ_{max} may be taken to be the applied working stress denoted by σ_{av}, and δ_a is the elastic buckling stress denoted by σ_E. Therefore, the safety check using Eq. (5.1a) can be rewritten as follows:

$$\gamma_D \sigma_{av} \leq \frac{\sigma_E}{\gamma_C} \quad \text{or} \quad \eta_E = \gamma_C \gamma_D \frac{\sigma_{av}}{\sigma_E} \leq 1, \tag{5.8a}$$

where η_E = elastic buckling usage factor; σ_{av} = applied working stress; σ_E = elastic buckling stress; and γ_C, γ_D = partial safety factors accounting for the uncertainties associated with strength and loads, respectively.

Under multiple-load components, an elastic buckling interaction relationship can be established as a function of the applied working stress (load) components and the corresponding elastic buckling stress components. For a rectangular plate element surrounded by support members at all (four) edges, the maximum number of load components may be six: (1) axial stress σ_x in the x direction; (2) axial stress σ_y in the y direction; (3) edge shear stress τ; (4) in-plane bending stress σ_{bx} in the x direction; (5) in-plane bending stress σ_{by} in the y direction; and (6) lateral pressure p, as shown in Figure 5.4. In this case, the elastic buckling interaction criterion may be expressed as follows:

$$\Gamma_B = \Gamma_B \left(\sigma_x, \sigma_{xE}, \sigma_y, \sigma_{yE}, \tau, \tau_E, \sigma_{bx}, \sigma_{bxE}, \sigma_{by}, \sigma_{byE} \right) \leq 1, \tag{5.8b}$$

Table 5.3. *Buckling coefficients for a simply supported rectangular plate under single types of loads for* a/b ≥ 1

Load type	k
σ_x	$k_x = \left[\dfrac{a}{m_o b} + \dfrac{m_o b}{a}\right]^2$
	where m_o is the buckling half-wave number for the plate in the x direction, which is the minimum integer satisfying $a/b \leq \sqrt{m_o(m_o + 1)}$. For practical use, the half-wave number m may be taken as $m_o = 1$ for $1 \leq a/b \leq \sqrt{2}$, $m = 2$ for $\sqrt{2} < a/b \leq \sqrt{6}$, and $m = 3$ for $\sqrt{6} < a/b \leq 3$. If $a/b > 3$, the buckling coefficient can be approximated to $k_x = 4$.
σ_y	$k_y = \left[1 + \left(\dfrac{b}{a}\right)^2\right]^2$
τ	$k_\tau \approx 4\left(\dfrac{b}{a}\right)^2 + 5.34$, for $\dfrac{a}{b} \geq 1$ $\left(k_\tau \approx 5.34\left(\dfrac{b}{a}\right)^2 + 4.0,\ \text{for}\ \dfrac{a}{b} < 1\right)$
σ_{bx}	$k_{bx} \approx 23.9$
σ_{by}	$k_{by} \approx \begin{cases} 23.9, & \text{for } 1 \leq \dfrac{a}{b} \leq 1.5 \\ 15.87 + 1.87\left(\dfrac{a}{b}\right)^2 + 8.6\left(\dfrac{b}{a}\right)^2, & \text{for } \dfrac{a}{b} > 1.5 \end{cases}$

where σ_{xE} = elastic buckling stress for σ_x; σ_{yE} = elastic buckling stress for σ_y; τ_E = elastic buckling stress for τ; σ_{bxE} = elastic buckling stress for σ_{bx}; and σ_{byE} = elastic buckling stress for σ_{by}. Note that the effect of lateral pressure load, p, needs to be accounted for in the determination of the elastic buckling stress components.

5.5.1 Elastic Plate Buckling

Under single stress components, the elastic rectangular plate buckling stress can be given by the classic Bryan equation, as follows:

$$\sigma_E = k\frac{\pi^2 E}{12(1 - \nu^2)}\left(\frac{t}{b}\right)^2, \tag{5.9}$$

where σ_E = elastic plate buckling stress; b = plate breadth in the short direction; t = plate thickness; E = elastic modulus; ν = Poisson ratio; and k = buckling coefficient, which can be taken as indicated in Table 5.3 depending on the load type, for the case when all (four) edges are supported simply.

The buckling coefficient depends on the boundary conditions and cutouts as well as load types. For elastic buckling strength expressions of plates under single-load components taking into account the effects of different boundary conditions (e.g., all clamped edges; partly clamped edges plus partly simply supported edges; elastically restrained edges) and cutouts, see Paik and Thayamballi (2003).

The effect of lateral pressure on plate buckling needs to be considered. When a plate in a continuous stiffened panel is subjected to lateral pressure, the plate edges can approach the condition of being clamped depending on the thickness of the plate and the pressures involved. Also, lateral pressure loading may beneficially disturb occurrence of the inherent plate buckling pattern. In such cases, the buckling strength of long plate elements making up a continuous stiffened panel under lateral pressure may be greater than that without lateral pressure loading.

For practical design purposes, a correction factor is sometimes used to take into account the effect of lateral pressure on the plate buckling strength, the factor being applied by multiplication with the buckling strength calculated for the plate without lateral pressure loads. A similar approach also may be used to account for other effects that may be neglected in classical buckling theory; for example, initial deformations and residual stresses.

Fujikubo et al. (1998) proposed plate compressive buckling strength correction factors to account for the effect of lateral pressure, by the regression analysis of finite-element method solutions for long plate elements (i.e., with a/b ≥ 2) in a continuous stiffened panel, as follows:

$$C_{px} = 1 + \frac{1}{576} \left(\frac{pb^4}{Et^4} \right)^{1.6}, \quad \text{for} \quad \frac{a}{b} \geq 2, \tag{5.10a}$$

$$C_{py} = 1 + \frac{1}{160} \left(\frac{b}{a} \right)^{0.95} \left(\frac{pb^4}{Et^4} \right)^{1.75}, \quad \text{for} \quad \frac{a}{b} \geq 2, \tag{5.10b}$$

$$C_{px} = C_{py} = 1.0 \quad \text{for} \quad a/b \approx 1, \tag{5.10c}$$

where C_{px} and C_{py} are correction factors of the elastic compressive buckling strength in the x and y direction, respectively, to account for the effects of lateral pressure. p is the magnitude of net lateral pressure loads.

For nearly square plates (a/b ≈ 1) under combined axial compression and lateral pressure, a half-wave deflection may occur from the very beginning; thus, the bifurcation buckling phenomenon may not appear as the axial compressive loads increase. Further, the increase of buckling strength due to the rotational restraints and the decrease of buckling strength due to a half-wave deflection caused by lateral pressure may offset each other. For square plates, therefore, $C_{px} = C_{py} = 1.0$ may be approximated.

For convenience, the effect of lateral pressure on the elastic buckling strength for shear or in-plane bending is often neglected. Given such an assumption, Eq. (5.9) can then be rewritten to approximately account for the effect of lateral pressure, except for the case of edge shear or in-plane bending, as follows:

$$\sigma_E = kC_p \frac{\pi^2 E}{12 (1 - \nu^2)} \left(\frac{t}{b} \right)^2, \tag{5.11}$$

where $C_p = C_{px}$ or $C_{py} = $ buckling strength correction factor due to lateral pressure actions that can be determined as described previously, depending on the load type together with the plate aspect ratio.

Under a total of six types of multiple-load components, the elastic plate buckling interaction criterion may be given as a function of the individual elastic buckling stress components, which should take into account the effect of lateral pressure loads; for example, using Eq. (5.10), when involved, as follows (Paik and Thayamballi 2003):

$$\Gamma_B = \left[\frac{\sigma_x}{C_1 C_4 \sigma_{xE} \left\{ 1 - \left(\frac{\tau}{C_3 C_6 \tau_E} \right)^{\alpha_{11}} \right\}} \right]^{\alpha_1} + \left[\frac{\sigma_y}{C_2 C_5 \sigma_{yE} \left\{ 1 - \left(\frac{\tau}{C_3 C_6 \tau_E} \right)^{\alpha_{12}} \right\}} \right]^{\alpha_2} \leq 1, \tag{5.12}$$

where

$$C_1 = 1 - \left(\frac{\sigma_{bx}}{C_7 \sigma_{bxE}}\right)^2; \quad C_2 = \left\{1 - \left(\frac{\sigma_{bx}}{C_7 \sigma_{bxE}}\right)^{\alpha_4}\right\}^{1/\alpha_3}; \quad C_3 = \left\{1 - \left(\frac{\sigma_{bx}}{C_7 \sigma_{bxE}}\right)^2\right\}^{0.5};$$

$$C_4 = \left\{1 - \left(\frac{\sigma_{by}}{\sigma_{byE}}\right)^{\alpha_6}\right\}^{1/\alpha_5}; \quad C_5 = \left\{1 - \left(\frac{\sigma_{by}}{\sigma_{byE}}\right)^{\alpha_8}\right\}^{1/\alpha_7}; \quad C_6 = \left\{1 - \left(\frac{\sigma_{by}}{\sigma_{byE}}\right)^2\right\}^{1/2};$$

$$C_7 = \left\{1 - \left(\frac{\sigma_{by}}{\sigma_{byE}}\right)^{\alpha_{10}}\right\}^{1/\alpha_9};$$

$$\alpha_1 = \alpha_2 = 1, \quad \text{for } 1 \le \frac{a}{b} \le \sqrt{2};$$

$$\left.\begin{array}{l} \alpha_1 = 0.0293\left(\dfrac{a}{b}\right)^3 - 0.3364\left(\dfrac{a}{b}\right)^2 + 1.5854\left(\dfrac{a}{b}\right) - 1.0596 \\[2mm] \alpha_2 = 0.0049\left(\dfrac{a}{b}\right)^3 - 0.1183\left(\dfrac{a}{b}\right)^2 + 0.6153\left(\dfrac{a}{b}\right) + 0.8522 \end{array}\right\}, \quad \text{for } \frac{a}{b} > \sqrt{2};$$

$$\alpha_3 = \alpha_4 = 1.50\left(\frac{a}{b}\right) - 0.30, \quad \text{for } 1 \le \frac{a}{b} \le 1.6;$$

$$\left.\begin{array}{l} \alpha_3 = -0.625\left(\dfrac{a}{b}\right) + 3.10 \\[2mm] \alpha_4 = 6.25\left(\dfrac{a}{b}\right) - 7.90 \end{array}\right\}, \quad \text{for } 1.6 < \frac{a}{b} \le 3.2;$$

$$\left.\begin{array}{l} \alpha_3 = 1.10 \\ \alpha_4 = 12.10 \end{array}\right\}, \quad \text{for } 3.2 < \frac{a}{b};$$

$$\left.\begin{array}{l} \alpha_5 = 0.930\left(\dfrac{a}{b}\right)^2 - 2.890\left(\dfrac{a}{b}\right) + 3.160 \\[2mm] \alpha_6 = 1.20 \end{array}\right\}, \quad \text{for } 1 \le \frac{a}{b} \le 2;$$

$$\left.\begin{array}{l} \alpha_5 = 0.066\left(\dfrac{a}{b}\right)^2 - 0.246\left(\dfrac{a}{b}\right) + 1.328 \\[2mm] \alpha_6 = 1.20 \end{array}\right\}, \quad \text{for } 2 < \frac{a}{b} \le 5;$$

$$\left.\begin{array}{l} \alpha_5 = 1.117\left(\dfrac{a}{b}\right) - 3.837 \\[2mm] \alpha_6 = -0.167\left(\dfrac{a}{b}\right) + 2.035 \end{array}\right\}, \quad \text{for } 5 < \frac{a}{b} \le 8;$$

$$\left.\begin{array}{l} \alpha_5 = 5.10 \\ \alpha_6 = 0.70 \end{array}\right\}, \quad \text{for } 8 < \frac{a}{b};$$

$$\left.\begin{array}{l} \alpha_7 = 1.0; \\[2mm] \alpha_8 = \dfrac{1}{6.5}\left(14.0 - \dfrac{a}{b}\right) \end{array}\right\}, \quad \text{for } 1 \le \frac{a}{b} \le 7.5;$$

$$\alpha_7 = \alpha_8 = 1.0, \quad \text{for } 7.5 < \frac{a}{b};$$

$$\left.\begin{array}{l} \alpha_9 = 0.050\left(\dfrac{a}{b}\right) + 1.080 \\[2mm] \alpha_{10} = 0.268\left(\dfrac{a}{b}\right) - 1.248\left(\dfrac{b}{a}\right) + 2.112 \end{array}\right\}, \quad \text{for } 1 \le \frac{a}{b} \le 3;$$

$$\left.\begin{array}{l} \alpha_9 = 0.146 \left(\dfrac{a}{b}\right)^2 - 0.533 \left(\dfrac{a}{b}\right) + 1.515 \\[4mm] \alpha_{10} = 0.268 \left(\dfrac{a}{b}\right) - 1.248 \left(\dfrac{b}{a}\right) + 2.112 \end{array}\right\}, \quad \text{for } 3 < \dfrac{a}{b} \le 5;$$

$$\left.\begin{array}{l} \alpha_9 = 3.20 \left(\dfrac{a}{b}\right) - 13.50 \\[4mm] \alpha_{10} = -0.70 \left(\dfrac{a}{b}\right) + 6.70 \end{array}\right\}, \quad \text{for } 5 < \dfrac{a}{b} \le 8;$$

$$\left.\begin{array}{l} \alpha_9 = 12.10 \\[2mm] \alpha_{10} = 1.10 \end{array}\right\}, \quad \text{for } 8 < \dfrac{a}{b};$$

$$\alpha_{11} = \begin{cases} -0.160 \left(\dfrac{a}{b}\right)^2 + 1.080, \quad \left(\dfrac{a}{b}\right) + 1.082, \quad \text{for } 1 \le \dfrac{a}{b} \le 3.2 \\[4mm] 2.90, \quad \text{for } \dfrac{a}{b} > 3.2; \end{cases}$$

$$\alpha_{12} = \begin{cases} 0.10 \left(\dfrac{a}{b}\right) + 1.90, \quad \text{for } 1 \le \dfrac{a}{b} \le 2 \\[4mm] 0.70 \left(\dfrac{a}{b}\right) + 0.70, \quad \text{for } 2 < \dfrac{a}{b} \le 6 \\[4mm] 4.90, \quad \text{for } 6 < \dfrac{a}{b}. \end{cases}$$

5.5.2 Elastic Stiffener Web Buckling

The elastic buckling stress of a stiffener or girder web must be greater than the applied axial compressive stress by an adequate safety factor so that the elastic stiffener web buckling can be avoided prior to buckling of the attached plating. For a stiffener web surrounded by attached plating and stiffener flange, and two transverse frames, as shown in Figure 5.5, the elastic buckling stress can be calculated as follows (Paik and Thayamballi 2003):

$$\sigma_E^W = k_w \frac{\pi^2 E}{12 \left(1 - \nu^2\right)} \left(\frac{t_w}{h_w}\right)^2, \tag{5.13}$$

where σ_E^W = elastic buckling strength of stiffener web; and k_w = elastic buckling strength coefficient of stiffener web. To account for the effect of welding residual stress, the web buckling stress computed from Eq. (5.13) may be reduced by the compressive residual stress in the stiffener web. The approximate formulations of k_w in Eq. (5.13) are given by

$$k_w = \begin{cases} C_1 \zeta_p + C_2, \quad \text{for } 0 \le \zeta_p \le \eta_w \\[4mm] C_3 - \dfrac{1}{C_4 \zeta_p + C_5}, \quad \text{for } \eta_w < \zeta_p \le 60, \\[4mm] C_3 - \dfrac{1}{60 C_4 + C_5}, \quad \text{for } 60 < \zeta_p \end{cases} \tag{5.14}$$

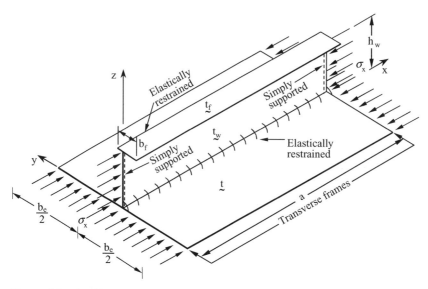

Figure 5.5. A stiffener web surrounded by attached plating, stiffener flange, and transverse frames under axial compressive loads.

where

$$\eta_w = -0.444\zeta_f^2 + 3.333\zeta_f + 1.0;$$

$$C_1 = -0.001\zeta_f + 0.303;$$

$$C_2 = 0.308\zeta_f + 0.427;$$

$$C_3 = \begin{cases} -4.350\zeta_f^2 + 3.965\zeta_f + 1.277, & \text{for } 0 \leq \zeta_f \leq 0.2 \\ -0.427\zeta_f^2 + 2.267\zeta_f + 1.460, & \text{for } 0.2 < \zeta_f \leq 1.5 \\ -0.133\zeta_f^2 + 1.567\zeta_f + 1.850, & \text{for } 1.5 < \zeta_f \leq 3.0 \\ 5.354, & \text{for } 3.0 < \zeta_f; \end{cases}$$

$$C_4 = \begin{cases} -6.70\zeta_f^2 + 1.40, & \text{for } 0 \leq \zeta_f \leq 0.1 \\ \dfrac{1}{5.10\zeta_f + 0.860}, & \text{for } 0.1 < \zeta_f \leq 1.0 \\ \dfrac{1}{4.0\zeta_f + 1.814}, & \text{for } 1.0 < \zeta_f \leq 3.0 \\ 0.0724, & \text{for } 3.0 < \zeta_f; \end{cases}$$

$$C_5 = \begin{cases} -1.135\zeta_f + 0.428, & \text{for } 0 \leq \zeta_f \leq 0.2 \\ -0.299\zeta_f^3 + 0.803\zeta_f^2 - 0.783\zeta_f + 0.328, & \text{for } 0.2 < \zeta_f \leq 1.0 \\ -0.016\zeta_f^3 + 0.117\zeta_f^2 - 0.285\zeta_f + 0.235, & \text{for } 1.0 < \zeta_f \leq 3.0 \\ 0.001, & \text{for } 3.0 < \zeta_f; \end{cases}$$

where $\zeta_p = \dfrac{GJ_p}{h_w D_w}$; $\zeta_f = \dfrac{GJ_f}{h_w D_w}$; $J_f = \dfrac{b_f t_f^3}{3}$ = torsion constant of stiffener flange; $D_w = \dfrac{Et_w^3}{12(1-\nu^2)}$ = bending rigidity of stiffener web; $G = \dfrac{E}{2(1+\nu)}$; t_w = stiffener web thickness;

h_w = stiffener web height, that is, without inclusion of both flange thickness and attached plate thickness (see Figure 5.2); b_f = flange breadth; t_f = flange thickness; E = elastic modulus; ν = Poisson ratio; $J_p = \frac{b_e t^3}{3}$; t = thickness of attached plating; and b_e = effective width of attached plating.

The effective width of the attached plating depends on the applied stresses, but it is often given by the Faulkner approximation, as follows:

$$b_e = \begin{cases} b, & \text{for } \beta \leq 1 \\ b\left(\dfrac{2}{\beta} - \dfrac{1}{\beta^2}\right), & \text{for } \beta > 1, \end{cases} \tag{5.15}$$

where $\beta = \frac{b}{t}\sqrt{\frac{\sigma_Y}{E}}$; and σ_Y = material yield stress.

For flat-bar stiffeners, Eq. (5.14) will become much simpler because $\zeta_f = 0$; the computed results are well approximated by

$$k_w = \begin{cases} 0.303\zeta_p + 0.427, & \text{for } 0 \leq \zeta_p \leq 1 \\ 1.277 - \dfrac{1}{1.40\zeta_p + 0.428}, & \text{for } 1 < \zeta_p \leq 60 \\ 1.2652, & \text{for } 60 < \zeta_p. \end{cases} \tag{5.16}$$

Figures 5.6(a) and 5.6(b) show the variation of the elastic buckling coefficients for flat-bar or angle/T-stiffener web as a function of the web aspect ratio a/h_w and the torsional rigidity of plating, respectively. Figure 5.6 shows that with an increase in the torsional rigidity of plating, the web buckling coefficient increases significantly. Therefore, accounting for such effects can be important, particularly in cases where stiffener web buckling is a possibility. The effects of the web aspect ratio on the buckling strength of the stiffener web, however, can be ignored in many practical cases.

5.5.3 Elastic Tripping of Stiffener

In design, it is usually desired that the elastic torsional-flexural buckling or tripping of stiffener under axial compressive loads does not occur before the buckling of attached plating. For nonsymmetric angle stiffeners, a closed-form expression for the elastic tripping stress σ_E^T can be obtained as follows (Paik and Thayamballi 2003):

$$\sigma_E^T = \min_{m=1,2,3\ldots} \left| \frac{C_2 + \sqrt{C_2^2 - 4C_1 C_3}}{2C_1} \right|, \tag{5.17}$$

where $C_1 = (b_e t + h_w t_w + b_f t_f)I_p - S_f^2$;

$$C_2 = -I_p\left[EI\left(\frac{m\pi}{a}\right)^2 - \frac{qa^2}{12}\frac{S_1}{I_y}\left(1 - \frac{3}{m^2\pi^2}\right) \right]$$

$$- (b_e t + h_w t_w + b_f t_f)\left[G(J_w + J_f) + EI_z h_w^2\left(\frac{m\pi}{a}\right)^2 - \frac{qa^2}{12}\frac{S_2}{I}\left(1 - \frac{3}{m^2\pi^2}\right) \right]$$

$$+ 2S_f\left[EI_{zy}h_w\left(\frac{m\pi}{a}\right)^2 - \frac{qa^2}{12}\frac{S_3}{I_y}\left(1 - \frac{3}{m^2\pi^2}\right) \right];$$

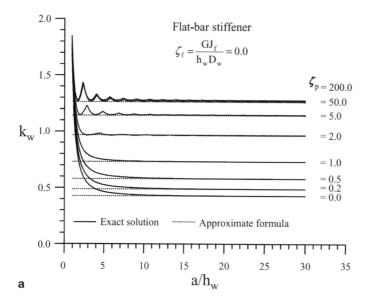

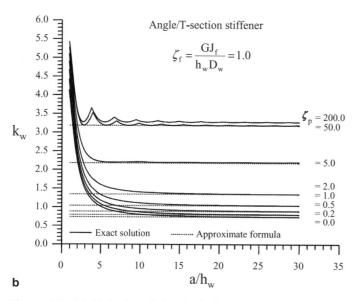

Figure 5.6. (a) Variation of the elastic buckling strength coefficient for a flat-bar stiffener web as a function of the web aspect ratio and torsional rigidity of plating. (b) Variation of the elastic buckling strength coefficient for angle or T-section stiffener web as a function of the web aspect ratio and torsional rigidities of plating or stiffener flange.

$$
C_3 = \left[EI \left(\frac{m\pi}{a} \right)^2 - \frac{qa^2}{12} \frac{S_1}{I_y} \left(1 - \frac{3}{m^2\pi^2} \right) \right]
$$
$$
\times \left[G \left(J_w + J_f \right) + EI_z h_w^2 \left(\frac{m\pi}{a} \right)^2 - \frac{qa^2}{12} \frac{S_2}{I_y} \left(1 - \frac{3}{m^2\pi^2} \right) \right]
$$
$$
- \left[EI_{zy} h_w \left(\frac{m\pi}{a} \right)^2 - \frac{qa^2}{12} \frac{S_3}{I_y} \left(1 - \frac{3}{m^2\pi^2} \right) \right]^2 ;
$$

$$S_f = -\frac{t_f b_f^2}{2};$$

$$S_1 = -\left(z_p - h_w\right) t_f b_f - b_e t z_p - h_w t_w \left(z_p - \frac{h_w}{2}\right);$$

$$S_2 = -\left(z_p - h_w\right) t_f \left(h_w^2 b_f + \frac{b_f^3}{3}\right) - h_w^3 t_w \left[\frac{1}{3}z_p - \frac{h_w}{4}\right];$$

$$S_3 = \left(z_p - h_w\right) \frac{b_f^2 t_f}{2};$$

$$I_y = \frac{b_e t^3}{12} + A_p z_p^2 + \frac{t_w h_w^3}{12} + A_w \left(z_p - \frac{t}{2} - \frac{h_w}{2}\right)^2 + \frac{b_f t_f^3}{12} + A_f \left(z_p - \frac{t}{2} - h_w - \frac{t_f}{2}\right)^2;$$

$$I_z = A_p y_o^2 + A_w y_o^2 + A_f \left(y_o^2 - b_f y_o + \frac{b_f^2}{3}\right);$$

$$I_{zy} = A_p z_p y_o + A_w \left(z_p - \frac{t}{2} - \frac{h_w}{2}\right) y_o + A_f \left(z_p - \frac{t}{2} - h_w - \frac{t_f}{2}\right)\left(y_o - \frac{b_f}{2}\right);$$

I_p = polar moment of inertia of stiffener about toe, given by

$$I_p = \frac{t_w h_w^3}{3} + \frac{t_w^3 h_w}{3} + \frac{b_f^3 t_f}{3} + \frac{b_f t_f^3}{3} + A_f h_w^2;$$

$$z_p = \frac{0.5 A_w \left(t + h_w\right) + A_f \left(0.5t + h_w + 0.5 t_f\right)}{b_e t + h_w t_w + b_f t_f};$$

$$y_{oe} = \frac{b_f^2 t_f}{2 \left(b_e t + h_w t_w + b_f t_f\right)};$$

J_w = torsion constant for the web, given by

$$J_w = \frac{1}{3} t_w^3 h_w \left(1 - \frac{192}{\pi^5} \frac{t_w}{h_w} \sum_{n=1,3,5}^{\infty} \frac{1}{n^5} \tanh \frac{n\pi h_w}{2 t_w}\right);$$

J_f = torsion constant for the flange, given by

$$J_f = \frac{1}{3} t_f^3 b_f \left(1 - \frac{192}{\pi^5} \frac{t_f}{b_f} \sum_{n=1,3,5}^{\infty} \frac{1}{n^5} \tanh \frac{n\pi b_f}{2 t_f}\right);$$

$A_p = b_e t; A_w = h_w t_w; A_f = b_f t_f; b_e$ = as defined in Eq. (5.15);

q = equivalent line pressure ($q = pb$);

p = lateral pressure;

b = breadth of attached plating; and

m = tripping half-wave number of the stiffener.

For symmetric T-stiffeners, a closed-form expression of the elastic tripping stress σ_E^T can be obtained as follows:

$$\sigma_E^T = \min_{m=1,2,3\ldots} \left| -\frac{a^2 G \left(J_w + J_f\right) + E I_f h_w^2 m^2 \pi^2}{I_p a^2} + \frac{q a^2}{12} \frac{S_4}{I_y I_p} \left(1 - \frac{3}{m^2 \pi^2}\right) \right|, \qquad (5.18)$$

where $S_4 = -(z_p - h_w) t_f \left(h_w^2 b_f + \frac{b_f^3}{12}\right) - h_w^3 t_w \left[\frac{1}{3} z_p - \frac{h_w}{4}\right];$

$$I_p = \frac{t_w h_w^3}{3} + \frac{t_w^3 h_w}{12} + \frac{b_f t_f^3}{3} + \frac{b_f^3 t_f}{12} + A_f h_w^2; \quad I_f = \frac{b_f^3 t_f}{12}.$$

Figure 5.7. A stiffener flange with three simply supported edges and one free edge.

5.5.4 Elastic Stiffener Flange Buckling

The stiffener flange must normally not buckle before the stiffener web or plating between stiffeners. The elastic buckling stress of the stiffener flange under axial compressive loads for the case shown in Figure 5.7 can be calculated by

$$\sigma_E^F = k_f \frac{\pi^2 E}{12\left(1 - \nu^2\right)} \left(\frac{t_f}{b_f^*}\right)^2, \tag{5.19}$$

where σ_E^F = elastic buckling stress of stiffener flange; $k_f = 0.425 + (\frac{b_f^*}{a})^2$; $b_f^* = b_f$ for nonsymmetric angle stiffeners; $b_f^* = 0.5 b_f$ for symmetric T-stiffeners; E = elastic modulus; a = length of flange; b_f = breadth of flange; t_f = thickness of flange; and ν = Poisson ratio.

5.6 Permanent Set Deflection Limits: Under Impact-Pressure Actions

While in service, ship-shaped offshore structures can be subjected to impact-pressure actions arising from green water, bow slamming, or sloshing, as described in Sections 4.10–4.12 of Chapter 4. Impact-pressure actions can cause excessive deflection of structural components.

Although the characteristics of most actions in ship-shaped offshore units may be dynamic or impact in nature, they can be idealized as quasistatic action problems when the action duration is long enough compared to the natural period of the structure as described in Section 4.10 of Chapter 4. When the duration time of actions is relatively short compared to the natural period of the structure, however, the actions must, in principle, be dealt with in the impact or impulsive domains. Note that not all cases of green water, bow slamming, or sloshing necessarily lead to impact-pressure pulses; nonimpact pressure variations are more common and can usually be dealt with in a quasistatic manner.

In terms of SLS design for impact-pressure actions with a short impact duration time, the following procedure might be used:

Step (a): Identify design profile of impact-pressure actions in terms of pressure versus time history, as described in Sections 4.10–4.12 of Chapter 4. At least two action parameters: (1) the peak pressure p, and (2) the pressure duration time τ for the design environmental conditions must be characterized by relevant vessel-motion analysis using site-specific environmental data.

Step (b): Calculate the structural damage (or permanent set of lateral deflection) due to the impact-pressure actions with the two parameter values determined in step (a).

Step (c): Optimize the structure so that the structural damage amount calculated from step (b) does not exceed the acceptable limiting value that may be prescribed according to the structure type, location, and severity of consequences; this value is usually taken as a few times the plate thickness, say, 1.5t (t = plate thickness).

The two action parameters – that is, peak pressure and pressure duration time – can be determined using the procedures described in Section 4.10 of Chapter 4. The type of local damage of structural components under impact-pressure actions that needs to be calculated in step (b) is typically that of permanent set deflection. Note that due to repeated actions of impact pressure, structural components can also suffer low-cycle fatigue cracking or rupture, which, where relevant, must be considered separately in the design process.

In Sections 5.6.1–5.6.3, closed-form formulations of permanent set deflections of stiffened plate structures under impact-pressure actions are presented. The formulae have been implemented by the ALPS/ULSAP (2006) computer program. Some illustrative examples are also shown, involving comparisons among ALPS/ULSAP calculations, nonlinear impact finite-element analyses, and experimental results. The calculation procedure noted in steps (a)–(c) is somewhat different from the current practice used, for example, by classification societies adopting the damage equivalent quasistatic pressure concept.

It is considered that permanent set deflection on a basic part of ship-shaped offshore structures must be evaluated at three levels: (1) a plate level between support members, (2) a stiffened panel level between transverse frames, and (3) a grillage level as an entire cross-stiffened panel. Also, the formulae need to account for the effects of strain rate in association with impact actions, as necessary.

5.6.1 Plates between Support Members

Under the impact-pressure-action applications, plating is likely to deform between support members, which are generally designed to provide enough support to plating and not to fail before plating. Also, the adjacent plating is assumed to deflect in the same direction of pressure loading, implying that the rotational restraints along the support member boundaries are large. In such a case, we may assume that the plate is clamped along its four edges. We assume also that the material obeys the Tresca-type yield criterion, and shear forces do not affect yielding.

The collapse mode of plating is presumed as indicated in Figure 5.8. The material is rigid and perfectly plastic and the loaded plate is divided into a number of rigid sections separated by straight-line plastic hinges, as shown in Figure 5.8. In this case, the maximum permanent set deflection w_p of the plate under impact-pressure action, but without the consideration of strain-rate effect, is given as the following lower-bound solution (Chen 1993):

$$\frac{w_p}{t} = \sqrt{2\frac{\alpha}{A_2}\lambda + \left(\frac{A_1}{A_2}\right)^2} - \frac{A_0}{A_2} - \frac{A_1}{A_2}, \tag{5.20}$$

where a = plate length; b = plate breadth; t = plate thickness; $\lambda = \frac{\mu V_o^2 b^2}{4M_p t}$; $V_o = \frac{P_c \tau}{\mu}$; $M_p = \frac{\sigma_Y t^2}{4}$; $\mu = \rho t$; $\tan \phi = \sqrt{3 + \alpha^2} - \alpha$; $\alpha = \frac{b}{a}$; $A_0 = \frac{3}{2\sin\phi\cos\phi} + \frac{1}{\alpha} - \tan\phi$;

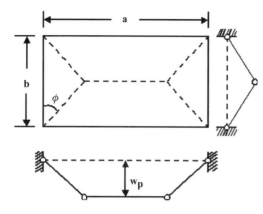

Figure 5.8. Presumed collapse mode for the plate clamped at all (four) edges.

$A_1 = \frac{1}{3\sin\phi\cos\phi} + \frac{2}{\tan\phi} + \frac{2}{\alpha}$; $A_2 = 4\left(\frac{1}{\sin\phi\cos\phi} + \frac{1}{\tan\phi} + \frac{4}{\alpha} - 3\tan\phi\right)$; σ_Y = material yield stress; P_e = effective peak pressure as defined in Section 4.10 of Chapter 4; ρ = mass per unit of volume, which is given by $\rho = \frac{\gamma}{g} = 7.85\,(\text{N} \times \sec^2/\text{mm}^4)$ for steel; γ = density ($\gamma = 7850\,\text{kg/m}^3$ for steel); and g = acceleration of gravity = $9.8\,\text{m/s}^2$.

For impact-pressure actions, the effect of strain rate is of significance. When the strain-rate effect is accounted for, the permanent set deflection formula should become a function of the strain rate. In this case, Eq. (5.20) is rewritten as a nonlinear function of the strain rate by replacing σ_Y (static yield stress) in Eq. (5.20) with σ_{Yd} (dynamic yield stress), as follows:

$$f_1(\lambda, w_p) - \left(\frac{w_p}{t}\right) = 0, \tag{5.21}$$

where $f_1(\lambda, w_p)$ = nonlinear function for plating as variables of $\lambda = \frac{P_e^2\tau^2 b^2}{\sigma_{Yd}\rho t^2}$ and w_p; $\sigma_{Yd} = \{1 + (\frac{V_o}{2w_p}C)^{1/Q}\}\sigma_Y$; C, Q = coefficients of the so-called Cowper–Symonds equation, which are given by C = 40.4 and Q = 5 for mild steel and C = 3,200 and Q = 5 for high-tensile steel (Paik and Thayamballi 2003; Jones 2006).

It is realized that the permanent set of the plate deflection under impact-pressure actions may not exceed the deflection value at the fundamental (natural) period of plating under dynamic pressure loads (Paik and Shin 2006). Therefore, the following condition must be satisfied:

$$w_p \le w_p^*, \tag{5.22}$$

where w_p = permanent set of plate deflection under impact-pressure actions; and w_p^* = permanent set deflection at the impact direction time equal to the natural period (T) of plating.

The natural period of steel plating under dynamic lateral pressure is calculated, approximately, as follows (KR 1997):

$$T = \frac{1}{f_n}, \tag{5.23}$$

where $f_n = \frac{\lambda_n}{2\pi b^2}\sqrt{\frac{D}{\rho t}}$ for square plates, with $\lambda_n = 19.74$ for simply supported edges, and $\lambda_n = 35.98$ for clamped edges; $f_n = \frac{\alpha_r\pi}{2b^2}\sqrt{\frac{D}{\rho t}}$ for rectangular plates with α_r as depicted in Figure 5.9; $D = \frac{Et^3}{12(1-\nu^2)}$; a, b, t, ρ = as defined in Eq. (5.20); E = elastic modulus; and ν = Poisson ratio.

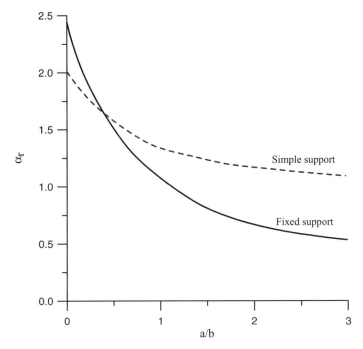

Figure 5.9. Coefficient α_r for determining the natural period of a rectangular plate.

5.6.2 Longitudinally Stiffened Panels between Transverse Frames

When the pressure pulse is not very large, the relatively heavier transverse frames in a grillage – that is, the overall cross-stiffened gross panel – may not fail until the uniaxially stiffened panel between transverse frames fail. In this case, the uniaxially stiffened panel between the two adjacent transverse frames may be modeled by a plate-stiffener combination clamped at both ends as representative of the panel, as shown in Figure 5.1.

Jones (1997) derived a closed-form expression for the permanent set of a beam deflecting under impact pressure when the strain-rate effect is not accounted for, as follows [for symbols not defined below, refer to Eq. (5.21)]:

$$\frac{w_p}{t_{eq}} = \frac{1}{2}\left[\left(1 + \frac{3\lambda}{4}\right)^{1/2} - 1\right], \tag{5.24}$$

where $\lambda = \frac{\mu V_o^2 a^2}{16 M_p t_{eq}}$; $M_p = \frac{\sigma_o b t_{eq}^2}{4}$; σ_o = flow stress taking into account the strain-hardening effect, which may be taken as $\sigma_o = \frac{\sigma_Y + \sigma_T}{2}$; σ_Y = yield stress; and σ_T = ultimate tensile stress; and t_{eq} = equivalent thickness, which is given by $t_{eq} = \frac{bt + h_w t_w + b_f t_f}{b}$, with the nomenclature presented in Figure 5.10.

When the strain-rate effect is accounted for, Eq. (5.24) becomes a nonlinear function of the strain-rate effect, by replacing σ_o (static flow stress) in Eq. (5.24) with σ_{od} (dynamic flow stress), as follows [for symbols not specified below, see Eq. (5.21)]:

$$f_2(\lambda, w_p) - \left(\frac{w_p}{t}\right) = 0, \tag{5.25}$$

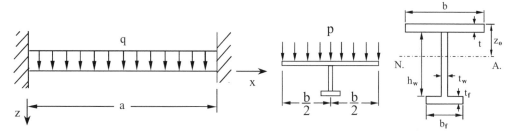

Figure 5.10. A plate-stiffener combination model clamped at both ends, representing a uni-axially stiffened panel between transverse frames and subject to an impact-line load, q = pb.

where $f_2(\lambda, w_p)$ = nonlinear function for interframe panel as variables of λ and w_p; $\lambda = \frac{\mu V_o^2 a^2}{16 M_p t_{eq}}$; $M_p = \frac{\sigma_{od} b t_{eq}^2}{4}$; $\sigma_{od} = \{1 + (\frac{V_o}{2 w_p C})^{1/Q}\} \sigma_o$; and C, Q = as defined in Eq. (5.22).

5.6.3 Cross-Stiffened Plate Structures

When the pressure pulse is very large and/or the transverse frames are relatively weak, the transverse frames may be postulated to fail together with longitudinal stiffeners and as plating. In this case, the cross-stiffened panel or grillage may be idealized as an orthotropic plate.

The permanent set of the orthotropic panel deflection may be calculated, approximately, by a method similar to that for plating described in Section 5.6.1, but with the equivalent plate thickness for an orthotropic plate, which may be given by smearing – that is, uniformly distributing – all (both longitudinal and transverse) support members into the plating in terms of volume of the structure, as follows [for symbols not defined below, refer to Eq. (5.21) and the nomenclature of Figure 5.1]:

$$t_{eq} = \frac{V}{LB},\tag{5.26}$$

where V = total volume of the cross-stiffened plate structure.

Although Eqs. (5.20) and (5.21) will be used for calculating the permanent set deflection, the related parameters including impact-energy parameter λ will be calculated from the following equations:

$$\lambda = \frac{\mu V_o^2 B^2}{4 M_p t_{eq}}, \quad \alpha = \frac{B}{L}, \text{ for neglecting the strain-rate effect}\tag{5.27a}$$

$$\lambda = \frac{P_e^2 \tau^2 B^2}{\left\{1 + \left(\dfrac{V_o}{2 w_p C}\right)^{1/Q}\right\} \sigma_Y \rho t_{eq}^2}, \text{ for considering the strain-rate effect}\tag{5.27b}$$

where V_o = as defined in Eq. (5.20); and $M_p = \frac{\sigma_Y b t_{eq}^2}{4}$.

Note that the permanent set deflection, w_p of the cross-stiffened plate structure under impact-pressure action, should not be greater than the permanent set deflection for the case with the pressure duration time equal to the natural period of the panel, as indicated in Eq. (5.22).

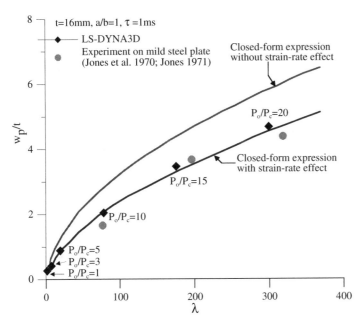

Figure 5.11. Comparison of permanent set deflections among DYNA3D simulations, ALPS/ULSAP and experiments, for steel plating surrounded by support members; shown as a function of the ratio of applied impact pressure P_o ($= P_e$) to the quasistatic pressure collapse load P_c [λ = as defined in Eq. (5.20)].

5.6.4 Illustrative Examples

Impact-pressure actions arising in service from sloshing, slamming, and green water can cause moderate to severe structural damage in ship-shaped offshore units. Therefore, avoiding structural damage due to impact pressure is one of the tasks required for safe structural design in such cases.

Pertinent simplified and efficient methods useful for this purpose have been described above and in Sections 4.10–4.12 of Chapter 4. These methods noted have also been implemented by the ALPS/ULSAP (2006) computer program, where the nonlinear functions – for example, Eq. (5.21) or Eq. (5.25) – for the permanent set deflection of panels are solved by the bisection method; typically, the number of iterations required to find a solution using this method is fewer than 15, with an associated computing time of less than 1 second.

The ALPS/ULSAP methods can predict the permanent set deflections of plating (between stiffeners), interframe stiffened panels (between transverse frames), and grillages under impact-pressure actions once the two parameters, the impact peak pressure value and its duration time, are known. Predictions may be made with and without strain-rate effects.

Application examples related to predicting the permanent set deflection of steel-stiffened plate structures under impact-pressure actions are now demonstrated by the ALPS/ULSAP methods together with a comparison of results by more refined direct DYNA3D nonlinear dynamic finite-element simulations (DYNA3D 2004) and also relevant existing experimental results as shown in Figure 5.11. It is apparent that the effect of strain rate is of significance on the permanent set deflection of plating under impact-pressure actions. Also, the ALPS/ULSAP solutions are seen to be in good agreement with the DYNA3D simulations and experimental results.

As another example, the permanent set deflection of a cross-stiffened plate structure under impact-pressure actions is now analyzed. The dimensions of the structure with the nomenclature indicated in Figure 5.1 are L = 15,300mm; B = 3,760mm; and t = 16mm. The number of longitudinal stiffeners is 3, and their type is a T-bar with 520mm × 12mm (web) and 150mm × 20mm (flange). The number of transverse frames is 2, and their type is a T-bar with 2,730mm × 18mm (web) and 450mm × 45mm (flange). The material yield stress is $\sigma_Y = 315$ N/mm^2 and the elastic modulus is E = 205,800 N/mm^2. The mass density is $\rho = 7,850$ kg/m^3.

Figure 5.12 shows comparisons of permanent set deflection predictions obtained by ALPS/ULSAP and through the use of the more elaborate direct DYNA3D nonlinear FEM considering plating, longitudinally stiffened panels, and grillages (cross-stiffened panels) under impact-pressure loads, with variation in the peak impact-pressure values and the impact-pressure duration time. Figure 5.12 shows that the permanent set deflections are significantly affected by impact-duration time as well as the peak-pressure value. Also, ALPS/ULSAP solutions are in good agreement with nonlinear FEM results over relatively wide ranges of impact-pressure values and duration times of practical applicability.

5.7 Intact Vessel Stability

A ship-shaped offshore unit must be designed to meet buoyancy requirements, and it will float at the proper attitude and remain generally upright during normal use or during an accident and while incurring heavy-weather effects. This involves the problems of gravitational stability and related criteria for judging the adequacy of the vessel's stability, accounting for internal loading and external upsetting hazards.

This section deals briefly with the vessel's stability criteria in the intact condition. The stability criteria in damaged conditions – for example, due to unintended flooding – will be treated in association with accidental limit-state design, described in Chapter 8.

The stability design procedures used for a moored ship-shaped offshore system today are similar to those used to design a trading ship (Lewis 1988). Calm-water concepts – that is, in the absence of wind and current – are used. The vessel's stability is governed primarily by the nature of equilibrium between weight (gravity forces), buoyancy, and center of gravity, among other factors. The inherent stability is judged both on its own terms (i.e., through required experience with proven threshold values) and for stability against upsetting forces and moments, such as predefined, service-proven, wind-heeling moments.

Winds can generate large external actions that affect the ship-shaped offshore unit. The curves of righting and wind-heeling moments similar to Figure 5.13 must be prepared in this case. The righting-moment and wind-heeling-moment curves are plotted in the most critical cases. A full range of draughts including those in transit (tow) conditions must be considered. In calculating the wind-heeling moments, the maximum variable deck and equipment loads must, in principle, be accounted for in the most unfavorable condition and position.

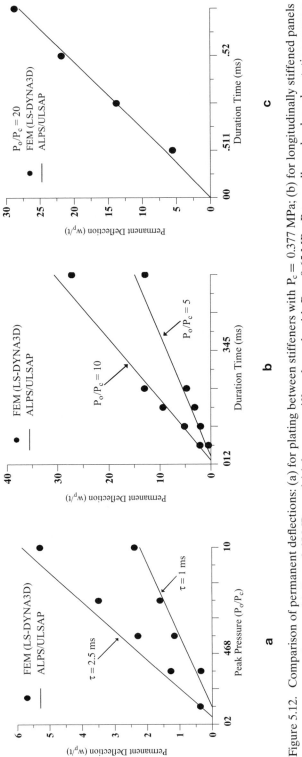

Figure 5.12. Comparison of permanent deflections: (a) for plating between stiffeners with $P_c = 0.377$ MPa; (b) for longitudinally stiffened panels between transverse frames with $P_c = 0.65$ MPa; and (c) for cross-stiffened panels with $P_c = 0.65$ MPa; $P_c =$ collapse loads under static pressure; $P_o = P_e =$ applied peak pressure value; $w_p =$ permanent set deflection of plating; $t =$ plate thickness; and $\tau =$ peak pressure duration time.

135

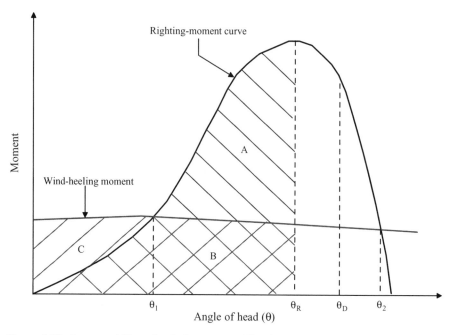

Figure 5.13. Intact stability criteria for a moored floating production system with the definition of heading angle illustrated in Figures 7.15 and 7.16 of Chapter 7, following HSE (2001b; courtesy of HSE).

To solve for intact stability, all of the following conditions must be satisfied, according to a set of criteria from (HSE 2001b):

$$A + B \geq 1.4\,(B + C)\,, \theta_1 \leq 15°, \theta_2 \geq 30°, GM \geq 0.3m, GZ \geq 0.5GM_o \sin\theta, \tag{5.28a}$$

where A, B, C = areas illustrated in Figure 5.13; θ_1 = the static angle of heel due to wind; θ_2 = the second intercept of wind-heeling and righting moment curves; GZ = righting lever; GM_o = the minimum permissible GM (metacentric height); θ = the angle in the range of 0 to minimum of θ_D or θ_M or 15 degrees; θ_D = the angle of first down-flooding; and θ_M = the angle of maximum righting lever.

In calculating A, B, and C, the following conditions must be satisfied:

$$\theta_R \leq \theta_D, \theta_R \leq \theta_2, \tag{5.28b}$$

where θ_R = the angle to which the areas A, B, and C are evaluated.

The curves of wind-heeling moments can be drawn using in part the distribution of wind forces that may be estimated from Eq. (4.5) in Chapter 4 or by similar means. Note that the wind forces are significantly affected by the wind speed, which needs to be determined for the site-specific environmental data as described in Chapter 4. In cases where those speeds are less than statutorily mandated wind speeds, the latter must be used. The wind speeds for normal operating, transit (tow), intermediate conditions, and severe storm conditions must be considered. The area of all surfaces exposed to wind due to heeling or trim should be included when calculating the projected areas in the vertical plane. Topsides and other deck structures can significantly affect the wind-heeling moments as the vessel is heeled; therefore, these must be included specifically in the calculations.

When assessing the wind-heeling moments, the moment arm is measured in distance in vertical direction, from the center of pressure of all exposed surfaces to the center of lateral resistance of the underwater body of the vessel. When a dynamic positioning system is available, the wind-heeling-moment curves will be prepared for the difference in force between the maximum applied thrust and the total wind force acting in the center of lateral resistance. We assume, in such a case, that the maximum thrust is applied at the thruster elevation. A complete range of cases, with varying heeling angles, should be evaluated to obtain the wind-heeling moment curves. It is sometimes assumed that the wind-heeling moment curve for a ship-shaped offshore unit may take the form of a cosine function (HSE 2001b). Wind-tunnel testing, as described in Appendix 5, is highly desirable as part of the process for assessing wind effects on the vessel hull and also for assessing the interactions of wind forces between topsides modules (HSE 2000).

As an intact stability criterion, a limiting-value static-heeling angle to simulate wind may be prescribed for any condition, say, 15 degrees maximum. In addition, the ratio of the area under the righting-moment curve to that under the wind-heeling-moment curve, as illustrated in Figure 5.13, should not be less than a limiting value, which may be about 1.4 for surface and self-elevating offshore units and 1.3 for column-stabilized offshore units (HSE 2001b). These areas will be measured from the upright position to a heeling angle, which is the lesser of either the angle of first down-flooding or the second intercept of the wind-heeling and righting-moment curves.

There are also limits on metacentric heights usually specified in the intact condition. Also, the vessel must be in stable equilibrium for a range of large angles of heel after damage, say, up to 30 degrees; see Section 8.3 of Chapter 8 for a more detailed discussion. Project specifications for FPSOs usually refer to international standards or requirements as indicated in Appendix 6; for example, the IMO SOLAS requirements with regard to intact and damage stability for oceangoing vessels.

5.8 Vessel Station-Keeping

Position-keeping and motion control are both important to meet functional and operational requirements of ship-shaped offshore units. The mooring system, thrusters, dynamic positioning system, or their combinations are useful means for station-keeping of a floating structure in the various environmental conditions involving wind, waves, and currents. The types, features, and related design considerations for mooring systems are discussed in Chapter 9; this section is concerned with the SLS assessment and criteria for station-keeping.

The analysis of the fluid-dynamic behavior for the structural parts above sea level subject to winds is much more complicated than that for the submerged parts subject to waves and currents. Although for submerged parts one may be confident with refined theoretical and numerical simulations of the submerged parts validated by model test results, it is usually necessary to rely more on model test data for the parts above sea level, as described in Appendix 5.

Moorings are generally the primary means of maintaining a floating offshore unit on station in all design conditions, including the maximum design storm conditions. Also, they are relied on even more to restrain the structure within certain excursion limits for operational purposes. These limits may be specified as a percentage of water depth, say, 5–8 percent for a case of intact spread mooring; and a few percent more with a lost mooring line. The integrity and performance of moorings are also important to other systems such as the risers connected to subsea equipment.

Mooring forces, which are related to SLS design for station-keeping with mooring systems, should be evaluated under all pertinent operational and extreme design environmental conditions by model testing as well as theoretical and numerical simulations as appropriate (Barltrop 1998). In addition, the effects of mooring forces transferred to the offshore unit from the mooring lines must be evaluated for structural design. The extreme stresses and the load-carrying capacity of the mooring lines must also be evaluated. Applicable criteria similar to Eq. (5.1b) will need to be satisfied in order to prevent the sudden failure of any single mooring line, which may, in turn, cause excessive loads in the remaining lines and also lead to increased loads on anchors.

Thrusters are often arranged to assist the mooring system in reducing the mooring forces and to provide for better heading control. The mooring analysis, in such cases, needs to take into account not only the beneficial effects of any thruster-related reduction in mooring-line tensile forces but also the possible detrimental effects of a sudden failure of the thruster. This is important to avoid excessive mooring line tensions and excursions. A similar consideration is important when thrusters need to be taken out of the commission for the purposes of maintenance and repair. As part of SLS limits, some operational restrictions may, in such situations, be put in place instead of strengthening the mooring system capability.

In mooring analysis, the site-specific environmental data must be used with the relevant return period storm, as we described in Section 4.13 of Chapter 4. Also, as we describe in Section 5.10 on motion exceedance, the lateral excursion, for heeling or trim, must be considered in the process of meeting the requirements of motion limits for operability of production systems, risers, gangways between fixed platforms, and floating units.

A more detailed description of mooring analysis and related considerations can be found in the standard books, for example, Barltrop (1998) and Faltinsen (1999). Guidelines on station-keeping of floating production offshore units can be found in HSE (2001c) and most classification society rules and guidelines. These documents strongly recommend that one should evaluate the strength capacity of mooring lines – for example, chain and wire rope – by full-scale testing under axial tensile loads wherever possible as well.

It is important to establish a relevant scheme for the periodic inspection of the mooring system components and mooring lines while in service in addition to procedures for their deployment or redeployment. A specific strategy is required to maintain the integrity of the mooring system and for its changeout, particularly when the life of mooring system components is shorter than the overall expected life of the offshore deployment. Inspection plans for damage, wear, corrosion, or fracture of components of the mooring system need to be developed, and procedures for replacement of such components must be in place as well (Brown et al. 2005).

5.9 Vessel Weathervaning and Heading Control

For the correct heading of a moored offshore unit in all design operating conditions, a dynamic positioning system (and occasionally tugs) may be used together with turret-mooring systems that maintain it in position. In Chapter 9, we describe in detail one advantage of a turret-mooring system in comparison to a spread-mooring system: it allows the vessel to rotate and adopt the optimum orientation in response to environmental conditions such as wind, waves, and currents.

The rotation of the vessel about the turret is "weathervaning." Two types of weathervaning may be relevant: *free weathervaning* and *partial weathervaning*. In free weathervaning, the vessel can freely rotate through 360 degrees; however, in the case of partial weathervaning, the rotations are restrained to within a more limited extent, ±270 degrees (i.e., to one and also the other directions), for instance. Partial weathervaning is sometimes implemented for a turret using a drag-chain attachment.

Heading control of the vessel is important in terms of mitigating the vessel's motions and environmental actions, including mooring forces. The issues related to bow slamming and green water are also closely related to the vessel's heading into waves, as described in Sections 4.11 and 4.12 of Chapter 4.

Although severe storms tend to be unidirectional, it is not rare that in medium to severe cases, the wind direction can change 30 degrees or more. An abrupt change in heading may significantly increase roll motion before the vessel's heading can be brought back to face the incoming wave. The vessel can develop sufficiently adverse roll motions that may even affect the operation of machinery, such as the vessel's power generators, even if the vessel's heading to a wind-driven wave is at a modest angle or if there is a beam swell with waves close to the natural period in roll.

To make accurate theoretical and numerical predictions of a vessel's heading into waves, appropriate joint distributions for the relevant metocean parameters must be identified or, at least, appropriate design values that are neither too extreme nor too insignificant must be determined and used together considering uncertainties in one's ability to accurately predict responses for complex scenarios. Model testing may then be strongly recommended to reduce some of the uncertainties involved and to obtain better predictions of weathervaning and its effects (HSE 2000); see also Appendix 5 of this book.

Where a dynamic positioning system is arranged with a mooring system, the analysis of both systems should account for the effects and interactions of each system. Therefore, the possible failure of a mooring line must be considered in the design of thrusters because sudden mooring line failure may lead to changes in heading, mooring loads, and perhaps the overloading of the dynamic positioning system. Similarly, the mooring analysis must take into account the possible failure of any dynamic positioning system. Also, even in the event of failure of a main generating unit, sufficient power to maintain the position of the unit must be available.

It is usual that the dynamic positioning system for heading control incorporates a device that is able to detect and record any position loss and also give a visual and/or audible warning when such position loss exceeds certain threshold values. Useful information on dynamic positioning systems and their implementation can be found in various guidelines provided by NMD (1986), NPD (1999), and IMCA

(1999), as well as classification society rules and guidelines; see Appendix 6 of this book.

Some of these guidelines might have been developed in other contexts – for example, for driving support vessels without production equipment such as risers. In such cases, additional factors must also be considered to control the maximum allowable excursions relative to production risers, for instance. Also, the need for the necessary means and procedures to safely disconnect the risers in certain cases may be an important design consideration.

5.10 Vessel Motion Exceedance

Most production equipment has certain operational limits associated with it, including motion velocities and accelerations. In some cases, vessel motions can lead to sloshing in equipment, which if not properly considered in design can cause failure of the equipment; a case in point is the separator failure described by Bradley and Sanders (2000). Therefore, it is important that for a ship-shaped offshore unit the related motions and excursions be limited to prescribed values in order to meet the operational requirements of the unit and that all related effects be accounted for in design.

Excessive motions, accelerations, and excursions of a floating unit can lead to increased downtime because of exceedence of limits associated with functions such as drilling, production, mooring, and station-keeping and because of a possible drop in efficiency, or even failure, in some cases of structure and equipment associated with those functions. Of particular importance are offshore installations of essentially land-based technology, such as gas depropanisers. Production facilities with large vessel motions can experience problems related to the oil–water and gas–liquid separation processes and bow slamming and green water. Problems due to bow slamming and green water are discussed in Sections 4.11 and 4.12 in Chapter 4.

Lateral excursions and roll motions, in particular, may cause the serviceability problems for production equipment and risers. In addition, in drilling operations there is a prescribed heave motion limit (usually about 4m in amplitude), again related to how much flexibility one can expect from the riser system. Most operations involving personnel transfer are motion-sensitive. A related factor is that offshore vessels have a relatively higher center of gravity due to their topsides equipment than trading tankers, which again can imply larger vessel motions than a trading tanker.

The matter of vessel motion exceedance is also closely related to heading control and station-keeping, as we described in previous sections of this chapter. Ship-shaped offshore units are likely to be subjected to large lateral motions, particularly due to beam seas. The beam sea condition can occur not only in particular cases of the unit moored with a spread-mooring system but also with a weathervaning system when significant currents act in different directions to the wind and waves. An example of this is discussed by da Silva and dos Reis Correia (2004) in the context of shuttle tanker offloading operations for turret-moored F(P)SOs in the *Campos* basin in Brazil, where they state that, in some cases, winds and local waves come from a specific direction and the swells come from a direction that is 90 degrees removed, which can cause large roll motions for the F(P)SO.

Motion and hydrodynamic analysis can be made for particular situations using the refined methods available today, except for certain details such as input to the behavior in a roll, which is significantly affected by nonlinear viscous roll damping. In such cases, experiments need to be done to collect the specific data needed. We recommend that the experimental results be checked against the analysis tools used to obtain the data.

One of the cost-effective methods to increase the roll damping and reduce the roll motions is to increase the length and size of the bilge keels (HSE 2005). In such cases, one may resort to model tests to determine the optimum location and size of bilge keels. The strength and fatigue capacity of bilge keels also need to be confirmed as part of the design.

In addition to these functional degradations of the offshore units due to vessel motions, excessive vessel motions will prevent crew members from performing their tasks effectively. Therefore, it is important to evaluate the effects of motion on human performance, health, and safety as well. In fact, in some cases, it is possible that human limitations may be more severe than those related to machinery. Several classes of human-performance degradation associated with vessel motions are considered (Colwell 1989; HSE 1999; ABS 2002):

- Postural stability
- Motion-sickness incidence
- Motion-induced fatigue
- Vibration effects on the human body

We now recognize that postural stability is an issue related to low-frequency vibrations typically below 1 Hz and that human postural stability is maintained by a complex musculo-skeletal system with various controls. The effect of motion sickness has been investigated extensively, both on land and at sea. For example, Reason (1978) and Pingree (1989) studied the causes and effects of motion sickness on human performance. Fatigue of crew members is also a factor affecting the operational effectiveness during prolonged periods aboard vessels at sea. It is a concern that such fatigue may lead to errors in critical tasks.

The low-frequency motions and vibrations, which are regarded to be below 0.5–1.0 Hz, are generally the causes of motion-sickness and postural-control issues. High-frequency vibrations can cause mechanical resonance in humans that leads to discomfort and lack of concentration. In addition, crew habitability needs such as indoor climate and lighting are also relevant considerations in modern-day ship-shaped floating offshore units. A primary reason for the consideration of these factors is that human error is of great concern with regard to accident prevention.

An excellent review on the national and international standards regarding the effects of vessel motion on human performance and on limiting values is presented in HSE (1999). See ABS (2002) for general considerations and limiting criteria for habitability design.

5.11 Vibration and Noise

Vibration on floating offshore units must be controlled for the same reasons it is controlled on merchant trading ships; these include protection of personnel health,

Table 5.4. *Vibration-limit categories for offshore installations, following HSE (2001d)*

Category	Description
I	Restricted area (less than 4 minutes exposure) vibration limits. Short exposure to levels above these limits may create a health hazard and cause difficulty in walking. These high levels of vibration usually cause such alarm and discomfort that persons affected will intuitively leave the affected area. Vibration levels above these limits are treated as "prohibited."
II	Just acceptable locally to equipment, although vibration limits for machinery itself may be more restrictive than these levels. Annoyance and discomfort may be experienced.
III	Recommended design vibration limits for all general work areas. Vibration levels are easily detectable but not uncomfortable.
IV	Recommended design vibration limits for office, control rooms, and similar areas.
V	Recommended design vibration limits for sleeping, recreation, and similar areas in living accommodations. These vibration levels are just detectable.

ability of personnel to perform designated tasks, and provision of an acceptable rest and recreation environment in the accommodation areas. Similarly, noise must be controlled to minimize the risk of hearing damage to personnel in work areas; to ensure that warning signals are heard; to allow communication by speech, telephone, and radio; to maintain working efficiency; and to provide an acceptable rest and recreation environment in the living accommodations (e.g., HSE 2001d; ISO 6954).

The technologies for and approaches to vibration, noise analysis, and control for offshore units are similar to those of trading ships and include avoidance of resonance and limitations on amplitudes, velocities, accelerations, and acceptable noise levels.

In ships and offshore facilities, vibrations may be present due to machinery, thrusters, propulsion engines, and propellers. Compared to vessel motions, these vibrations are usually of a higher frequency. Bow slamming can also induce higher-frequency transient vibrations, called "whipping." One type of steady-state vibration that is admittedly rare both in oceangoing ships and in ship-shaped offshore units is "springing" related to the two-noded mode of vibration of the hull girder. In FPSOs, it is often only the local vibration effects that are pertinent because large machinery, such as the main engines and the propeller, are either not there or are not operational.

Considerations related to vibration and noise control are necessary during the selection and specification of various equipment and pieces of machinery. Vendors may be expected to supply appropriate testing and documentary evidence of the vibration and noise levels that their equipment and machinery are to meet. Also, the vibration and noise limits that apply to an offshore unit will be included as part of the technical specifications for the unit.

In HSE (2001d), vibration limits are divided into four categories, ranging from recommended vibration limits in living accommodation (category V) to prohibited vibration limits (greater than category I), as indicated in Table 5.4. Sample vibration limits for human exposure are presented in Figure 5.14 for the vertical direction and in Figure 5.15 for the horizontal direction.

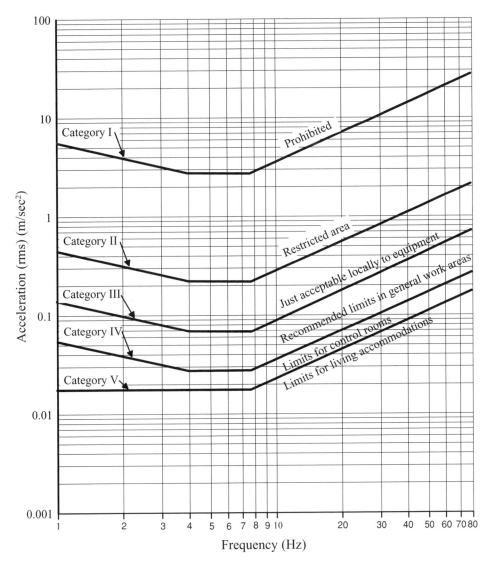

Figure 5.14. Vibration limits for the vertical direction (HSE 2001d; courtesy of HSE).

Related to noise, the guidelines found in HSE (2001d) recommend pertinent noise limits depending on various locations on offshore installations, such as specific work areas, sleep and recreation areas, and areas of living accommodations. These limits range from 45 to 70 dBA.

5.12 Mooring Line Vortex-Induced Resonance Oscillation

A unique source of vibration that has been experienced in some offshore units is related to the vortex shedding in mooring lines or risers under heavy current conditions. The condition is in part related to current velocity, which can be significant even in deep water in certain locations, as described in Section 4.6 of Chapter 4.

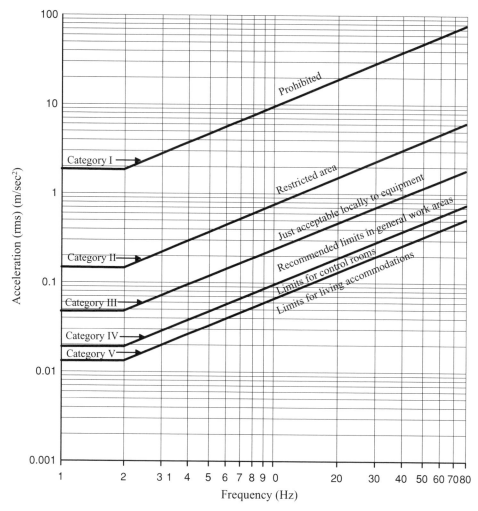

Figure 5.15. Vibration limits for the horizontal direction (HSE 2001d; courtesy of HSE).

When the fluid separates and starts to flow around a circular cylinder, the result-ing wake pattern may be symmetrical at first, but this subsequently changes to an asymmetric wake pattern, with the vortices developing and shedding on either side of the cylinder in turn. This phenomenon is called *vortex shedding*, and it can cause oscillatory forces to be applied on the cylinder. If there is resonance between these oscillatory forces and the structure of the cylinder, the structural displacements grow and can, in extreme cases, disturb the function of the cylinder, which could perhaps be a mooring line or a riser.

For a vertically oriented cylinder in a current, the vortex-shedding period is inversely proportional to the current velocity. The oscillations of interest occur in the cylinder when the natural frequency of the cylinder becomes identical to the vortex-shedding frequency. It is interesting to note that in a certain range of the cur-rent velocity, lock-in of the vortex-shedding frequency to the natural frequency of the cylinder may take place, meaning that the vortex-shedding frequency becomes almost identical to the natural period of the cylinder. This phenomenon is also referred to as *resonance* or *synchronization*.

In contrast to the resonance oscillations in the lock-in region noted previously, galloping is another phenomenon of oscillations for the structure that occurs when the vortex-induced hydrodynamic forces in effect cause a sufficiently large negative damping of the oscillations. The galloping type of instability oscillations can occur for lower current velocities than those for lock-in.

Consideration may also need to be given to fatigue damage due to vortex-induced vibrations where appropriate (Petruska et al. 2002). A detailed description of the problems on vortex-induced resonance oscillations of cylinders such as risers in currents may be found in Faltinsen (1990).

5.13 Corrosion Wastage

Because of localized corrosion wastage (e.g., pitting, grooving) as well as general (uniform) corrosion, leakage can take place in oil or watertight boundaries, which can potentially lead to undesirable pollution, cargo mixing, or gas accumulation in enclosed spaces. The effects of general corrosion and localized corrosion must in general be taken into account in corrosion management, as described in Chapter 10.

As corrosion-wastage limiting values triggering plate replacement, some classification societies recommend local plate thickness reductions of about 20–25 percent from considerations of strength (Paik et al. 2006). Margins for general corrosion are routinely added in structural design and appropriate corrosion protection, including coatings, impressed current cathodic protection (ICCP), and anodes, is used where necessary. Pitting is treated somewhat differently in design in that no specific steel thickness margins are added. In either case, periodic inspection with close-up surveys and gauging are relied upon to detect threshold levels of corrosion and replace the affected plating when necessary. In certain types of floating offshore units that must remain on site for extended periods compared to dry-docking available to trading tankers every five years, corrosion is a factor that requires particularly enhanced consideration. Related data collection and ongoing studies are also necessary for the long term. Chapter 10 presents a more detailed description for corrosion management and protection.

REFERENCES

ABS (2002). *Guide for crew habitability on offshore installations*. American Bureau of Shipping, Houston, May.

ALPS/ULSAP (2006). A computer program for ultimate limit state assessment for stiffened panels. Proteus Engineering (http://www.proteusengineering.com), Stevensville, MD.

Barltrop, N. D. P. (1998). *Floating structures: A guide for design and analysis*. The Centre for Marine and Petroleum Technology (CMPT). Herefordshire, England: Oilfield Publications Limited.

Bradley, R., and Sanders, C. J. (2000). *Operating experiences with the captain FPSO*. Offshore Technology Conference, OTC 12055, Houston, May.

Brown, M. G., Hall, T. D., Marr, D. G., English, M., and Snell, R. O. (2005). *Floating production mooring integrity JIP – Key findings.* Offshore Technology Conference, OTC 17499, Houston, May.

Chen, W. (1993). "A new bound solution for quadrangular plates subjected to impulsive loads." *Proceedings of the 3rd International Offshore and Polar Engineering Conference*, Vol. 4, pp. 701–708, Singapore, June 6–11.

Chen, L., and Atsuta, C. (1977). *Theory of beam-columns.* New York: McGraw-Hill.

Colwell, J. L. (1989). *Human factors in the naval environment: A review of motion sickness and biodynamic problem.* (DREA Technical Memorandum 89/220), Canadian National Defence Research Establishment, Dartmouth, September.

da Silva, S., and dos Reis Correia, C. (2004). "Shuttle tankers basic requirements for Brazilian waters." *Proceedings of OMAE Specialty Symposium on FPSO Structural Integrity, OMAE–FPSO'04-0071*, Houston, August 30– September 2.

DYNA3D (2004). User's manual (Version 960). Livermore Software, California.

Eurocode (1992). *Design of steel structures.* (Eurocode 3), British Standards Institution, London.

Faltinsen, O. M. (1990). *Sea loads on ships and offshore structures.* Cambridge, UK: Cambridge University Press.

Fujikubo, M., Yao, T., Varghese, B., Zha, Y., and Yamamura, K. (1998). "Elastic local buckling strength of stiffened plates considering plate/stiffener interaction and lateral pressure." *Proceedings of the International Offshore and Polar Engineering Conference*, Montreal, Vol. 4, pp. 292–299.

HSE (1999). *Human factors review of vessel motion standards.* (Offshore Technology Report, OTO 1999/036), Health and Safety Executive, UK.

HSE (2000). *Review of model testing requirements for FPSOs.* (Offshore Technology Report, OTO 2000/123), Health and Safety Executive, UK.

HSE (2001a). *Loads.* (Offshore Technology Report, OTO 2001/013), Health and Safety Executive, UK.

HSE (2001b). *Stability.* (Offshore Technology Report, OTO 2001/049), Health and Safety Executive, UK.

HSE (2001c). *Station-keeping.* (Offshore Technology Report, OTO 2001/050), Health and Safety Executive, UK.

HSE (2001d). *Noise and vibration.* (Offshore Technology Report, OTO 2001/068), Health and Safety Executive, UK.

HSE (2005). *Findings of expert panel engaged to conduct a scoping study on survival design of floating production storage and offloading vessels against extreme metocean condition.* (Research Report, No. 357), Health and Safety Executive, UK.

Hughes, O. F. (1983). *Ship structural design – A rationally-based, computer-aided optimization approach.* The Society of Naval Architects and Marine Engineers, NJ.

Hughes, O. F., and Caldwell, J. B. (1991). *Marine structures – Selected topics, examples and problems: Vol. I Plate Bending.* (Supplement to Chapter 9 of *Ship Structural Design*), The Society of Naval Architects and Marine Engineers, NJ.

IMCA (1999). *Guidelines on the design and operation of DP vessels.* International Marine Contractors Association, M103, UK.

Jones, N. (1971). "A theoretical study of the dynamic plastic behavior of beams and plates under finite-deflections." *International Journal of Solids and Structures*, 7: 1007–1029.

Jones, N. (1973). "Slamming damage." *Journal of Ship Research*, 17(2): 80–86.

Jones, N. (1977). "Damage estimate for plating of ships and marine vehicles." *Proceedings of PRADS'77*, Tokyo, Japan, pp. 121–128.

Jones, N. (1997). *Structural impact.* Cambridge: Cambridge University Press.

Jones, N. (2006). "Some recent developments in the dynamic inelastic behaviour of structures." *Ships and Offshore Structures*, 1(1): 37–44.

Jones, N., Liu, T. G., Zheng, J. J., and Shen, W. Q. (1991). "Clamped beam grillages struck transversely by a mass at the center." *International Journal of Impact Engineering*, 11(3): 379–399.

Jones, N., Uran, T. O., and Tekin, S. A. (1970). "The dynamic plastic behavior of fully clamped rectangular plates." *International Journal of Solids and Structures*, 6: 1499–1512.

KR (1997). *Control of ship vibration and noise.* Korean Register of Shipping, Daejeon, Korea.

Lewis, E. V. (1988). *Principles of naval architecture.* Jersey City, NJ: The Society of Naval Architects and Marine Engineers.

Mano, M., Okumoto, Y., and Takeda, Y. (2000). *Practical design of hull structures.* Senpaku Gijutsu Kyoukai, Tokyo, Japan, September.

NMD (1986). *Guidelines on dynamic positioning systems.* Norwegian Maritime Directorate, Norway.

NPD (1999). *Guidelines for the specification and operation of dynamically positioned diving support vessels.* Norwegian Petroleum Directorate, Norway.

Paik, J. K., Brennan, F., Carlsen, C. A., Daley, C., Barbatov, Y., Ivanov, L., Rizzo, C. M., Simonsen, B. C., Yamamoto, N., and Zhuang, H. Z. (2006). *Condition assessment of aged ships. Report of the International Ships and Offshore Structures Congress, Specialist Committee V.6*, Southampton, UK: Southampton University Press.

Paik, J. K., and Hughes, O. F. (2006). Ship structures. In *Computational analysis of complex structures*, R. E. Melchers and R. Hough, eds. Reston, VA: American Society of Civil Engineers.

Paik, J. K., and Shin, Y. S. (2006). "Structural damage and strength criteria for ship stiffened panels under impact pressure actions arising from sloshing, slamming, and green water loading." *Ships and Offshore Structures*, 1(3): 249–256.

Paik, J. K., and Thayamballi, A. K. (2003). *Ultimate limit state design of steel-plated structures.* Chichester, UK: John Wiley & Sons.

Petruska, D. J., Zimmermann, C. A., Krafft, K. M., Thurmond, B. F., and Duggal, A. (2002). *Riser system selection and design for a deepwater FSO in the Gulf of Mexico.* Offshore Technology Conference, OTC 14154, Houston, May.

Pingree, B. J. W. (1989). "Motion commotion – A seasickness update." *Journal of the Royal Naval Medical Services*, 75: 75–84.

Reason, J. T. (1978). "Motion sickness: Some theoretical and practical considerations." *Applied Ergonomics*, 9(3): 163–167.

Szilard, R. (2004). *Theories and applications of plate analysis – Classical, numerical, and engineering methods.* Chichester, UK: John Wiley & Sons.

Timoshenko, S. P., and Gere, J. M. (1961). *Theory of elastic stability.* New York: McGraw-Hill.

Timoshenko, S., and Woinowsky-Krieger, S. (1981). *Theory of plates and shells.* New York: McGraw-Hill.

CHAPTER 6

Ultimate Limit-State Design

6.1 Introduction

As described in Chapters 3 and 5 of this book, limit states are classified into four types: serviceability limit states (SLS), ultimate limit states (ULS), fatigue limit states (FLS), and accidental limit states (ALS). The ULS for ship-shaped offshore structures include the following:

- Structural instability of part or all of the global structure resulting from buckling and collapse of its structural components
- Attainment of the maximum load-carrying capacity of the structure or its components by any combination of buckling, yielding, rupture, and fracture
- Significant in-flooding and loss of watertight integrity of the hull due to extreme actions under harsh environmental conditions
- Loss of static equilibrium in part or for all of the global structure considered as a rigid body; that is, capsizing or overturning

This chapter presents ULS design principles and criteria, together with useful practices, applicable for ship-shaped offshore structures. Core technologies for ULS design and assessment of the buckling and plastic collapse of structural components (plates, stiffened panels, support members) and progressive hull girder collapse are addressed.

6.2 Design Principles and Criteria

The structural design criteria for the ULS are primarily based on buckling collapse or ultimate strength. To be safe in the ULS, the design criterion can be expressed following Eq. (3.10) in Chapter 3:

$$C_d - D_d > 0, \tag{6.1a}$$

where C_d is design capacity (strength) and D_d is design demand (actions or action effects). The subscript d denotes the *design value*, which considers the uncertainties associated with capacity or demand. In ULS design, C_d indicates the ultimate strength and D_d is the extreme working load or stress in consistent units. When the structure is subjected to multiple-load components, C_d and D_d need to be expressed as the corresponding interaction functions taking into account the effects of combined actions.

Eq. (6.1a) may be rewritten in the form of a conventional structural safety check as follows:

$$\eta = \frac{C_d}{D_d} > 1, \tag{6.1b}$$

where η = measure of structural adequacy that must be greater than unity to be safe.

Using the partial safety factor approach, Eqs. (6.1a) and (6.1b) can be rewritten because $C_d = C_k/\gamma_C$ and $D_d = \gamma_D D_k$, as follows:

$$\frac{C_k}{\gamma_C} - \gamma_D D_k > 0 \tag{6.1c}$$

$$\eta = \frac{1}{\gamma_C \gamma_D} \frac{C_k}{D_k} > 1, \tag{6.1d}$$

where C_k, D_k = characteristic values for capacity and demand, respectively; and γ_C, γ_D = partial safety factors associated with capacity and demand, respectively, both of which are defined to be greater than unity. The partial safety factors must be obtained by probabilistic analysis involving associated uncertainties.

The ULS design criterion of ship-shaped offshore unit hulls under vertical bending moments are similar to trading ships and may be expressed as follows:

$$\frac{M_u}{\gamma_u} \geq \gamma_{sw} M_{sw} + \gamma_w M_w, \tag{6.1e}$$

where M_u = ultimate bending moment; M_{sw} = still-water bending moment; M_w = wave-induced bending moment; and $\gamma_u, \gamma_{sw}, \gamma_w$ = partial safety factors for M_u, M_{sw}, and M_w, respectively.

Note that for trading tankers, M_w in the design condition may be determined from Eq. (3.28) of Chapter 3, and M_{sw} can be calculated from Eq. (3.25), or approximately from Eq. (3.26). The calculation method of M_u is described in Section 6.7. For ULS check, the recent IACS common structural design rules for trading double-hull oil tankers (IACS 2005) suggest using $\gamma_u = 1.1$, $\gamma_{sw} = 1.0$, and $\gamma_w = 1.3$ in the sagging condition as long as M_{sw}, M_w, and M_u are determined from the methods documented in the rules.

This section focuses on the determination of the characteristic value C_k for ULS of structural components and vessel hulls. For ULS calculations of ship-shaped offshore structures, gross thickness (i.e., as-built thickness) is usually applied, although the net thickness (i.e., as-built thickness minus a nominal corrosion margin or allowance) is used in many cases for modern trading tanker structural design (IACS 2005). This is often simply a matter of practice.

In usual operational condition of vessels, tensile strains of structural components at gross yielding may be small enough such that no fracture may occur. However, for offshore units operating in cold waters or for aged vessel structures, the structural material is more likely to become brittle and/or the fracture strain of the structural components may become smaller. In such cases, the structural components may experience brittle or ductile fracture; thus, this type of failure must also be considered.

6.3 Actions and Action-Effects Analysis

For ULS design of structural components, the demand term in Eq. (6.1) is typically measured by stresses in terms of single components or their combinations. On the other hand, the demand of vessel hull structures – that is, at the global system level – is given by hull girder loads such as vertical bending moments, horizontal bending moments, shearing forces, or torsional moments or their combinations.

For ship-shaped offshore structures, extreme conditions for the environment and operation must be identified based on tow and site-specific environmental data (e.g., waves, wind, currents) together with an operational plan (e.g., loading, offloading), as described in Chapter 4. The resulting actions, in terms of pressures or forces, can be determined by vessel motion analysis.

Hull girder actions (e.g., hull girder bending moments), similar to those in trading tanker designs, can be obtained by the integration of actions (e.g., pressures) along the entire vessel. Actions of topsides and risers, as well as the vessel hull, must also be identified by their interaction (see Section 9.3 of Chapter 9). Action effects (e.g., stresses, internal resistive moments) in individual structural components can be calculated by structural analysis, typically by applying the finite-element method, as described in Section 5.3 of Chapter 5.

With hull girder bending moments known, bending stresses over the vessel hull cross section of ship-shaped offshore units are often calculated at the early design stage by classical beam theory, as follows:

$$\sigma = \frac{M}{I/z} = \frac{M}{Z},$$ (6.2)

where σ = hull girder bending stress; M = applied bending moment; I = moment of inertia over hull cross section with regard to the elastic neutral axis; z = distance from the neutral axis to the location of stress calculations; and $Z = I/z$ = section modulus.

It is interesting to note that the allowable working stress-based design methods for longitudinal strength of trading tankers consider that the hull girder bending stress σ must not exceed the allowable stress; that is, $175/K$ (N/mm^2) as indicated in Eq. (3.1) of Chapter 3. On the other hand, the ULS design methods consider that the ultimate hull girder strength denoted by M_u must be greater than the extreme (maximum) bending moment applied by an appropriate factor of safety.

As a matter of convenience, the site environment is often categorized into two groups: *harsh conditions* and *benign conditions*. The harsh environment may be further subdivided into *harsh* (e.g., North Sea) and *hurricane* (e.g., Gulf of Mexico). Some classification societies (e.g., DNV 2000) indicate that the environment is benign if the following criterion is satisfied for a ship-shaped offshore structure (MacMillan 2001):

$$M_{wb}\gamma_{fi}\gamma_{nc} \leq 1.17M_{wr} + 0.17M_s,$$ (6.3)

where M_{wb} = 100-year return period ($10^{-8.7}$ value) wave-bending moment; M_{wr} = IACS rule wave-induced bending moment (10^{-8} value); M_s = design still-water bending moment; γ_{fi} = partial load coefficient that may be taken as 1.15 for ULS design; and γ_{nc} = nonlinear correction factor that may be taken as 1.1 in sagging and 0.9 in

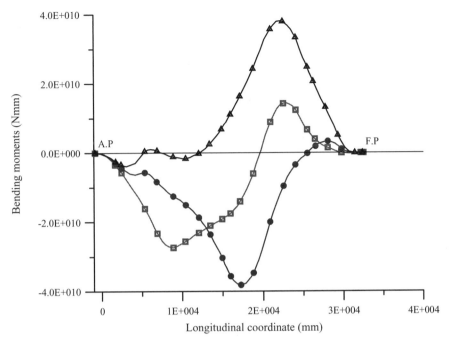

Figure 6.1. Sample variations of still-water bending moment along the vessel length of an FPSO during loading and offloading conditions.

hogging unless otherwise specified or known. In general, wave environments with a significant wave height of less than about 8–9m usually may be considered benign in the same context, that is, in comparison to unrestricted service for trading tankers.

For design, some direct calculations are required for the harsh or hurricane environments. The classification society requirements applicable to trading tanker designs, considering unrestricted service as a baseline as described in Section 3.5 of Chapter 3, may be applied for benign environments. In the latter cases, the wave and environmental loads of the site, for example, would be estimated by multiplying the unrestricted service load by a predefined environmental severity factor (ESF), obtained from correspondence with directly calculated values in past cases. Where applicable, both harsh environments and hurricane environments should be considered applying the ULS requirements. Passing storms should be considered during tow in many regions.

Note that the still-water hull girder bending moment configurations can be significantly changed during offloading and, in fact, can have very different features from those of tankers. Figure 6.1 shows sample calculations of still-water hull girder bending moments for an FPSO during offloading. Direct calculations are essential to determine extreme hull girder bending moments during loading and offloading.

6.4 Structural Component Configuration

Figure 6.2 shows a basic part of a ship-shaped offshore structure. The main types of structural components used for ship-shaped offshore units are stiffened plate panels, but tubular members are typically used for other types of offshore platforms; for

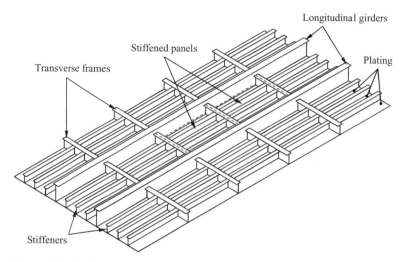

Figure 6.2. Basic part of a ship-shaped offshore structure.

example, jackets and semisubmersibles. Tubular joint-related ULS methods can be found in, for example, NORSOK N004 (1999).

The plate panel is usually reinforced by beam members (stiffeners) in the longitudinal, or transverse, direction. Figure 6.3 shows typical members used for the stiffening of plate panel in ship-shaped offshore structures. To improve the stiffness and strength of plate panels, increasing the stiffener dimensions is usually more efficient than simply increasing the plate thickness.

The stiffened panels are likely to be subjected to large lateral loads or out-of-plane bending. In any event, for lateral support, the stiffened panels are supported by stronger beam members, called plate girders or web frames. The deep web of a plate girder is also often stiffened vertically and/or horizontally. In contrast, box-type support members, which consist of plate panels together with diaphragms, or transverse floors at a relevant spacing, are used for construction of land-based steel bridges.

Although plating primarily sustains in-plane loads, support members resist out-of-plane (lateral) actions and bending. Both girders and frames may be called major support members. For strength analysis of stiffened plate structures, stiffeners, or some support members, together with their associated plating, may be modeled as beams, columns, or beam-columns. This occurs, for example, in a grillage model for

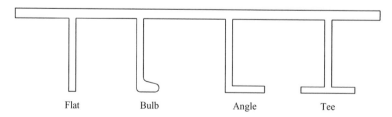

Figure 6.3. Various types of stiffeners used for plate panels.

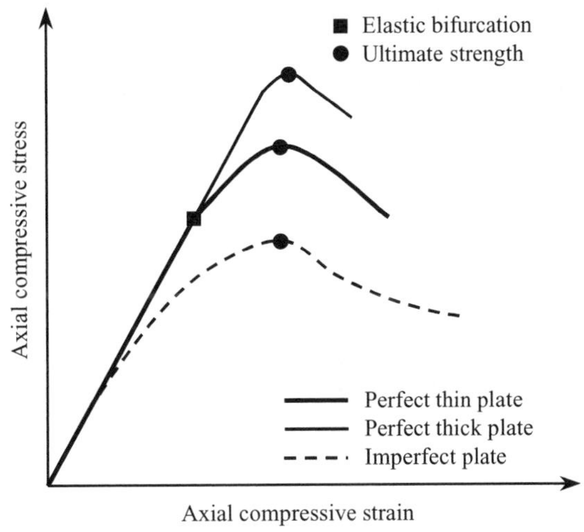

Figure 6.4. A schematic of the ultimate strength behavior of steel plates under predominantly compressive loads.

topsides structure, where the plating between beams and frames may be allowed to buckle by design, but where plate ultimate strength is not exceeded. A similar idealization is also often used for purposes of evaluating the compressive strength of plate and stiffener combinations, wherein plate buckling occurs prior to the failure of stiffener together with an effective width of plating. Note that a design in which a stiffener fails prior to the plating to which it is attached is generally undesirable. Similarly, it is a requirement that major support members (which can be stiffened plate panels) do not fail prior to the plate and/or stiffened plate structure to which they provide support.

6.5 Ultimate Strength of Plates

A number of ULS methods for plates can be found in the literature (Paik and Thayamballi 2003). This section presents some selected and useful methods to determine the characteristic value C_k in Eq. (6.1) for ULS of plates between support members or stiffeners. For a more elaborate description of plate ULS methods, see Paik and Thayamballi (2003).

6.5.1 Fundamentals

When axial tensile loads are predominant, plates fail by gross yielding, meaning that the plate ultimate strength is equal to the yield strength, in this case. Plates can buckle under predominantly compressive loads and reach the ULS by subsequent plastic collapse. The ultimate strength behavior of plates under compression is usually classified into five regimes: (1) prebuckling, (2) buckling, (3) post-buckling, (4) collapse (ultimate strength), and (5) post-collapse. Figure 6.4 shows a schematic of the plate collapse behavior under predominantly axial compressive loads.

In the prebuckling regime, the structural response between loads and displacements is usually linear, and the structural component will be stable. As the predominant compressive stress reaches a critical value, buckling occurs. In contrast to columns, where buckling is usually synonymous with ultimate strength, plates buck-led in the elastic regime still may be stable, in the sense that further loading can be sustained until the ultimate strength is reached, even if the in-plane stiffness significantly decreases after the inception of buckling. Because little residual strength of a plate is retained after buckling occurs in the inelastic regime, inelastic buckling is sometimes considered to be the ULS of the plate.

As the applied loads increase, the plate eventually reaches the ULS, due to expansion of the yielded region. The in-plane stiffness of the collapsed plate numerically takes a "negative" value in the postultimate regime, meaning there is a high degree of instability. A plate with initial imperfections starts to deflect from the very beginning as the compressive loads increase and, therefore, a bifurcation buckling phenomenon does not appear. The ultimate strength of imperfect structures is normally smaller than that of perfect structures.

The ultimate strength behavior of plates usually depends on a variety of influential factors such as geometric and material properties, loading characteristics, initial imperfections (i.e., initial deflections and residual stresses), and boundary conditions. This section presents useful ULS methods for rectangular plates, where a = plate length; b = breadth; t = thickness; E = elastic modulus; σ_Y = matrial stress; and v = Poisson ratio.

6.5.2 Closed-Form Expressions

One of the most practical methods to determine the plate ULS is to use closed-form expressions or design formulations. This section presents some selected formulations, where the average working stresses indicated in Figure 5.4 in Chapter 5 are denoted by $\sigma_x = \sigma_{xav}$, $\sigma_y = \sigma_{yav}$, and $\tau = \tau_{av}$. The lateral pressure distribution is considered to be uniform over the plate surface and is denoted by p.

As we described in Chapter 3, classification society rules have used the critical buckling strength-based design (CBSD) approach. The critical buckling strength is calculated by a plasticity correction of elastic buckling strength. The Johnson–Ostenfeld formula, Eq. (3.2) in Chapter 3, is typically used for such a plasticity correction. The critical buckling strength is approximately the ultimate strength for relatively thick plates without imperfections (e.g., fabrication-related initial imperfections, cutouts); that is, when the following condition is satisfied:

$$\sigma_E \geq p_r \sigma_F, \tag{6.4}$$

where σ_E = elastic plate buckling stress; $\sigma_F = \sigma_Y$ for normal stress and $\sigma_F = \tau_Y = \sigma_Y/\sqrt{3}$ for shear stress; σ_Y = material yield stress; and p_r = coefficient accounting for the plasticity sensitivity, which is typically taken as $p_r = 0.5$ to 0.6.

In the CBSD approach, the critical buckling strength or pseudo-ultimate strength can then be calculated from Eq. (3.1) once the elastic buckling stress is determined. Under single-stress components, the elastic rectangular plate buckling stress can be given by Eq. (5.9) or Eq. (5.11) in Chapter 5, considering the effect of lateral pressure load p.

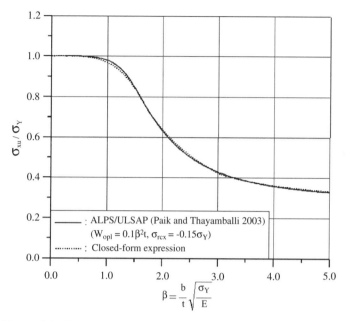

Figure 6.5. The validity of Eq. (6.5) by a comparison with the results of ALPS/ULSAP (2006) for a plate with an average level of initial imperfections (w_{opl} = maximum plate initial deflection; and σ_{rcx} = compressive residual stress in the x direction).

Reliable empirical formulations for plate ULS assessment are available. For $a/b \geq 1$, the ultimate strength of the imperfect plating (i.e., with an average level of initial imperfections), under uniaxial compression in the x direction alone, may be predicted as a function of the plate slenderness ratio β (Paik et al. 2004), as follows:

$$\frac{\sigma_{xu}}{\sigma_Y} = \begin{cases} -0.032\beta^4 + 0.002\beta^2 + 1.0 & \text{for} \quad \beta \leq 1.5 \\ 1.274/\beta & \text{for} \quad 1.5 < \beta \leq 3.0 \\ 1.248/\beta^2 + 0.283 & \text{for} \quad \beta > 3.0, \end{cases} \tag{6.5}$$

where σ_{xu} = ultimate compressive stress in the x direction; $\beta = \frac{b}{t}\sqrt{\frac{\sigma_Y}{E}}$ = plate slenderness ratio; σ_Y = material yield stress; b = plate breadth; t = plate thickness; and E = elastic modulus.

However, the ultimate compressive strength formula for the imperfect plating (i.e., with an average level of initial imperfections), under uniaxial compression in the y direction alone, may be predicted as a function of the plate aspect ratio and the slenderness ratio β (Paik et al. 2004), as follows:

$$\frac{\sigma_{yu}}{\sigma_Y} = \frac{b}{a}\frac{\sigma_{xu}}{\sigma_Y} + \frac{0.475}{\beta^2}\left(1 - \frac{b}{a}\right) \quad \text{for} \quad \frac{a}{b} \geq 1, \tag{6.6}$$

where σ_{yu} = ultimate compressive stress in the y direction; a = plate length; β = as defined in Eq. (6.5); and σ_{xu} = ultimate strength of the plating, under uniaxial compression in the x direction alone, as defined in Eq. (6.5).

Note that Eqs. (6.5) and (6.6) were empirically derived by regression analysis of experimental and numerical solutions. Figures 6.5 and 6.6 confirm the validity of the two empirical formulae, Eqs. (6.5) and (6.6), respectively.

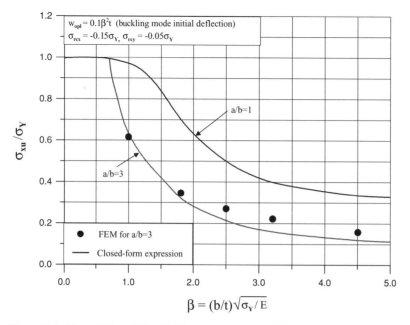

Figure 6.6. The validity of Eq. (6.6) by a comparison with nonlinear finite-element analyses for a plate with an average level of initial imperfections (w_{opl} = maximum plate initial deflection; σ_{rcx} = compressive residual stress in the x direction; and σ_{rcy} = compressive residual stress in the y direction).

An empirical formula for the plate ultimate shear strength may be given as follows (Paik and Thayamballi 2003):

$$\frac{\tau_u}{\tau_Y} = \begin{cases} 1.324\left(\dfrac{\tau_E}{\tau_Y}\right) & \text{for } 0 < \dfrac{\tau_E}{\tau_Y} \le 0.5 \\[2ex] 0.039\left(\dfrac{\tau_E}{\tau_Y}\right)^3 - 0.274\left(\dfrac{\tau_E}{\tau_Y}\right)^2 + 0.676\left(\dfrac{\tau_E}{\tau_Y}\right) + 0.388 & \text{for } 0.5 < \dfrac{\tau_E}{\tau_Y} \le 2.0 \\[2ex] 0.956 & \text{for } \dfrac{\tau_E}{\tau_Y} > 2.0, \end{cases}$$

$$(6.7)$$

where τ_u = ultimate shear strength of plates; τ_Y = shear yield stress, as defined in Eq. (6.4); and τ_E = elastic shear buckling stress, which is given in Eq. (5.9) in Chapter 5, that is, by $\tau_E = k_\tau \frac{\pi^2 E}{12(1-\nu^2)}\left(\frac{t}{b}\right)^2$, with $k_\tau \approx 4\left(\frac{b}{a}\right)^2 + 5.34$ for $\frac{a}{b} \ge 1$ or $k_\tau \approx 5.34\left(\frac{b}{a}\right)^2 + 4.0$ for $\frac{a}{b} < 1$.

Note that Eq. (6.7) implicitly considers that the plate includes an average level of initial imperfections (i.e., post-weld initial deflection and residual stresses). Figure 6.7 plots Eq. (6.7) by comparing it to relevant nonlinear finite-element solutions for steel plates with initial imperfections. In Figure 6.7, the dotted line indicates the critical shear buckling stress that is obtained by the Johnson–Ostenfeld plasticity correction of the elastic buckling stress, and the solid line indicates the result of Eq. (6.7).

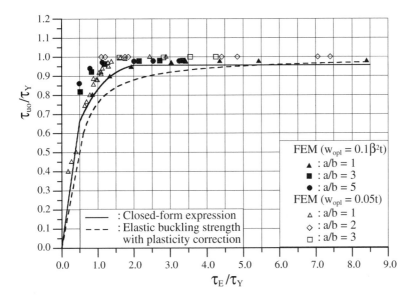

Figure 6.7. The ultimate strength versus the elastic bifurcation buckling stress of a plate under edge shear (Paik and Thayamballi 2003) (w_{opl} = maximum plate initial deflection).

The static collapse load, for rectangular plates subject to a uniformly distributed lateral pressure, can be found between the lower and upper bounds, as follows (Jones 1975):

$$\frac{8M_P}{b^2}\left(1+\alpha+\alpha^2\right) \le p_c \le \frac{24M_P}{b^2}\frac{1}{\left(\sqrt{3+\alpha^2}-\alpha\right)^2}, \text{ for simply supported plates,}$$

(6.8a)

$$\frac{16M_P}{b^2}\left(1+\alpha^2\right) \le p_c \le \frac{48M_P}{b^2}\frac{1}{\left(\sqrt{3+\alpha^2}-\alpha\right)^2}, \text{ for clamped plates,}$$ (6.8b)

where $M_P = \sigma_Y t^2/4$ is the plastic bending moment per unit of breadth that the plate cross section may carry; p_c = plastic collapse strength, which may be approximated as the ultimate strength p_u; and $\alpha = b/a$.

Eqs. (6.8a) and (6.8b) have been derived using the upper- and lower-bound theorems of plasticity for plates made of rigid plastic material and assuming that the plating obeys the Tresca yield criterion. The effect of shear on the yielding has been neglected because the plates are considered relatively thin.

Of interest, Eqs. (6.8a) and (6.8b) can be simplified for a square plate because $\alpha = 1$, as follows:

$$\frac{24M_P}{b^2} \le p_c \le \frac{24M_P}{b^2}, \qquad \text{for simply supported plates,}$$ (6.8c)

$$\frac{32M_P}{b^2} \le p_c \le \frac{48M_P}{b^2}, \text{ for clamped plates.}$$ (6.8d)

Eqs. (6.8c) and (6.8d) show that the lower and upper limits can coincide for simply supported plates, but the lower and upper limits can differ significantly; that is, they are in the ratio 2:3 for clamped plates. Fox (1974) has shown that the collapse pressure load equals $42.85 M_P/b^2$ for clamped plates.

In this regard, an upper limit p_{cr} of the ultimate lateral pressure load for simply-supported plates may be given pessimistically as follows:

$$p_{cr} = \frac{6t^2 \sigma_Y}{b^2} \frac{1}{\left(\sqrt{3 + \alpha^2} - \alpha\right)^2}. \tag{6.8e}$$

In predicting the plate ultimate lateral pressure loads, the computed value therefore should not generally be greater than the upper limit p_{cr} of Eq. (6.8e). Note that the rigid plastic theory formulae given herein do not account for membrane effects and, hence, predict the critical lateral pressure pessimistically. For practical design purposes, however, the collapse pressure load p_c is often regarded as the ultimate strength of plates under lateral pressure loads, but, again, it is desirable to account for the membrane effect. We describe methods to account for the membrane effect in Sections 6.5.3–6.5.5.

Under combined edge shear and lateral pressure, the plate ultimate strength interaction relationship (Paik and Thayamballi 2003) may be given by

$$\left(\frac{\tau_{av}}{\tau_u}\right)^{1.5} + \left(\frac{p}{p_u}\right)^{1.2} = 1, \tag{6.9}$$

where τ_u = plate ultimate strength under edge shear alone as defined in Eq. (6.7); and p_u = plate ultimate strength under lateral pressure alone as defined in Eq. (6.8) as $p_u \approx p_c$ but with $p_u \leq p_{cr}$.

Although various types of the plate ultimate strength interaction relationships with multiple working stress components have been suggested in the literature (Paik and Thayamballi 2003), most of them may be generalized to the following form (taking a negative sign for compressive stress and a positive sign for tensile stress):

$$\left(\frac{\sigma_{xav}}{\sigma_{xu}}\right)^{c_1} + \alpha \left(\frac{\sigma_{xav}}{\sigma_{xu}}\right)\left(\frac{\sigma_{yav}}{\sigma_{yu}}\right) + \left(\frac{\sigma_{yav}}{\sigma_{yu}}\right)^{c_2} + \left(\frac{\tau_{av}}{\tau_u}\right)^{c_3} = 1, \tag{6.10a}$$

where $\sigma_{xu}, \sigma_{yu}, \tau_u$ = ultimate strengths for axial stress σ_{xav} in the x direction, axial stress σ_{yav} in the y direction, and edge shear τ_{av}, respectively; and α, c_1, c_2, c_3 = coefficients. For axial tensions in the x or y direction, $\sigma_{xu} = \sigma_Y$, for σ_{xav}, and $\sigma_{yu} = \sigma_Y$, for σ_{yav}, will be used.

In Eq. (6.10a), σ_{xu}, σ_{yu}, and τ_u must take into account the effect of lateral pressure, when lateral pressure is a constant value and is applied to σ_{xu}, σ_{yu}, and τ_u simultaneously. In this case, σ_{xu} and σ_{yu} may be approximated by using Eq. (5.11) in Chapter 5 together with Eq. (3.2) in Chapter 3, and τ_u can be determined as a solution of

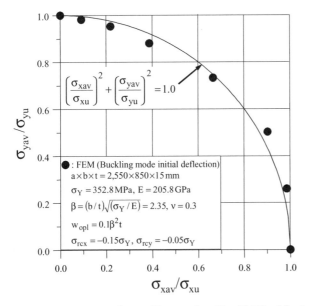

Figure 6.8. A comparison of interaction Eq. (6.11) with plate-compressive ultimate strength data obtained by finite-element analyses(w_{opl}, σ_{rcx}, σ_{rcy} = as defined in Figure 6.6).

Eq. (6.9) with regard to τ_{av} with known lateral pressure loads p. Also, the coefficients may be taken as follows:

$$c_1 = c_2 = c_3 = 2, \alpha = 0, \quad \text{when both } \sigma_{xav} \text{ and } \sigma_{yav} \text{ are compressive (negative)},$$

$$\tag{6.10b}$$

$$c_1 = c_2 = c_3 = 2, \alpha = -1, \quad \text{when either } \sigma_{xav} \text{ or } \sigma_{yav}, \text{ or both, are tensile (positive)}.$$

$$\tag{6.10c}$$

Under biaxial compressive loads, Eq. (6.10) simplifies to

$$\left(\frac{\sigma_{xav}}{\sigma_{xu}}\right)^2 + \left(\frac{\sigma_{yav}}{\sigma_{yu}}\right)^2 = 1. \tag{6.11}$$

Figure 6.8 compares the ultimate strength interaction relationships for imperfect plating under biaxial compressive loads, as given by Eq. (6.11), with verification results obtained by nonlinear finite-element analysis in a set of cases.

6.5.3 Analytical Methods

Analytical methods are accurate and efficient tools for the plate ultimate strength calculations. In these methods, the membrane stresses inside the plate are computed typically by solving the nonlinear governing differential equations of elastic large deflection plate theory, and it is assumed that the plate will collapse if the membrane stress reaches a critical value (e.g., the yield stress) or when a relevant criterion, as a function of membrane stresses, is satisfied.

The elastic large deflection behavior of plates with initial deflections is governed by two differential equations, one representing the equilibrium condition and the other representing the compatibility condition (Marguerre 1938), as follows:

$$\Phi = D\left(\frac{\partial^4 w}{\partial x^4} + 2\frac{\partial^4 w}{\partial x^2 \partial y^2} + \frac{\partial^4 w}{\partial y^4}\right)$$
$$- t\left[\frac{\partial^2 F}{\partial y^2}\frac{\partial^2(w+w_o)}{\partial x^2} + \frac{\partial^2 F}{\partial x^2}\frac{\partial^2(w+w_o)}{\partial y^2} - 2\frac{\partial^2 F}{\partial x \partial y}\frac{\partial^2(w+w_o)}{\partial x \partial y} + \frac{p}{t}\right] = 0, \quad (6.12a)$$

$$\frac{\partial^4 F}{\partial x^4} + 2\frac{\partial^4 F}{\partial x^2 \partial y^2} + \frac{\partial^4 F}{\partial y^4}$$
$$- E\left[\left(\frac{\partial^2 w}{\partial y \partial x}\right)^2 - \frac{\partial^2 w}{\partial x^2}\frac{\partial^2 w}{\partial y^2} + 2\frac{\partial^2 w_o}{\partial x \partial y}\frac{\partial^2 w}{\partial x \partial y} - \frac{\partial^2 w_o}{\partial x^2}\frac{\partial^2 w}{\partial y^2} - \frac{\partial^2 w}{\partial x^2}\frac{\partial^2 w_o}{\partial y^2}\right] = 0,$$

$$(6.12b)$$

where w = added deflection (i.e., deflection due to applied actions); w_o = initial deflection; t = plate thickness; E = material elastic modulus; ν = Poisson ratio; p = lateral pressure; $D = \frac{Et^3}{12(1-\nu^2)}$ = plate bending rigidity; and F = Airy stress function.

By using the Airy stress function F, the stress components at a certain location inside the plate may be calculated as follows:

$$\sigma_x = \frac{\partial^2 F}{\partial y^2} - \frac{Ez}{1-\nu^2}\left(\frac{\partial^2 w}{\partial x^2} + \nu\frac{\partial^2 w}{\partial y^2}\right), \quad (6.13a)$$

$$\sigma_y = \frac{\partial^2 F}{\partial x^2} - \frac{Ez}{1-\nu^2}\left(\frac{\partial^2 w}{\partial y^2} + \nu\frac{\partial^2 w}{\partial x^2}\right), \quad (6.13b)$$

$$\tau = \tau_{xy} = -\frac{\partial^2 F}{\partial x \partial y} - \frac{Ez}{2(1+\nu)}\frac{\partial^2 w}{\partial x \partial y}, \quad (6.13c)$$

where σ_x, σ_y = normal stress components in the x and y directions; τ = shear stress; and z = coordinate in the plate thickness direction with $z = 0$ in the midthickness.

Also, the corresponding strain components at a certain location inside the plate are given by

$$\varepsilon_x = \frac{\partial u}{\partial x} + \frac{1}{2}\left(\frac{\partial w}{\partial x}\right)^2 + \frac{\partial w}{\partial x}\frac{\partial w_o}{\partial x} - z\frac{\partial^2 w}{\partial x^2}, \quad (6.14a)$$

$$\varepsilon_y = \frac{\partial v}{\partial y} + \frac{1}{2}\left(\frac{\partial w}{\partial y}\right)^2 + \frac{\partial w}{\partial y}\frac{\partial w_o}{\partial y} - z\frac{\partial^2 w}{\partial y^2}, \quad (6.14b)$$

$$\gamma_{xy} = \frac{\partial u}{\partial y} + \frac{\partial v}{\partial x} + \frac{\partial w}{\partial x}\frac{\partial w}{\partial y} + \frac{\partial w_o}{\partial x}\frac{\partial w}{\partial y} + \frac{\partial w}{\partial x}\frac{\partial w_o}{\partial y} - 2z\frac{\partial^2 w}{\partial x \partial y}, \quad (6.14c)$$

where ε_x, ε_y, γ_{xy} = strain components corresponding to stress components σ_x, σ_y, τ, respectively; and u, v = axial displacements in the x or y direction.

Each strain component in Eq. (6.14) can be expressed as a function of stress components, as follows:

$$\varepsilon_x = \frac{1}{E}(\sigma_x - \nu\sigma_y), \tag{6.15a}$$

$$\varepsilon_y = \frac{1}{E}(\sigma_y - \nu\sigma_x), \tag{6.15b}$$

$$\gamma_{xy} = \frac{2(1+\nu)}{E}\tau_{xy}. \tag{6.15c}$$

The governing differential equations of plate behavior, that is, Eq. (6.12), can be solved directly under prescribed loading and boundary conditions; for example, using the Galerkin method (Paik and Thayamballi 2003). In solving the nonlinear governing differential equations, Eqs. (6.12a) and (6.12b), the added deflection w and initial deflection w_o can be assumed as follows:

$$w = \sum_{m=1}\sum_{n=1} A_{mn}\, f_m(x)g_n(y), \tag{6.16a}$$

$$w_o = \sum_{m=1}\sum_{n=1} A_{omn}\, f_m(x)g_n(y), \tag{6.16b}$$

where $f_m(x)$ and $g_n(y)$ are functions that satisfy the boundary conditions for the plate. A_{mn} and A_{omn} are unknown *added deflection amplitudes* and known *initial deflection amplitudes*, respectively.

Upon substituting Eq. (6.16) into Eq. (6.12) and solving for the stress function F, the particular solution F_P can be expressed as

$$F_P = \sum_{r=1}\sum_{s=1} K_{rs}\, p_r(x)\, q_s(y), \tag{6.17}$$

where the coefficients K_{rs} will be second-order functions with regard to the unknown deflection amplitudes A_{mn}.

Including the applied loading, the complete stress function F can be given by

$$F = F_H + \sum_{r=1}\sum_{s=1} K_{rs}\, p_r(x)\, q_s(y), \tag{6.18}$$

where F_H is the homogeneous solution of the stress function that satisfies the applied load condition.

To compute the unknown amplitudes A_{mn}, one may use the Galerkin method for the equilibrium equation, Eq. (6.12a), resulting in the following equation:

$$\iiint \Phi f_r(x)\, g_s(y)\, dvol = 0, r = 1, 2, 3, \ldots, s = 1, 2, 3, \ldots. \tag{6.19}$$

Substituting Eqs. (6.16) and (6.18) into Eq. (6.19) and performing the integration over the whole volume of the plate, a set of third-order simultaneous

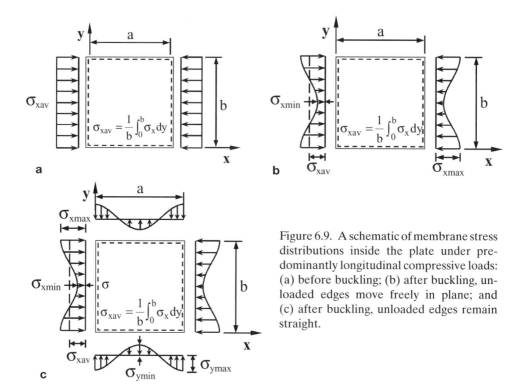

Figure 6.9. A schematic of membrane stress distributions inside the plate under predominantly longitudinal compressive loads: (a) before buckling; (b) after buckling, unloaded edges move freely in plane; and (c) after buckling, unloaded edges remain straight.

equations, with regard to the unknown amplitudes A_{mn}, will be obtained. The nonlinear stress distribution inside the plate can then be obtained from Eq. (6.13) with A_{mn}.

Figure 6.9 shows a typical example of the axial membrane stress distribution inside a plate under predominantly longitudinal compressive loading before and after buckling occurs. It is important to realize that the membrane stress distribution in the loading x direction can become nonuniform as the plate deflects from many causes, including buckling, initial deflection, and lateral pressure loading. The membrane stress distribution in the y direction also becomes nonuniform as long as the unloaded plate edges remain straight. However, no membrane stresses will develop in the y direction if the unloaded plate edges move freely in plane as long as no axial loading is applied in the y direction.

As is apparent from Figure 6.9, the maximum compressive membrane stresses, $\sigma_{x\,max}$ and $\sigma_{y\,max}$, are developed around the plate edges that remain straight, but the minimum membrane stresses, $\sigma_{x\,min}$ and $\sigma_{y\,min}$, occur in the middle of the plate where a membrane tension field is formed by the plate deflection because the plate edges remain straight.

The location of the maximum compressive stresses depends on the residual stresses. If no residual stresses exist, the maximum compressive stresses will develop along the edges. However, when there are residual stresses, the maximum compressive stresses may be located inside the plate at the limits that are the tensile residual stress block breadths from the plate edges (Paik and Thayamballi 2003).

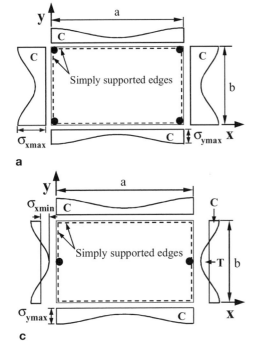

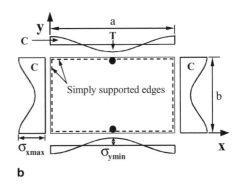

Figure 6.10. Three possible locations for the initial plastic yield at the plate edges under combined loads: (a) plasticity at corners; (b) plasticity at longitudinal edges; and (c) plasticity at transverse edges. (•) expected yielding locations; T: tension; C: compression.

With an increase in the plate deflection, the upper and/or lower fibers inside the middle of the plate will initially yield by bending. However, as long as it is possible to redistribute the applied loads to the straight plate boundaries by membrane action, the plate will not collapse. Collapse will occur when the most stressed boundary locations yield because the plate cannot keep the boundaries straight any further. This results in a rapid increase of lateral plate deflection.

Because of the nature of combined membrane axial stresses in the x and y directions, three possible locations for initial yield at edges are generally considered: plate corners, longitudinal edges, and transverse edges; see Figure 6.10. The stress status for the two edge locations, that is, at each longitudinal or transverse edge, can be expected to be the same as long as the longitudinal or transverse axial stresses are uniformly applied, that is, without in-plane bending. Depending on the predominant half-wave mode in the long direction, the location of the possible plasticity can vary at the long edges because the location of the minimum membrane stresses can be different; however, it is always the mid-edges in the short direction.

Yielding can be assessed by using the Mises–Henckey yield criterion. Therefore, the three resulting ultimate strength criteria for the most probable yield locations may be found as follows:

(1) Yielding at corners:

$$\left(\frac{\sigma_{x\,max}}{\sigma_Y}\right)^2 - \left(\frac{\sigma_{x\,max}}{\sigma_Y}\right)\left(\frac{\sigma_{y\,max}}{\sigma_Y}\right) + \left(\frac{\sigma_{y\,max}}{\sigma_Y}\right)^2 + \left(\frac{\tau_{av}}{\tau_Y}\right)^2 = 1. \qquad (6.20a)$$

(2) Yielding at longitudinal edges:

$$\left(\frac{\sigma_{x\,max}}{\sigma_Y}\right)^2 - \left(\frac{\sigma_{x\,max}}{\sigma_Y}\right)\left(\frac{\sigma_{y\,min}}{\sigma_Y}\right) + \left(\frac{\sigma_{y\,min}}{\sigma_Y}\right)^2 + \left(\frac{\tau_{av}}{\tau_Y}\right)^2 = 1. \qquad (6.20b)$$

(3) Yielding at transverse edges:

$$\left(\frac{\sigma_{x\,min}}{\sigma_Y}\right)^2 - \left(\frac{\sigma_{x\,min}}{\sigma_Y}\right)\left(\frac{\sigma_{y\,max}}{\sigma_Y}\right) + \left(\frac{\sigma_{y\,max}}{\sigma_Y}\right)^2 + \left(\frac{\tau_{av}}{\tau_Y}\right)^2 = 1. \qquad (6.20c)$$

Because the maximum or minimum membrane stresses of plates are expressed as functions of applied stress components, as well as initial deflections and welding residual stresses, Eqs. (6.20a)–(6.20c) are nonlinear functions. The smallest value among the solutions of these functions with regard to applied stress components will become the plate ultimate strength. This theory has been added to ALPS/ULSAP (2006).

6.5.4 Semianalytical Methods

In these methods, the geometrical nonlinearity-related behavior of plates – that is, elastic large deflection behavior – is analyzed by direct solutions of nonlinear governing differential equations and material nonlinearities (i.e., plasticity) are evaluated by numerical techniques to account for the effect of progressive plasticity expansion with increase in the applied loads.

Paik et al. (2001) developed a semianalytical method using this concept. For this purpose, it is assumed that the added deflection w_i and stress function F_i at the end of the ith step of load increment are calculated by

$$w_i = w_{i-1} + \Delta w, \qquad (6.21a)$$

$$F_i = F_{i-1} + \Delta F, \qquad (6.21b)$$

where Δw and ΔF are the increments of deflection or stress functions, respectively, where the prefix Δ indicates the increment of the variable.

The incremental forms of governing differential equations of large deflection plate theory, Eqs. (6.12a) and (6.12b), are derived as follows:

$$\Delta\Phi = D\left(\frac{\partial^4 \Delta w}{\partial x^4} + 2\frac{\partial^4 \Delta w}{\partial x^2 \partial y^2} + \frac{\partial^4 \Delta w}{\partial y^4}\right)$$

$$-t\left[\frac{\partial^2 F_{i-1}}{\partial y^2}\frac{\partial^2 \Delta w}{\partial x^2} + \frac{\partial^2 \Delta F}{\partial y^2}\frac{\partial^2(w_{i-1} + w_o)}{\partial x^2} + \frac{\partial^2 F_{i-1}}{\partial x^2}\frac{\partial^2 \Delta w}{\partial y^2}\right.$$

$$\left.+ \frac{\partial^2 \Delta F}{\partial x^2}\frac{\partial^2(w_{i-1} + w_o)}{\partial y^2} - 2\frac{\partial^2 F_{i-1}}{\partial x \partial y}\frac{\partial^2 \Delta w}{\partial x \partial y} - 2\frac{\partial^2 \Delta F}{\partial x \partial y}\frac{\partial^2(w_{i-1} + w_o)}{\partial x \partial y} + \frac{\Delta p}{t}\right] = 0,$$

$$(6.22a)$$

$$\frac{\partial^4 \Delta F}{\partial x^4} + 2\frac{\partial^4 \Delta F}{\partial x^2 \partial y^2} + \frac{\partial^4 \Delta F}{\partial y^4}$$
$$- E\left[2\frac{\partial^2(w_{i-1} + w_o)}{\partial x \partial y}\frac{\partial^2 \Delta w}{\partial x \partial y} - \frac{\partial^2(w_{i-1} + w_o)}{\partial x^2}\frac{\partial^2 \Delta w}{\partial y^2} - \frac{\partial^2 \Delta w}{\partial x^2}\frac{\partial^2(w_{i-1} + w_o)}{\partial y^2}\right] = 0,$$

(6.22b)

where the terms of very small quantities, with an order higher than the second-order of the increments Δw and ΔF, have been neglected.

At the end of the $(i-1)$th step of load increment, the deflection w_{i-1} and the stress function F_{i-1} are obtained by

$$w_{i-1} = \sum_{m=1}\sum_{n=1} A_{mn}^{i-1} f_m(x)g_n(y),$$

(6.23a)

$$F_{i-1} = F_H^{i-1} + \sum_{i=1}\sum_{j=1} K_{ij}^{i-1} p_i(x)q_j(y),$$

(6.23b)

where A_{mn}^{i-1} and K_{ij}^{i-1} are the known coefficients and F_H^{i-1} is a homogeneous solution for the stress function satisfying the applied loading condition. The welding-induced residual stresses can be included in the stress function F_H^{i-1} as initial stress terms.

The deflection increment Δw associated with the load increment at the ith step can be assumed as follows:

$$\Delta w = \sum_{k=1}\sum_{l=1} \Delta A_{kl} f_k(x)g_l(y),$$

(6.24)

where ΔA_{kl} is the increment of unknown added deflection amplitude.

Substituting Eqs. (2.23) and (6.24) into Eq. (6.22), the stress function increment ΔF can be obtained by

$$\Delta F = \Delta F_h + \sum_{i=1}\sum_{j=1} \Delta K_{ij} p_i(x)q_j(y),$$

(6.25)

where ΔK_{ij} are linear (i.e., first-order) functions in the unknown coefficients ΔA_{kl}. ΔF_H is a homogeneous solution for the stress function increment that satisfies the applied load condition.

To compute the unknown coefficients ΔA_{kl}, the Galerkin method can be applied to Eq. (6.22a):

$$\iiint \Delta \Phi f_r(x)g_s(y)\,dvol = 0, r = 1, 2, 3, \ldots, s = 1, 2, 3, \ldots.$$

(6.26)

By substituting Eqs. (6.23), (6.24), and (6.25) into Eq. (6.26) and performing the integration over the entire volume of the plate, a set of linear simultaneous equations for the unknown coefficients ΔA_{kl} will be obtained. Solving these simultaneous linear equations is not difficult. Having obtained ΔA_{kl}, we can calculate Δw from Eq. (6.24), ΔF from Eq. (6.25), $w_i = (w_{i-1} + \Delta w)$ from Eq. (6.23a), and $F_i = (F_{i-1} + \Delta F)$ from Eq. (6.23b).

Thus far, the differential equations governing the elastic large deflection behavior of plates have been formulated and are solved analytically. But the effects of

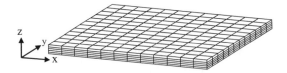

Figure 6.11. Example subdivision of plate mesh regions used for treatment of plasticity. (Note that geometric nonlinearity is dealt with analytically.)

plasticity have not been included. It is not straightforward to formulate governing differential equations representing both geometric and material nonlinearities simultaneously, although it is not impossible. A major source of difficulty is that an analytical treatment of plasticity with increase in the applied loads is very difficult. Even if such treatment were possible, it would not be straightforward to solve the resulting equations analytically. An easier alternative is to deal with progress of the plasticity numerically.

In the present method, the progress of plasticity, with an increase in the applied loads, is treated by a numerical approach. For this purpose, the plate is subdivided into a number of mesh regions in the three directions similar to the conventional finite-element method, as depicted by Figure 6.11. The average membrane stress components for each mesh region can be calculated at every step of load increment. Yielding for each mesh region is checked by using the relevant yield criterion, that is, the Mises–Henckey yield condition:

$$\sigma_x^2 - \sigma_x \sigma_y + \sigma_y^2 + 3\tau^2 = \sigma_Y^2. \tag{6.27}$$

As the applied loads increase, the stiffness matrices for the plate are recalculated. The fiber associated with yielded regions is removed and not included in the plate stiffness equation. By repeating the procedure and increasing the applied loads, the elastic/plastic large deflection behavior for the plate can be obtained. This theory has been added to ALPS/SPINE (2006).

6.5.5 Nonlinear Finite-Element Methods

Today, it is considered that the nonlinear finite-element method is the most powerful tool to investigate the elastic/plastic large deflection response characteristics of plates, as long as modeling, in terms of material, geometry, loading condition, boundary condition, and other influential parameters, is adequate.

For practical design purposes, the elastic/perfectly plastic material model, that is, without considering the strain-hardening effect, is typically applied for finite-element analyses. Figure 6.12 shows the stress–strain curves of steel material with varying the strain-hardening characteristics, while keeping the yield stress identical. Figure 6.13 shows the effect of strain hardening on the elastic/plastic large deflection behavior of the considered plate under uniaxial compression in the x direction, as obtained by the nonlinear finite-element method. Because of the strain-hardening effect, the plate ultimate compressive strength can be larger than that obtained by neglecting it. For pessimistic strength assessment of steel plates, however, an elastic/perfectly plastic material model, that is, without strain hardening, may be considered enough. But note that this material model may not be adequate for accidental limit-state design purposes, as described in Chapter 8 of this book. A more realistic material

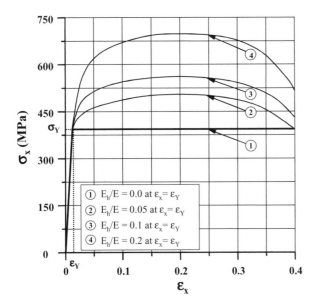

Figure 6.12. The stress–strain curves of a high-tensile steel with varying strain-hardening characteristics (E_h = tangent modulus of stress–strain relation after yielding).

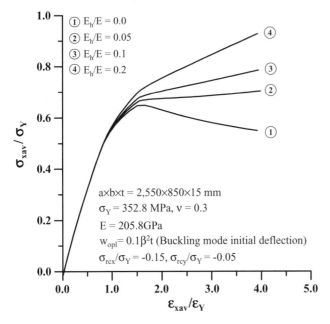

Figure 6.13. The effect of strain hardening on the elastic/plastic large deflection behavior of a plate under uniaxial compressive loads as obtained by the nonlinear finite-element analysis using the stress–strain curves defined in Figure 6.12 (E_h = tangent modulus of stress–strain relation after yielding).

model, taking into account the strain-hardening and strain-softening effects, must be considered in the accidental limit-state design. Also, selecting a fine-enough mesh is of importance in nonlinear finite-element analyses. For this purpose, a convergence study is usually required, varying the mesh size.

6.5.6 Illustrative Examples

As an illustrative example, a simply supported rectangular plate under combined biaxial compression, or under combined longitudinal axial compression and edge shear, is now considered. The plate dimensions are: a (length) = 2,400mm; b (breadth) = 800mm; t (thickness) = 20mm; σ_Y (yield stress) = 235 N/mm^2; E (elastic modulus) = 205,800 N/mm^2; and ν (Poisson ratio) = 0.3.

Figure 6.14 shows a comparison of ALPS/ULSAP calculations with nonlinear finite-element analysis and selected classification society rules formulae solutions. In the case of the classification society rules, initial imperfections (e.g., initial deflection, welding residual stresses) are implicit in the formulae, presumably also at an average level, although this cannot be said with certainty. ALPS/ULSAP calculations and the nonlinear finite-element method can deal with initial imperfections as parameters of influence. For the ALPS/ULSAP (2006) and DNV PULS (Steen et al. 2004) calculations, the maximum plate initial deflection $w_{opl}/t = 0.1\beta^2$ ($w_{opl} = 3.66$mm) with a buckling mode initial deflection shape was assumed, where β = as defined in Eq. (6.5), and welding residual stresses were not considered to exist. Lloyd's Register (LR) ship rules (SR) and new ship rules (NSR) and American Bureau of Shipping (ABS) rules are also compared.

Figure 6.14 shows that ALPS/ULSAP gives reasonably accurate predictions of the plate ULS compared with more refined finite-element-analysis solutions. The classification design formulae studied appear to give optimistic results when transverse axial compression is predominant and provide reasonably accurate solutions when longitudinal axial compression is predominant. PULS calculations appear to be overestimations when compared with ALPS/ULSAP solutions and more refined nonlinear finite-element analysis for the biaxial compressive loading cases. Both PULS and ALPS appear to be in very good agreement with the more refined nonlinear finite-element analysis for combined longitudinal axial compression and edge shear loading cases.

6.6 Ultimate Strength of Stiffened Plate Structures

In this section, various methods are presented to determine the characteristic value C_k in Eq. (6.1) for ULS of stiffened panels or grillages under combined in-plane and lateral pressure loads, as shown in Figure 6.15. Although more elaborate descriptions of these stiffened panel ULS methods may be found in Paik and Thayamballi (2003), this section presents only a summary.

6.6.1 Fundamentals

When subjected to predominantly axial tension, a stiffened panel may fail by gross yielding. On the other hand, a stiffened panel under predominantly compressive actions may show a variety of failure modes until the ultimate strength is reached.

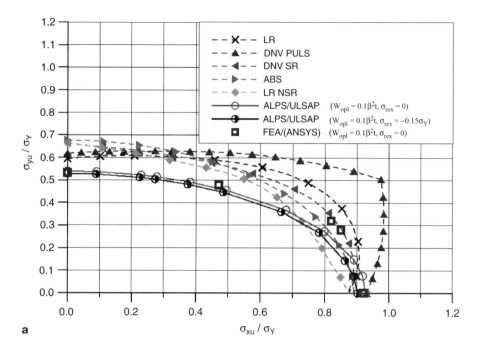

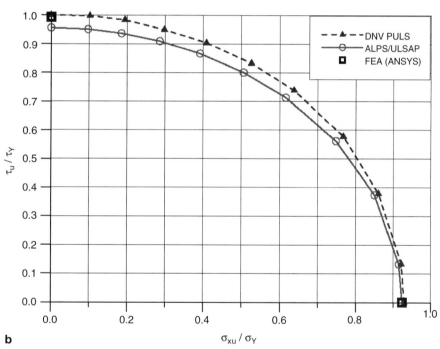

Figure 6.14. ULS interaction curves for a simply supported steel plate under combined biaxial compression or under combined longitudinal compression and edge shear, as obtained by design class formulae, DNV PULS, ALPS/ULSAP, and nonlinear finite-element analysis (ANSYS 2005): (a) under biaxial compression σ_x and σ_y; (b) under longitudinal axial compression σ_x and edge shear τ.

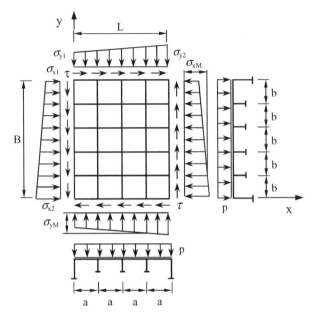

Figure 6.15. A stiffened plate structure under combined in-plane and out-of-plane loads.

The primary modes of overall failure for a stiffened plate structure can be classified into the following six modes (Paik and Thayamballi 2003):

- Mode I: overall collapse of plating and stiffeners as a unit
  - Mode I-1: Mode I for uniaxially stiffened panels; see Figure 6.16(a)
  - Mode I-2: Mode I for cross-stiffened panels; see Figure 6.16(b)
- Mode II: biaxial compressive collapse; see Figure 6.17
- Mode III: beam-column type collapse; see Figure 6.18(a)
- Mode IV: local buckling of stiffener web; see Figure 6.18(b)
- Mode V: tripping of stiffener; see Figure 6.19
- Mode VI: gross yielding

Mode I typically represents the collapse pattern when the stiffeners are relatively weak. In this case, the stiffeners can buckle together with plating as a unit; the overall buckling behavior is perhaps initially elastic. The stiffened panel can normally

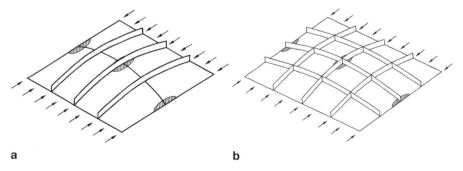

a   b

Figure 6.16. (a) Mode I-1: overall collapse of a uniaxially stiffened panel (shaded area represents yielded region). (b) Mode I-2: overall collapse of a cross-stiffened panel (shaded area represents yielded region).

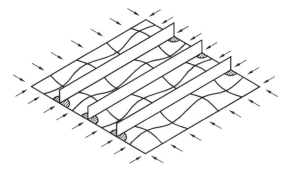

Figure 6.17. Mode II: biaxial compressive collapse (shaded area represents yielded region).

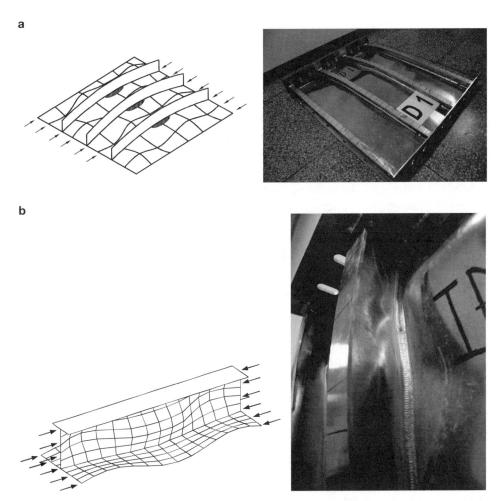

Figure 6.18. (a) Mode III: beam-column-type collapse (shaded area represents yielded region). (b) Mode IV: collapse by local buckling of stiffener web.

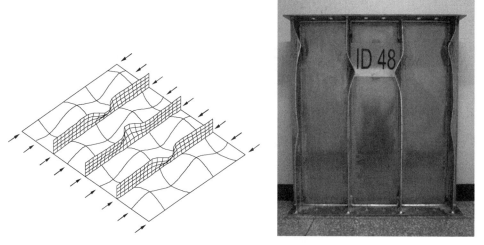

Figure 6.19. Mode V: collapse by inelastic tripping of stiffener.

sustain further loading even after overall buckling in the elastic regime occurs. The ultimate strength is reached eventually by formation of a large yield region inside the panel and/or along the panel edges. In Mode I, the collapse behavior of a uniaxially stiffened panel termed Mode I-1 is slightly different from that of a cross-stiffened panel termed Mode I-2. Mode I-1 is, in fact, initiated by the beam-column-type failure, but Mode I-2 failure resembles that of an "orthotropic plate."

Mode II represents the collapse pattern where the panel collapses by yielding along the plate-stiffener intersection at panel edges, with no stiffener failure. This type of collapse can be important, in some cases, when the panel is subjected to biaxial compressive loads and/or when the plating is stocky.

Mode III indicates a failure pattern in which the ultimate strength is reached by the yielding of the plate-stiffener combination at midspan. Mode III failure typically occurs when the dimensions of the stiffeners are of intermediate properties, that is, neither weak nor very strong.

Modes IV and V typically arise from the stiffener-induced failure, when the ratio of stiffener web height to stiffener web thickness is large, and/or when the type of the stiffener flange is inadequate to remain straight so that the stiffener web buckles or twists sideways. Mode V can occur when the ultimate strength is reached subsequent to the lateral-torsional buckling (also called "tripping") of stiffener. Mode IV represents a failure pattern in which the panel collapses by local compressive buckling of the stiffener web.

Mode VI typically takes place when the panel slenderness is relatively small or the panel is relatively stocky, and/or when the panel is predominantly subjected to the axial tensile loading so that neither local nor overall buckling occurs until the panel cross section yields entirely.

Although Figures 6.16–6.19 illustrate each collapse pattern separately and it is possible to design structures that exhibit such distinct behavior, it is also possible that some collapse modes may interact in real cases and occur simultaneously. For practical design purposes, however, we usually consider that the collapse of stiffened panels occurs at the lowest value among the various ultimate loads calculated, considering each of the collapse patterns separately.

6.6.2 Closed-Form Expressions

In the literature, closed-form expressions for the ULS of a stiffened panel for the various collapse modes discussed in Section 6.6.1 are available. In the case of problems of multiple-load components, the relevant relationship for ultimate strength interaction is established using single-load component results as the starting point.

For all types of collapse modes noted (i.e., Modes I to VI), the ultimate strength interaction equations for a stiffened plate structure under multiple stress components (i.e., axial loads in the longitudinal (x) and transverse (y) directions, edge shear, in-plane bending moment in the x and y directions, and lateral pressure) will have similar expressions but with different single ultimate strength components, depending on the load type and the collapse mode, as follows:

$$\left(\frac{\sigma_{xM}}{\sigma_{xu}}\right)^{c_1} - \alpha\left(\frac{\sigma_{xM}}{\sigma_{xu}}\right)\left(\frac{\sigma_{yM}}{\sigma_{yu}}\right) + \left(\frac{\sigma_{yM}}{\sigma_{yu}}\right)^{c_2} + \left(\frac{\tau}{\tau_u}\right)^{c_3} = 1, \tag{6.28}$$

where σ_{xM}, σ_{yM}, τ = references to working stresses obtained by action-effect analysis as defined in Figure 6.15; $\alpha = 0$, when both σ_{xM} and σ_{yM} are compressive (negative), and $\alpha = 1$, when either σ_{xM} or σ_{yM} or both are tensile (positive); c_1, c_2, c_3 = coefficients, which may be taken as $c_1 = c_2 = c_3 = 2$; σ_{xu} = ultimate stress of the panel under axial load in the x direction considering the effect of lateral pressure; and σ_{yu} = ultimate stress of the panel under axial load in the y direction considering the effect of lateral pressure; and τ_u = ultimate stress of the panel under edge shear loads considering the effect of lateral pressure.

$\sigma_{xu} = \sigma_{Yeq}$ may be taken when σ_{xM} is tensile and, similarly, $\sigma_{yu} = \sigma_{Yeq}$ is taken when σ_{yM} is tensile, where $\sigma_{Yeq} = \frac{\sigma_{Yp}bt+\sigma_{Yw}h_wt_w+\sigma_{Yf}b_ft_f}{bt+h_wt_w+b_ft_f}$ is an equivalent yield stress to be used when the material yield stress of plating is different from that of stiffeners.

In Eq. (6.28), σ_{xu}, σ_{yu}, and τ_u will be calculated taking into account the effects of initial imperfections and lateral pressure loads for all six collapse modes, Modes I to VI, separately. Therefore, a set of six different ultimate strength interaction equations for Eq. (6.28) will exist. The solution of Eq. (6.28), with regard to any reference working stress, will be the ultimate strength for the corresponding collapse mode, where the ratio of all load components involved may be considered constant, implying that the number of the unknown load parameter is 1. The minimum value among the six candidates for ultimate strength is considered the ultimate strength of the panel: the stiffened panel will reach the ULS in the corresponding collapse mode.

A detailed description of the derivation and other details of closed-form expressions is presented by Paik and Thayamballi (2003). This method has been added to ALPS/ULSAP (2006).

6.6.3 Analytical Methods

The overall buckling behavior of the panel can be analyzed by solving the two non-linear governing differential equations of large deflection orthotropic plate theory, which are *the equilibrium equation* and *the compatibility equation* (e.g., Troitsky 1976). This theory and related solutions are very useful for ULS calculations based on Mode I collapse.

By taking into account the effect of initial deflections, the two governing differential equations for the cross-stiffened panel (i.e., having stiffeners in both x and y directions) can be written as follows:

$$
D_x \frac{\partial^4 w}{\partial x^4} + 2H \frac{\partial^4 w}{\partial x^2 \partial y^2} + D_y \frac{\partial^4 w}{\partial y^4}
$$

$$
- t \left[\frac{\partial^2 F}{\partial y^2} \frac{\partial^2 (w + w_o)}{\partial x^2} - 2 \frac{\partial^2 F}{\partial x \partial y} \frac{\partial^2 (w + w_o)}{\partial x \partial y} + \frac{\partial^2 F}{\partial x^2} \frac{\partial^2 (w + w_o)}{\partial y^2} + \frac{p}{t} \right] = 0,
$$

(6.29a)

$$
\frac{1}{E_y} \frac{\partial^4 F}{\partial x^4} + \left(\frac{1}{G_{xy}} - 2 \frac{\nu_x}{E_x} \right) \frac{\partial^4 F}{\partial x^2 \partial y^2} + \frac{1}{E_x} \frac{\partial^4 F}{\partial y^4}
$$

$$
- \left[\left(\frac{\partial^2 w}{\partial x \partial y} \right)^2 - \frac{\partial^2 w}{\partial x^2} \frac{\partial^2 w}{\partial y^2} + 2 \frac{\partial^2 w_o}{\partial x \partial y} \frac{\partial^2 w}{\partial x \partial y} - \frac{\partial^2 w_o}{\partial x^2} \frac{\partial^2 w}{\partial y^2} - \frac{\partial^2 w}{\partial x^2} \frac{\partial^2 w_o}{\partial y^2} \right] = 0,
$$

(6.29b)

where w_o and w = initial and added deflection functions for the orthotropic plate, respectively; F = Airy stress function; and E_x and E_y = elastic moduli of the orthotropic plate in the x and y directions, respectively. G_{xy} = elastic shear modulus of the orthotropic plate that can be approximated as follows:

$$
G_{xy} = \frac{E_x E_y}{E_x + \left(1 + 2 \sqrt{\nu_x \nu_y} \right) E_y} \approx \frac{\sqrt{E_x E_y}}{2 \left(1 + \sqrt{\nu_x \nu_y} \right)}.
$$

(6.30)

Here, D_x and D_y in Eq. (6.29) are the flexural rigidities of the orthotropic plate in the x and y directions, respectively. H is the effective torsional rigidity of the orthotropic plate. Detailed definition of the constants (D_x, D_y, H) and also the solution method of Eq. (6.29) are presented in Paik and Thayamballi (2003).

Once the Airy stress function F and the added deflection w are known, the stresses inside the panel can be calculated as follows:

$$
\sigma_x = \frac{\partial^2 F}{\partial y^2} - \frac{E_x z}{1 - \nu_x \nu_y} \left(\frac{\partial^2 w}{\partial x^2} + \nu_y \frac{\partial^2 w}{\partial y^2} \right),
$$

(6.31a)

$$
\sigma_y = \frac{\partial^2 F}{\partial x^2} - \frac{E_y z}{1 - \nu_x \nu_y} \left(\frac{\partial^2 w}{\partial y^2} + \nu_x \frac{\partial^2 w}{\partial x^2} \right),
$$

(6.31b)

$$
\tau = -\frac{\partial^2 F}{\partial x \partial y} - 2 G_{xy} z \frac{\partial^2 w}{\partial x \partial y},
$$

(6.31c)

where σ_x and σ_y are the axial stresses in the x and y directions, respectively; τ is the edge shear stress; and z is the axis in the plate thickness direction with $z = 0$ at the mid-thickness.

Eq. (6.20) is then applied to determine the ULS of the stiffened panel, but with membrane stresses given by Eq. (6.31). This orthotropic plate theory has also been added to ALPS/ULSAP (2006) for the Mode I ultimate strength calculation of stiffened panels.

Membrane stress components in Eq. (6.31) can alternatively be obtained by solving the following nonlinear governing differential equations (Paik and Lee 2005) [for symbols not specified below, see Eq. (6.12)]:

$$\Phi = D \left(\frac{\partial^4 w}{\partial x^4} + 2 \frac{\partial^4 w}{\partial x^2 \partial y^2} + \frac{\partial^4 w}{\partial y^4} \right) - t \left[\frac{\partial^2 F}{\partial y^2} \frac{\partial^2 (w + w_o)}{\partial x^2} + \frac{\partial^2 F}{\partial x^2} \frac{\partial^2 (w + w_o)}{\partial y^2} \right.$$

$$\left. - 2 \frac{\partial^2 F}{\partial x \partial y} \frac{\partial^2 (w + w_o)}{\partial x \partial y} \right] + \sum_{ii=1}^{n_{sx}} \left[E I_{ii} \frac{\partial^4 w}{\partial x^4} - A_{ii} \left(\frac{\partial^2 F}{\partial y^2} - \nu \frac{\partial^2 F}{\partial x^2} \right) \frac{\partial^2 (w + w_o)}{\partial x^2} \right]_{y=y_{ii}}$$

$$+ \sum_{jj=1}^{n_{sy}} \left[E I_{jj} \frac{\partial^4 w}{\partial y^4} - A_{jj} \left(\frac{\partial^2 F}{\partial x^2} - \nu \frac{\partial^2 F}{\partial y^2} \right) \frac{\partial^2 (w + w_o)}{\partial y^2} \right]_{x=x_{jj}} - p = 0, \tag{6.32a}$$

$$\frac{\partial^4 F}{\partial x^4} + 2 \frac{\partial^4 F}{\partial x^2 \partial y^2} + \frac{\partial^4 F}{\partial y^4} - E \left[\left(\frac{\partial^2 w}{\partial y \partial x} \right)^2 - \frac{\partial^2 w}{\partial x^2} \frac{\partial^2 w}{\partial y^2} + 2 \frac{\partial^2 w_o}{\partial x \partial y} \frac{\partial^2 w}{\partial x \partial y} \right.$$

$$\left. - \frac{\partial^2 w_o}{\partial x^2} \frac{\partial^2 w}{\partial y^2} - \frac{\partial^2 w}{\partial x^2} \frac{\partial^2 w_o}{\partial y^2} \right] = 0, \tag{6.32b}$$

where n_{sx}, n_{sy} = number of stiffeners in the x or y direction; A = cross-sectional area of stiffener; I = moment of inertia of stiffener; and x_{jj}, y_{ii} = location coordinates of transverse or longitudinal stiffeners, respectively.

By using the Airy stress function, the stress components at a certain location inside the panel may be expressed as follows:

$$\sigma_x = \frac{\partial^2 F}{\partial y^2} - \frac{Ez}{1 - \nu^2} \left(\frac{\partial^2 w}{\partial x^2} + \nu \frac{\partial^2 w}{\partial y^2} \right), \tag{6.33a}$$

$$\sigma_y = \frac{\partial^2 F}{\partial x^2} - \frac{Ez}{1 - \nu^2} \left(\frac{\partial^2 w}{\partial y^2} + \nu \frac{\partial^2 w}{\partial x^2} \right), \tag{6.33b}$$

$$\tau = \tau_{xy} = -\frac{\partial^2 F}{\partial x \partial y} - \frac{Ez}{2(1 + \nu)} \frac{\partial^2 w}{\partial x \partial y}. \tag{6.33c}$$

The stiffened panel governing differential equations of Eq. (6.32) does not reflect local buckling of stiffener webs (Mode IV) or tripping of stiffeners (Mode V), but these equations can be used to accurately analyze for the elastic large deflection behavior for Modes I, II, and III. These equations can be solved directly under applied loading and boundary conditions, for example, using the Galerkin method. Again, once the membrane stresses inside the stiffened panel are calculated, the panel ULS can be determined applying Eq. (6.28).

6.6.4 Semianalytical Methods

A method, similar to that described in Section 6.5.4 for plates, can be applied to solve the governing differential equations, Eq. (6.32), for stiffened panels (Paik and Lee

2005). First, it is assumed that the load is applied incrementally. At the end of the $(i-1)$th step of load increment, the deflection and stress function can be denoted by w_{i-1} and F_{i-1}, respectively. In the same manner, the deflection and stress function at the end of the ith step of load increment are denoted by w_i and F_i, respectively. Therefore, the accumulated (total) deflection w_i and stress function F_i at the end of the ith step of load increment are calculated by

$$w_i = w_{i-1} + \Delta w, \tag{6.34a}$$

$$F_i = F_{i-1} + \Delta F, \tag{6.34b}$$

where Δw and ΔF are the increments of deflection or stress function, respectively, where the prefix Δ indicates the increment for the variable.

Similar to Eq. (6.22), the incremental forms of governing differential equations, Eq. (6.32), can be given by

$$\Delta\Phi = D\left(\frac{\partial^4 \Delta w}{\partial x^4} + 2\frac{\partial^4 \Delta w}{\partial x^2 \partial y^2} + \frac{\partial^4 \Delta w}{\partial y^4}\right) - t\left[\frac{\partial^2 F_{i-1}}{\partial y^2}\frac{\partial^2 \Delta w}{\partial x^2} + \frac{\partial^2 \Delta F}{\partial y^2}\frac{\partial^2 (w_{i-1}+w_o)}{\partial x^2}\right.$$

$$\frac{\partial^2 F_{i-1}}{\partial x^2}\frac{\partial^2 \Delta w}{\partial y^2} + \frac{\partial^2 \Delta F}{\partial x^2}\frac{\partial^2 (w_{i-1}+w_o)}{\partial y^2} - 2\frac{\partial^2 F_{i-1}}{\partial x \partial y}\frac{\partial^2 \Delta w}{\partial x \partial y} - 2\frac{\partial^2 \Delta F}{\partial x \partial y}\frac{\partial^2 (w_{i-1}+w_o)}{\partial x \partial y}\right]$$

$$+ \sum_{ii=1}^{n_{sx}}\left[EI_{ii}\frac{\partial^4 \Delta w}{\partial x^4} - A_{ii}\left(\frac{\partial^2 F_{i-1}}{\partial y^2} - \nu\frac{\partial^2 F_{i-1}}{\partial x^2}\right)\frac{\partial^2 \Delta w}{\partial x^2}\right.$$

$$\left. - A_{ii}\left(\frac{\partial^2 \Delta F}{\partial y^2} - \nu\frac{\partial^2 \Delta F}{\partial x^2}\right)\frac{\partial^2 (w_{i-1}+w_o)}{\partial x^2}\right]_{y=y_{ii}}$$

$$+ \sum_{jj=1}^{n_{sy}}\left[EI_{jj}\frac{\partial^4 \Delta w}{\partial y^4} - A_{jj}\left(\frac{\partial^2 F_{i-1}}{\partial x^2} - \nu\frac{\partial^2 F_{i-1}}{\partial y^2}\right)\frac{\partial^2 \Delta w}{\partial y^2}\right.$$

$$\left. - A_{jj}\left(\frac{\partial^2 \Delta F}{\partial x^2} - \nu\frac{\partial^2 \Delta F}{\partial y^2}\right)\frac{\partial^2 (w_{i-1}+w_o)}{\partial y^2}\right]_{x=x_{jj}}$$

$$+ \Delta Q = 0, \tag{6.35a}$$

$$\frac{\partial^4 \Delta F}{\partial x^4} + 2\frac{\partial^4 \Delta F}{\partial x^2 \partial y^2} + \frac{\partial^4 \Delta F}{\partial y^4} - E\left[2\frac{\partial^2 (w_{i-1}+w_o)}{\partial x \partial y}\frac{\partial^2 \Delta w}{\partial x \partial y}\right.$$

$$\left. - \frac{\partial^2 (w_{i-1}+w_o)}{\partial x^2}\frac{\partial^2 \Delta w}{\partial y^2} - \frac{\partial^2 \Delta w}{\partial x^2}\frac{\partial^2 (w_{i-1}+w_o)}{\partial y^2}\right] = 0, \tag{6.35b}$$

where the terms of very small quantities with an order higher than the second order of the increments Δw and ΔF have been neglected.

Similar to Eq. (6.26), the Galerkin method can be applied to determine the unknown amplitudes of assumed deflection functions. Then, the membrane stress distribution inside the stiffened panel can be calculated. The differential equations

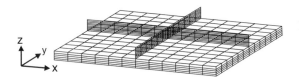

Figure 6.20. Example subdivision of mesh regions for plating and stiffeners in a stiffened panel.

governing the elastic large deflection response of stiffened panels have been formulated and solved analytically, although the effects of plasticity have not been included. To include plasticity effects, a numerical approach similar to the approach described in Section 6.5.4 can be employed.

For this purpose, the panel is subdivided into a number of mesh regions in the three directions, as shown in Figure 6.20. The average membrane stress components for each mesh region can be calculated at every step of load increment. Yielding for each mesh region is checked for both plate part and stiffeners by using the following yield criteria:

$$\sigma_x^2 - \sigma_x\sigma_y + \sigma_y^2 + 3\tau^2 \geq \sigma_{Yp}^2, \text{ for plate parts,} \tag{6.36a}$$

$$\sigma_{sx} \geq \sigma_{Ys}, \sigma_{sy} \geq \sigma_{Ys}, \qquad \text{for stiffeners,} \tag{6.36b}$$

where σ_{Yp} = yield stress of plate; σ_{Ys} = yield stress of stiffeners; σ_{sx} = working stress of stiffener in the x direction; and σ_{sy} = working stress of stiffener in the y direction.

We assume, in the solution process, that the plate parts are composed of a number of membrane strings (or fibers) in the two (i.e., x and y) directions. Each fiber has a number of layers in the z direction. The stiffener is regarded as an assembly of membrane fibers in their length direction, which also have a number of layers in the z direction. The boundary condition of each fiber would satisfy the panel edge condition as well. In fact, due to the membrane action of the fibers, occurrence of the additional panel deflection to some extent may be disturbed with a further increase in the applied loads. However, if any local region in the fiber is yielded, the fiber regarded as a "string" will be cut such that the membrane action stops.

Details of this theory are presented in Paik and Lee (2005) and have been added to ALPS/SPINE (2005).

6.6.5 Nonlinear Finite-Element Methods

Nonlinear finite-element methods are powerful tools for the ULS assessment of stiffened plate structures, as well as other types of structures, although they may require large computing times. In this regard, the elastic/perfectly plastic model is often applied for ULS assessment. But note that this type of material modeling may not be adequate for accidental limit-state design purposes, as we describe in Chapter 8, because in such cases strain-hardening and strain-softening effects must also be considered.

Another issue related to the use of nonlinear finite-element methods for elastic/plastic large deflection analysis of stiffened plate structures arises in the selection of the analysis extent and the associated mesh size. Although a fine-enough mesh size for analysis can be determined based on the convergence study as described in Section 6.5.5, several options may exist in terms of the extent of analysis for stiffened

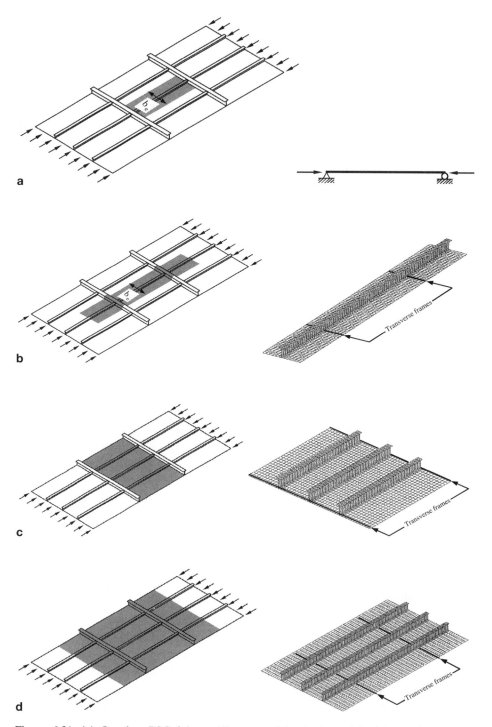

Figure 6.21. (a) One-bay PSC (plate-stiffener combination) model with attached plating. (b) Two-bay PSC model with attached plating. (c) One-bay SP (stiffened panel) model between transverse frames. (d) Two-bay SP model.

plate structures. For instance, the following options can be considered for a continuously stiffened panel, subject to predominantly longitudinal axial compressive loads:

- One-bay plate-stiffener combination model; see Figure 6.21(a)
- Two-bay plate-stiffener combination model; see Figure 6.21(b)
- One-bay stiffened panel model between transverse frames; see Figure 6.21(c)
- Two-bay stiffened panel model; see Figure 6.21(d)

The PSC (plate-stiffener combination) model adopts a stiffener with the attached plating as representative of the stiffened panel, but the SP (stiffened panel) model takes the entire panel in the breadth direction. The PSC model may behave as a beam-column, but the SP model can reflect the effect of continuity more precisely. The PSC model is often used for the elastic/plastic large deflection analyses of a stiffened panel under predominantly uniaxial loads in the stiffener length direction, with or without lateral pressure loads, typically when the stiffeners are medium or large in stiffness and strength.

The one-bay model considers the interframe panel, that is, between two adjacent transverse frames. This model is available when the effect of rotational restraints along the transverse frames is negligible, that is, the support conditions at the transverse frames can be assumed to be simply supported. The two-bay model, however, takes the extent of analysis as a half length of the two adjacent interframe panels and a full length of the center panel. The two-bay model is useful when the effect of rotational restraints along the transverse frames is important, and it may be used with a symmetric boundary condition applied along the center lines of the two adjacent interframe panels.

Therefore, the one-bay PSC model is the simplest one and useful for theoretical calculations, but this model cannot represent the rotational restraints along the transverse frames. To resolve this problem, the two-bay PSC model can be used where the lateral deflections along transverse frames are restricted and their rotational restraints are accounted for more realistically. Because the two-bay modeling technique can automatically take into account the longitudinal rotation effect of the longitudinal stiffeners at the transverse frames, where one panel deflects down while the adjacent panels buckle up in the continuous plate structure supported by transverse frames, and the stiffener cross section remains upright at the transverse frames, as shown in Figure 6.22.

Directions of lateral pressure loads and/or column type initial deflections of stiffeners can govern the panel collapse patterns and result in the plate-induced failure (PIF) or stiffener-induced failure (SIF), as illustrated in Figure 6.22. As would be expected, the panel ultimate strength for PIF can be quite different from that for SIF. This means that the nonlinear finite-element modeling should correctly define the directions of lateral pressure loads and also of initial deflections of both plating and stiffeners when they are present.

Under application of multiple-load components, including biaxial loads and shear, the PSC model cannot be used. In this case, the SP model must be used. A one-bay SP model cannot account for the effect of rotational restraints along transverse frames. A two-bay SP model is then required for resolving this problem. The two-bay SP model

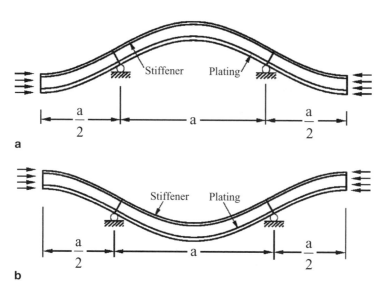

Figure 6.22. A schematic representation of the plate-induced failure (PIF) and stiffener-induced failure (SIF) patterns in the two-bay model: (a) PIF or compression in plate side (CIP); (b) SIF or compression in stiffener side (CIS).

is of course more refined and appropriate than either two-bay PSC model or one-bay SP model in terms of the resulting accuracy, even under uniaxial compressive loads.

We emphasize that the nonlinear finite-element method solutions may not be very accurate, and the solutions can even be completely incorrect if the finite-element modeling is not adequate.

6.6.6 Illustrative Examples

ALPS/ULSAP (2006) has been integrated with MAESTRO (2006) software for the purpose of ULS assessment and structural optimization of ships and ship-shaped offshore structures. Figure 6.23 shows a sample display related to the ALPS/ULSAP calculations with the MAESTRO modeler and post-processing system.

With the nomenclature of the stiffened panel shown in Figure 5.1 of Chapter 5, a comparison of ALPS/ULSAP, with nonlinear finite-element analyses and DNV PULS (Steen et al. 2004) solutions, is shown in Figure 6.24. The longitudinally stiffened panel between transverse frames is considered to be simply supported along the boundary and subject to biaxial compressive actions with or without lateral pressure actions. The dimensions of the object panel are: a (panel length) = 5,120mm; B (panel breadth) = 9,100mm; and t (panel thickness) = 20mm.

The number of stiffeners in the panel considered is 9 and the stiffener type is T-bar with the dimensions of 598.5mm × 12mm + 200mm × 20mm. The panel material is a high-tensile steel with yield stress of 315 N/mm². An average level of initial deflection is considered with the so-called hungry-horse shape (maximum plate initial deflection = 4.55mm), and we assume that no residual stresses exist. The column-type initial deflection of stiffeners is taken to be 5.12mm. The lateral pressure applied is 0.2531N/mm², which is equivalent to a water head of 25m.

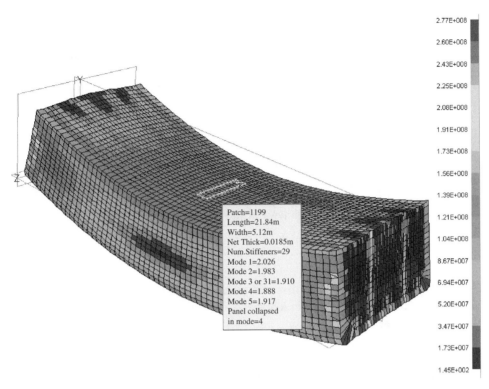

Figure 6.23. A sample display related to ULS calculations by ALPS/ULSAP for an FPSO hull structure in sagging.

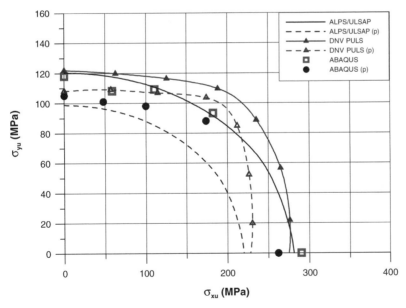

Figure 6.24. ULS interaction curves for a longitudinally stiffened panel under biaxial compression and lateral pressure actions (p denotes the case with lateral pressure actions).

Figure 6.24 shows that ALPS/ULSAP and DNV PULS give reasonably accurate predictions of the stiffened panel ULS under biaxial compressive loads compared with a more refined nonlinear finite-element analysis. ALPS/ULSAP provides somewhat pessimistic ULS solutions when with large lateral pressure actions; however, DNV PULS appears to alway give somewhat optimistic ULS predictions in the cases considered.

6.7 Ultimate Strength of Vessel Hulls

In this section, various methods are presented to determine the characteristic value C_k in Eq. (6.1) for ULS of ship-shaped offshore hull girders; that is, for purposes of considering ultimate strength behavior at the global-system level.

6.7.1 Fundamentals

A vessel hull in the intact condition will normally sustain applied hull girder loads smaller than the design hull girder loads and, in normal service, it may not suffer any structural damage such as buckling and plastic collapse. However, the hull girder loads are uncertain both because of the nature of site-specific environments and because of a possibly unusual operation associated with loading or offloading of cargo, for example, and perhaps due to human error. In rare cases, applied hull girder loads may therefore exceed their design loads. Because aging vessel hulls may have suffered structural deterioration due to corrosion, fatigue cracking, local denting, and related weakening, their structural resistance will play a role as well.

As applied hull girder loads increase beyond their design loads, structural components of the vessel hull may buckle in compression and yield in tension. The vessel's hull can normally carry loads beyond the onset of buckling or yielding if limited to a small enough number of structural components, but the structural effectiveness of any such failed component will decrease, and its individual structural stiffness can even become "negative," with their internal stress being redistributed to adjacent intact components. Such events will weaken the structure as to its ability to carry additional loads. The most highly compressed component will, deterministically speaking, collapse earlier, and the stiffness of the overall hull will decrease gradually. As loads continue to increase, buckling and plastic collapse of more structural components may occur progressively, until the ULS is reached for the hull girder as a whole. Thin-walled structures consisting of components prone to buckling in the elastic regime can fail without a large amount of plastic deformation.

When the structural safety of a vessel's hull is considered, the ultimate hull girder strength must then be accurately evaluated. Simple expressions for the calculation of the ultimate hull strength should be derived and used for the ready formulation of failure functions in reliability analysis and for the early stages of structural design.

6.7.2 Closed-Form Expressions

Useful closed-form expressions of ultimate hull girder strength have been suggested in the literature (e.g., Paik and Thayamballi 2003). The related existing studies may be classified into three types: the linear (knock-down factor) approach, the empirical approach, and the analytical approach.

In the linear approach, the behavior of the hull up to collapse of the compression flange (i.e., upper deck in sagging or outer bottom in hogging) is assumed to be linear, and the ultimate moment capacity of the hull is basically expressed as the ultimate strength of the compression flange multiplied by the elastic section modulus with a simple correction for buckling and yielding. In the empirical approach, an expression for strength is derived on the basis of experimental or numerical data for hull models. The analytical approach is based on a presumed stress distribution over the hull section, from which the moment of resistance of the hull is calculated theoretically, taking into account buckling in the compression flange and yielding in the tension flange.

The linear (knock-down factor) approach is quite simple, but its accuracy may not be always good because after buckling of the compression flange, the behavior of the hull is then no longer necessarily linear and the neutral axis changes position. Empirical formulations may provide reasonable solutions for a conventional vessel's hulls, but one must be careful when using empirical formulations for new or unusual hull types, and characteristics are usually derived on the basis of a limited database. However, analytical formulations can be applied with somewhat greater certainty to new or general hulls because they include section geometry and other effects more precisely.

Using the linear approach, the ultimate hull girder strength may be given by

$$M_u = kZ\sigma_u, \tag{6.37}$$

where σ_u = ultimate stress of the compressed hull flange (e.g., upper deck in sagging or outer bottom in hogging for vertical bending) or an outer side shell for horizontal bending; Z = the relevant section modulus of the hull at the compressed flange (deck in sagging and bottom in hogging for vertical bending; port or starboard side shell for horizontal bending); and k = coefficient accounting for the shift of neutral axis or other factors (usually obtained by correlation with more sophisticated analyses).

The ultimate stress of the compressed hull flange can be determined by the methods described in Section 6.6. As long as the k value can be obtained properly, Eq. (6.37) will give a good estimate of the vessel's hull girder ultimate strength.

Caldwell (1965) is a pioneer in the use of analytical methods for ultimate hull girder strength calculations. Paik and Mansour (1995) revived these methods with some advances. Caldwell (1965) made an assumption that at ULS, all material over the hull cross section in compression reaches its buckling plastic collapse state, and all material in tension reaches fully plastic state. However, it has been recognized that the vessel hull can reach the ULS even before the material in tension yields fully or the material in compression collapses entirely. As a result, the analytical method used with Caldwell's hypothesis of bending stress distribution at ULS may overestimate the ultimate hull girder strength.

Paik and Mansour (1995) developed an advanced analytical method (i.e., the Paik stress hypothesis) by modifying the Caldwell approach, where the bending stress distribution over a vessel's hull cross section at ULS (see Figure 6.25). In this case, the compression flange has collapsed and the tension flange has yielded

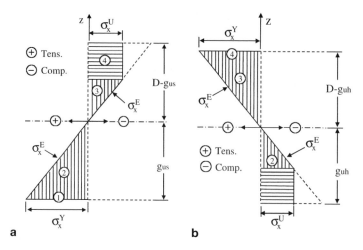

Figure 6.25. Bending stress distribution over the hull cross section at ULS, originally suggested by Paik and Mansour (1995): (a) sagging; (b) hogging.

at ULS, but the mid-height materials in the vicinity of the neutral axis is still intact (i.e., unfailed).

The detailed derivation of closed-form expressions for ultimate hull girder strength based on the Paik stress hypothesis is presented in Paik and Thayamballi (2003) and Paik (2004). The formulation can be given by

$$M_u = \sum_{i=1}^{N} \sigma_i A_i \left(z_i - g_u \right),\tag{6.38}$$

where $g_u = \frac{\sum \sigma_i A_i z_i}{\sum \sigma_i A_i}$ = distance from the base (reference) position to the neutral axis of the vessel's hull cross section; z_i = distance from the base (reference) position to the neutral axis of the ith structural component; σ_i = stress for the ith structural component adopting the stress hypothesis by Paik and Mansour (1995), as shown in Figure 6.25, with a negative sign in the compressed part and a positive sign in the tensioned part considering the hogging condition or the sagging condition, as the case may be; and A_i = cross-sectional area of the ith structural component.

For the use of Eq. (6.38), the ultimate stresses of individual structural components can be obtained from the methods described in Sections 6.5 and 6.6. The summation of Eq. (6.38) will be performed for the total number N of structural components that are effective for resisting the applied hull girder loads. Eqs. (6.37) and (6.28) are, in principle, applicable to the horizontal bending case as well as the vertical bending case.

The ultimate hull girder strength under shearing forces at hull cross section can be approximated by considering that the contribution of horizontal structural components to resistance against shearing forces can be neglected, as follows:

$$F_u = \sum_{i=1}^{N} A_{si} \tau_{ui},\tag{6.39}$$

where F_u = ultimate hull girder strength under shearing forces; A_{si} = cross-sectional area of the ith vertical component plating (excluding stiffeners); and τ_{ui} = ultimate shear stress of vertical component plating between stiffeners that may be computed from Eq. (6.7). The summation of Eq. (6.39) will be performed for the total number N of all vertical structural components.

Under simultaneous action of multiple hull girder load components, such as vertical bending, horizontal bending, and shearing force, the ultimate hull girder strength interaction relationship (Paik and Thayamaballi 2003) is given by

$$\left(\frac{M_V}{M_{Vu}F_{VR}}\right)^{c_1} + \left(\frac{M_H}{M_{Hu}F_{HR}}\right)^{c_2} - 1 = 0, \qquad (6.40a)$$

where $F_{VR} = \{1 - (F/F_u)^{c_4}\}^{1/c_3}$; $F_{HR} = \{1 - (F/F_u)^{c_6}\}^{1/c_5}$; M_{Vu}, M_{Hu}, F_u = ultimate hull girder strength under vertical bending moment (hogging or sagging) alone, horizontal bending moment (hogging or sagging) alone, and shearing force alone, respectively. The solutions of Eq. (6.40a) with regard to M_V, M_H, and F are the ultimate hull girder strength under combined hull girder loads.

In Eq. (6.40a), c_1–c_6 are constants that take into account the effect of load combination types as follows:

$$\left(\frac{M_V}{M_{Vu}}\right)^{c_1} + \left(\frac{M_H}{M_{Hu}}\right)^{c_2} = 1, \qquad (6.40b)$$

$$\left(\frac{M_V}{M_{Vu}}\right)^{c_3} + \left(\frac{F}{F_u}\right)^{c_4} = 1, \qquad (6.40c)$$

$$\left(\frac{M_H}{M_{Hu}}\right)^{c_5} + \left(\frac{F}{F_u}\right)^{c_6} = 1. \qquad (6.40d)$$

Paik et al. (1996) propose using $c_1 = 1.85$, $c_2 = 1.0$, $c_3 = 2.0$, $c_4 = 5.0$, $c_5 = 2.5$, and $c_6 = 5.5$ regardless of the vessel type and direction of bending (i.e., hogging or sagging). Gordo and Guedes Soares (1997) suggest using $c_1 = c_2 = 1.50 \sim 1.66$ for trading tankers and container ships. Ozguc et al. (2005) suggest using $c_1 = 2.0$ and $c_2 = 1.45$ for hogging and $c_2 = 1.35$ for sagging.

Figure 6.26 compares the ultimate hull girder strength interactions from these three suggestions. It is seen from Figure 6.26 that the Paik formula may give the most pessimistic solutions; however, it is based on numerical solutions of progressive collapse analyses, taking into account the effects of an average level of initial imperfections (i.e., initial deflections and welding residual stresses) in all structural components making up the vessel hulls.

6.7.3 Progressive Hull Collapse Analysis: Idealized Structural Unit Method

6.7.3.1 Background of Idealized Structural Unit Method

Progressive hull girder collapse analysis is highly desirable for the determination of the ultimate hull girder strength (Paik 2004). This is because the closed-form expressions presented in Section 6.7.2, although useful, cannot accurately take

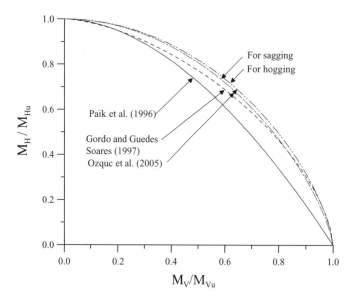

Figure 6.26. Ultimate hull girder strength interaction formulations for trading tanker hulls; taking the same interaction form for both hogging and sagging in Paik et al. (1996) and Gordo and Guedes Soares (1997).

into account the progressive failures of structural components, and their interactions, in particular cases, as the applied hull girder loads increase until the ULS is reached.

Applying analytical schemes for the progressive hull girder collapse analysis of vessel structures is usually not straightforward. Computer-aided numerical simulations will then be powerful tools for this purpose. However, conventional nonlinear finite-element methods (FEM) are not so successful for practical design purposes because they require a lot of computational effort. The idealized structural unit method (ISUM) is now recognized as an accurate and efficient tool for nonlinear analyses of large sized structures including ship-shaped offshore structures, as well as trading ships (Paik and Thayamballi 2003).

Ueda and Rashed (1974, 1984) are the pioneers of ISUM developments. Their first effort on ISUM developments was to perform the progressive collapse analysis of a transverse framed ship structure idealized as an assembly of the deep girder ISUM elements. In an almost parallel development to Ueda and Rashed (1974), Smith (1977) applied a similar approach for the progressive collapse analysis of a ship hull under vertical bending. Smith modeled the ship hull as an assembly of plate-stiffener combinations, that is, stiffeners with attached plating (or beam-column elements) representing stiffened panels.

The method developed by Smith (1977) is sometimes called the "Smith method." However, the Smith method adopts the same techniques as ISUM in terms of theoretical formulations and approaches; therefore, it can also be classified as a type of ISUM. In fact, the concept of ISUM is considerably more general, involving structural modeling techniques and ULS behavior formulations using large idealized structural units called "ISUM elements," which can be formulated based on theoretical, numerical, and experimental results.

For the progressive collapse analyses, ISUM elements should be developed in advance. Indeed, the main task involved in developing ISUM elements is to efficiently characterize possible nonlinearities of the target large structural element behavior under prescribed conditions, considering the geometric and material properties and loading and boundary conditions that are involved in the nonlinear behavior of the object structure. The result is something like the education of pupils in school by teachers where, as they say, "better teachers produce better pupils." The same is true for the development of ISUM elements and the related accuracy of results.

Depending on the analysis purpose, using ISUM elements with less complex characteristics can provide a good enough level of the computational accuracy, although ISUM elements with more sophisticated characteristics, of course, may need to be used for other purposes. For instance, if structural components of any object structure are subjected to predominantly uniaxial actions alone, then the corresponding ISUM elements can be idealized so that only the uniaxial loading condition is accounted for in the element properties and computations. However, if the structural components are subjected to multiple actions or action effects, the effect of combined loads on structural failure behavior should be taken into account in the development of ISUM elements. This means that even if the geometric shape of any two ISUM elements is identical or similar, its usage or characteristics can vastly differ.

For the development of ISUM elements, an analytical, numerical, or experimental approach can be used by "educating" the element regarding the expected nonlinear behavior of that type of structural element, under the prescribed conditions, in advance. The characteristics of the developed ISUM elements can be formulated in either explicit or implicit forms.

Individual developers of the ISUM elements may employ somewhat different approaches from each other to idealize and to formulate the actual nonlinear behavior of the structural components. Also, the features of existing ISUM elements can be advanced continually to accommodate more factors of influence or to make them more sophisticated.

In the following sections, ISUM theories useful for the progressive collapse analysis of general types of plated structures are described. These have been added to computer software called ALPS/GENERAL (2006) and integrated with the pre- and post-processing systems of MAESTRO (2006). The special version of ALPS/GENERAL is ALPS/HULL (2006) for progressive hull girder collapse analysis. Some illustrative examples on the progressive hull girder collapse analysis are presented, together with a comparison of ALPS/HULL analyses and pertinent experimental results.

6.7.3.2 ISUM Structural Modeling

Unlike the conventional nonlinear FEM that models the object structure using a large number of finite elements and varying degrees of mesh fineness depending on the stress gradients that may be present, ISUM idealizes a large structural component making up the structure as one ISUM element with a few nodal points.

For the nonlinear analysis of structures, it is apparent that various types of ISUM elements are necessary to make a complete structural model. For instance, ship-shaped offshore structures are primarily composed of plating and support

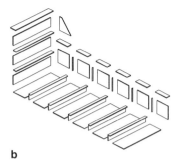

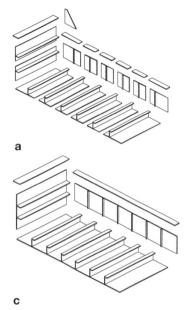

a

b

c

Figure 6.27. Various types of ISUM idealizations for a basic part of plated structures: (a) structural idealizations as an assembly of plate-stiffener combination elements–the so-called Smith method; (b) structural idealization as an assembly of plate-stiffener separation elements; (c) structural idealization as an assembly of stiffened panel elements.

members (e.g., stiffeners, girders, frames), and they can be idealized as assemblies of various types of ISUM elements such as plate-stiffener combination elements, plate-stiffener separation elements (i.e., plate elements for plating and beam-column elements for support members), and stiffened panel elements or their combinations.

Figure 6.27 shows some typical types of ISUM modeling for plated structures. Plate-stiffener combination models idealize any stiffened plate structure as an assembly of stiffeners together with attached plating; see Figure 6.27(a). Plate-stiffener separation modeling technique uses plate elements for plating between support members, but support members are modeled as beam-column elements without attached plating; see Figure 6.27(b). Stiffened panel element modeling takes the entire stiffened panel as one ISUM element; see Figure 6.27(c). As necessary, multiple types or combinations of such ISUM elements can be used for modeling any global structure of interest.

ISUM modeling of a vessel hull structure as an assembly of plate-stiffener separation elements, as shown in Figure 6.27(b), can be the most efficient and accurate modeling strategy for the progressive collapse analysis purpose (Paik et al. 2005). In the case shown, two types of ISUM elements, *plate element* for plating between support members and *beam-column element* for support members (stiffeners), are used to model the structure.

6.7.3.3 ISUM Plate Element

This section presents the formulations for the ISUM plate element used for ALPS/HULL modeling of plating between support members. It is important to realize that plate elements of vessel hulls can be subjected to combined stress components such as longitudinal axial stresses, transverse axial stresses, and edge shear stresses together with lateral pressure. Even for vessel hulls under the vertical bending

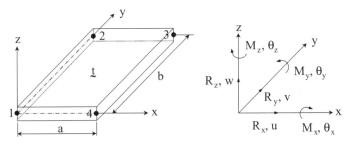

Figure 6.28. The local coordinate system, nodal forces, and displacements for the ISUM plate element.

moment alone, the vertical structural components – for example, side shell plating and longitudinal bulkhead plating – can be subjected to edge shear stresses as well as axial bending stresses.

(1) Nodal Forces and Nodal Displacements. Figure 6.28 shows a particular ISUM rectangular plate element used for the present ALPS/HULL computation. The element has a nodal point at each corner. The combined in-plane and out-of-plane deformation behavior for the ISUM plate element can be expressed by the nodal force vector $\{R\}$ and the displacement vector $\{U\}$ with 6 degrees of freedom at each corner nodal point that is taken to be located in the midthickness of the element, as follows:

$$\{R\} = \left\{R_{x1}\,R_{y1}\,R_{z1}\,M_{x1}\,M_{y1}\,M_{z1}\ldots R_{x4}\,R_{y4}\,R_{z4}\,M_{x4}\,M_{y4}\,M_{z4}\right\}^{T}, \tag{6.41a}$$

$$\{U\} = \left\{u_1v_1w_1\theta_{x1}\theta_{y1}\theta_{z1}\ldots u_4v_4w_4\theta_{x4}\theta_{y4}\theta_{z4}\right\}^{T}, \tag{6.41b}$$

where R_x, R_y, R_z = translational nodal forces in the x, y, and z directions, respectively; M_x, M_y = out-of-plane bending moments with regard to the x and y directions, respectively; M_z = torsional moment with regard to the z direction; u, v, w = translational displacements in the x, y, and z directions, respectively; θ_x $(= -\partial w/\partial y)$, θ_y $(= \partial w/\partial x)$, θ_z = rotations with regard to the x, y, and z directions, respectively; and $\{\ \}^{T}$ = transpose of the vector. A digit in the subscript indicates the node number of the rectangular element.

(2) Strain versus Displacement Relationship. The strain versus displacement relationship of the element is given in the Cartesian coordinate system, as follows:

$$\varepsilon_x = \frac{\partial u}{\partial x} - z\frac{\partial^2 w}{\partial x^2} + \frac{1}{2}\left\{\left(\frac{\partial u}{\partial x}\right)^2 + \left(\frac{\partial v}{\partial x}\right)^2\right\} + \frac{1}{2}\left(\frac{\partial w}{\partial x}\right)^2, \tag{6.42a}$$

$$\varepsilon_y = \frac{\partial v}{\partial y} - z\frac{\partial^2 w}{\partial y^2} + \frac{1}{2}\left\{\left(\frac{\partial u}{\partial y}\right)^2 + \left(\frac{\partial v}{\partial y}\right)^2\right\} + \frac{1}{2}\left(\frac{\partial w}{\partial y}\right)^2, \tag{6.42b}$$

$$\gamma_{xy} = \left(\frac{\partial u}{\partial y} + \frac{\partial v}{\partial x}\right) - 2z\frac{\partial^2 w}{\partial x\partial y} + \left\{\left(\frac{\partial u}{\partial x}\right)\left(\frac{\partial u}{\partial y}\right) + \left(\frac{\partial v}{\partial x}\right)\left(\frac{\partial v}{\partial y}\right)\right\} + \left(\frac{\partial w}{\partial x}\right)\left(\frac{\partial w}{\partial y}\right), \tag{6.42c}$$

where ε_x, ε_y, and γ_{xy} are the generalized strain components for a plane stress state.

The first term on the right-hand side of the equation represents the small deformation in-plane strain. The second term denotes the small deformation out-of-plane strain. The third and fourth terms are nonlinear strain components due to large in-plane and out-of-plane deformations, respectively. It is necessary to take into account the effect of out-of-plane large deformations because the plate element considered is very large compared with conventional finite elements. The incremental expressions of Eq. (6.42a) are given by

$$\Delta\varepsilon_x = \frac{\partial\Delta u}{\partial x} - z\frac{\partial^2\Delta w}{\partial x^2} + \left(\frac{\partial u}{\partial x}\right)\left(\frac{\partial\Delta u}{\partial x}\right) + \left(\frac{\partial v}{\partial x}\right)\left(\frac{\partial\Delta v}{\partial x}\right) + \left(\frac{\partial w}{\partial x}\right)\left(\frac{\partial\Delta w}{\partial x}\right)$$

$$+ \frac{1}{2}\left\{\left(\frac{\partial\Delta u}{\partial x}\right)^2 + \left(\frac{\partial\Delta v}{\partial x}\right)^2\right\} + \frac{1}{2}\left(\frac{\partial\Delta w}{\partial x}\right)^2, \tag{6.43a}$$

$$\Delta\varepsilon_y = \frac{\partial\Delta v}{\partial y} - z\frac{\partial^2\Delta w}{\partial y^2} + \left(\frac{\partial u}{\partial y}\right)\left(\frac{\partial\Delta u}{\partial y}\right) + \left(\frac{\partial v}{\partial y}\right)\left(\frac{\partial\Delta v}{\partial y}\right) + \left(\frac{\partial w}{\partial y}\right)\left(\frac{\partial\Delta w}{\partial y}\right)$$

$$+ \frac{1}{2}\left\{\left(\frac{\partial\Delta u}{\partial y}\right)^2 + \left(\frac{\partial\Delta v}{\partial y}\right)^2\right\} + \frac{1}{2}\left(\frac{\partial\Delta w}{\partial y}\right)^2, \tag{6.43b}$$

$$\Delta\gamma_{xy} = \left(\frac{\partial\Delta u}{\partial y} + \frac{\partial\Delta v}{\partial x}\right) - 2z\frac{\partial^2\Delta w}{\partial x\partial y} + \left(\frac{\partial u}{\partial x}\right)\left(\frac{\partial\Delta u}{\partial y}\right) + \left(\frac{\partial u}{\partial y}\right)\left(\frac{\partial\Delta u}{\partial x}\right)$$

$$+ \left(\frac{\partial y}{\partial x}\right)\left(\frac{\partial\Delta v}{\partial y}\right) + \left(\frac{\partial v}{\partial y}\right)\left(\frac{\partial\Delta v}{\partial x}\right) + \left(\frac{\partial w}{\partial x}\right)\left(\frac{\partial\Delta w}{\partial y}\right)$$

$$+ \left(\frac{\partial w}{\partial y}\right)\left(\frac{\partial\Delta w}{\partial x}\right) + \left(\frac{\partial\Delta u}{\partial x}\right)\left(\frac{\partial\Delta u}{\partial y}\right) + \left(\frac{\partial\Delta v}{\partial x}\right)\left(\frac{\partial\Delta v}{\partial y}\right)$$

$$+ \left(\frac{\partial\Delta w}{\partial x}\right)\left(\frac{\partial\Delta w}{\partial y}\right), \tag{6.43c}$$

where the prefix Δ denotes an infinitesimal increment of the variable.

For convenience in the formulations of the element, the nodal displacement vector $\{U\}$ is now split into three components: the in-plane component $\{S\}$, the out-of-plane component $\{W\}$, and the component for the rotations about the axis z. Thus, Eq. (6.43a) can be rewritten by the matrix form using the vectors, $\{S\}$ and $\{W\}$, as follows:

$$\{\Delta\varepsilon\} = [B_p]\{\Delta S\} - z[B_b]\{\Delta w\} + [C_p][G_p]\{\Delta S\} + [C_b][G_b]\{\Delta w\}$$

$$+ \frac{1}{2}[\Delta C_p][G_p]\{\Delta S\} + \frac{1}{2}[\Delta C_b][G_b]\{\Delta w\} = [B]\{\Delta U\}, \tag{6.44}$$

where $\{\Delta\varepsilon\} = \left\{\Delta\varepsilon_x \Delta\varepsilon_y \Delta\gamma_{xy}\right\}^T = $ increment of strain vector; $\{U\} = \{SW\}^T = $ nodal displacement vector; $\{S\} = \{u_1v_1u_2v_2u_3v_3u_4v_4\}^T = $ in-plane displacement vector;

$\{W\} = \{w_1 \theta_{x1} \theta_{y1} w_2 \theta_{x2} \theta_{y2} w_3 \theta_{x3} \theta_{y3} w_4 \theta_{x4} \theta_{y4}\}^T =$ out-of-plane displacement vector; and $[B] =$ strain versus displacement matrix.

$$\left\{\frac{\partial u}{\partial x} \frac{\partial v}{\partial y} \frac{\partial u}{\partial y} + \frac{\partial v}{\partial x}\right\}^T = [B_p]\{S\}; \quad \left\{\frac{\partial^2 w}{\partial x^2} \frac{\partial^2 w}{\partial y^2} 2\frac{\partial^2 w}{\partial x \partial y}\right\}^T = [B_b]\{W\};$$

$$\left\{\frac{\partial u}{\partial x} \frac{\partial v}{\partial x} \frac{\partial u}{\partial y} \frac{\partial v}{\partial y}\right\}^T = [G_p]\{S\}; \quad \left\{\frac{\partial w}{\partial x} \frac{\partial w}{\partial y}\right\}^T = [G_b]\{W\};$$

$$[C_p] = \begin{bmatrix} \dfrac{\partial u}{\partial x} & \dfrac{\partial v}{\partial x} & 0 & 0 \\[2mm] 0 & 0 & \dfrac{\partial u}{\partial y} & \dfrac{\partial v}{\partial y} \\[2mm] \dfrac{\partial u}{\partial y} & \dfrac{\partial v}{\partial y} & \dfrac{\partial u}{\partial x} & \dfrac{\partial v}{\partial x} \end{bmatrix}; \quad [C_b] = \begin{bmatrix} \dfrac{\partial w}{\partial x} & 0 \\[2mm] 0 & \dfrac{\partial w}{\partial y} \\[2mm] \dfrac{\partial w}{\partial y} & \dfrac{\partial w}{\partial x} \end{bmatrix}.$$

(3) Stress versus Strain Relationship. The membrane stress increments $\{\Delta\sigma\}$ due to strain increments $\{\Delta\varepsilon\}$ can be calculated for a plane stress state, as follows:

$$\{\Delta\sigma\} = [D]\{\Delta\varepsilon\}, \tag{6.45}$$

where $\{\Delta\sigma\} = \{\Delta\sigma_x \Delta\sigma_y \Delta\tau_{xy}\}^T =$ increment of average membrane stress components for a plane stress state; and $[D] =$ stress–strain matrix, which can be determined as a function of various parameters of influence including geometric and material properties, applied stresses, and failure status (e.g., buckling, plastic collapse), among others.

In the present element formulation, the $[D]$ matrix is derived in closed-form expressions by an analytical method, that is, by solving nonlinear plate governing differential equations, as a function of geometric and material properties, initial deflection, welding residual stresses, and applied stresses for two different failure states: the pre- and post-ultimate limit states (Paik and Thayamballi 2003).

In the pre- or post-ultimate strength regimes, the $[D]$ matrices can be expressed as follows:

$$[D^B] = \frac{1}{A_1 B_2 - A_2 B_1}\begin{bmatrix} B_2 & -A_2 & 0 \\ -B_1 & A_1 & 0 \\ 0 & 0 & 1/C_1 \end{bmatrix}, \tag{6.46a}$$

$$[D]^U = \begin{bmatrix} D_1 & 0 & 0 \\ 0 & D_2 & 0 \\ 0 & 0 & D_3 \end{bmatrix}, \tag{6.46b}$$

where $[D]^B = [D]$ matrix in the pre-ultimate strength regime; $[D]^U = [D]$ matrix in the post-ultimate strength regime; $A_1 = \frac{1}{E}\frac{\partial \sigma_{x\,max}}{\partial \sigma_{xav}}$; $A_2 = \frac{1}{E}(\frac{\partial \sigma_{x\,max}}{\partial \sigma_{yav}} - v)$; $B_1 = \frac{1}{E}(\frac{\partial \sigma_{y\,max}}{\partial \sigma_{xav}} - v)$; $B_2 = \frac{1}{E}\frac{\partial \sigma_{y\,max}}{\partial \sigma_{yav}}$; $C_1 = \frac{1}{G_e}(1 - \frac{\tau_{av}}{G_e}\frac{\partial G_e}{\partial \tau_{av}})$; $D_1 = -\frac{\sigma_{x\,max}^u}{2}\frac{\sigma_{xcr}}{E\varepsilon_{xav}^2}$; $D_2 = -\frac{\sigma_{y\,max}^u}{2}\frac{\sigma_{ycr}}{E\varepsilon_{yav}^2}$; $D_3 \approx 0$; $E =$ elastic modulus; $v =$ Poisson ratio; σ_{xav}, σ_{yav},

τ_{av} = average membrane stress components; $\sigma_{x\,max}$, $\sigma_{y\,max}$ = maximum membrane stresses in the x or y direction; $\sigma_{x\,min}$, $\sigma_{y\,min}$ = minimum membrane stresses in the x or y direction; G_e = effective shear modulus; and ε_{xav}, ε_{yav} = average membrane strains in the x or y direction.

The effective shear modulus can be determined using a concept similar to the effective width of plating buckled under axial compression but buckled under edge shear, as suggested by Paik (1995).

In Eq. (6.46), maximum or minimum membrane stresses as shown in Figure 6.10 can be calculated by solving the two nonlinear governing differential equations (6.12) and taking into account the effects of influential parameters, including loading, boundary conditions, and initial imperfections. The maximum or minimum stresses, as well as other membrane stress components, will be varied with applied loads (or average membrane stresses).

For a detailed description of deriving the [D] matrix indicated in Eq. (6.46), see Paik and Thayamballi (2003). Note that Fourier series functions are adopted as plate deflection functions in the derivation of the [D] matrix formulations – that is, to solve the nonlinear governing differential equations (6.12) – although the displacement (shape) functions of the plate element for determining the working stress and strain components will be given by polynomial functions, as presented later in Eq. (6.51).

When no failure has occurred in the perfect plate element (i.e., without initial imperfections), the [D] matrix in Eq. (6.46) will become

$$[D] = \frac{E}{1 - \nu^2} \begin{bmatrix} 1 & \nu & 0 \\ \nu & 1 & 0 \\ 0 & 0 & \dfrac{1 - \nu}{2} \end{bmatrix}. \tag{6.47}$$

(4) Tangent Stiffness Equation. By applying the principle of virtual work, the element tangent stiffness equation can be derived as follows:

$$\{\Delta R\} = [K]\{\Delta U\}, \tag{6.48}$$

where $[K] = [K_p] + [K_b] + [K_g] + [K_\sigma]$ = tangent stiffness matrix of the plate element.

In the [K] matrix noted previously, the first and second terms represent the stiffness matrices related to the in-plane and the out-of-plane small deformations, respectively. The third term is the initial deformation stiffness matrix, which consists of three terms representing the geometric nonlinear effects associated with the in-plane and out-of-plane deformations and their interactions. The fourth term is the initial stress stiffness matrix that is produced by the initial stresses for the element, in which a term related to their interactions does not appear. Each term mentioned can be developed in more detail as follows:

$$[K_p] = \begin{bmatrix} [K_1] & 0 \\ 0 & 0 \end{bmatrix}, \quad [K_b] = \begin{bmatrix} 0 & 0 \\ 0 & [K_2] \end{bmatrix},$$

$$[K_g] = \begin{bmatrix} [K_3] & [K_4] \\ [K_4]^T & [K_5] \end{bmatrix}, \quad [K_\sigma] = \begin{bmatrix} [K_6] & 0 \\ 0 & [K_7] \end{bmatrix}, \tag{6.49}$$

where $[K_1] = \int_V [B_p]^T [D] [B_p] \, d \, vol$; $[K_2] = \int_V [B_b]^T [D]^e [B_b] z^2 \, d \, vol$;

$$[K_3] = \int_V [G_p]^T [C_p]^T [D] [B_p] \, d \, vol + \int_V [B_p]^T [D] [C_p] [G_p] \, d \, vol$$

$$+ \int_V [G_p]^T [C_p]^T [D]^e [C_p] [G_p] \, d \, vol;$$

$$[K_4] = \int_V [B_p]^T [D]^E [C_b] [G_b] \, d \, vol + \int_V [G_p]^T [C_p]^T [D]^E [C_b] [G_b] \, d \, vol;$$

$$[K_5] = \int_V [G_b]^T [C_b]^T [D] [C_b] [G_b] \, d \, vol; \quad [K_6] = \int_V [G_p]^T [\sigma_p] [G_p] \, d \, vol;$$

$$[K_7] = \int_V [G_b]^T [\sigma_b] [G_b] \, d \, vol; \quad [\sigma_p] = \begin{bmatrix} \sigma_x & 0 & \tau_{xy} & 0 \\ 0 & \sigma_x & 0 & \tau_{xy} \\ \tau_{xy} & 0 & \sigma_y & 0 \\ 0 & \tau_{xy} & 0 & \sigma_y \end{bmatrix}; \quad [\sigma_b] = \begin{bmatrix} \sigma_x & \tau_{xy} \\ \tau_{xy} & \sigma_y \end{bmatrix}.$$

The stiffness matrix components for the rotations with regard to the z axis may normally be set to zero, but this can in some cases produce numerical instability in the computation of the structural stiffness equation. To get a stabilizing effect in the numerical computation, the stiffness matrix components for the displacement component θ_z can be added to the stiffness matrix, Eq. (6.49). The stiffness equation for the displacement component θ_z may be given by (Zienkiewicz 1977):

$$\begin{Bmatrix} M_{z1} \\ M_{z2} \\ M_{z3} \\ M_{z4} \end{Bmatrix} = \alpha E \, A t \begin{bmatrix} 1 & -\frac{1}{2} & -\frac{1}{2} & -\frac{1}{2} \\ -\frac{1}{2} & 1 & -\frac{1}{2} & -\frac{1}{2} \\ -\frac{1}{2} & -\frac{1}{2} & 1 & -\frac{1}{2} \\ -\frac{1}{2} & -\frac{1}{2} & -\frac{1}{2} & 1 \end{bmatrix} \begin{Bmatrix} \theta_{z1} \\ \theta_{z2} \\ \theta_{z3} \\ \theta_{z4} \end{Bmatrix}, \tag{6.50}$$

where t = plate thickness; A = surface area of the element; α = constant, which may normally be taken to be a very small value, for example, 5.0×10^{-5}.

(5) Displacement (Shape) Function. To attain a uniform state of shear stresses inside the element, a nonlinear function is in the present finite-element method assumed for the in-plane displacements (u, v), and a polynomial function is assumed for the out-of-plane displacement (w), which is expressed in terms of twelve parameters. Thus we have

$$u = a_1 + a_2 x + a_3 y + a_4 xy + \frac{b_4}{2}(b^2 - y^2), \tag{6.51a}$$

$$v = b_1 + b_2 x + b_3 y + b_4 xy + \frac{a_4}{2}(a^2 - x^2), \tag{6.51b}$$

$$w = c_1 + c_2 x + c_3 y + c_4 x^2 + c_5 xy + c_6 y^2 + c_7 x^3 + c_8 x^2 y$$

$$+ c_9 xy^2 + c_{10} y^3 + c_{11} x^3 y + c_{12} xy^3, \tag{6.51c}$$

where $a_1, a_2, \cdots, c_{12}$ = unknown coefficients, which are expressed in terms of nodal displacements {U}.

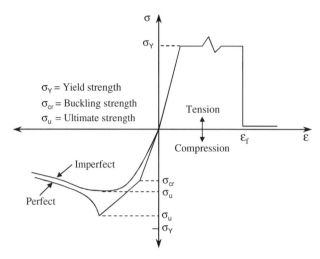

Figure 6.29. A schematic of one type of failure behavior of the present ISUM plate element.

For the rectangular plate element with length of a and breadth of b, the coefficients of the displacement functions can be obtained by substituting local coordinates and displacements at the corresponding nodes into Eq. (6.51), and eventually the stiffness matrix [K] of Eq. (6.48) can be calculated at each incremental loading step. Note that the shape function indicated in Eq. (6.51) is suitable for calculating the average membrane stresses and strains; however, the maximum or minimum membrane stress calculations may need to use different shape functions, such as sinusoidal functions, upon solving Eq. (6.12).

(6) Failure State Considerations. Figure 6.29 represents the schematic of one type of failure behavior considered for developing the present ISUM plate element, where only the failure behavior under predominantly axial loading is shown, although the present ISUM plate element can deal with combined stresses including longitudinal axial stress, transverse axial stress, edge shear, and lateral pressure.

As the axial compressive stress increases, for instance, the in-plane stiffness of the imperfect plate element, for example, with initial deflections, decreases from the very beginning of loading and eventually reaches the ULS. In the post-ultimate strength regime, the internal compressive stress continues to decrease as long as the axial compressive displacement increases. As noted, the stress–strain relations in the pre- and post-ultimate strength regimes can be derived by analytical approaches (Paik and Thayamballi 2003). The ULS criteria can also be formulated considering the various possible parameters of influence (e.g., geometric and material properties, combined loads, initial imperfections, local damages) by analytical approaches.

On the other hand, when the applied action effects are predominantly tensile, the stress–strain relation may follow the typical linear elastic behavior until gross yielding is reached. For practical ULS assessment, the strain-hardening effect is often neglected so that somewhat pessimistic results are obtained, but the full set of material behavior including strain hardening and strain softening must be taken into account for accidental limit-state analyses, discussed in Chapter 8.

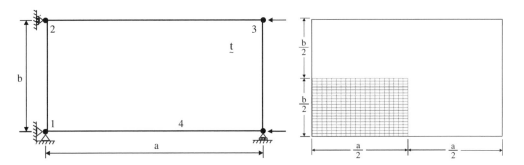

Figure 6.30. Modeling of the plate by ALPS/GENERAL (left) and ANSYS (right).

In the present element formulation, it is assumed that fracture takes place if the equivalent tensile strain reaches a prescribed value of critical (fracture) strain where relevant.

(7) Post-Ultimate Strength Behavior. The tangent stiffness equation of the plate element in the post-ultimate regime is given by Eq. (6.48) but with the [D] matrix of Eq. (6.46b).

(8) Benchmark Study of the Plate Element. A number of benchmark studies have been previously undertaken by a comparison with ISUM plate element analyses, more refined FEM results and physical test data (Paik & Thayamballi 2003); a very fundamental case of the benchmark study is now presented in the following as an example.

An imperfect rectangular plate under uniaxial compressive actions is considered. The dimension of the plate is $a \times b \times t = 1,000 \times 1,000 \times 15$(mm), Young's modulus $= 205,800$ N/mm^2, yield stress $= 352.8$ N/mm^2, and Poisson's ratio $= 0.3$. Initial deflection of the plate is $w_o = 0.05t \cdot \sin\left(\frac{\pi x}{a}\right) \sin\left(\frac{\pi y}{b}\right)$ and no residual stress is considered to exist. The plate is considered to be simply supported along all (four) edges. Figure 6.30 shows the analysis models by ALPS/HULL and nonlinear FEM (ANSYS 2006). For the ANSYS FEM analysis, a quarter of the plate is taken as the extent of the analysis.

Figure 6.31 compares the progressive collapse behavior of the plate under axial compressive actions. In the FEM analysis, two cases are considered, one for plate edges kept straight and the other for unloaded edges moving in plane freely. Since the developed ISUM plate presumes that all edges are simply supported keeping them straight, the ISUM solutions are in good agreement with the FEM results for the former type of edge condition. The plate-edge conditions are more likely to be kept straight in continuous plated structures.

6.7.3.4 ISUM Beam-Column Element

For ISUM modeling, small support members (stiffeners) are to be modeled as beam-column elements, where the nodal points may be located at the intersections between attached plating and support members, as indicated in Figure 6.32. For strong support members with deep webs, flanges of the support members may be idealized as beam-column elements, and the deep webs may be modeled as plate elements because their plate-like behavior is of interest.

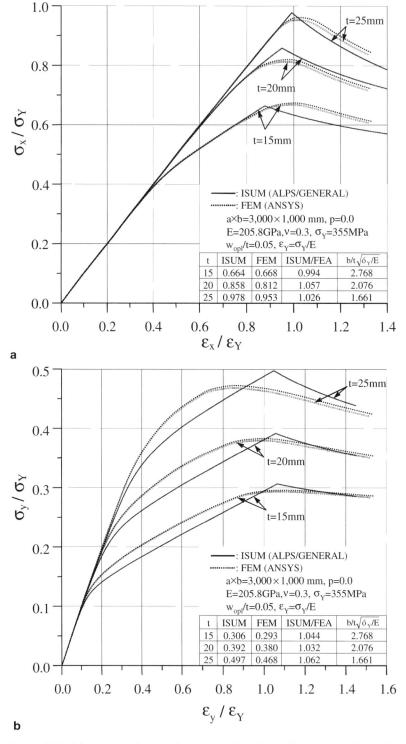

Figure 6.31. (a) A comparison of the progressive plate collapse behavior, under uniaxial compression in the x direction obtained by ISUM and FEM. (b) A comparison of the progressive plate collapse behavior under uniaxial compression in the y direction, obtained by ISUM and FEM.

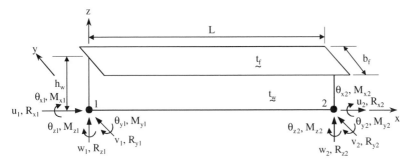

Figure 6.32. The ISUM beam-column element without attached plating (•: nodal points).

(1) Nodal Forces and Nodal Displacements. The element shown in Figure 6.32 has two end nodal points. Each node has 6 degrees of freedom; thus, the nodal force vector {R} and nodal displacement vector {U} are given by [for symbols, unless specified below, see Eq. (6.41)]

$$\{R\} = \left\{ R_{x1}\, R_{y1}\, R_{z1}\, M_{x1}\, M_{y1}\, M_{z1}\, R_{x2}\, R_{y2}\, R_{z2}\, M_{x2}\, M_{y2}\, M_{z2} \right\}^T, \qquad (6.52a)$$

$$\{U\} = \left\{ u_1 v_1 w_1 \theta_{x1} \theta_{y1} \theta_{z1} u_2 v_2 w_2 \theta_{x2} \theta_{y2} \theta_{z2} \right\}^T. \qquad (6.52b)$$

(2) Strain versus Displacement Relationship. The relationship between strain and displacement of the beam-column element should be formulated to reflect the sideways displacements as well as vertical and axial displacements, as follows:

$$\varepsilon = \frac{\partial u}{\partial x} + \frac{1}{2}\left(\frac{\partial v}{\partial x}\right)^2 + \frac{1}{2}\left(\frac{\partial w}{\partial x}\right)^2 - y\frac{\partial^2 v}{\partial x^2} - z\frac{\partial^2 w}{\partial x^2}, \qquad (6.53)$$

where ε = axial strain of the element.

The incremental form of Eq. (6.53) is given by

$$\Delta\varepsilon = \frac{\partial \Delta u}{\partial x} + \frac{\partial v}{\partial x}\frac{\partial \Delta v}{\partial x} + \frac{1}{2}\left(\frac{\partial \Delta v}{\partial x}\right)^2 + \frac{\partial w}{\partial x}\frac{\partial \Delta w}{\partial x} + \frac{1}{2}\left(\frac{\partial \Delta w}{\partial x}\right)^2 - y\frac{\partial^2 \Delta v}{\partial x^2} - z\frac{\partial^2 \Delta w}{\partial x^2}, \qquad (6.54)$$

where the prefix Δ denotes the infinitesimal increment.

(3) Stress versus Strain Relationship. The relationship between axial stress and strain of the beam-column element may be expressed as follows:

$$\Delta\sigma = \frac{L}{A}k\Delta\varepsilon, \qquad (6.55)$$

where L = element length; A = element cross-sectional area; and k = axial stiffness.

The axial stiffness k of the element may vary with various factors such as failure status, loading condition, and possible degradation (e.g., initial imperfections). Although detailed derivation of the k expressions is presented in Paik and

Thayamballi (2003), the following selected results may be relevant for purposes of the present analysis when axial compressive actions are predominant:

$$k = \begin{cases} k_E & \text{for pre-ultimate limit-state regime} \\ k_U & \text{for post-ultimate limit-state regime,} \end{cases} \tag{6.56}$$

where $k_E = \left[\frac{L}{EA} + \frac{\pi^2(\delta_o + w_{q\,max})^2}{2LP_E(1 - P/P_E)^3}\right]^{-1}$; $k_U = -\left\{\frac{4M_P^2}{LP^3}[1 - (\frac{P}{P_P})^4]\right\}^{-1}$; L = element length; δ_o = maximum initial deflection; $w_{q\,max} = \frac{5qL^4}{384EI}$; q = applied lateral line load; E = material elastic modulus; I = moment of inertia of the element cross section with respect to the intersection at the nodal point; P = axial compressive force with positive sign; P_E = elastic buckling force of the element as one of flexural buckling, flexural-torsional buckling (tripping), or local web buckling; M_P = full plastic bending moment of the element cross section, which can be given as $M_P = \frac{t_w h_w^2}{4}\sigma_{Yw}$ for rectangular cross section, that is, without flange, where h_w = web height; t_w = web thickness; σ_{Yw} = web yield stress; and P_P = full plastic axial force that can be given by $P_P = A\sigma_{Yeq}$, where σ_{Yeq} = equivalent yield stress.

When axial tensile forces are predominant or when no failures occur in the perfect element, that is, without initial imperfections, k in Eq. (6.56) becomes

$$k = \frac{EA}{L}. \tag{6.57}$$

(4) Displacement (Shape) Function. The displacement of the beam-column element can be given by

$$u = a_1 + a_2 x, \tag{6.58a}$$
$$v = b_1 + b_2 x + b_3 x^2 + b_4 x^3, \tag{6.58b}$$
$$w = c_1 + c_2 x + c_3 x^2 + c_4 x^3, \tag{6.58c}$$

where $a_1, a_2, b_1, \ldots, c_4$ = unknown constants that can be determined as functions of the nodal displacements.

(5) Tangent Stiffness Equation. Applying the virtual energy principle, the tangent stiffness equation of the beam-column element can be obtained as follows:

$$\{\Delta R\} = [K]\{\Delta U\}, \tag{6.59}$$

where $[K]$ = tangent stiffness matrix, which is given by

$$[K] = \begin{bmatrix} K_{uu} & K_{uv} & K_{uw} \\ & K_{vv} & K_{vw} \\ \text{sym.} & & K_{ww} \end{bmatrix};$$

$$[K_{uu}] = S\int [U_{a1}]^T [U_{a1}]\, d\,vol; [K_{uv}] = S\int [U_{a1}]^T [C_v][U_{b1}]\, d\,vol$$

$$-Sy\int [U_{a1}]^T [U_{b2}]\, d\,vol;$$

$$[K_{uw}] = S \int [U_{a1}]^T [C_w] [U_{c1}] \, d\,vol - Sz \int [U_{a1}]^T [U_{c2}] \, d\,vol;$$

$$[K_{vv}] = S \int [U_{b1}]^T [C_v]^T [C_v] [U_{b1}] \, d\,vol - Sy \int [U_{b1}]^T [C_v]^T [U_{b2}] \, d\,vol$$

$$+ \int \sigma [U_{b1}]^T [U_{b1}] \, d\,vol - Sy \int [U_{b2}]^T [C_v] [U_{b1}] \, d\,vol$$

$$+ Sy^2 \int [U_{b2}]^T [U_{b2}] \, d\,vol;$$

$$[K_{vw}] = S \int [U_{b1}]^T [C_v]^T [C_w] [U_{c1}] \, d\,vol - Sz \int [U_{b1}]^T [C_v]^T [U_{c2}] \, d\,vol$$

$$- Sy \int [U_{b2}]^T [C_w] [U_{c1}] \, d\,vol + Syz \int [U_{b2}]^T [U_{c2}] \, d\,vol;$$

$$[K_{ww}] = S \int [U_{c1}]^T [C_w]^T [C_w] [U_{c1}] \, d\,vol - Sz \int [U_{c1}]^T [C_w] [U_{c2}] \, d\,vol$$

$$+ \int \sigma [U_{c1}]^T [U_{c1}] \, d\,vol - Sz \int [U_{c2}]^T [C_w] [U_{c1}] \, d\,vol$$

$$+ Sz^2 \int [U_{c2}]^T [U_{c2}] \, d\,vol;$$

$$\frac{\partial u}{\partial x} = [U_{a1}] \{U_n\}; \quad \frac{\partial v}{\partial x} = [U_{b1}] \{V_n\}; \quad \frac{\partial w}{\partial x} = [U_{c1}] \{W_n\};$$

$$\frac{\partial^2 v}{\partial x^2} = [U_{b2}] \{V_n\}; \quad \frac{\partial^2 w}{\partial x^2} = [U_{c2}] \{W_n\};$$

$$[C_v] = \frac{\partial v}{\partial x}; [C_w] = \frac{\partial w}{\partial x};$$

$$\{U_n\} = \{u_1 \theta_{x1} u_2 \theta_{x2}\}^T; \{V_n\} = \{v_1 \theta_{z1} v_2 \theta_{z2}\}^T; \{W_n\} = \{w_1 \theta_{y1} w_2 \theta_{y2}\}^T;$$

where σ = axial stress; $S = \frac{kL}{A}$; k = as defined in Eqs. (6.56) and (6.57); and $\int d\,vol$ = integration over the element volume.

(6) Failure State Considerations. Figure 6.33 represents the idealized behavior of the beam-column element for the purpose of ultimate strength analysis. In contrast to plate elements, this beam-column element would not have reserve strength once buckling occurs; thus, the element reaches the ULS immediately after buckling. Again, the effect of strain hardening is usually neglected in the ULS formulations of the element, but it can, of course, be accounted for where necessary.

Possible types of collapse modes for beam-column elements such as beam-column-type collapse, web buckling, lateral-torsional buckling, and gross yielding, as previously described in Section 6.6, must be included in the element formulations.

(7) Post-Ultimate Strength Behavior. The tangent stiffness equation of the beam-column element in the post-ultimate regime is given by Eq. (6.59) but with a different

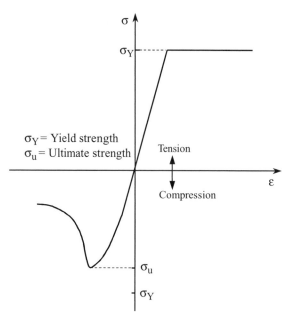

Figure 6.33. Idealized structural behavior of the ISUM beam-column element for ULS analysis.

axial stiffness, which is a function of applied loads. For the calculation of axial stiffness of the beam-column element in the post-ultimate regime, see Paik and Thayamballi (2003).

(8) Benchmark Study of the Beam-Column Element. The ISUM theory of the beam-column element was also added to ALPS/GENERAL (2006) computer program. To verify the accuracy of the ISUM theory presented, a benchmark study is now performed. For this purpose, an unsymmetric I-section beam-column, which is a part of a stiffened plate structure as a representative indicated in Figure 5.1 of Chapter 5, is analyzed by ISUM (ALPS/GENERAL 2006) and FEM (ANSYS 2006) until and after the ultimate strength is reached.

With the nomenclature indicated in Figure 5.2 of Chapter 5, the sectional dimension of the beam-column considered is: $b = b_e = 850$mm, $t = 15$mm, $h_w = 150$mm, $t_w = 11$mm, $b_f = 90$mm, and $t_f = 16$mm. The length of the beam-column is a = 4,250mm. The Young modulus (E) is 205.8GPa, yield stress (σ_Y) is 315MPa, and Poisson ratio (v) is 0.3.

For ALPS/GENERAL analysis, the beam-column is modeled by one ISUM beam-column element simply supported at both ends. For ANSYS nonlinear FEM analysis, the two-bay PSC model indicated in Figure 6.21(b) is used to account for the rotational restraint effect along the transverse frames.

The larger flange has an initial deflection of $W_{opl} = 0.05t$ with local plate buckling mode, and the global initial deflection of the beam-column is 0.0015a for both the vertical direction and sideways of the beam-column. The ultimate strength behavior of the beam-column can be influenced by the pattern of initial deflection. Therefore, two types of initial deflection pattern are considered in the analysis; compression in

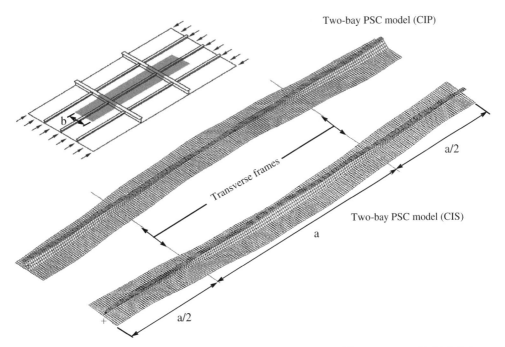

Figure 6.34. ANSYS finite-element modeling of the structure with two types of initial deformation pattern (amplification factor of 20).

plate side (CIP) and compression in stiffener side (CIS), as shown in Figure 6.34. In CIP mode of initial deflection, larger flange (plate) is compressed, although smaller flange is compressed in CIS mode of initial deflection. No residual stress is considered to exist.

Figure 6.35 shows the progressive collapse analysis results for the beam-column obtained by ISUM and FEM. It is seen that ISUM solutions are in good agreement with more refined nonlinear FEM results until and after the ultimate strength is reached.

6.7.3.5 Illustrative Examples

(1) Test Hull Models under Vertical Bending. Experimental results for two-test hull models, that is, a frigate ship hull model tested by the United Kingdom Royal Navy under sagging moment (Dow 1991) and unidirectional double-hull girder system tanker hull models tested by the United States Navy under both sagging and hogging moments (Bruchman et al. 2000), are now analyzed using ALPS/HULL code, which is the special version of ALPS/GENERAL used for progressive hull girder collapse analysis.

Figures 6.36(a) and 6.36(b) show the ALPS/HULL model with von Mises stress distribution at ULS of the frigate test hull under sagging and hogging, respectively. The applied stress components at ULS of some selected plate elements are also presented in Figure 6.36, showing that the effects of transverse axial stresses and edge shear stresses in side shell plating may not be negligible.

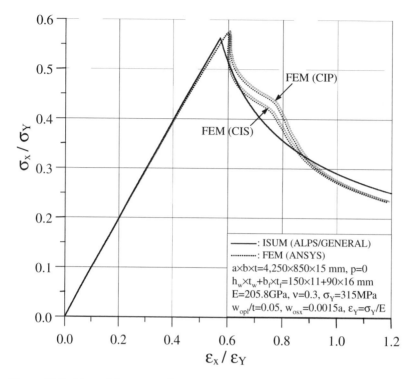

Figure 6.35. The axial compressive stress versus strain curves of the structure with two types of initial deformation pattern, obtained by ISUM and FEM.

For ALPS/HULL simulations, the test ship hull is modeled as an assembly of plate-stiffener separation elements; that is, plate elements for plating between stiffeners and beam-column elements for support members. Although the level of initial plate deflection is kept at $w_{opl}/t = 0.1$, where w_{opl} = maximum initial deflection and t = plate thickness, the compressive residual stress level in plating is varied by $\sigma_{rc}/\sigma_Y = 0.0$ or 0.05, where σ_{rc} = compressive residual stress and σ_Y = material yield stress. The critical fracture strain is assumed to be 25 percent.

Table 6.1 indicates computing times required for ALPS/HULL progressive hull girder collapse analyses using a laptop computer with Pentium (M) processor. Figure 6.37 shows the collapse mode distribution for the individual plate elements (plating) and beam-column elements (support members) at ULS of the frigate test ship hull under sagging and the comparison of the progressive collapse behavior under vertical bending moments as obtained by ALPS/HULL and as obtained from the experiment.

In this case, it is observed that buckling collapse took place at the deck panels and gross yielding occurred at bottom panels until the ULS is reached, while the midheight part of the hull still remains relatively intact. When the critical strain for fracture at the plate or

Table 6.1. *Computing times required for ALPS/HULL progressive hull girder collapse analyses using a laptop computer with Pentium (M) processor*

Test model or ship hull	Computing time
Frigate hull test model	6.5 seconds
Double-skin tanker hull test model	5.5 seconds
FPSO hull	52 seconds
Shuttle tanker hull	45 seconds

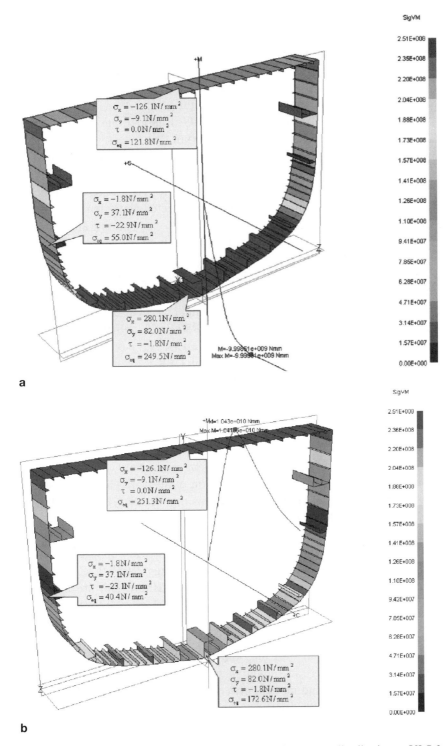

Figure 6.36. ALPS/HULL models together with von Mises stress distribution at ULS for the frigate test hull: (a) under sagging; (b) under hogging.

beam-column element level was set as 25 percent, no fracture took place until and after the hull reached the ULS, but it is possible that fracture can occur at much lower levels of critical fracture strain in some cases, for example, 5 percent, for damaged or aged structures and/or under operation in cold water (Paik 1994; Drouin 2006). It is evident from Figure 6.37 that ALPS/HULL analysis is in good agreement with the experimental results for the frigate test ship hull.

Figures 6.38(a) and 6.38(b) show the ALPS/HULL model together with the von Mises stress distribution at ULS for the double-skin tanker test hull under sagging and hogging, respectively. The test hull is composed of only plate elements and so the ALPS/HULL also models the test hull as an assembly of ISUM plate elements only. Table 6.1 indicates the computing time required for ALPS/HULL simulations.

The applied stress components at ULS for some selected plate elements are shown in Figure 6.38, indicating that edge shear stresses in side shell plating are usually very large in this type of unique structural system and cannot, of course, be neglected in the progressive hull girder collapse analysis even under vertical bending. Figure 6.39(a) shows the collapse modes distribution for individual plate elements at ULS of the double-skin tanker under sagging, and Figure 6.39(b) shows the comparison of the progressive collapse behavior under vertical bending moments as obtained by the experiments and also the ALPS/HULL and the ULTSTR computer programs, the last being developed by the U.S. Navy (e.g., Bruchman et al. 2000). It is seen from Figure 6.39(b) that until ULS is reached under the sagging condition, buckling collapse took place mostly at the upper-half part, while gross yielding occurred at the lower-half part. We also see that no intact elements remain at ULS in this case, somewhat in contrast to the typical ship hull case shown previously, which was composed of stiffened panels together with support members.

Note that this type of unique hull girder system can be very efficient and advantageous in terms of expecting a full contribution of all plate elements against the hull girder collapse process. Figure 6.39 confirms that ALPS/HULL and ULTSTR analyses are in reasonably good agreement with the experimental results.

(2) An FPSO Hull under Vertical Bending. The progressive collapse analysis of a FPSO hull under vertical bending is now performed using ALPS/HULL. The general arrangement and midship section drawings of this offshore unit are indicated in Figure 1.17 of Chapter 1. Figure 6.40 shows ALPS/HULL models for the FPSO hull between two transverse frames or between two transverse bulkheads, both being composed of plate-stiffener separation ISUM elements. If transverse frames are not sufficiently strong so that they may fail before interframe stiffened panels or if combined hull load effects, including shearing forces at the vessel hull section, are significant, one cargo hold model between two transverse bulkheads should be used. For the present example, however, one sliced hull section model between two transverse frames is taken under the assumption that the transverse frames are strong enough under the vertical bending moment.

Figures 6.41 and 6.42 represent the axial stress distribution and the collapse mode distribution over the hull cross section at ULS under sagging or hogging, respectively. Figure 6.43 shows the resulting vertical bending moment versus curvature curve of

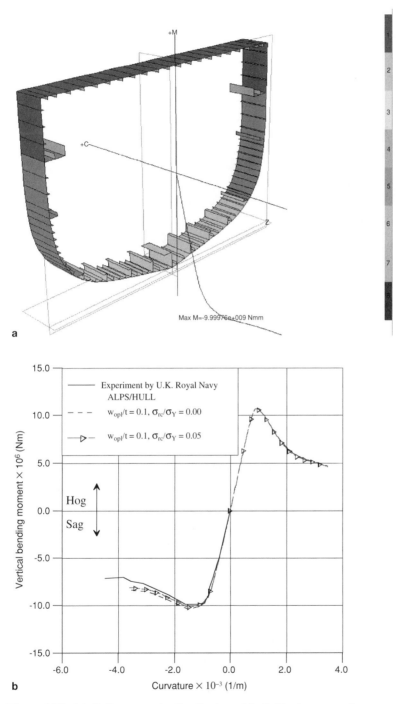

Figure 6.37. (a) Collapse mode distribution of individual structural components at ULS as obtained by ALPS/HULL (Legend: Mode 1 = overall collapse; Mode 2 = biaxial compressive collapse; Mode 3 = beam-column type collapse; Mode 4 = web stiffener buckling collapse; Mode 5 = stiffener tripping collapse; Mode 6 = gross yielding; Mode 7 = ultimate tensile yielding; Mode 8 = fracture). (b) Vertical bending versus curvature curves for the frigate ship-hull test model as obtained by ALPS/HULL and the experiment.

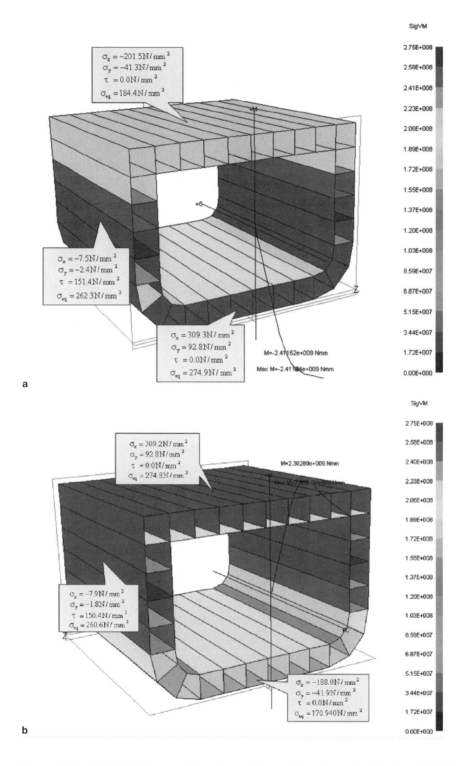

Figure 6.38. ALPS/HULL models together with von Mises stress distribution at ULS for the double-skin tanker test hull: (a) under sagging; (b) under hogging.

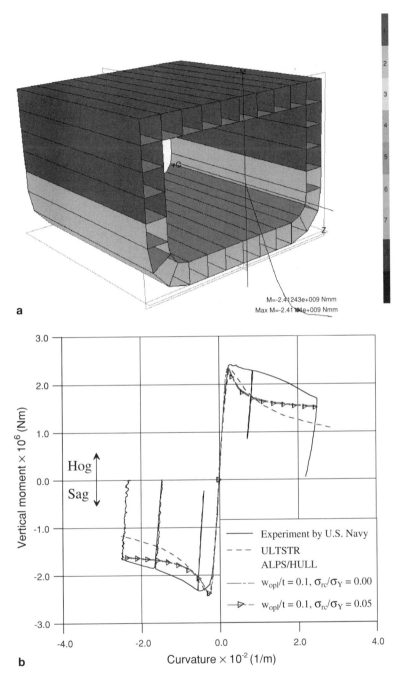

Figure 6.39. (a) Collapse mode distribution for individual structural components of the unidirectional double-skin tanker hull test model at ULS under sagging as obtained by ALPS/HULL. (b) The resulting vertical bending versus curvature curves for the unidirectional double-skin tanker hull test model as obtained by ALPS/HULL, ULTSTR, and the experiment.

the FPSO hull in sagging or hogging. An average level of initial imperfections was applied for the analysis. The safety factor of the FPSO hull can then be defined as the ratio of the computed ultimate hull girder strength to the applied maximum vertical bending moment.

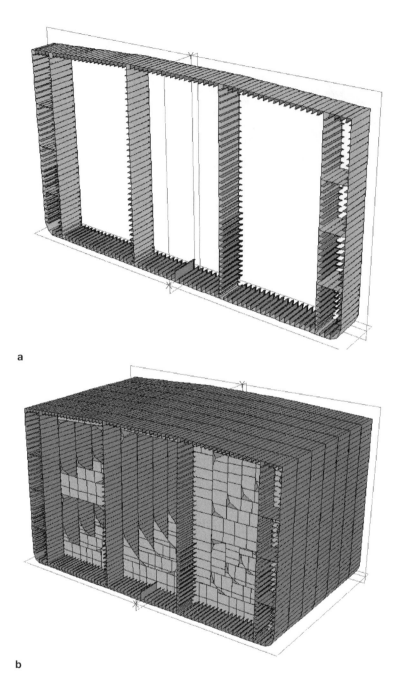

a

b

Figure 6.40. ALPS/HULL models for progressive collapse analysis of an FPSO hull: (a) one sliced hull section model between two transverse frames; (b) one cargo hold model between two transverse bulkheads.

(3) A Shuttle Tanker Hull under Combined Vertical and Horizontal Bending. The progressive collapse analysis of a double-skin shuttle tanker hull under combined vertical and horizontal bending moments is now undertaken by ALPS/HULL analysis. It is expected that the plate elements will be subjected to combined biaxial

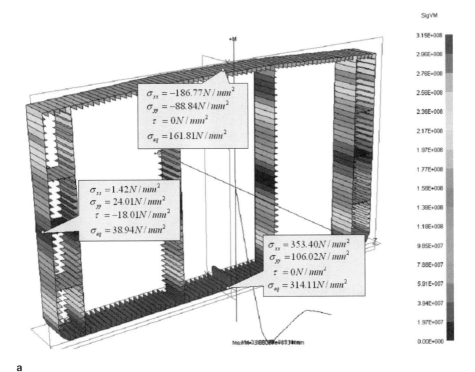

a

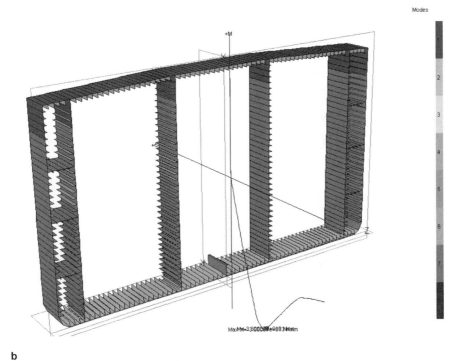

b

Figure 6.41. The stress distribution and the collapse mode distribution over the hull cross section of the FPSO at ULS under sagging, as obtained by ALPS/HULL: (a) von Mises stress distribution; (b) collapse mode distribution.

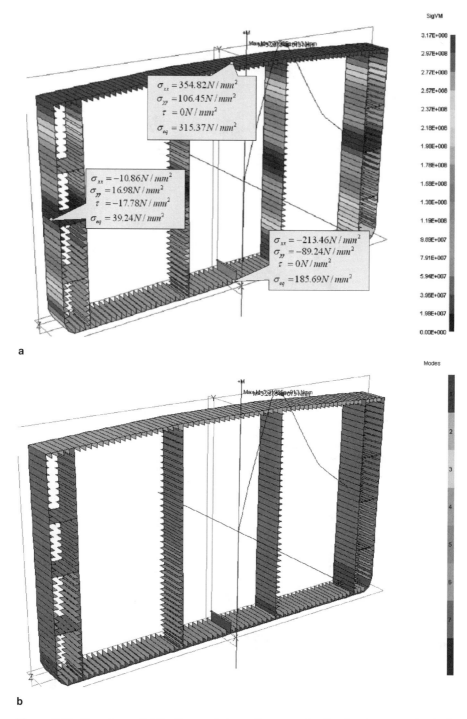

Figure 6.42. The stress distribution and the collapse mode distribution over the hull cross section of the FPSO at ULS under hogging, as obtained by ALPS/HULL: (a) von Mises stress distribution; (b) collapse mode distribution.

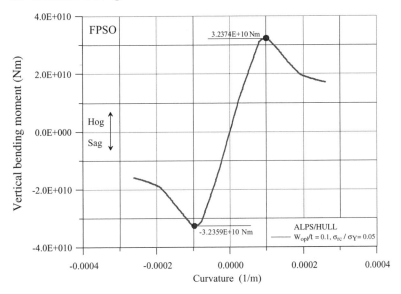

Figure 6.43. The vertical bending versus curvature curve for an FPSO hull, as obtained by ALPS/HULL.

stresses and edge shear stress in this case. Figure 6.44 shows the ALPS/HULL modeling of the tanker. The plate-stiffener separation modeling technique was applied. An average level of initial imperfections in the form of initial distortions and welding residual stresses was assumed in each structural component. Table 6.1 indicates the computing time required for ALPS/HULL progressive collapse analyses.

For the present illustrative examples, three different levels of horizontal bending moments together with pure vertical bending are varied to investigate the effects of horizontal bending on the progressive hull girder collapse analysis. The ratio of the horizontal hull cross-sectional rotation (θ_H) to the vertical hull cross-sectional rotation (θ_V) is kept constant until and after the ULS is reached. Because the in-plane stiffness of individual structural components varies with applied stresses in conjunction with the failure status, the subsequent ratio of the horizontal bending moment (M_H) to the vertical bending moment (M_V) is not the same. The present ALPS/HULL simulations show that the effect of positive horizontal bending is the same to that of negative horizontal bending, which follows because the ship hull section is symmetrical with respect to the ship center line and no structural damage on any one side alone is considered.

The von Mises stresses for the individual structural components at ULS of the ship hull are also shown in Figure 6.44, when $\theta_H/\theta_V = 0.1$ or $M_H/M_V \approx 0.2$. The applied stresses for some selected plate elements are also presented in Figure 6.44. The results show that both transverse axial stresses and edge shear stresses, as well as longitudinal axial bending stresses, may not be negligible and should be carefully taken into account in the progressive collapse analysis in this case. This clearly indicates that any ISUM plate elements that have been developed by considering only longitudinal axial stresses can not be applied reliably for this particular type of problem.

Figure 6.45 shows the failure status of individual structural components at the ULS of the ship hull under combined vertical and horizontal bending when $\theta_H/\theta_V = 0.1$ or

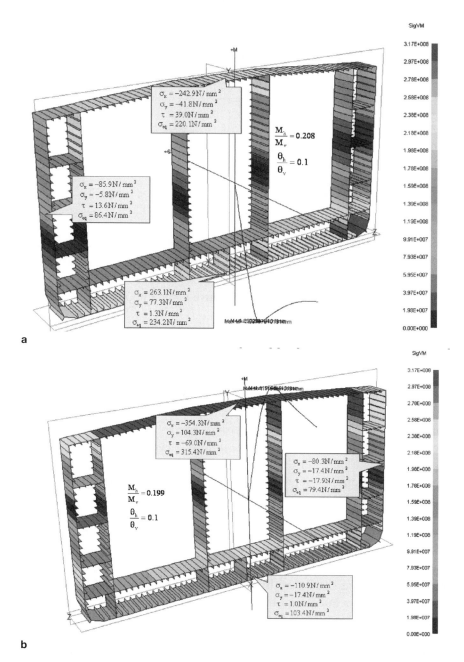

Figure 6.44. ALPS/HULL models together with von Mises stress distribution at ULS for the double-skin shuttle tanker hull when $\theta_H/\theta_V = 0.1$ or $M_H/M_V \approx 0.2$: (a) under combined vertical bending (sagging) and horizontal bending; (b) under combined vertical bending (hogging) and horizontal bending.

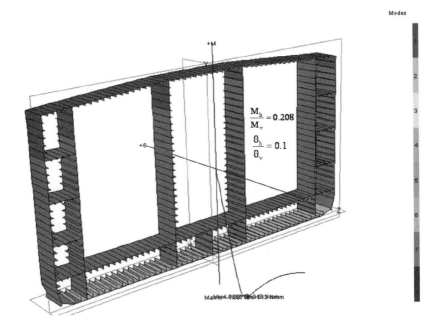

a

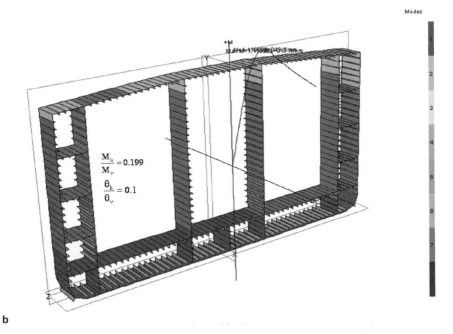

b

Figure 6.45. Collapse mode distribution for the shuttle tanker structural components at ULS when $\theta_H/\theta_V = 0.1$ or $M_H/M_V \approx 0.2$: (a) under combined vertical bending (sagging) and horizontal bending; (b) under combined vertical bending (hogging) and horizontal bending.

Table 6.2. *Effects of horizontal bending on ultimate hull girder strength of the shuttle tanker*

M_V	θ_H/θ_V	M_H/M_V	M_{Vu}/M_{Vu}^*
Sag	0.10	0.208	0.975
	0.15	0.309	0.951
	0.30	0.721	0.839
Hog	0.10	0.199	0.970
	0.15	0.329	0.967
	0.30	0.697	0.894

Notes: M_{Vu} = ultimate hull girder strength under combined vertical and horizontal bending; M_{Vu}^* = ultimate hull girder strength under pure vertical bending.

$M_H/M_V \approx 0.2$. Buckling collapse of some structural components that are subjected to additional compressive loads due to horizontal bending is found to be accelerated compared to those under pure vertical bending. Subsequently, it is also worth noting that the collapse mode distribution of structural components is not symmetrical on either side of the ship center line or with respect to the ship depth direction.

Figure 6.46 shows the effects of horizontal bending moments on the progressive hull girder collapse behavior for the shuttle tanker. Table 6.2 indicates the ultimate hull girder strength reduction due to horizontal bending. For the present specific example where the ship breadth is quite wide, it is observed that horizontal bending does not reduce the ultimate strength of ship hulls under predominantly vertical bending significantly. For horizontal bending with about 30 percent of the maximum

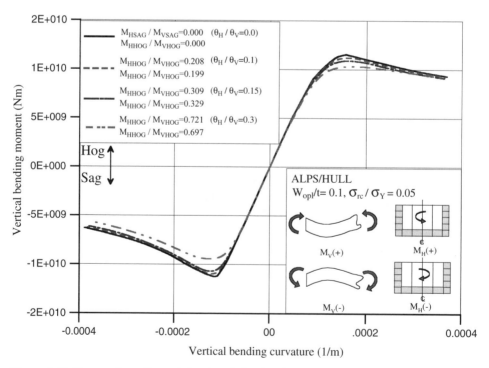

Figure 6.46. Progressive collapse behavior of the shuttle tanker hull under combined vertical and horizontal bending, as obtained by ALPS/HULL progressive collapse analyses.

vertical bending, the ultimate hull girder strength reduction is about 5 percent when compared to the pure vertical bending case.

Of course, the greater the magnitude of horizontal bending, the larger the expected ultimate hull girder strength reduction. In practice, whether the horizontal bending moment is significant would depend on the ship characteristics including hull form, operating condition, and sea-state parameters. Unless known otherwise, it is, of course, only prudent to account for the effect of horizontal bending in addition to vertical bending in any progressive hull girder collapse analysis.

REFERENCES

ALPS/GENERAL (2006). A computer program for progressive collapse analysis of general types of plated structures. Proteus Engineering, Stevensville, MD.

ALPS/HULL (2006). A computer program for progressive collapse analysis of ship hulls. Proteus Engineering, Stevensville, MD (http://www.proteusengineering.com).

ALPS/SPINE (2006). A computer program for elastic-plastic large deflection analysis of stiffened plate structures by the incremental energy method. Proteus Engineering, Stevensville, MD (http://www.proteusengineering.com).

ALPS/ULSAP (2006). A computer program for ultimate limit-state assessment of stiffened panels. Proteus Engineering, Stevensville, MD (http://www.proteusengineering.com).

ANSYS (2006). User's manual (Version 10.0). ANSYS Inc., Canonsburg, PA.

Bruchman, D. D., Kihl, D. P., and Adamchak, J. C. (2000). *Evaluation of the effect of construction tolerances on vessel strength.* Ship Structure Committee Report, SSC-411, Washington, DC.

Dow, R. S. (1991). "Testing and analysis of 1/3-scale welded steel frigate model." *Proceedings of the International Conference on Advances in Marine Structures*, Dunfermline, Scotland, pp. 749–773.

DNV (2000). *Structural design of offshore ships.* (Offshore Standards, DNV-OS-C102), Det Norske Veritas, Oslo, January.

Drouin, P. (2006). "Brittle fracture in ships – A lingering problem." *Ships and Offshore Structures*, 1(3): 229–233.

DYNA3D (2004). User's manual (Version 960), Livermore Software, CA.

Fox, E. N. (1974). "Limit analysis for plates: The exact solution for a clamped square plate of isotropic homogeneous material obeying the square yield criterion and loaded by a uniform pressure." *Philosophical Transactions of the Royal Society of London, Series A (Mathematical and Physical Sciences)*, 277: 121–155.

Gordo, J. M., and Guedes Soares, C. (1997). "Interaction equation for the collapse of tankers and containerships under combined bending moments." *Journal of Ship Research*, 41(3): 230–240.

IACS (2005). *Common structural rules for double hull oil tankers.* International Association of Classification Societies (http://www.jtprules.com), London, December.

Jones, N. (1975). "Plastic behavior of beams and plates." In *Ship structural design concepts.* Cambridge, MD: Cornell Maritime Press, pp. 747–778.

MacMillan, A. (2001). *Effective FPSO/FSO hull structural design.* Offshore Technology Conference, OTC 13212, Houston, May.

MAESTRO (2006). Version 8.7.2. Proteus Engineering, Stevensville, MD (http://www.proteusengineering.com).

Marguerre, K. (1938). "Zur theorie der gekreummter platte grosser formaenderung." *Proceedings of the 5th International Congress for Applied Mechanics*, Germany.

NORSOK N004 (1999). *Design of steel structures.* Norwegian Standards, Norway.

Ozguc, O., Das, P. K., and Barltrop, N. D. P. (2005). *Rational interaction design equations for the ultimate longitudinal strength of tankers, bulk carriers, general cargo, and container ships under coupled bending moment.* Department of Naval Architecture and Marine Engineering, Universities of Glasgow and Strathclyde, Glasgow.

Paik, J. K. (1994). "Tensile behavior of local members on ship hull collapse." *Journal of Ship Research*, 38(3): 239–244.

Paik, J. K. (1995). "A new concept of the effective shear modulus for a plate buckled in shear." *Journal of Ship Research*, 39(1): 70–75.

Paik, J. K. (2004). "A guide for the ultimate longitudinal strength assessment of ships." *Marine Technology*, 41(3): 122–139.

Paik, J. K., and Lee, M. S. (2005). "A semi-analytical method for the elastic-plastic large deflection analysis of stiffened panels under combined biaxial compression/tension, biaxial in-plane bending, edge shear and lateral pressure loads." *Thin-Walled Structures*, 43: 375–410.

Paik, J. K., Lee, J. M., Shin, Y. S., and Wang, G. (2004). "Design principles and criteria for ship structures under impact pressure actions arising from sloshing, slamming and green water." *SNAME Transactions*, 112: 292–313.

Paik, J. K., and Mansour, A. E. (1995). A simple formulation for predicting the ultimate strength of ships. *Journal of Marine Science and Technology*, 1(1): 52–62.

Paik, J. K., and Thayamballi, A. K. (2003). *Ultimate limit state design of steel-plated structures.* Chichester, UK: John Wiley & Sons.

Paik, J. K., Thayamballi, A. K., and Che, J. S. (1996). "Ultimate strength of ship hulls under combined vertical bending, horizontal bending, and shearing forces." *SNAME Transactions*, 104: 31–59.

Paik, J. K., Thayamballi, A. K., and Lee, J. M. (2004). "Effect of initial deflection shape on the ultimate strength behavior of welded steel plates under biaxial compressive loads." *Journal of Ship Research*, 48(1): 45–60.

Paik, J. K., Thayamballi, A. K., Lee, S. K., and Kang, S. J. (2001). "A semi-analytical method for the elastic-plastic large deflection analysis of welded steel or aluminum plating under combined in-plane and lateral pressure loads." *Thin-Walled Structures*, 39: 125–152.

Paik, J. K., Hughes, O. F., Hess, P. E., and Renaud, C. (2005). "Ultimate limit state design technology for aluminum multi-hull ship structures." *SNAME Transactions*, 113: 270–305.

Saitoh, T., Yoshikawa, T., and Yao, H. (1995). "Estimation of deflection of steel panel under impulsive loading." *Journal of the Japan Society of Mechanical Engineers*, 61: 2241–2246.

Smith, C. S. (1977). "Influence of local compressive failure on ultimate longitudinal strength of ship's hull." *Proceedings of the International Symposium on Practical Design in Shipbuilding*, Tokyo, pp. 73–79.

Steen, E., Byklum, E., Vilming, K. G., and Ostvold, T. K. (2004). "Computerized buckling models for ultimate strength assessment of stiffened ship hull panels." *Proceedings of the 9th International Symposium on Practical Design of Ships and Floating Structures*, Vol. 1, pp. 235–242.

Troitsky, M. S. (1976). *Stiffened panels: Bending, stability, and vibrations.* Amsterdam, The Netherlands: Elsevier Scientific Publishing.

Ueda, Y., and Rashed, S. M. H. (1974). "An ultimate transverse strength analysis of ship structures." *Journal of the Society of Naval Architects of Japan*, 136: 309–324 (in Japanese).

Ueda, Y., and Rashed, S. M. H. (1984). "The idealized structural unit method and its application to deep girder structures." *Computers and Structures*, 18(2): 277–293.

Zienkiewicz, O. C. (1977). *The finite element method.* New York: McGraw-Hill.

CHAPTER 7

Fatigue Limit-State Design

7.1 Introduction

As we discussed in Chapters 3 and 5, limit states are classified into four categories: serviceability limit states (SLS), ultimate limit states (ULS), fatigue limit states (FLS), and accidental limit states (ALS). This chapter presents FLS design principles and criteria together with selected engineering practices applicable for the structure of ship-shaped offshore units.

Under the action of repeated loading, fatigue cracks may in time be initiated in the stress concentration areas of ship-shaped offshore structures, and indeed have been reported by Hoogeland et al. (2003) and Newport et al. (2004), among others. In general, the fatigue damage at a crack initiation site is affected by many factors, such as material properties (e.g., elastic modulus, ultimate tensile stress); high local stresses (e.g., stress concentration, residual stresses); size of components; nature of stress variation (e.g., stress variation during the loading and off-take cycles, number of wave-induced stress range cycles); and environmental and operational factors including corrosion and performance of coatings. Potential flaws (e.g., poor materials, porosity, slag inclusions, undercuts, lack of fusion, incomplete weld root penetration) and misalignments can also significantly increase stress concentration and initial defects at welds.

To achieve greater fatigue durability in a structure, therefore, stress concentrations, flaws, and structural degradation, including corrosion and fatigue effects, must either be avoided or minimized or, more commonly, their levels and effects either in design, construction, and/or service must be monitored and effectively controlled to acceptable levels. The effect of stress concentrations intentionally present is assessed at the design stage in order to ensure that the fatigue life of the structure is longer than the design service life with an adequate factor of safety. Construction defects are, in practice, to be monitored and controlled by appropriate construction standards. However, to the extent that such standards are often generic to a type of vessel and based mainly on what can be economically achieved, they may need to be selectively enhanced in many cases depending on the individual structural characteristics.

The in-service inspection and maintenance regime to be used is related to fatigue design and construction. In usual cases, it may be more economical in design to allow a certain (small) level of possibility of fatigue damage, as long as the structure can perform its function during and until repairs are made after the fatigue symptoms are detected. This is basically an economic consideration, in that to almost completely eliminate fatigue in such cases leads to a heavier structure with consequent additional initial cost and also loss in payload. In some other cases, however, one may choose to be less tolerant of fatigue damage if it is inconvenient or difficult to inspect the structure, interrupt production, or, in some sense, the related consequences of fatigue cracking are very significant.

Hull structures of ship-shaped offshore units as described in Section 1.5 in Chapter 1 are quite similar to those of trading tankers in terms of overall geometric dimensions and properties but differ as to the characteristics of action effects to which they are subject. Furthermore, ship-shaped offshore units have unique structural details at the connections between hull and topsides facilities and between hull and risers or mooring lines. Examples include topsides supports, pipe racks, flare tower foundation, and also supports to oil offloading lines, risers, and mooring lines connected to the vessel.

Therefore, traditional FLS design approach used for trading tanker structures, where certain dominant profiles of environmental and operational actions are considered as the main fatigue actions, may not be fully adopted for ship-shaped offshore units. Therefore, to be more accurate, first-principles methods must be employed to greater degrees in such cases.

It is important also to point out that traditional FLS design methods, typically applied in classification society rules for the routine fatigue design of trading tankers, must not normally be applied unchanged to offshore structures, primarily because such procedures often tend to contain various adjustment factors (some obvious and some not) usually obtained by a calibration exercise designed to prevent large departures from existing ranges of vessel experience and steel weight.

This chapter focuses on FLS design practices for ship-shaped offshore structures using first-principles methods involving direct seakeeping analysis of a vessel's motions and actions due to varying loads. In the common case of fatigue evaluation for wave-induced effects, this is the *spectral fatigue method*. The related spectra can be established either in closed-form (as we will see later in this chapter) or by cycle-counting techniques based on time history of the effects concerned. The fatigue life is evaluated using the S–N curves based on small specimen constant amplitude fatigue test data. This is the commonly used and accepted method, although in special cases other possibilities also exist, for example: (a) use of variable amplitude fatigue test data, (b) use of large specimen test data, and (c) use of fatigue crack growth and fracture mechanics methodologies.

For time-variant residual strength assessment of structures with fatigue cracking damage, it is necessary to identify the crack propagation characteristics; therefore, this chapter also presents and discusses time-variant fatigue crack propagation models. The effect of fatigue cracking on ultimate strength of steel plates and vessel hulls is discussed in Chapter 11.

7.2 Design Principles and Criteria

FLS represents the fatigue crack occurrence of structural details due to stress concentration and damage accumulation under the action of repeated loading. In the relatively common context of use of S–N curves derived from small specimen fatigue test data, the related state of failure is often assumed to correspond roughly to the

Table 7.1(a). *Sample safety factors for fatigue limit-state design of ship-shaped offshore structures*

Structure	Hull			Topsides
Location	All structures excluding side connections	Side shell connections	Noninspectable areas (e.g., off-vessel mooring components, I-tubes)	Uniform throughout
Safety factor	1 for North Atlantic condition; 3 for combined transit and onsite condition	2 for North Atlantic condition; 4 for combined transit and onsite condition	10	2

initiation of a through-thickness crack at a particular location. We appreciate, however, that for practical purposes a crack that is even so initiated may not be visually observed until it is longer. In the same vein, surface cracks are even more difficult to observe without a specialized means such as dye penetration or magnetic particle testing.

In any event, it is worth pointing out that there exists a certain amount of ambiguity as to what the FLS failure of the real structure physically correlates to the fatigue data used in design. For this and many other reasons, the FLS design in a particular case is carried out so that it is ensured that the structure has an adequate fatigue life that is longer than the design service life by an appropriate factor of safety. Also, the predicted fatigue life is required input for purposes of planning efficient inspection programs during the operation of the structure.

The design fatigue life for structural details of ship-shaped offshore units is normally specified by the operator or owner, but that of trading tankers is usually specified by bodies such as classification societies. For new-build ship-shaped offshore structures, the fatigue life may often be taken anywhere from 20 to 60 years or longer (including safety factors), and it is typically taken as 20–25 years for trading tankers.

In new-build ship-shaped offshore structures, the fatigue safety factors might vary from 1 to 3 or more, and occasionally even 10, depending on the maintenance philosophy to be employed in service and on the potential consequences of fatigue failure at a given location and the potential consequences of associated downtime. The shorter the design fatigue life, the smaller the inspection intervals need to be if a crack problem-free operation is to be assured. Table 7.1(a) indicates sample safety factors used for FLS design of ship-shaped offshore structures in practice. Table 7.1(b) indicates fatigue safety factors for hull interface structures suggested by ABS (2004).

The inability to dry-dock is also a factor in some owners specifying fatigue safety factors greater than those for trading tankers that are able to dry-dock every

Table 7.1(b). *Fatigue safety factors for hull interface structures, after ABS (2004)*

Severity	Inspectable and field repairable	
	Yes	No
Noncritical	3	5
Critical	5	10

5 years for extensive inspection and repairs. It will be appreciated that in the case of a tanker conversion to a ship-shaped offshore unit, fatigue safety factors closer to those for the trading tanker may usually be economically necessary; in such cases, it is also common that an extensive structural integrity monitoring program will be employed in service, at least in harsh environments. Tanker conversions, however, are typically targeted for shorter times of onsite service than their new-build counterparts.

The FLS assessment and design should, in principle, be undertaken for every suspect location of fatigue cracking that includes welded joints and local areas of stress concentrations and for all relevant loads. Although wave-induced actions are primary sources of fatigue, the effects due to the following loads may need to be considered depending on the design and circumstances:

- Functional loads including those related to loading and offtake of cargo
- Wind loads; for example, the effect of vortex-induced vibrations and vortex shedding
- Slamming loads
- Sloshing loads
- Local load effects arising from mooring and riser systems

Procedures and criteria related to slamming and sloshing effects for FLS purposes are not generally well defined, and neither are these two load effects amenable to a closed form spectral fatigue method. In any event, it appears in practice that these conditions are primarily evaluated for strength, even if fatigue is sometimes said to be suspected in related failures. The dry-docking condition, any docking condition afloat, and any damage condition, although relevant for strength assessment, normally do not need to be considered for FLS. The calculations should address all transit conditions, for example, tow to the field or to a shipyard for repair and all onsite operating conditions.

The structural design criteria for FLS are usually based on cumulative fatigue damage under repeated fluctuation of loading, as measured by the Palmgren–Miner cumulative damage accumulation rule for purposes of using the S–N curve approach (S = fluctuating stress, N = number of stress cycle). A particular value of the Miner sum (in principle, unity) is taken to be synonymous with the formation or initiation of a crack and is calculated as follows:

$$\frac{n_1}{N_1} + \frac{n_2}{N_2} + \cdots + \frac{n_i}{N_i} + \cdots + \frac{n_k}{N_k} = 1, \qquad (7.1)$$

where n_i = number of cycles at the ith stress range level; N_i = fatigue life at the ith stress range level; and k = number of stress range levels. The structure is designed so that when analyzed for fatigue, a reduced target damage sum results, implying that cracks will not form with a given degree of certainty or safety factor.

In applying the S–N curve approach, the following three steps are required: (1) define the histogram of cyclic stress ranges; (2) select the relevant S–N curve; and (3) calculate the cumulative fatigue damage.

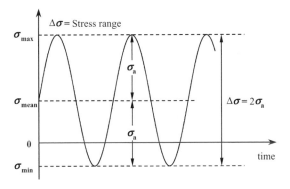

Figure 7.1. Cyclic stress range versus time.

7.2.1 Cyclic Stress Ranges

In the fatigue damage assessment of welded structural details, our primary concern is the ranges of cyclic maximum and minimum stresses rather than mean stresses, as shown in Figure 7.1. When the maximum and minimum stresses are denoted by σ_{max} and σ_{min}, respectively, the mean stress is given by $\sigma_{mean} = (\sigma_{max} + \sigma_{min})/2$ and the stress (or loading) ratio is defined by $R = \sigma_{min}/\sigma_{max}$. The stress range is defined by $\Delta\sigma = \sigma_{max} - \sigma_{min} = 2\sigma_a$, where σ_a is called the "stress amplitude."

In principle, the fatigue characteristics are affected by the R value as well as $\Delta\sigma$, but the effect of mean stress is typically neglected for welded structures for two interrelated reasons: (a) the likely presence of near yield magnitude tensile residual stresses in vicinity of a weld, which tends to make the entire stress range damaging for fatigue purposes; and (b) the typical use of S–N curves based on the R = 0 data. Note that in nonwelded cases, R effects may be significant, but in ship-shaped offshore structures, it is typically the welded details that usually govern for FLS purposes – except possibly cutouts.

In the context of spectral-analysis-based FLS design of ship-shaped offshore units for wave-induced effects, the vessel motion and load response amplitude operators (RAOs) are obtained by the seakeeping analysis for a range of wave frequencies and headings as well as for representative cases of each of the vessel loading cases identified.

Then, the global and local structural analyses with relevant finite-element models are used to obtain the wave-induced cyclic stress range transfer function for each wave heading angle and frequency. Note that the load transfer functions are complex numbers; hence, there actually are two stress analysis cases that must be gone through before the stress transfer function value can be established for a given heading, speed, frequency, and load case. This can be quite an intensive process of computation. There are variations of this process in which stress influence functions for loads are used.

Using scatter diagrams of waves and swell that represent a number of sea states describing short-crested seas in terms of joint probability of occurrence of wave height and period, the spectral density of stress range is calculated by integrating the product of the wave energy spectrum and the modulus-squared of the stress range spectrum for each combination of significant wave height and period; therefore, the variance of stress range is obtained for the short-term sea state. A series of such

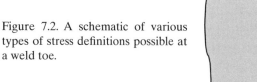

Figure 7.2. A schematic of various types of stress definitions possible at a weld toe.

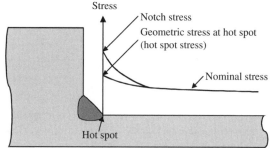

short-term response analyses are made, and the results are accumulated to obtain a total stress-range distribution (cyclic stress ranges) in the long term that represents the number of cycles for each combination of wave height and period for the computation period considered; for example, the time at sea or years on site.

For using the S–N curve approach, three types of stresses may be relevant: nominal stress, hot spot stress, and notch stress, depending on which type of stress the S–N curve to be used is based on. Figure 7.2 shows a schematic representation of various types of stresses.

The hot spot stress is defined as the largest value of the extrapolation to the weld toe of the geometric stress distribution immediately outside the region affected by the geometry of the weld. The notch stress represents the total stress at the weld toe including the geometric stress and the stress due to the presence of the weld itself. Note that the hot spot stress shown in Figure 7.2 is a general case; quite often in the derivation of S–N curves, its distribution in the vicinity of the weld toe is linearized. In any event, one important point is that application of S–N curves requires consistent determination of stresses (stress ranges) of a particular type and obtained in a particular manner. Although seemingly trivial, such considerations are important because the fatigue damage predictions are proportional to the stress range cubed at least; therefore, any stress-related errors and approximations are also significantly magnified.

Nominal stress is considered in the sectional area by neglecting stress concentrations due to structural joint configuration as well as notch at weld toe. Structural ineffectiveness related to the reduced effective plate breadth due to stress diffusion, or stress gradients due to cutouts, whose effects are not implicit in the S–N curve in the vicinity of the structural detail (represented by geometric stress concentration factors), must be accounted for.

Typically, the nominal stress consists of two components: global stress and local bending stress, the former being associated with the whole hull girder or primary structural components and the latter with secondary structural behavior such as those of plates and stiffeners between web frames, and the plate itself between stiffeners. When the nominal stress in a particular structure is used in conjunction with nominal stress-based S–N curves, it must be multiplied by the appropriate geometric stress concentration factors. The hot spot stress, on the other hand, implicitly represents the geometric stress (also called structural stress) at weld toe accounting for the effects of the geometric stress concentration due to structural discontinuities and presence of attachments, for instance, but excludes the localized notch stress increase due to the presence of the weld itself. In applying a hot spot stress-based S–N curve, it is the hot spot stress that would be calculated for application purposes, while the notch stress enhancement effect is implicit in the S–N lives.

The effect of fabrication-related imperfections (e.g., misalignment) must be additionally considered. However, it is often the assumption that standardized FLS procedures based on nominal stress-based S–N curves in particular have the effect of fabrication imperfections implicit in them (e.g., in the safety factors used); therefore, in such cases it is only the level of fabrication imperfections exceeding standard construction standards that are usually separately considered; such need for evaluation may occur in the reanalysis of a particular known misalignment situation to decide if it must be repaired.

The relation between nominal stress range and hot spot stress range may be given by

$$\Delta\sigma_h = K\Delta\sigma_n, \qquad (7.2a)$$

where $\Delta\sigma_h$ = hot spot stress range; $\Delta\sigma_n$ = nominal stress range; and K = stress concentration factor (also called the K-factor). The K-factor is sometimes further subdivided as follows:

$$K = K_g K_e K_a K_n, \qquad (7.2b)$$

where K_g = K-factor related to the geometry of the detail; K_e = K-factor related to eccentricity for plate butt-weld connections and cruciform joints, for example; K_a = K-factor related to angular mismatch for plate connections; and K_n = K-factor for asymmetrical stiffeners on laterally loaded panels applicable when the stress is calculated by classical beam theory without stiffener rotation effects.

This chapter focuses on hot spot stress-based FLS assessment and design methods. In Section 7.7, we describe how the hot spot stress can be obtained in particular cases by appropriate finite-element analysis of structural details and the extrapolation of the applicable stress is concerned to the weld toe in a certain predefined manner. The use of notch stresses is not common but also may occur, for instance, in using strain versus life curves for the prediction of low-cycle fatigue effects.

7.2.2 S–N Curves

For practical FLS assessment using the S–N curve approach, the relevant S–N curves must be developed for various types of weld joints. To do this, fatigue tests are carried out on physical models of specific structural details. For convenience, the fatigue tests are usually performed so that the test models are subjected to cyclic stress ranges of uniform amplitude, although actual stress-cycle amplitude in service will in most cases be irregular and nonuniform. As surmised from Eq. (7.1), however, the fatigue damage accumulation can be calculated once individual stress levels and their numbers are quantified to the extent that *Miner's rule* holds true for the situation considered. This appears to be largely true for structural effects due to a stationary relatively narrow band Gaussian sea state, for example.

Based on the fatigue test results conducted usually in air and at room temperature, the number of stress cycles until a crack initiates (or other predefined failure criteria including, for example, a small specimen breaking apart or a certain percentage

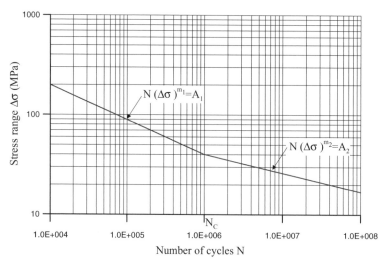

Figure 7.3. A schematic of a typical bilinear S–N design curve.

increase in compliance) can be obtained. Through a series of such tests for a variety of stress ranges $\Delta\sigma$, the S–N curves for the particular structural details may be obtained and plotted. It is of some interest that typical S–N design curves used for ship-shaped offshore structures may consist of two segments, as shown in Figure 7.3. The curves in the figure are usually expressible by curve fitting of the test results plotted on a log–log scale, as follows:

$$N(\Delta\sigma)^{m_1} = A_1, \quad \text{for } N < N_c, \tag{7.3a}$$

$$N(\Delta\sigma)^{m_2} = A_2, \quad \text{for } N \geq N_c, \tag{7.3b}$$

where $\Delta\sigma$ = corresponding stress range (i.e., nominal or hot spot); N = number of stress cycles with constant stress range $\Delta\sigma$, until failure; m_1, m_2 = negative inverse slopes of the S–N curve; $\log A_1 = \log a_1 - 2s_1$; $\log A_2 = \log a_2 - 2s_2$; a_1, a_2 = life intercepts of the mean S–N curve; s_1, s_2 = standard deviations of $\log N$, used to account for scatter in S–N data for design purposes; $\Delta\sigma_c$ = stress range at which the slope of S–N curve changes or at $N = N_c$; and N_c = number of cycles at which the slope of S–N curve changes.

Figure 7.3 shows a two-segment design S–N curve, obtained from the mean by subtracting approximately 2 standard deviations of the data scatter in fatigue life given the stress range. Although this may not be obvious, it should also be stated that the low-stress high-cycle regime curve is often not based on particular data but on a generic hypothesis of how the curve would behave in that regime. The real data is often in the first zone of the S–N curve alone. This is partly due to testing economics reasons; the second zone slope change is also common when only justified by better equivalence to failure under variable amplitude loading – the Haibach hypothesis. The usual practice is that the two-segment design S–N curve is applied for cases of structural details in air and when effectively protected from corrosion primarily by coatings.

Where the primary corrosion protection system is cathodic protection, the first zone is simply extrapolated to the long-life regime. Under freely corroding conditions, a penalty factor of two to three must be used on the fatigue lives predicted using in-air

S–N curves. For a good exposition of these various assumptions and considerations, as well as fatigue testing in general, and the derivation and application of S–N curves for design, see Gurney (1979).

As necessary, the S–N curves for particular cases in design can of course be obtained from recognized FLS design guidance notes and classification society procedures rather than from direct fatigue testing. Regarding the S–N curves, relatively extensive and clear guidelines appear to be available in some cases; see offshore standards codes and recommended practices of DNV in particular (e.g., DNV 1998, 2002, 2006; and the cross references therein). Comparisons of some of the available codes and standards regarding S–N curves are made in HSE (2001), Wang and Cheng (2003), and Wang et al. (2005), among others. Needless to say, although the longtime existence of specific structural codes and related software may give another impression, fatigue-life predictions of offshore structures remain a subject that typically requires specialist involvement and advice.

In many guidelines, the S–N curves for different connection types of interest in ship-shaped offshore structures may be grouped into four types: (a) those for flame cut edges; (b) those corresponding to weld toe stress that is parallel to weld toe; (c) those for weld toe stress perpendicular to weld toe; and (d) those for fillet weld root failure in shear. A fifth type, that for base metal, is often not of interest. A hot spot stress-based curve corresponding to (c) is usually included in addition to any nominal stress-based S–N curves for situations involving (c) and others.

7.2.3 Fatigue Damage Accumulation

The fatigue damage accumulation using the S–N curve approach can be made in nondimensional form using the Miner hypothesis once the distribution of the long-term stress range has been discretized into a relevant stress range versus number of cycles histogram. For a histogram with a number of constant amplitude stress range levels $\Delta\sigma_i$ each with a number of stress fluctuations n_i, the fatigue damage for a two zone S–N curve shown may be calculated as follows:

$$D = \sum_{i=1}^{k} \frac{n_i}{N_i} = \sum_{i=1}^{k_1} \frac{n_i}{A_1}(\Delta\sigma_i)^{m_1} + \sum_{j=1}^{k_2} \frac{n_j}{A_2}(\Delta\sigma_j)^{m_2}, \tag{7.4}$$

where D = accumulated fatigue damage; $k = k_1 + k_2 =$ total number of stress levels; k_1 = number of stress levels for $\Delta\sigma < \Delta\sigma_c$; k_2 = number of stress levels for $\Delta\sigma \geq \Delta\sigma_c$; n_i = number of stress cycles in stress level i; and N_i = number of cycles until failure at the ith constant amplitude stress range $\Delta\sigma_i$. In specific cases, closed-form approximations to this expression for accumulated fatigue damage are also possible.

To be safe, the calculated fatigue damage accumulation must be smaller than the target cumulative fatigue damage for design, as follows:

$$D \leq D_{cr}, \tag{7.5}$$

where D_{cr} = target fatigue damage sum or usage factor that is inverse value of design fatigue safety factor. As previously stated, the factor of safety required in particular cases will depend on the long-term reliability required, the feasibility of dry-docking, the physical consequence of failure, the ease of inspection, the ease of repair, and related aspects of cost including production interruption.

Table 7.2. *Recommended target Miner sum for new-build FPSO structures (BV 2004)*

	Degree of accessibility for inspection, maintenance, and repair		
Consequence of failure	Not accessible[2]	Underwater[2]	Dry inspection[3]
Critical[1]	0.1	0.25	0.5
Noncritical	0.2	0.50	1.0

Notes: [1]Includes loss of life, uncontrolled pollution, collision, sinking, other major damage to the installations, and major production loss. [2]Includes areas that can be inspected in dry or underwater conditions but require dry-docking for repair. [3]Includes areas that can be inspected in dry conditions but with extensive preparation and heavy impact on operation.

Table 7.2 from BV (2004) indicates one example of recommended practice in this regard; in this case, for new-build FPSOs by tabulating the recommended target Miner sum for a given life.

7.3 Practices for Spectral-Analysis-Based FLS Design

FLS design methods may be classified into three types: (1) deterministic, (2) semiprobabilistic, or (3) spectral-analysis methods (HSE 1997). One of the simplest of deterministic methods is based on an extreme dynamic response, for example, one that occurs once in 20–25 years. In conjunction with an assumed number, for example, 5×10^6 of wave cycles per year, the extreme dynamic response is used to define a log-linear exceedance diagram of stresses. The fatigue damage is then estimated after obtaining the number of cycles corresponding to various stress ranges. This method is generally used for a very preliminary estimation of fatigue damage in simple cases of wave-induced fatigue effects. However, it should be realized that the predicted fatigue life using this method can be significantly affected by loads, local stress determination, and the S–N curve. Fricke et al. (2002) made a comparative study on deterministic fatigue strength assessment procedures of various classification societies. The predicted fatigue lives for a structural detail of a Panamax container vessel longitudinal coaming considered as an example had large differences ranging from 1.8 to 20.7 years.

The semiprobabilistic method, the next higher level of approximation, divides each sea state into a number of constituent waves of various wave heights and periods. It is assumed that the total response can be characterized by the sum of the responses to the constituent waves as individual waves (Holmes et al. 1978). Although this method is more refined than the deterministic method, it has its limitations arising primarily from the approximate calculation of stress ranges as a function of the wave height – for example, in the form $C_1 H^{C_2}$ often used, where C_1 and C_2 are coefficients and H is wave height.

The coefficients in such an equation must be predetermined based on experience with structures using calculations by more detailed and sophisticated methods such as the spectral fatigue method in the frequency domain.

The spectral fatigue method is a preferred frequency domain analysis approach that can handle a number of factors related to fatigue life prediction in a more detailed and direct manner than what is practical by the other two methods. For instance, the springing response as well as slamming on the structure can be dealt with in

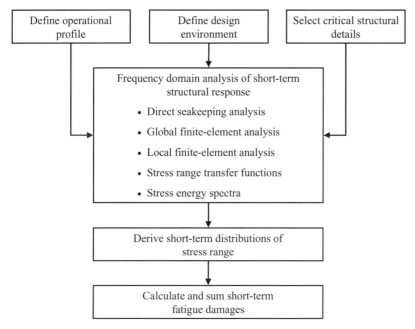

Figure 7.4. A spectral-analysis-based FLS assessment procedure for ship-shaped offshore units.

terms of natural frequency as some part of the spectrum of exciting forces. Combined effects between waves and swell and directional spreading effects can also be treated. Useful past studies related to FLS assessment of ship-shaped offshore units also tend to reinforce the view that the spectral-analysis-based method is perhaps the most reliable approach (see, e.g., Francois et al. 2000; Lotsberg 2000; Bultema et al. 2000; Oh et al. 2003; Tam and Wu 2005).

For FLS assessment of ship-shaped offshore units, some important aspects that must be considered include the following (API RP 2FPS 2001):

- The effects of weathervaning for turret-moored vessels, also taking into account vessel drift and occurrence of waves in off-head seas orientations
- The probability distribution of the environmental parameters to accurately reflect the actual site and mooring conditions
- Load conditions during operations
- The effects of corrosion on the stress range
- Fatigue damage incurred during transit or tow
- Frequency and form effects in corrosion fatigue if relevant

Although API RP 2FPS (2001) and some other guidelines notes for design of offshore units – for example, DNV-RP-C206 (DNV 2006) – provide for adopting the traditional FLS assessment methods of recognized classification society rules used for trading vessel structures with the right adjustments, this chapter focuses mainly on the more refined methods using the direct calculations using the spectral fatigue analysis.

Figure 7.4 shows an FLS assessment procedure using spectral analysis. The main tasks involved in this procedure, in order to determine the cyclic stress ranges and associated number of cycles at structural details of interest, are the determination

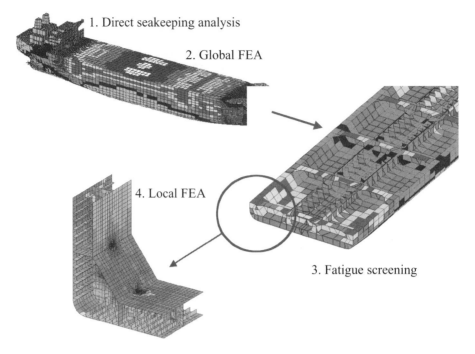

Figure 7.5. Various stages of structural analyses for FLS assessment.

of seakeeping analysis, global structural analysis, and local structural analysis, as illustrated in Figure 7.5. Once the stress ranges and associated number of cycles have been determined consistent with a particular type of S–N curve, calculations related to fatigue damage accumulation are made and compared to target allowable, hence completing the FLS assessment.

With principal dimensions, weight distribution, and other particulars of the structure known, the route or site-specific design environmental and operational data will be used to perform the seakeeping analysis. In principle, noncollinear environments in terms of combination of wind, waves, swell, and currents must be accounted for, as shown in Figure 7.6, but the effects of some of the less important parameters at the specific site, such as temperature variations, might be neglected depending on the case involved.

For calculation purposes, route- and site-specific scatter diagrams need to be defined for waves and swell. Design operational conditions need to be defined, reflecting loading and offloading of oil or gas products, in addition to transit conditions for

Figure 7.6. A weathervaning turret-moored ship-shaped offshore unit subject to noncollinear environmental actions.

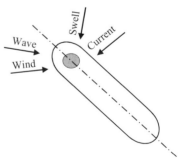

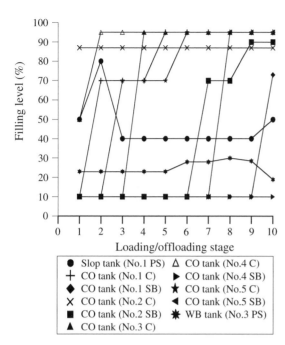

Figure 7.7. Example of filling level variations of oil tanks and water ballast tanks due to loading and offloading of an actual FPSO (CO = cargo oil, C = center tank, PS = port side tank, SB = starboard side tank, and WB = water ballast).

towing or relocating the new-build vessel from the construction yard to the operating site. For convenience, three types of operational conditions are often considered: (1) ballast, (2) fully loaded conditions, (3) and partially loaded conditions. The associated time in each condition must be determined. For example, the time distribution of each loading condition may be estimated in a typical case to be 25 percent for ballast condition, 25 percent for fully loaded condition, and 50 percent for partially loaded condition.

Because the transit is carried out in ballast condition, the time duration of the ballast condition may be further subdivided into 24 percent for operational ballast condition and 1 percent for transit ballast condition. Note that the filling levels of oil tanks are usually significantly changed due to loading and offloading, as shown in Figure 7.7, and their changes will affect the fatigue behavior as well.

The seakeeping analysis is carried out to evaluate the actions arising from the environmental and operating conditions. Mass information of individual structures and other items are used to develop the global finite-element model for the overall structural analysis; such a global finite-element model would include the entire installation including hull, topside modules, flare tower, piperack, accommodation, helideck, crane pedestal, caissons, fire pumps, and fluid transfer line. Additional local models and related finite-element analysis would be performed for the more detailed stress information necessary.

Wave headings in the range of 0 to 360 degrees may need to be considered for a spread-moored unit and for units during tow. For a turret-moored unit, the wave headings may be from head or following seas, and a limited angular sector about these main headings as long as the weathervaning system works successfully. However, even in this case, a greater range of wave headings must be considered to reflect the effect of short-crested swell waves. For instance, some ten wave headings from –90 degrees (port beam) to 90 degrees (starboard beam) with an interval of 18 degrees

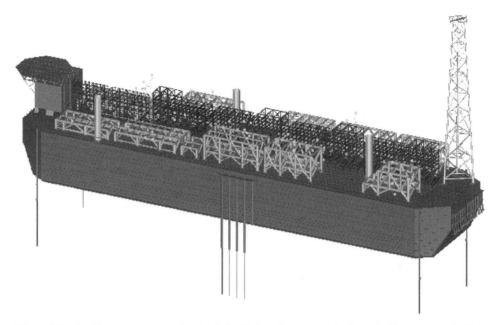

Figure 7.8. An illustrative example of a global finite-element analysis model for a moored ship-shaped offshore unit (Oh et al. 2003; courtesy of Society of Petroleum Engineers/Offshore Technology Conference).

may be considered for onsite operation conditions of a vessel that is symmetrical about its centerline. This would normally be sufficient to describe the directional distribution of energy in short-crested waves. Note that the wave-energy spreading function during tow could be different from that on site; for example, transit conditions might use a cosine squared rather than cosine-power-4 spreading function. The entire range of wave periods must usually be considered in any case, for example, from 1 to 40 seconds.

Although it is recognized that the traditional potential theory is accurate enough to analyze the first-order vessel motion, it does not take into account the effect of nonlinear roll damping that is primarily characterized by roll angle, wave heading, and wave frequency. Therefore, a strategy to account for the nonlinear roll damping effects must be developed, usually based on model test data, and included in the potential theory calculations as an adjustment. This aspect is discussed further in Section 7.4.

Once the actions are determined from the hydrodynamic and vessel motion analysis, the global structural analysis is performed by applying the calculated actions to the global finite-element model to evaluate the action effects required (e.g., stresses and displacements). Figure 7.8 shows an illustrative example of a global finite-element analysis model for a ship-shaped offshore unit moored on site (Oh et al. 2003). Note that all topsides structures, process equipment, and mooring lines as well as hull parts are included in the global finite-element analysis model.

Relevant boundary conditions will need to be applied for the global structural-analysis model in order to remove rigid-body motions. For instance, the locations of transverse bulkheads may be considered to be supported with springs. Typical types of finite elements used for the structural analysis of trading tankers can also be used

Table 7.3. *Typical structural details and sources of fatigue action for ship-shaped offshore units*

Area type	Structural details	Sources of fatigue action
Specific area for ship-shaped offshore units	Topside module supports; flare tower foundation; riser porches; caissons; crane pedestals; mooring foundations; deck penetrations; helideck-to-deck connections; turret structure	Hull girder bending actions; variation of side shell pressure; deck deformation; riser actions; mooring actions; topsides inertia actions; crane actions; wind actions; temperature actions
Longitudinal hull part, similar to normal trading tankers	Doubling plates; bracket toes and heels; rat holes and erection butts; deck openings; longitudinal girders; structural terminations	Hull girder bending and shear; wave-pressure actions; pressures from internal fluid; topsides actions; stresses due to loading and offloading
Transverse hull part, similar to normal trading tankers	Shear lugs and cutouts; hopper corners; transverse frames and gussets; transverse bulkheads	Wave-pressure actions; pressures from internal fluid; topsides actions; differential pressure actions

for ship-shaped offshore units (Paik and Hughes 2006). The adequacy of the global finite-element modeling and analysis can be checked through various means, such as the correct balancing of mass when compared to the loading manual, checks of excitation, checks of reaction forces (which must be zero) at support springs, and checks of the stress level at main deck or outer bottom under pure vertical bending moment application.

After the global structural analysis is completed, the local structural analysis for structural details of interest is performed by applying the boundary displacements obtained from the global structural analysis to the local finite-element model together with any intervening local loads. The fatigue stress ranges are consistent with the type of S–N curve being used, for example, the hot spot stress ranges are to be determined for the structural details.

For FLS design of ship-shaped offshore units, critical joints and details that are considered include the following (API RP 2FPS 2001; DNV 2006):

- Integration locations of the mooring system with the hull structure
- Main hull locations at deck, bottom, side, and longitudinal bulkheads
- Main hull longitudinal stiffener connections to transverse web frames and bulkheads
- Significant openings in the main hull and elsewhere
- Local stiffener end connections on transverse frames
- Flare tower and its attachment to hull
- Riser interfaces
- Major process-equipment supports

Table 7.3 summarizes typical structural details and locations for fatigue assessment and the sources of related fatigue actions. Figures 7.9 and 7.10 show typical structural details that are of interest regarding fatigue crack occurrence in

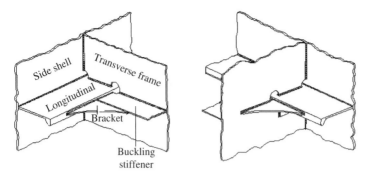

Figure 7.9. Example of critical areas of fatigue cracking in hull part; shown here are intersections between stiffener and brackets welded to the flange of the longitudinal.

hull part. Stress concentration areas specific to ship-shaped offshore units include topside supports (Figure 7.11), piperack supports (Figure 7.12), riser connections (Figure 7.13), and oil offloading line connections (Figure 7.14). Useful studies for FLS assessment and design of such structural details are found in the literature, for example, welded pipe penetrations in stiffened plate structures (Lotsberg 2004) and side longitudinals in floating offshore structures (Lotsberg and Landet 2005).

7.4 Seakeeping Analysis

As a complement to Section 7.3, this section presents more elaborate description of the seakeeping analysis. The main objective of the seakeeping analysis is to first determine the vessel motion and hydrodynamic load response amplitude operators (RAOs) for a range of wave frequencies and headings and for each of the vessel-load cases identified. The hydrodynamic actions obtained from the analysis will then be used as input data for the global structural analysis.

In the analysis, 6 degrees of freedom rigid body vessel motions are typically considered. The wave-load RAOs include those for hull girder actions, external wave-pressure actions, internal tank pressure actions due to accelerations, inertia forces associated with masses of structural components, and significant items of process

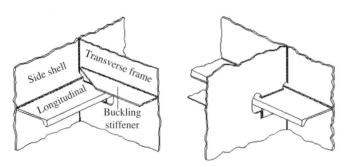

Figure 7.10. Example of critical areas of fatigue cracking in hull part; shown here are buckling stiffeners and longitudinals welded to transverse frame at web plate only.

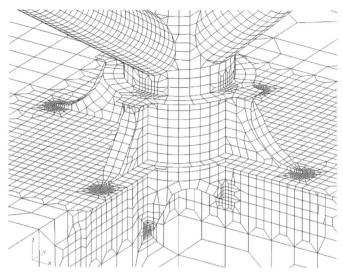

Figure 7.11. An illustrative example for critical areas for fatigue at a topside support (Oh et al. 2003; courtesy of Society of Petroleum Engineers/Offshore Technology Conference).

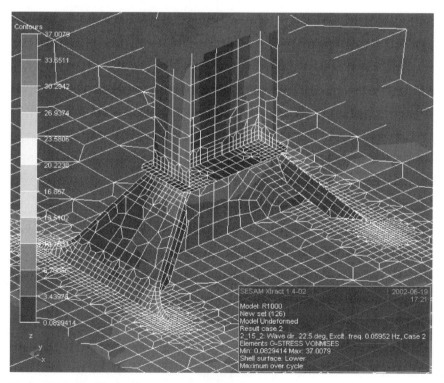

Figure 7.12. An illustrative example of critical areas for fatigue at piperack supports (Oh et al. 2003; courtesy of Society of Petroleum Engineers/Offshore Technology Conference).

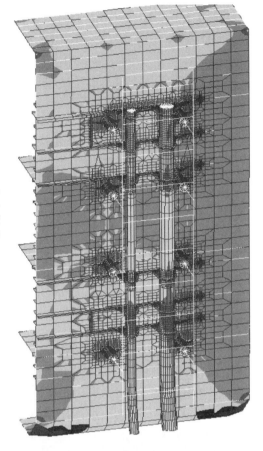

Figure 7.13. An illustrative example of critical areas for fatigue at riser line connections (Oh et al. 2003; courtesy of Society of Petroleum Engineers/Offshore Technology Conference).

equipment. The effect of the mooring system must be accounted for in the analysis as necessary. In addition, the mooring system is also subject to its own separate FLS assessment. Figures 7.15 and 7.16 schematically show the wave heading profiles for a turret- and a spread-moored ship-shaped offshore unit, respectively.

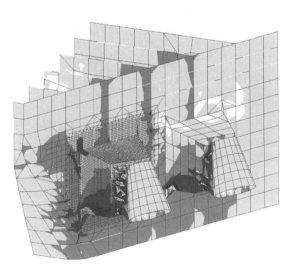

Figure 7.14. An illustrative example of critical areas for fatigue at oil offloading line connections (Oh et al. 2003; courtesy of Society of Petroleum Engineers/Offshore Technology Conference).

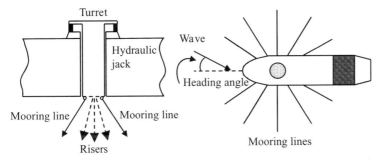

Figure 7.15. Wave heading for a turret-moored ship-shaped offshore installation.

For the seakeeping analysis, a substantial number of regular wave periods (or frequencies) must be considered. In principle, the range of wave periods must cover all relevant response transfer functions, such as vessel motions, sectional loads, pressures, and drift forces. Note that some 25–30 wave periods are necessary to describe the transfer functions smoothly. The shortest wavelength (or lowest wave period) considered may correspond to about 10 percent of the vessel length (DNV 2006).

Vessel motion-induced accelerations and inertia actions play an important role in FLS assessment. The g (acceleration of gravity) component induced by the vessel pitch and roll motion must be included when determining accelerations in longitudinal and transverse directions of the vessel. Large-amplitude vessel motions due to large and very infrequent waves may be neglected for the FLS assessment; however, they are primary action sources for ULS assessment because it is the stress ranges at lower action levels with intermediate wave amplitudes that are more likely to contribute to the fatigue damage accumulation.

Note that existing three-dimensional linear-potential theory tends to underestimate the roll damping at roll resonance and, subsequently, the vessel roll motion is likely to be overestimated. In this regard, a relevant strategy must be considered to reflect the nonlinear roll damping effect within the calculations. Typically, vortex-induced damping (eddy-making) near sharp knuckles and drag of the hull-skin friction and normal forces or flow separation from bilge keels must be accounted for in the roll-motion analysis. Some useful methods to deal with nonlinear roll-damping

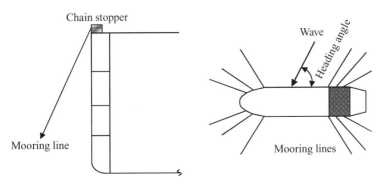

Figure 7.16. Wave heading for a spread-moored ship-shaped offshore installation.

effects may be found in the literature (e.g., DNV 2006). Most of these methods are based on empirical approaches developed by curve fitting of test data obtained from physical models.

7.5 Stress Range Transfer Functions

This section is also a complement to Section 7.3. For FLS assessment, the usual linear superposition of the response of all regular wave components making up irregular seas is adopted, leading to a frequency-domain analysis. Therefore, the resulting stress may be obtained as the sum of all contributing action effects. In using the linear frequency-domain-analysis approach based on small amplitude wave hydrodynamics, the effect of large waves may be inaccurately represented, although this is not usually a problem because the fatigue damage accumulation is mostly associated with the more moderate waves for which the theory works better.

The transfer function represents the response to a sinusoidal wave with unit amplitude for given wave frequencies, heading angles, and speeds. Generally, linear potential theory is typically applied to calculate the transfer functions. A substantial number of wave frequencies and headings (and speeds if relevant) should then be considered to obtain a complete set of transfer functions. Masses of all structural components and cargo must be considered for motions in and about the vertical, longitudinal, and transverse directions.

The short-term wave response distribution for FLS assessment can be estimated using the site-specific wave scatter diagram identified. Upon calculating the transfer functions, the Pierson–Moskowitz wave spectrum (i.e., Eq. (4.11) in Chapter 4) or another appropriate wave spectral density may be used to obtain the response spectrum. The stress range response may usually be assumed to be Rayleigh distributed within each short-term sea-state condition. The stress range distribution for a given sea state and heading direction is then determined on the basis of the response spectrum. The fatigue damage within each sea state and heading direction may be calculated in closed form when the stress range distribution within each short-term condition is known analytically, for example, to be Rayleigh-distributed.

If a swell component is present, the response to swell may also be evaluated in a similar way to that of wind and waves but with different spectrum; for example, the JONSWAP spectrum (i.e., Eq. (4.14) in Chapter 4) with appropriate factors may be used. Where the responses to wind and waves and to swell can be considered independent, the combined effect may then be evaluated by adding the appropriate variances.

Low-frequency responses associated with the riser or mooring system can contribute to the fatigue damage. Therefore, it is usually recommended that the riser or mooring system be included in the global structural analysis. However, if the low-frequency response due to the riser or mooring system has been evaluated separately from the global analysis, the stress range due to the low-frequency riser or mooring forces needs to be combined with that from wave and swell effects.

Also, in contrast to a situation where fatigue damage is accumulated sea state by sea state, the long-term stress range distribution can be calculated as a weighted sum of all sea states and heading directions, and the fatigue damage can then be calculated using the long-term distribution of stress ranges; see, for example, DNV-RP-C206 (DNV 2006).

7.6 Global Structural Analysis

Again, this section is a complement to Section 7.3. The global structural analysis model must include all structures and process equipment modules, as shown in Figure 7.8. The actions obtained from the hydrodynamic actions and vessel motion analysis will then be applied to the global finite-element structural analysis from which the nodal displacements applicable for the boundary conditions of the local finite-element analysis for the structural details are calculated. The global structural analysis will be performed for each wave frequency and heading angle in a relevant range to obtain the wave-induced cyclic stress range transfer functions. If a forward speed exists, such as in a transit case, it is an additional parameter to be considered.

Note that the global and local system action effects in association with FLS assessment may be affected by dynamic vessel actions as well as hydrodynamic actions. Typical dynamic action sources include slamming, sloshing, green water, and springing, as described in Chapter 4. As a complement to Section 7.3, this section provides some considerations related to the mechanism of dynamic action sources that affect the fatigue damage accumulation.

Slamming may result in a dynamic transient action (i.e., damped oscillatory response) of the vessel structure. In ship-shaped offshore units, the most frequent locations exposed to slamming are bow flare and accommodation structure if located in the fore part of the vessel, in particular with a turret-moored system. For spread-moored installations, stern part may also be exposed to slamming. Slamming can potentially induce the fatigue actions of the vessel hull through whipping.

In general, the frequency of occurrence and severity of slamming are governed by the vessel draft, hull geometrical form, site environment, and wave heading and vessel speed, when present. The effects of slamming-induced actions must be accounted for in terms of hull girder actions (e.g., bending moments, shear forces), local strength, and through limitations on reduction of ballast draft.

For many types of ship-shaped offshore units, partial filling of tanks is very often of a necessity (see Figure 7.7) and this potentially exposes the vessel structure to sloshing that may contribute to the fatigue damage accumulation. Major factors governing the occurrence of sloshing include tank size, tank filling levels, structural arrangements inside the tank (e.g., swash bulkheads, web frames), transverse metacentric height (GM), vessel draft, and natural periods of vessel and cargo in roll and pitch motions.

Because sloshing-induced actions are primarily characterized by the inertia forces induced by the liquid in the tank, it is recognized, based on analysis of several ship-shaped offshore units ranging in length from 100–260m, that a Weibull shape parameter h = 1.0 (i.e., the exponential distribution) may be appropriate for FLS assessment for sloshing both in the longitudinal and transverse sloshing modes (DNV 2006).

The need for inclusion of sloshing effects in FLS assessment is usually judged according to the fatigue damage accumulation level predicted solely by sloshing actions. For instance, fatigue damage accumulation due to sloshing actions may be calculated by assuming appropriate K-factors for the details and the Weibull shape parameter of h = 1.0; if it exceeds 0.1, then sloshing effects would be accounted for in the FLS assessment. Sloshing-induced stresses are typically then superimposed to stresses induced by other action sources, that is, without any specific correlation considerations.

Onsite loading and offloading cycles may also contribute to the fatigue damage accumulation due to low-cycle fatigue effects. If the number of loading and offloading cycles during the entire lifetime is less than a certain value – say, 1,600 – then this action may not affect total fatigue damage significantly (DNV 2006). Otherwise, low-cycle fatigue may

Table 7.4. *Loading sequence for low-cycle fatigue assessment at transverse bulkheads*

Load step	One tank	Adjacent tank
1	Empty	Empty
2	Full	Empty
3	Full	Full
4	Empty	Full

need to be considered because the effect of the global bending and/or local effects due to the continuous loading and offloading may be significant (HSE 2004).

A low-cycle fatigue assessment (see Section 7.10) may be required for transverse bulkheads between tanks in particular, which may experience a full load reversal according to the loading steps, as indicated in Table 7.4, where loading steps 2 and 4 may result in different stress ranges because of the loading sequence.

Low-cycle fatigue effects may also be relatively more significant where the loading patterns for a ship-shaped floating offshore structure as converted are very different from that of the original tanker insofar as longitudinal bulkheads are concerned. This occurs, for example, in the conversion of some tankers where the longitudinal bulkheads originally were not designed to the full tank load due to filling on one side, with the tank on the other side empty. Low-cycle fatigue effects have also been found to be of relatively greater significance where the remaining fatigue lives in a conversion are small for any reason.

7.7 Local Structural Analysis and Hot Spot Stress Calculations

Local structural analysis is carried out for selected joints and details for purposes of obtaining the stress range transfer function for each wave frequency and heading angle. As previously noted, structural details within interface to topside modules and riser or mooring systems need in particular to be analyzed in addition to the hull part.

Today, finite-element analysis (hereafter abbreviated as FEA) is typically applied for the structural analyses at either the local or global level. This section presents the FEA approaches to determine hot spot stress at structural details. Recent advances in the areas are summarized in this section; for greater detail, see Lotsberg (2006).

7.7.1 Definition of Hot Spot Stress

Although the nominal stress-based S–N curve approach has been well established for FLS assessment of many types of welded structures, such as ships and land-based structures, their use often requires one to accurately predict the related stress concentration factors (SCFs) at the structural details involved, as may be surmised from Eq. (7.2). In those cases where this is possible, whether by direct analysis or by parametric formulae, a nominal stress-based approach to fatigue-life prediction is also feasible and indeed may be preferred.

Fine-mesh FEA can accurately calculate local stresses for SCF prediction and also calculate hot spot stresses involving stress concentration characteristics at hot spot

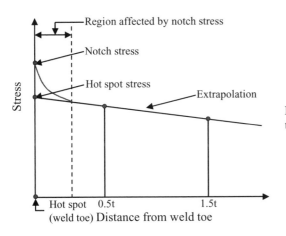

Figure 7.17. Extrapolation to derive the hot spot stress.

areas when the hot spot approaches are preferred. For purposes of the hot spot stress application, the hot spot stress prediction approach at a particular structural detail must be the same as that used to establish the hot spot S–N curve. Care is needed to define the characteristic fatigue stress in a particular case independent of the stress gradients that may be present through the plate thickness using the conventional shell-element modeling. Even so, the hot spot stress concept can be an efficient engineering methodology for FLS assessment of ship-shaped offshore structures.

To approximately account for the notch effect, the hot spot stress is derived by extrapolation of the geometric stress to the weld toe, as shown in Figures 7.2 and 7.17. It is evident that the stress used as basis for such an extrapolation should be outside that affected by the weld notch but close enough to pick up the stress gradients to geometry.

Some classification societies suggest that one may derive the hot spot stress from a linear extrapolation using the stress values calculated at points 0.5t and 1.5t (t = plate thickness) from the weld toe (or intersection line), when FEA with a finely defined mesh size at the hot spot region (e.g., t × t) is applied.

Although the previously mentioned technique is good enough in terms of computational accuracy, it has some disadvantages. For instance, the manual work to derive the hot spot stresses for various locations can be time-consuming. There is also often confusion as to whether the hot spot stresses must be derived from the Gaussian stresses in the finite element, as shown in Figure 7.18, or from the nodal stresses or some sort of an average stress, say at the midpoint of the element. Note that FEA programs do not always provide the required stresses at the required locations as computational results routinely.

In this regard, a simpler definition of the hot spot stress may be more attractive. For instance, the hot spot stress may be defined at a position 0.5t from the weld toe and, in some cases, this has been shown to lead to less scatter in the calculated results (Fricke and Säbel 2000; Storsul et al. 2004a). The distance measurement may be different depending on the finite-element type. For the FEA with shell elements, the distance to the hot spot stress readout point is measured from the mean intersection lines because neither the plate thickness nor the weld is normally included in the finite-element model. For FEA with solid elements, there can actually be less uncertainty in this regard because the distance to the hot spot stress readout point can be measured from a readily identifiable upper surface and weld toe because both the thickness

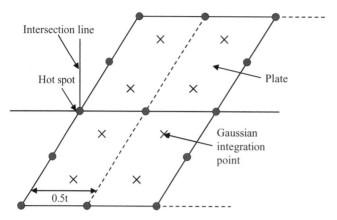

Figure 7.18. Hot spot stress calculations from the Gaussian stresses (•: nodal point; +: integration point).

and the weld will be included in the finite-element model. This is not, however, to suggest that solid elements can be used for practical purposes in all cases.

It is well worth noting that when shell elements are used to predict the hot spot stress, there indeed is an implicit assumption being made as to the fatigue effect of the associated through-thickness stresses because the shell element usually assumes a linear distribution of stresses through the shell.

Other approximations may also be suggested to calculate the hot spot stress. For instance, a linear extrapolation of the stresses to the hot spot (weld toe or intersection line) from the readout points at 0.5t and 1.5t can be made, as shown in Figure 7.17. The principal stress at the hot spot is then calculated from the extrapolated component values. Otherwise, the hot spot stress could be taken as the stress at the readout point 0.5t away from the hot spot and multiplied by, say, 1.1.

More recently, the concept of a "structural stress" as one consisting of axial and bending components through the plate thickness has been formalized by Battelle for use in fatigue strength prediction. In the proprietary Battelle approach, a hot spot stress that is obtained directly from the nodal forces calculated by FEA through well-defined automated procedures capable of consistently taking account of through-thickness axial and bending effects is used; available S–N data were reanalyzed on the same basis to define the S–N curve to be used. Fracture mechanics concepts can be employed to reduce the fatigue data scatter; that is, adjustment factors related to these can also be part of the structural stress definition in particular cases. A detailed description of the Battelle structural stress method may be found in Dong et al. (2001); for the subject of structural stresses, see also Niemi (2001) and Doerk et al. (2003).

In yet another development, it has been suggested that the stress range may be approximately redefined by considering the combination of membrane and bending stresses, as follows:

$$\Delta\sigma = \Delta\sigma_m + \alpha\Delta\sigma_b, \tag{7.6}$$

where $\Delta\sigma =$ total stress range; $\Delta\sigma_m =$ membrane component of stress range; $\Delta\sigma_b =$ bending component of stress range; and $\alpha =$ a "correlation" constant.

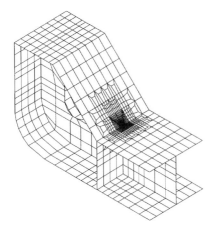

Figure 7.19. Stress concentration area in a hopper corner knuckle.

The correlation constant in Eq. (7.6) usually takes a value less than 1.0. This is in part because the load-shedding effect during crack growth with bending stresses present in addition to axial stresses is attempted to be accounted for. Note that the bending effect can be important in areas with a localized stress concentration, such as that seen in hopper corner knuckles in Figure 7.19. When stress variation along the weld is small, however, the bending effect in the FLS assessment may, of course, be negligible. When the bending effect needs to be considered, it has been suggested that the reduction factor for hopper corners, for instance, may be taken as $\alpha = 0.6$ or otherwise $\alpha = 0.0$. This approximation is said to be supported by fatigue test data developed under bending (Kang et al. 2002; Kang and Kim 2003; Petinov et al. 2006) and by fracture mechanics analyses (Lotsberg and Sigurdsson 2004). For a more detailed description, see Lotsberg (2006).

We know that the hot spot stress concept may not be the best approach for simple cruciform joints, simple T-joints in plated structures, or simple butt joints that are welded from one side only. In these cases, the conventional nominal stress concept is perhaps more reliable for FLS assessment. Also note that in such situations, fabrication-related misalignments will significantly affect the fatigue life; to the extent that such effects are not implicit in laboratory test data related to fatigue, they may need to be additionally considered in the FLS assessment for these joints through the use of an additional SCF usually estimated by simplified formulae.

7.7.2 Finite-Element Analysis Modeling

A linear-elastic finite-element structural analysis is usually performed to calculate hot spot stresses as well as nominal stress as required. Fabrication-related misalignments are usually not included in the FEA model. The extent of the FEA model is selected to be large enough so that the boundary uncertainties are minimized. For structural (stress) analysis, three types of hot spots at weld toe are relevant, as shown in Figure 7.20: at the weld toe on the plate surface at an ending attachment (point a), at the weld toe around the plate edge of an ending attachment (point b); and along the weld of an attached plate or weld toes on both the plate and attachment surface (point c).

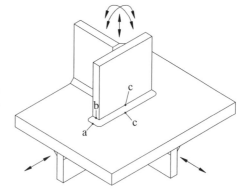

Figure 7.20. Types of hot spots at weld toe (i.e., a, b, or c).

Thin plate-shell elements or solid elements can be adopted as finite-element types, but all types of the adopted elements must adequately reflect the relatively steep stress gradients around the weld toe as well as plate bending deformation, although a linear approximation of stress through the plate thickness will be assumed in all types.

Figure 7.21 shows typical finite-element models at weld toe using shell or solid elements. Although the shell-element model is much simpler, the solid-element model is more relevant to reflect steep stress gradients.

The 4-node shell elements need to include internal degrees of freedom for improved in-plane behavior. Typically, the element size for a 4-node shell-element model needs to be as fine as keeping the t × t rule (t = plate thickness) in terms of element size. For an 8-node shell-element model, the element size can be up to 2t × 2t, as shown in Figure 7.22. An FEA model with a larger element size at the hot spot region may underestimate the stress level.

The three-dimensional solid-element model is more desirable for analyzing complex structural details, although the use of such elements can be much more time-consuming. For the model with solid elements, the element length of the two or three elements near weld toe may be selected to approximately equal the plate thickness. The transverse element size parallel to the plate breadth may also be the plate thickness, but it should not exceed the width of attachment, as shown in Figure 7.20.

A 20-node isoparametric solid-element model (i.e., with midside nodes at the edges) may be very useful for analysis of more complex cases in terms of geometric and loading condition. Figure 7.21 shows the finite-element model with 20-node solid elements compared to that of 8-node shell elements.

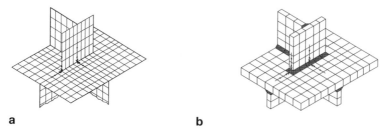

a   b

Figure 7.21. Schematic finite-element models at weld toe: (a) shell-element model; (b) solid-element model including weld toe.

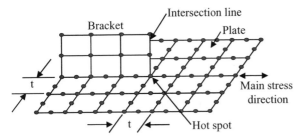

Figure 7.22. Finite-element model with 8-node shell elements.

For the 20-node hexahedral element model shown in Figure 7.23, only one element over the thickness may be enough to get a linear stress distribution, while at least four elements are necessary with the 8-node shell-element model. Another benefit of using three-dimensional solid elements is that the fillet weld can be more accurately included in the model so that local geometry and its stiffness can be properly accounted for in the analysis (Storsul et al. 2004b).

The element aspect ratio (i.e., the ratio of element length to element breadth) may also affect the computational accuracy and it should be smaller than 4 to prevent any significant errors.

The stress evaluation method may be different depending on the element types and sizes. The element size at a hot spot region typically follows the $t \times t$ rule (t = plate thickness), as noted previously. When the plate shell-element model is used, the surface stress is defined at the corresponding midside points. The stresses at midside nodes can be taken as the stresses at readout points $0.5t$ and $1.5t$ away from the weld toe.

When the solid-element model is used, the stress is first extrapolated from the Gaussian points to the surface. Then these stresses are interpolated linearly to the surface center or extrapolated to the edge of the elements if this is along the line for hot spot stress calculation. When the element size exceeds the $t \times t$ rule, a second-order polynomial approximation to the element stresses usually must be made.

7.8 Selection of S–N Curves

Depending on the types of stresses and structural details, relevant S–N curves must be developed by fatigue testing or selected from recognized standards and codes, for

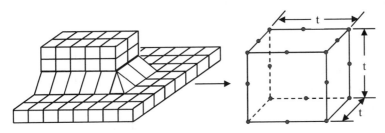

Figure 7.23. Finite-element model with 20-node solid elements.

Table 7.5. *Classification of welding connection types for application of S–N curves*

Type	Detail	Stress feature in S–N curves	Remark
B		Nominal stress multiplied by the stress concentration factor associated with local geometry such as cutouts	Machine gas cut or sheared material (drag lines, corners removed, visible imperfections, or corrosion pitting are not considered)
C		Nominal stress multiplied by the stress concentration factor associated with local geometry such as cutouts	Machine thermally cut edges
		Local stress parallel to the weld	Welded connection with stress parallel to the weld and butt weld ground flush with stress normal to the weld
D		Local stress normal to the weld	Welded connection with stress normal to the weld toe (Welds proven free from significant defects)
W		Engineering shear stress in the weld throat	Design of weld size (throat thickness)

Note: D-type S–N curves are usually used for the hot spot stress-based FLS assessment. For welding connection types other than those presented above, see DNV-RP-C206 (DNV 2006).

example, DNV-RP-C206 (DNV 2006). Selection of the design S–N curves may then be based on actual geometry of the detail at weld toe or root, potential flaws at weld and base material, and stress direction relative to the weld.

Table 7.5 indicates a commonly used classification scheme for S–N curves according to the critical area types noted in Section 7.2. This particular classification is based on the work of T. R. Gurney and S. J. Maddox at The Welding Institute. This work subsequently was reflected in various codes and standards including BS 5400 (1980) for bridges (http://www.bsi-global.com) and the now philosophically withdrawn UK DEN (Department of Energy) and HSE (Health and Safety Executive) prescriptive fatigue guidance for offshore platforms (http://www.hse.gov.uk).

However, it continues to be widely used in ships and offshore practices, for example, DNV-RP-C206 (DNV 2006). Another widely used classification or catalog of S–N curves, now increasingly common in Europe, is that from International

Table 7.6. *Examples of design S–N curves in air, following DNV-RP-C206 (DNV 2006)*

| S–N curve | N < N_c | | N ≥ N_c | | Fatigue limit at | p |
	$\log A_1$	m_1	$\log A_2$	m_2	N = N_c (MPa)	
B1	15.117	4.0	17.146	5.0	106.97	0.0
B2	14.885	4.0	16.856	5.0	93.59	0.0
C	12.592	3.0	16.320	5.0	73.10	0.15
C1	12.449	3.0	16.081	5.0	65.50	0.15
C2	12.301	3.0	15.835	5.0	58.48	0.15
D	12.164	3.0	15.606	5.0	52.63	0.20
E	12.010	3.0	15.350	5.0	46.78	0.20
F	11.855	3.0	15.091	5.0	41.52	0.25
F1	11.699	3.0	14.832	5.0	36.84	0.25
F3	11.546	3.0	14.576	5.0	32.75	0.25
G	11.398	3.0	14.330	5.0	29.24	0.25
W1	11.261	3.0	14.101	5.0	26.32	0.25
W2	11.107	3.0	13.845	5.0	23.39	0.25
W3	10.970	3.0	13.617	5.0	21.05	0.25

Note: $N_c = 10^7$; p = exponent for thickness effect.

Institute of Welding (http://www.iiw.org) (IIW 1996). The U.S. Ship Structure Committee (http://www.shipstructure.org) has also published several reports containing S–N curves potentially usable for ship-shaped offshore units.

In much of the practice today, the design S–N curves normally follow the mean-minus-two-standard-deviation lines (on a log-log scale) based on relevant fatigue test data, implying that a 97.6 percent probability of fatigue durability may be achieved if S–N curve and data scatter alone is the consideration, which it typically is not.

Tables 7.6 and 7.7 indicate representative design S–N curve fatigue parameters for structural components in air and in sea water with cathodic protection, respectively, following Eq. (7.3). Figures 7.24 and 7.25 plot the various S–N curves that can be used for offshore structures under typical wind and wave actions with the total number of cycles $N > 10^6$, while a different S–N curve would be used in the low-cycle fatigue region (DNV 2006). Additional S–N curves and also related discussion can be found in HSE (1999). For structural thickness greater than about 25mm, the S–N curve expressions of Eq. (7.3) are often modified to account for the "thickness effect," as follows:

$$\log N = \log A - m \log \left[\Delta\sigma \left(\frac{t}{t_r} \right)^p \right], \tag{7.7}$$

where A = A_1 or A_2; m = m_1 or m_2; t = plate thickness; t_r = reference thickness, (e.g., $t_r = 25$mm); and p = exponent for thickness effect.

7.9 Fatigue Damage Calculations

The fatigue damage accumulation is calculated by use of the Palmgren–Miner rule, Eq. (7.4). When a bilinear or two-sloped S–N curve is used, the fatigue damage

Table 7.7. *Examples of design S–N curves in sea water with cathodic protection, following DNV-RP-C206 (DNV 2006)*

S–N curve	N < N_c		N ≥ N_c		Fatigue limit at N = N_c (MPa)	p
	$\log A_1$	m_1	$\log A_2$	m_2		
B1	14.917	4.0	17.146	5.0	106.97	0.0
B2	14.685	4.0	16.856	5.0	93.59	0.0
C	12.192	3.0	16.320	5.0	73.10	0.15
C1	12.049	3.0	16.081	5.0	65.50	0.15
C2	11.901	3.0	15.835	5.0	58.48	0.15
D	11.764	3.0	15.606	5.0	52.63	0.20
E	11.610	3.0	15.350	5.0	46.78	0.20
F	11.455	3.0	15.091	5.0	41.52	0.25
F1	11.299	3.0	14.832	5.0	36.84	0.25
F3	11.146	3.0	14.576	5.0	32.75	0.25
G	10.998	3.0	14.330	5.0	29.24	0.25
W1	10.861	3.0	14.101	5.0	26.32	0.25
W2	10.707	3.0	13.845	5.0	23.39	0.25
W3	10.570	3.0	13.617	5.0	21.05	0.25

Note: $N_c = 10^7$; p = exponent for thickness effect.

accumulation for short-term responses that are postulated to follow a Weibull distribution can be calculated by

$$D = f_o T_d \sum_{k=1}^{k} \left[\frac{q^{m_1}}{A_1} \Gamma_1 \left(1 + \frac{m_1}{h}, \left(\frac{\Delta \sigma_c}{q}\right)^h\right) + \frac{q^{m_2}}{A_2} \Gamma_2 \left(1 + \frac{m_2}{h}, \left(\frac{\Delta \sigma_c}{q}\right)^h\right) \right], \quad (7.8)$$

where k = total number of stress levels, as defined in Eq. (7.4); f_o = long-term average response zero up-crossing frequency; h = Weibull stress range shape distribution

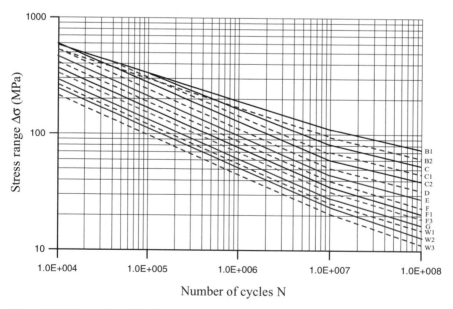

Figure 7.24. Examples of design S–N curves in air, after DNV-RP-C206 (DNV 2006; courtesy of DNV).

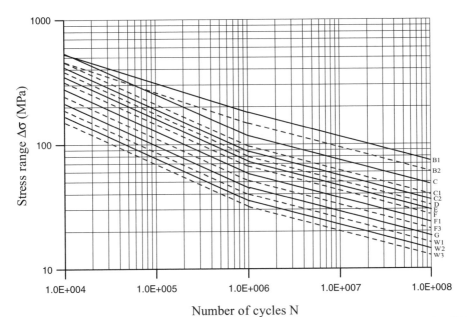

Figure 7.25. Examples of design S–N curves in seawater with cathodic protection, after DNV-RP-C206 (DNV 2006; courtesy of DNV).

parameter; q = Weibull stress-range scale-distribution parameter; $\Delta\sigma_c$ = as defined in Eq. (7.3); A_1, m_1 = S–N fatigue parameters, typically for $N < 10^7$; A_2, m_2 = S–N fatigue parameters, typically for $N \geq 10^7$; and Γ_1, Γ_2 = complementary incomplete gamma function and incomplete gamma function, respectively.

For the global hull response due to hull girder bending, as an example, the zero up-crossing frequency f_o in Eq. (7.8) may be approximated by

$$f_o = \frac{1}{4\log_{10}(L)}, \tag{7.9}$$

where L = vessel length in meter. This implies that for a vessel with L = 350m, $f_o \approx 0.098$ sec^{-1} is approximated. The origin of this often useful equation is believed to be from some early work by DNV (1998).

When the Weibull stress range shape distribution parameter h in Eq. (7.8) is known, the allowable stress range for a Miner sum of unity may be found as a function of the S–N curve types (or welding connection types). The corresponding allowable stress range at 10^{-8} probability level for a give S–N curve and Weibull parameter can be conveniently obtained from precalculated results shown in Figure 7.26 for structural components in air and Figure 7.27 for structural components in sea water with cathodic protection, DNV (2006). For these figures, a target fatigue life of 20 years, or number of cycles in order of 0.5×10^8, has been assumed.

Note also that the Weibull stress range scale distribution parameter q in Eq. (7.8) may be determined by

$$q = \frac{\Delta\sigma_o}{(\ln n_o)^{1/h}}, \tag{7.10}$$

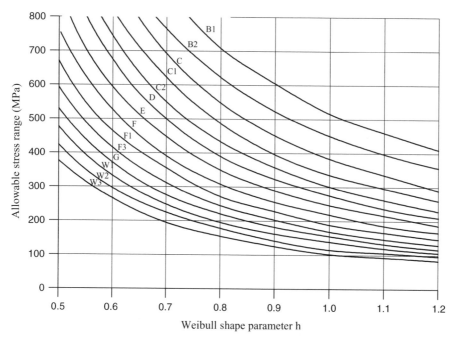

Figure 7.26. Example of allowable stress range at 10^{-8} probability level as a function of the Weibull shape distribution parameter and S–N curve types for structural components in air, after DNV-RP-C206 (DNV 2006; courtesy of DNV).

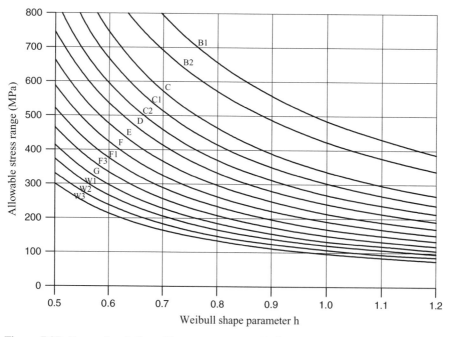

Figure 7.27. Example of allowable stress range at 10^{-8} probability level as a function of the Weibull shape distribution parameter and S–N curve types for structural components in sea water with cathodic protection, after DNV-RP-C206 (DNV 2006; courtesy of DNV).

where h = the shape parameter; $\Delta\sigma_o$ = the stress range that occurs once in the total number of cycles involved; and n_o = total number of cycles during the time period considered.

When a linear or one-sloped S–N curve is used, the fatigue damage accumulation formula can be simplified to

$$D = \frac{f_o T_d}{A_1} \sum_{k=1}^{k} q^{m_l} \Gamma\left(1 + \frac{m_l}{h}\right),$$ (7.11)

where $\Gamma()$ = gamma function; and k = total number of stress levels, as defined in Eq. (7.4).

In a case where the long-term stress ranges arise from a series of short-term sea states following the Rayleigh distribution, the fatigue damage accumulation can be calculated for the different loading conditions involved and a bilinear or two-sloped S–N curve application, as follows [for symbols not specified below, see Eq. (7.3)]

$$D = f_o T_d \sum_{i=1, j=1}^{I,J} r_{ij} \left[\frac{\left(2\sqrt{2m_{oij}}\right)^{m_1}}{A_1} \Gamma_1 \left(1 + \frac{m_1}{2}\left(\frac{\Delta\sigma_c}{2\sqrt{2m_{oij}}}\right)^2\right) \right.$$
$$\left. + \frac{\left(2\sqrt{2m_{oij}}\right)^{m_2}}{A_2} \Gamma_2 \left(1 + \frac{m_2}{2}\left(\frac{\Delta\sigma_c}{2\sqrt{2m_{oij}}}\right)^2\right) \right],$$ (7.12)

where r_{ij} = relative number of stress cycles in short-term conditions i, j; A_1, m_1 = S–N fatigue parameters typically for $N < 10^7$; A_2, m_2 = S–N fatigue parameters typically for $N \geq 10^7$; I = total number of sea-state conditions; J = total number of wave headings; and m_{oij} = zero spectral moment of stress response process.

In some FLS design cases, one part of the fatigue damage may arise from a particular process of fatigue actions, but another fatigue damage for the same hot spot area may be caused by yet another different process of fatigue actions. In this case, the combination of fatigue damages from two different fatigue processes must be considered in an adequate manner because a linear superposition may not necessarily provide conservative FLS assessment (Lotsberg 2005).

To combine fatigue damages from two different dynamic processes, DNV-RP-C206 (DNV 2006) suggests using the following equation when one-sloped S–N curve is applied for fatigue damage calculations:

$$D = D_1 \left(1 - \frac{n_2}{n_1}\right) + n_2 \left\{ \left(\frac{D_1}{n_1}\right)^{1/m} + \left(\frac{D_2}{n_2}\right)^{1/m} \right\}^m,$$ (7.13)

where D_1 = fatigue damage due to one dynamic process; D_2 = fatigue damage due to another dynamic process; n_1 = mean zero up-crossing frequency for one dynamic process; n_2 = mean zero up-crossing frequency for another dynamic process; and m = inverse slope of one-sloped S–N curve. It is considered that both D_1 and D_2 are also calculated applying the corresponding one-sloped S–N curves.

However, when two-sloped S–N curves are applied, for example, so that high-cycle fatigue is likely dominant, it is considered that the main contribution to overall fatigue damage will be from the region of $N \geq N_c$. In this regard, Eq. (7.13) may still be applicable for this case but with the inverse slope of two-sloped S–N curve at $N \geq N_c$, for example, m = 5.0 as defined in Tables 7.6 and 7.7 (Lotsberg 2005).

7.10 High-Cycle Fatigue versus Low-Cycle Fatigue

Typically, fatigue damage of offshore structures is likely due to high-cycle fatigue actions that are considered to have an associated number of loading cycles that are more than 10^4. It may be considered that fatigue initiated with a number of cycles less than 10^4 is low-cycle fatigue and vice versa for high-cycle fatigue.

Most guidance and practices for FLS assessment and design presented in codes and standards, as well as in this chapter, are associated with high-cycle fatigue. It is important to realize that action effects (e.g., stresses) related to low-cycle fatigue more likely involve plasticity at structural details (hot spot areas), but those due to high-cycle fatigue mostly remain in the elastic regime. Therefore, low-cycle fatigue analysis must apply the action effects obtained by taking account of nonlinear material behavior.

Offshore structures are normally designed for other types of limit states such as ULS so that a sufficient factor of safety is achieved for extreme environmental and operational conditions as well as in normal service. Although the effects of stresses due to local notches or at structural details (hot spot areas) are usually not accounted for in ULS design, it is considered that the ULS design results in a structure wherein the stress ranges during extreme actions are themselves limited to few in number.

However, it is to be noted that low-cycle fatigue must certainly be considered when dynamic actions with low-cycle fatigue are frequent. In fact, the loading and unloading cycle is quite frequent in FPSOs when compared to many trading tankers, and the resulting hull girder loading variations can also be significant, as shown in Figure 6.1 of Chapter 6. Similar scenarios may also need to be considered for trading tankers in particular cases (Urm et al. 2004).

7.11 Time-Variant Fatigue Crack Propagation Models

Fatigue cracking damage has been a primary source of costly repair work for aging structures. Such cracking damage has been found primarily in welded joints and local areas of stress concentrations; for example, at the weld intersections of longitudinals, frames, and girders. Initial defects may also be formed in the structure by fabrication procedure and may conceivably remain undetected over time. Under a cyclic loading or even monotonic extreme loading, cracks and defects may propagate and become larger with time.

Because cracks of a large enough size can conceivably lead to the catastrophic failure of the structure, it is essential to properly consider and establish relevant crack-tolerant design procedures for structures in addition to implementation of close-up survey and maintenance strategy. For reliability assessment of aging structures under extreme loads, it is often necessary to take into account a known (existing or premised or anticipated) crack on the ultimate limit-state analysis as a parameter

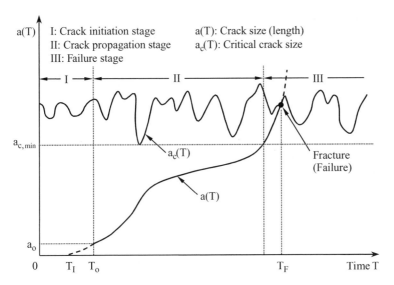

Figure 7.28. A schematic of crack initiation and growth for a steel structure with time.

of influence (Paik et al. 2003). To make such aims possible, it is required to develop a time-dependent fatigue crack propagation model that can predict the crack tip locations and size as the vessel ages.

Figure 7.28 shows a schematic of fatigue related cracking damage progress as a function of time (age) in steel structures. The fatigue damage progress can be separated into three stages: initiation (stage I), propagation (stage II), and failure (fracture) stage (stage III) (ISO 2394 1998).

It is theoretically assumed in the figure that no initial defects exist so that there is no cracking damage until time T_I. The initiation of fatigue cracking damage is affected by many factors, such as the stress ranges experienced during the load cycles, local stress concentration characteristics, and the number of stress range cycles, and it can be evaluated by appropriate fatigue analysis.

However, when any crack is detected in an existing structure at time T_o, it normally has a certain crack size (length), denoted by a_o called the initial crack size, which must be detectable.

Fatigue cracks propagate with time progressively in ductile material. They may, however, become quite unstable in brittle material. Crack propagation is affected by many parameters such as initial crack size, history of local nominal stresses, load sequence, crack retardation, crack closure, crack growth threshold, and stress intensity range in addition to stress intensity factor at the crack tip that depends on material properties and geometry. The fracture mechanics approach is often used to analyze the behavior of crack propagation.

The time-dependent cracking damage model may normally be composed of the following three models:

- A model for crack initiation assessment with a related detection threshold
- A model for crack growth assessment
- A model for failure assessment

Macrocrack initiation at a critical structural detail can be theoretically predicted using the S–N curve approach, as described previously. Cracks at critical joints and details can be detected during inspection when the crack size is large enough – usually, say, about 15–30mm, sometimes more. In terms of the integrity of aged ship-shaped offshore structures, it is often assumed that the crack with length of a_o at a critical joint or detail has initiated at T_o years.

Crack growth can be assessed by the fracture mechanics approach, which considers that one or more premised cracks of a small dimension exist in the structure and then predicts the details of their propagation including any coalescence and subsequent fracture. In this approach, a major task is to establish the relevant crack growth equations or "laws" as a function of time (year) and other relevant parameters.

The crack growth rate is often expressed as a function of the stress intensity factor at the crack tip, on the assumption that the yielded area around the crack tip is relatively small. The Paris–Erdogan law is often used for this purpose and is expressed as follows:

$$\frac{da}{dN} = C(\Delta K)^m, \tag{7.14}$$

where ΔK = stress intensity factor at the crack tip; C, m = constants to be determined based on tests; a = crack length; and N = number of stress cycles.

For steel structures with typical types of cracks, the stress intensity factor formulae are given in Broek (1986) and Paik and Thayamballi (2003). In ship-stiffened panels, cracks are often observed along the weld intersections between plating and stiffeners, for example. For a plate with cracking, ΔK may be given, when the stiffening effect is neglected, as follows:

$$\Delta K = F\Delta\sigma\sqrt{\pi a}, \tag{7.15}$$

where $\Delta\sigma$ = stress range (or double amplitude of applied fatigue stress); a = crack size (length); and F = geometric parameter depending on the loading and configuration of the cracked body. For plates with typical types of cracks and under axial tension as shown in Figure 7.29, F is given approximately as follows:

$$F = \left(\sec\frac{\pi a}{b}\right)^{1/2}, \quad \text{for a center crack, see Figure 7.29(a)}, \tag{7.16a}$$

$$F = 30.38\left(\frac{a}{b}\right)^4 - 21.71\left(\frac{a}{b}\right)^3 + 10.55\left(\frac{a}{b}\right)^2$$
$$- 0.23\left(\frac{a}{b}\right) + 1.12, \quad \text{for a crack on one side, see Figure 7.29(b)}, \tag{7.16b}$$

$$F = 15.44\left(\frac{a}{b}\right)^3 - 4.78\left(\frac{a}{b}\right)^2 + 0.43\left(\frac{a}{b}\right)$$
$$+ 1.12, \quad \text{for cracks on both sides, see Figure 7.29(c)}. \tag{7.16c}$$

Note that these simplified situations use idealized boundary conditions. The effect of stiffening is also neglected. More refined stress intensity factor solutions and methods of calculating improved stress intensity factors in particular situations exist, for

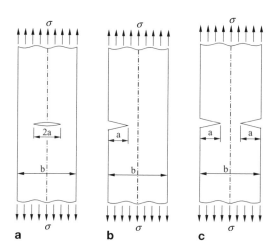

Figure 7.29. Typical crack locations in a plate under tensile stress: (a) center crack; (b) crack to one side; (c) crack to both sides.

example, by the nonlinear finite-element method or the Green function technique (Broek 1986).

In the time-dependent reliability analysis considering the growth of fatigue cracking, it is more efficient to express the crack growth behavior in a closed form. The crack length a(T) as a function of time T can then be calculated by integrating Eq. (7.14) with regard to the stress cycle N. In the integration of Eq. (7.14), it is often assumed that the geometric parameter F is constant; that is, assuming that the geometric parameter F is unchanged with the crack propagation. This assumption may be reasonable as long as the initial crack size a_o is small.

In such a case, the integration of Eq. (7.14), after substituting Eq. (7.15) into Eq. (7.14), results in

$$a(T) = \begin{cases} \left[a_o^{1-m/2} + \left(1 - \dfrac{m}{2}\right) C \left(\Delta\sigma F\sqrt{\pi}\right)^m (T - T_o)q \right]^{\frac{1}{1-m/2}}, & \text{for } m \neq 2, \\ a_o \exp\left[C\Delta\sigma^2 F^2\pi(T - T_o)q \right], & \text{for } m = 2, \end{cases}$$

(7.17)

where a_o = initial crack size; and a = subsequent crack size at time T.

It is interesting to note that in trading ships, the expected number of wave-load cycles occurs approximately once in every 6–10 seconds, and the total number of cycles after crack initiation may therefore be estimated by

$$N \approx (T - T_o) \times 365 \times 24 \times 60 \times 60/10 = q \times (T - T_o),$$

(7.18)

where $q \approx 365 \times 24 \times 60 \times 60/10$; T = structure age; and T_o = age at the initiation of the cracking with a = a_o.

Figure 7.30 shows a sample application of Eq. (7.17) by comparing it to a direct integration of Eq. (7.14), which accounts for the effect of crack growth on the geometric parameter F, that is, varied with time. Figure 7.30 shows that Eq. (7.17) slightly overestimates the fracture life as the crack propagates. This is expected because Eq. (7.17) was derived under the assumption that the geometric parameter F remains unchanged with time. However, it is considered that their difference for a small initial crack size is negligible, and, in such cases, Eq. (7.17) may provide a reasonable tool

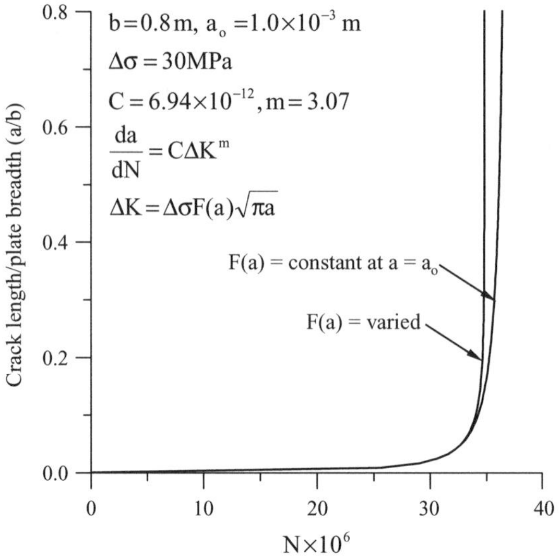

Figure 7.30. A comparison of Eq. (7.17) with a direct (numerical) integration of Eq. (7.14) for a small initial crack size.

for the crack-growth assessment. This is certainly not true for large initial crack sizes; in such cases, a more rigorous approach must be used.

REFERENCES

ABS (2004). *Guide for building and classing floating production installations*. American Bureau of Shipping, Houston, April.

API RP 2FPS (2001). *Recommended practice for planning, designing, and constructing floating production systems*. (Recommended Practices, 2FPS), American Petroleum Institute.

Bergan, P. G., and Lotsberg, I. (2004). "Advances in fatigue assessment of FPSOs." *Proceedings of International Conference on Offshore Mechanics and Arctic Engineering, OMAE-FPSO'04-0012*, Houston, August 30–September 2.

Broek, D. (1986). *Elementary engineering fracture mechanics*. The Netherlands: Martinus Nijhoff Publishers.

BS 5400 (1980). *Part 10: Code of practice for fatigue design*. British Standard, UK.

Bultema, S., van den Boom, H., and Krekel, M. (2000). *FPSO integrity: JIP on FPSO fatigue loads*. Offshore Technology Conference, OTC 12142, Houston, May.

BV (2004). *Hull structure of production, storage, and offloading surface units*. (Rule Notes, No. 497), Bureau Veritas, Paris, October.

DNV (1998). *Fatigue assessment of ship structures*. (Classification Notes, No. 30.7), Det Norske Veritas, Oslo.

DNV (2002). *Structural Design of Offshore Units (WSD Method)*. (Offshore Standards, DNV-OS-C201), Det Norske Veritas, Oslo.

DNV (2006). *Fatigue methodology for offshore ships*. (Recommended Practices, DNV-RP-C206), Det Norske Veritas, Oslo.

Doerk, O., Fricke, W., and Weissenborn, C. (2003). "Comparison of different calculation methods for structural stresses at welded joints." *International Journal of Fatigue*, 25: 359–369.

Dong, P., Hong, J. K., and Cao, C. (2001). *A mesh-insensitive structural stress procedure for fatigue evaluation of welded structures*. (IIW Document, No. XIII-1902–01), International Institute of Welding, USA.

Francois, M., Mo, O., Fricke, W., Mitchell, K., and Healy, B. (2000). *FPSO integrity: Comparative study of fatigue analysis methods.* Offshore Technology Conference, OTC 12148, Houston, May.

Fricke, W., Cui, W., Kierkegaard, H., Kihl, D., Koval, M., Mikkola, T., Parmentier, G., Toyosada, M., and Yoon, J. H. (2002). "Comparative fatigue strength assessment of a structural detail in a containership using various approaches of classification societies." *Marine Structures*, 15: 1–13.

Fricke, W., and Säbel, A. (2000). *Hot spot stress analysis of five structural details and recommendations for modeling, stress evaluation and design S–N curve.* (Report No. FF99.188), Germanischer Lloyd, Germany.

Gurney, T. R. (1979). *Fatigue of Welded Structures.* 2nd ed. Cambridge, UK: Cambridge University Press.

Holmes, P., Tickell, R. G., and Burrows, R. (1978). *Prediction of long term wave loading on offshore structures.* (Liverpool University Report, Nos. OT7823 and OT 7824 for OSFLAG Project 5), University of Liverpool, UK.

Hoogeland, M. G., van der Nat, C. G. J. M., and Kaminski, M. L. (2003). *FPSO fatigue assessment: Feedback from in-service inspections.* Offshore Technology Conference, OTC 15064, Houston, May.

HSE (1997). *A review of monohull FSUs and FPSUs.* (Offshore Technology Report, OTO 1997/800), Health and Safety Executive, UK.

HSE (1999). *Background to new fatigue guidance for steel joints and connections in offshore structures.* (Offshore Technology Report, OTH 1992/390), Health and Safety Executive, UK.

HSE (2001). *Comparison of fatigue provisions in codes and standards.* (Offshore Technology Report, OTO 2001/083), Health and Safety Executive, UK.

HSE (2004). *Review of low cycle fatigue resistance.* (Research Report, No. 207), Health and Safety Executive, UK.

IIW (1996). *Fatigue design of welded joints and components. Recommendations of IIW Joint Working Group XIII-1539–96/XV-845–96,* A. Hobbacher, ed., International Institute of Welding, Abington, UK: Abington Publishing.

ISO 2394 (1998). *General principles on reliability for structures.* International Standard, International Organization for Standardization, Geneva, June.

Kang, S. W., Kim, W. S., and Paik, Y. M. (2002). "Fatigue strength of fillet welded steel structure under out-of-plane bending." *Proceedings of International Welding/Joining Conference*, Korea.

Kang, S. W., and Kim, W. S. (2003). "A proposed S–N curve for welded ship structure." *Welding Journal*, 82(7): 161–169.

Lotsberg, I. (1998). *FPSO – Fatigue capacity, hot spot stress and S–N data: Background and planning.* (DNV Report, No. 98-3465), Det Norske Veritas, Oslo, December.

Lotsberg. I. (2000). *Background and status of the FPSO fatigue capacity JIP.* Offshore Technology Conference, OTC 12144, Houston, May.

Lotsberg, I. (2004). "Fatigue design of welded pipe penetrations in plated structures." *Marine Structures*, 17: 29–51.

Lotsberg, I. (2005). "Background for revision of DNV-RP-C203 fatigue analysis of offshore steel structures." *Proceedings of OMAE2005*, OMAE2005-67549, The 24th *International Conference on Offshore Mechanics and Artic Engineering (OMAE 2005)*, Halkidiki, Greece, June 12–17.

Lotsberg, I. (2006). "Fatigue design of plated structures using finite element analysis." *Ships and Offshore Structures*, 1(1): 45–54.

Lotsberg, I., and Landet, E. (2005). "Fatigue capacity of side longitudinals in floating structures." *Marine Structures*, 18: 25–42.

Lotsberg, I., and Sigurdsson, G. (2004). "Hot spot S–N curve for fatigue analysis of plated structures." *Proceedings of OMAE–FPSO 2004 – OMAE Specialty Symposium on FPSO Integrity (OMAE–FPSO'04-0014)*, Houston, August 30–September 2.

Newport, A., Basu, R., and Peden, A. (2004). "Structural modifications to the FPSO Kuito cargo tanks." *Proceedings of OMAE–FPSO 2004 – OMAE Specialty Symposium on FPSO Integrity, OMAE–FPSO'04-0085*, Houston, August 30–September 2.

Niemi, E. (2001). *Structural stress approach to fatigue analysis of welded components–Designer's guide.* (IIW Document, No. XV-1090-01), International Institute of Welding, USA.

Oh, M. H., Sim, W. S., and Shin, H. S. (2003). *Fatigue analysis of Kizomba 'A' FPSO using direct calculation based on FMS.* Offshore Technology Conference, OTC 15066, Houston, May.

Paik, J. K., and Hughes, O. F. (2006). "Ship structures." In *Computational analysis of complex structures.* Reston, VA: American Society of Civil Engineers.

Paik, J. K., and Thayamballi, A. K. (2003). *Ultimate limit state design of steel-plated structures.* Chichester, UK: John Wiley & Sons.

Paik, J. K., Wang, G., Thayamballi, A. K., Lee, J. M., and Park, Y. I. (2003). "Time-variant risk assessment of aging ships accounting for general/pit corrosion, fatigue cracking, and local denting damage." *SNAME Transactions*, 111: 159–197.

Petinov, S. V., Kim, W. S., and Paik, Y. M. (2006). "Assessment of fatigue strength of weld root in ship structure: An approximate procedure." *Ships and Offshore Structures*, 1(1): 55–60.

Storsul, R., Landet, E., and Lotsberg, I. (2004a). "Convergence analysis for welded details in ship-shaped structures." *Proceedings of OMAEFPSO 2004 – OMAE Specialty Symposium on FPSO Integrity, OMAE-FPSO'04-0016*, Houston, August 30–September 2.

Storsul, R., Landet, E., and Lotsberg, I. (2004b). "Calculated and measured stress at welded connections between side longitudinals and transverse frames in ship-shaped structures." *Proceedings of OMAE–FPSO 2004 – OMAE Specialty Symposium on FPSO Integrity, OMAE–FPSO'04-0017*, Houston, August 30–September 2.

Tam, G., and Wu, J. F. (2005). *A rational approach for the evaluation of fatigue strength of FPSO structures.* American Bureau of Shipping, Houston.

Urm, H. S., Yoo, I. S., Heo, J. H., Kim, S. C., and Lotsberg, I. (2004). "Low cycle fatigue strength assessment for ship structures." *Proceedings of the 9th Symposium on Practical Design of Ships and Other Floating Structures (PRADS 2004)*, Luebeck-Traemuende, Germany, September, pp. 774–781.

Wang, X., and Cheng, Z. (2003). "Sensitivity of fatigue assessment to the use of different reference S–N curves." *Proceedings of 22nd International Conference on Offshore Mechanics and Arctic Engineering (OMAE 2003), OMAE 2003-37238*, Cancun, Mexico, June 13–18.

Wang, X., Cheng, Z., Wirsching, P., and Sun, H. (2005). "Fatigue design factors and safety level implied in fatigue design of offshore structures." *Proceedings of 24th International Conference on Offshore Mechanics and Arctic Engineering (OMAE 2005), OMAE 2005-67488*, Halkidiki, Greece, June 12–17.

CHAPTER 8

Accidental Limit-State Design

8.1 Introduction

As discussed in Chapters 3 and 5, limit states are classified into four categories: serviceability limit states (SLS), ultimate limit states (ULS), fatigue limit states (FLS), and accidental limit states (ALS). This chapter presents ALS design principles and criteria together with some related practices applicable for ship-shaped offshore units.

ALS potentially leads to a threat of serious injury or loss of life, pollution, damage, and loss of property or significant financial expenditure. The intention of ALS design is to ensure that the structure is able to tolerate specified accidental events and, when accidents occur, subsequently maintains structural integrity for a sufficient period under specified (usually reduced) environmental conditions to enable the following risk mitigation and recovery measures to take place, as relevant:

- Evacuation of personnel from the structure
- Control of undesirable movement or motion of the structure
- Temporary repairs
- Safe refuge and firefighting in the case of fire and explosion
- Minimizing outflow of cargo or other hazardous material

Different types of accidental events may require different methodologies or different levels of refinement of the same methodology to analyze structural resistance or capacity during and following such events (demands). The ALS design is then necessarily an important part of design and operation in terms of risk assessment and management that consists of hazard identification, structural evaluation, and mitigation measure development for specific types of accidents, as we describe in Chapter 13.

The main focus of this chapter is on the prescriptive safety evaluation for accidental events such as unintended flooding (damage stability), collisions, dropped objects, fire, explosion, and progressive accidental hull girder collapse.

8.2 Design Principles and Criteria

The primary aim of the ALS design can be characterized by the following three broad objectives:

- To avoid loss of life in the structure or the surrounding area
- To avoid pollution of the environment
- To avoid loss of property or prevent significant financial expenditure

ALS considerations are necessary to achieve a design whose main safety functions are not impaired during any accidental event or within a certain time period after the accident to the necessary and acceptable degree. The ALS design criteria are normally based on limiting accidental consequences such as structural damage, health, and environmental pollution. Risk mitigation for such events will take account of not only design features but also operational measures, including ceasing production; it will also address crew member training.

Because the structural damage characteristics and the behavior of damaged structures depend on the types of accidents and risk perception is unique to individuals, societies, and circumstances, it is not straightforward to establish universally applicable ALS design criteria. Typically, for a given type of structure, design accidental scenarios and associated acceptance criteria must be chosen on the basis of particular risk assessment.

In selecting the design target ALS performance levels for such events, the approach is normally to tolerate a certain level of damage consistent with a greater aim such as survivability or minimized consequences; not to do so would result in an uneconomical structure. The main functions of the structure that should not be compromised during any accident event or within a certain time period after the accident may include the usability of escape ways, the integrity of shelter areas, and the integrity of global system structure and the environment.

For purposes of ALS design, the following three main aspects must be identified:

- Significant accident scenarios taking account of frequency of occurrence
- Structural and other evaluation methods of the accident consequences
- Relevant acceptance criteria

Accident scenarios must reflect accidental phenomena that affect the safety of the installation and the surrounding environment in an unfavorable fashion, but must also be credible. The largest credible accident possible of a particular type is often of interest. The frequency of occurrence of the corresponding accident must fall within an acceptable range. The structural evaluation methods should be adopted so that the accident consequences can be analyzed to the needed accuracy.

Although in some cases simplified approaches may often be enough, more sophisticated methodologies are in other cases necessary for analysis of the accident consequences that usually involve highly nonlinear aspects by their very nature. The acceptance criteria format depends on the accident situations to be avoided. Typical measures of the acceptance criteria include reserve stability, damage extent, quantity of oil outflow, and residual load-carrying capacity, for example. Required or limit values for accidental action effects (e.g., damage amount, material property change) and structural crashworthiness (e.g., energy absorption capability) are often used to represent the measure of safety level.

The ALS design format may, therefore, be a set of deterministic rules representing acceptable safety level or some given limits to the probability of occurrence to adverse events, or some specified bounds on the probability (likelihood) of consequences, or some combination of these. A deterministic ALS design format may be expressed in terms of limits of deformation or energy absorption capability until the critical consequence occurs, as follows:

$$w \leq w_a, \tag{8.1a}$$

$$E_k \gamma_k \leq \frac{E_r}{\gamma_r}, \tag{8.1b}$$

where w = factored accidental action effects (e.g., deformation, strain); w_a = allowable (factored) accidental action effects; E_k = characteristic value of kinetic energy loss due to accidental actions; E_r = characteristic value of energy absorption capability until a specified critical damage occurs; and γ_k, γ_r = partial safety factors, taking into account the uncertainties related to kinetic energy loss and energy absorption capability, respectively.

The partial safety factors used in Eq. (8.1b) may be chosen to represent one or more or perhaps even all of the following uncertainties:

- Natural variation of design variables
- Modeling uncertainties of the assessment method
- Return period of hazard event
- Societal factors including risk perception
- Consequences including economic factors

In contrast to deterministic ALS design criteria, the risk-based design format can be given by

$$R \leq R_a, \tag{8.2}$$

where $R = \sum_i F_i C_i$ = risk; F_i = frequency (or likelihood) of the ith failure event resulting in the consequence C_i; and R_a = acceptable risk level.

Risk-based criteria are more general in nature but usually more complex to apply than the prescriptive approaches. Risks to humans may be categorized into two main types:

- Individual fatality risks that are perhaps approximately the same as those typical for other occupational hazards
- Societal fatality risks associated with frequency of accidents and hazards

Any risk should not exceed a level defined as unconditionally intolerable, and the level of the consequences of any accident should be acceptable to the various stakeholders, primarily the owners, governments, and the public. To achieve these aims within a risk-based format to ALS, the well-known and general as low as reasonably practicable (ALARP) technique can be applied for risk assessment, discussed in Chapter 13. This chapter deals primarily with deterministic approaches to ALS, which may, however, also form part of a more general overall risk-based approach to design in many cases.

8.3 Damaged Vessel Stability: Accidental Flooding

If one or more internal spaces of a vessel are opened to the sea by structural damage, then cargo leakage and/or water ingress can potentially take place until stable equilibrium is established between these spaces and the sea. Accidental flooding

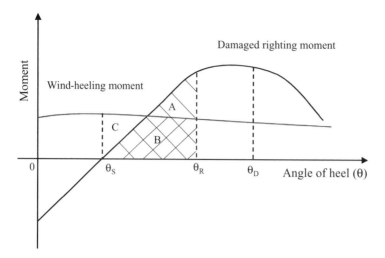

Figure 8.1. Damaged vessel stability criteria, after HSE (2001; courtesy of HSE).

and/or cargo leakage can cause significant changes in draft, trim, and heel. When such changes subsequently immerse nonwatertight portions of the vessel, stable equilibrium may not be attainable due to progressive flooding, and the vessel can sink either with or without capsizing.

In general, if the remaining maximum righting moment in terms of the metacentric height and other stability characteristics such as GZ is smaller than any existing heeling moment, capsizing could take place. Even for the case of symmetrical flooding with respect to the vessel centerline, some heeling forces usually exist due to unsymmetrical weights and/or wind; thus, capsizing may occur as long as the metacentric height, for example, is negative in the damaged condition. Also, the amount of reserve buoyancy present by design can be an important factor in such cases.

It is required under these circumstances to ensure that the offshore units survive any reasonable damage resulting in flooding. In particular, reserve stability in damaged conditions with unintended flooding must be sufficient to withstand, say, the wind-heeling moment imposed from any direction on the damaged unit.

The procedure for damage stability calculations of ship-shaped offshore units is similar to that for normal trading tankers (Lewis 1988); in particular, the effect of wind should not be overlooked. HSE (2001) reviews practical damage stability criteria for offshore units such as that illustrated in Figure 8.1. In this case, the curves of damaged righting moment and wind-heeling moment at the specific wind velocity are plotted as functions of heeling angle. For the areas depicted in Figure 8.1, the following criterion must be satisfied regarding accidental flooding in terms of damage stability:

$$A + B \geq B + C, \quad \theta_S \leq 15°, \tag{8.3a}$$

where A, B, C = shaded areas of the moment–heeling angle curves illustrated in Figure 8.1; θ_S = static angle of heel after damage without wind. In the calculations of the areas A, B, and C, the following conditions must also be satisfied:

$$\theta_R \leq \theta_D \quad \text{and} \quad \theta_R \leq \theta_2, \tag{8.3b}$$

where θ_2 = second intercept of wind-heeling and righting-moment curves; θ_D = angle of first down-flooding; and θ_R = angle from θ_S over which A, B, and C are evaluated.

In such calculations, the effects of mooring restraints are usually not accounted for, and a substantial number of heeling angles possible must be considered. Also, the effects of mitigation measures to reduce heel motion, such as pumping flooded water out or changes in ballasting of compartments, are not considered in the calculations in order to reflect the worst anticipated service condition after accidental flooding. In specific cases of damage stability assessment, it is highly desirable to perform wind-tunnel testing to better quantify the wind-heeling and righting moments (HSE 2000j).

8.4 Collisions

8.4.1 Fundamentals

Worldwide, there have been several collision accidents between ships and offshore oil and gas installations that sometimes have resulted in the total loss of the offshore units involved. The majority of the in-field vessel collisions is known to be with supply vessels and shuttle tankers, although there are also a few cases involving passing vessels, that is, involving a vessel that was not being operated in connection with the offshore installations (HSE 1999a).

Usually, a significant decrease in the frequency of in-field vessel impacts has been achieved in most cases by various efforts addressing the increased operator awareness of the associated hazards, the increased experience and training of the marine crews involved, and, in general, the insistent demand for higher safety standards.

It is, however, recognized that with increase in the use of ship-shaped offshore units worldwide, the probability of collision accidents, such as between a shuttle tanker and a ship-shaped offshore unit in a tandem offloading operation, as shown in Figure 8.2, cannot be avoided completely (Vinnem et al. 2003). A supply boat that routinely berths at the offshore installation can also cause a collision accident. Passing vessels may also pose a collision risk if the unit is close to or within a frequently sailed route. It is, therefore, only prudent to include such considerations in the ALS design of the ship-shaped offshore units where significant. It is equally important to manage collision risks and develop relevant safety measures for protecting offshore units because such accidents can lead to costly consequences in terms of loss of lives and damage to property and environment.

The reports of the International Ships and Offshore Structures Congress (ISSC) Specialist Committee on Collision and Grounding (Paik et al. 2003a; Wang et al. 2006) present a recent state-of-the art review on collision and grounding mechanics of trading ships, taking into account the probabilistic and physical nature of such accidents, which can also be relevant, at least in part, for offshore units. The effectiveness of structural arrangements for reducing or avoiding pollution due to leakage and for maintaining an adequate amount of residual strength of damaged structures is also discussed in the same ISSC reports.

It is important to realize that the collision risk profile of ship-shaped offshore units may be different from that of trading tankers. This is mainly because the former

Figure 8.2. Tandem offloading operation of an FPSO (Adhia et al. 2004; courtesy of Society of Petroleum Engineers/Offshore Technology Conference).

is located in a fixed position and is routinely visited by shuttle tankers as well as supply boats, in contrast to the latter. Therefore, the collision risk scenarios to be considered for the design of ship-shaped offshore structures must be different from those of trading tankers. However, in both cases, similar methods can be employed for the analysis of structural crashworthiness due to collisions because the collision mechanics are similar and the structure is also similar.

The damage from a collision with low kinetic energy may possibly be limited to permanent deformation without tearing or fracture of the side shell of the struck vessel structure. This low-energy collision is often termed a minor collision or contact. However, the collision with high kinetic energy can result in severe damage, including fracture of the struck structure that can cause cargo leakage and/or water ingress, posing a threat to the safety of the offshore installation and the protection of the environment. This high-energy collision is sometimes called a "major collision"; it obviously gets more attention than the minor collisions in terms of risk management of ship-shaped offshore units. Note, however, that the energy limit value defining low- or high-energy collision may depend on structural crashworthiness, as well as initial kinetic energy. In practice of ship-shaped offshore structures, a major collision can arise from the impact of a shuttle tanker or passing vessel, although a minor collision may be the result of the impact of a supply boat (Wang et al. 2003).

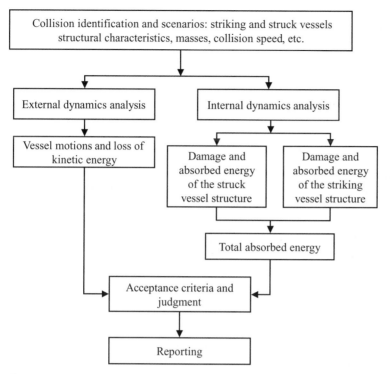

Figure 8.3. A prescriptive procedure for safety evaluation of a collision event.

8.4.2 Practices for Collision Assessment

Analysis of the accident mechanics for collisions can be classified into two parts: the external and internal mechanics. The external accident mechanics deal with the rigid-body global motion of the structures involved under the accidental actions. The internal accident mechanics include evaluation of the structural failure response during the accident. The internal mechanics of collision accidents are quite complex. Deformations many times larger than the structural thickness may occur, and the major part of energy absorption may take place by inelastic large "straining." Figure 8.3 shows a practical procedure for safety evaluation in a vessel-to-vessel collision event.

As the ALS design criterion related to minor collisions, Eq. (8.1a) could be used when the permanent deformation is calculated in the relevant collision scenario. For major collisions, however, Eq. (8.1b) is usually more adequate to apply. In this case, the kinetic energy loss associated with external collision mechanics and the energy absorption capability associated with internal collision mechanics need to be evaluated.

Collisions are dynamic by nature and can be characterized by their own general differential equations of motion. However, the collision event is also well described quasistatically by using the conservation of momentum principle as long as the impact duration is long enough compared to the natural period of vibration of the structure involved. For a collision between a ship-shaped offshore unit and a trading vessel (e.g., shuttle tanker, supply boat, passing vessel), the conservation of momentum

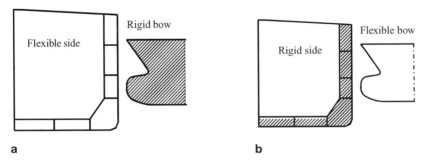

Figure 8.4. A schematic for the internal collision mechanics analysis of vessel structures: (a) internal collision mechanics analysis of the struck vessel structure; (b) internal collision mechanics analysis of the striking vessel structure.

principle yields the following equation when both striking and struck vessels are considered to move as one body after collision:

$$(M_A + \Delta M_A)V_A \sin \theta + (M_B + \Delta M_B)V_B = (M_A + \Delta M_A + M_B + \Delta M_B)V_T, \quad (8.4)$$

where M_A, M_B = masses of striking or struck vessel; V_A, V_B = velocities of striking or struck vessel at the start time of collision; $\Delta M_A, \Delta M_B$ = added masses of the striking or struck vessel; V_T = velocity of both striking and struck vessels as one body after collision; and θ = collision angle between striking and struck vessels.

The velocity of both striking and struck vessels as one body after collision can be readily obtained from Eq. (8.4), as follows:

$$V_T = \frac{(M_A + \Delta M_A)\,V_A \sin \theta + (M_B + \Delta M_B)\,V_B}{M_A + \Delta M_A + M_B + \Delta M_B}. \quad (8.5)$$

When M_B is much larger than M_A, $V_T \approx 0$ may be assumed, implying that both striking and struck vessels stop as one body after collision. The loss of kinetic energy after collision can be calculated as follows:

$$E_k = \frac{1}{2}(M_A + \Delta M_A)(V_A \sin \theta)^2 - \frac{1}{2}(M_A + \Delta M_A + M_B + \Delta M_B)V_T^2, \quad (8.6)$$

where E_k = kinetic energy loss as defined in Eq. (8.1b); and V_T = as defined in Eq. (8.5). When a ship-shaped offshore unit is considered to be in a fixed location, $\Delta M_B = 0$ and $V_B = 0$ can be assumed.

The energy absorption capability (i.e., E_r in Eq. (8.1b)) can be obtained by considering the internal collision mechanics. The strain-energy component dissipated by damage of the struck structure is usually computed by solving the internal collision mechanics problem for the struck structure when the striking structure regarded as a rigid body penetrates into the struck structure, as illustrated in Figure 8.4(a). However, the strain-energy component dissipated by damage of the striking structure is obtained by analyzing the internal collision mechanics problem for the striking structure when the striking structure crushes into a rigid vertical wall, as shown in Figure 8.4(b).

Once the relationship between collision force and penetration is obtained, until or after the critical damage resulting in a threat to the structural safety and the environment takes place, as illustrated in Figure 8.5, the energy capability absorbed

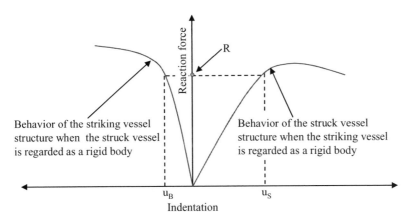

Figure 8.5. A schematic of energy absorption characteristics by both the bow structure of the striking vessel and the side structure of the struck vessel.

up to the occurrence of the critical damage can be calculated by integrating the area below the force versus penetration curve. The kinetic energy during the collision can be consumed to damage both the striking and struck structures. Note that during the collision, the reaction force denoted by R in Figure 8.5 must be the same in both striking and struck structures, where the indentation will be u_B at the striking vessel structure and u_S at the struck vessel structure. In this case, the total energy absorption is calculated as the sum of the two component absorptions.

8.4.3 Nonlinear Finite-Element Modeling Techniques

Although useful simplified methods have been developed in the literature to calculate the collision force penetration relations or the collision energy penetration relations (Paik et al. 2003a; HSE 2004), a great deal of progress has also been achieved in the use of nonlinear dynamic finite-element methods to collision problems that give more reliable and accurate simulation results of accident consequences. Indeed these advanced methods are now, arguably, mature enough to enter day-by-day design practice. Idealized structural unit method (ISUM) has also been useful for efficient simulation of structural crashworthiness due to collisions (Paik and Pedersen 1996, Paik et al. 1999). ALPS/SCOL (2006) is a computer program for analysis of collision and grounding mechanics using ISUM.

Related to nonlinear dynamic finite-element methods, two types of algorithms (i.e., implicit and explicit) are relevant based on the time integration techniques used. In the explicit algorithm, both internal and external forces are summed at each finite-element nodal point and nodal acceleration is computed by dividing total forces by nodal mass, but the implicit algorithm employs a traditional finite-element solver by applying the nodal force increments to calculate the nodal displacement increments. The explicit algorithm may solve the stiffness equation at time t by direct time integration using the central difference technique, as follows:

$$a_t = \frac{v_{t+0.5\Delta t} - v_{t-0.5\Delta t}}{\Delta t}, \quad v_t = \frac{u_{t+0.5\Delta t} - u_{t-0.5\Delta t}}{\Delta t}, \tag{8.7}$$

where a_t = acceleration at time t; v_t = velocity at time t; u_t = displacement at time t; and Δt = time increment (step).

However, the implicit algorithm solves the stiffness equation at time $t + \Delta t$ where the time step size may be selected by the program user, while the maximum time step size in the explicit algorithm is controlled properly. For dynamic structural simulations involving impact and crashworthiness, the explicit algorithm-based programs are perhaps more useful. The following are some examples of nonlinear finite-element programs that can analyze the structural crashworthiness due to accidental actions; there may be distributors other than the major ones noted here:

ABAQUS distributed by HKS (www.hks.com)

ADINA distributed by ADINA R&D, Inc. (www.adina.com)

ANSYS distributed by ANSYS, Inc. (www.ansys.com)

LS-DYNA distributed by Livemore Software Technology Corporation (www.lstc.com)

NASTRAN, MARC distributed by MSC Software Corporation (www.mscsoftware.com)

PAM-CRASH distributed by ESI Group (www.esi-group.com)

RADIOSS distributed by RADIOSS Consulting Corporation (www.radioss.com)

The Belytschko–Tsay-type-four-node plate-shell elements are typically used to analyze the dynamics of thin-walled structures such as ships and ship-shaped offshore structures in collision applications, and these are known to be more appropriate than the older triangular elements. To properly capture structural crashworthiness characteristics, the finite-element mesh size should be fine enough.

For many reasons, however, it is not always the case that very fine mesh modeling can be adopted. For example, very large complex structures need a huge number of finite elements so that it is not easy to execute the numerical simulations with such large number of elements. Convergence studies varying the mesh size and the number of elements often need to be undertaken to define relevant mesh size for practical purposes. In this regard, it will be very helpful for nonlinear finite-element structural modeling if the relevant optimal mesh size can be readily determined in important cases without an extensive convergence study. Some related guidelines and considerations are now discussed.

In collisions or grounding, crushing, fracture (tearing or cutting), and inelastic large straining are major failure modes. Among them, the crushing mechanism requires very fine mesh size to accurately reflect a folded configuration. Figure 8.6 shows a typical crushing pattern of thin-walled structures under compressive actions. It is evident from Figure 8.6 that more than eight (rectangular plate-shell type) finite elements may be necessary to capture the folding pattern within a half length H of one structural fold. It is interesting to note that relevant mesh size to properly capture crushing behavior of thin-walled structures can be determined depending on the expected half length of one structural fold.

A number of analytical formulae are now available in the literature to predict the expected length of one structural fold of thin-walled structures (Paik and Wierzbicki 1997), although they have been developed for different crushing patterns applicable to different structural geometries. For example, Wierzbicki and Abramowicz (1983)

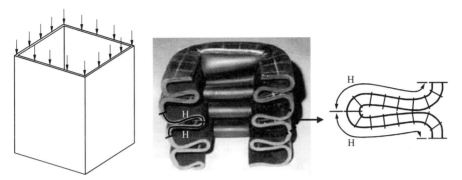

Figure 8.6. A thin-walled structure crushed under predominantly axial compressive actions, cut at its midsection.

proposed the following formula to predict the fold length for thin-walled structures under crushing actions:

$$H = 0.983b^{2/3}t^{1/3}, \tag{8.8}$$

where b = breadth of plating between stringers or support members; t = plate thickness; and H = a half-fold length; see Figure 8.6.

Equation (8.8) can be approximately applied for all types of structural cross sections such as L-, T-, or X-sections once b and t are known. The fold length can be changed by stiffeners, among other factors. Some experimental observations on the folding length change due to stiffeners of thin-walled structures are presented by Paik et al. (1996).

It is considered that fracture takes place when the following criterion in terms of strain is satisfied:

$$\varepsilon_{eq} \geq \varepsilon_{fc}, \tag{8.9}$$

where ε_{eq} = equivalent maximum tensile strain at a finite element; and ε_{fc} = critical fracture strain specified for the finite-element simulations.

The stress-based criteria can, of course, be used for judgment of fracture occurrence (Lehmann and Yu 1998; Zhu and Atkins 1998). In the finite-element simulation, a fractured element is usually removed from the object structure at the next incremental loading step. A double set of nodes is sometimes introduced to simulate the crack propagation such that the elements are allowed to separate once the fracture criteria are fulfilled (Amdahl and Stornes 2001), if the path of crack propagation can be known or presumed in advance.

The stress–strain relationship of material used for the finite-element modeling can significantly affect the resulting simulations for structural crashworthiness. The true stress-versus-strain characteristics of material can usually be obtained by the transformation of engineering (or nominal) stress-versus-strain relationship, as follows:

$$\sigma_t \approx \sigma_e(1 + \varepsilon_e); \varepsilon_t = \ln(1 + \varepsilon_e), \tag{8.10}$$

where σ_t = true stress, which allows approximately for Poisson-ratio thinning; ε_t = true strain representing integral of change in length over instantaneous length;

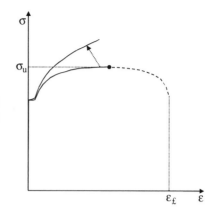

Figure 8.7. A schematic of material model I – the traditional approach to transform the engineering stress–strain curve to the true stress–strain curve.

σ_e = engineering stress, which is given as force divided by unstressed area; and ε_e = engineering strain, which represents change in length divided by initial length.

When the detailed material coupon test results do not exist, Eq. (8.10) can be approximately developed, for example, by using the Ramberg–Osgood model (Ramberg and Osgood 1941), as follows:

$$\varepsilon_e = \frac{\sigma_e}{E} + \varepsilon_{eo}\left(\frac{\sigma_e}{\sigma_{\varepsilon_{eo}}}\right)^n, \tag{8.11}$$

where E = elastic modulus at the origin of the stress–strain curve; ε_{eo} = residual engineering strain corresponding to the elastic limit; $\sigma_{\varepsilon_{eo}}$ = engineering stress at $\varepsilon_e = \varepsilon_{eo}$; and n = exponent constant, which may be taken as $n = \ln \alpha / \ln \sigma_c$ for $\alpha \geq 1$; σ_c = stress at $\varepsilon_e = \varepsilon_{eo}/\alpha$; α = test constant, which is often taken as $\alpha = 2$ for steel.

Note, however, that the Ramberg–Osgood model represents the stress–strain relation up to the ultimate stress, and no characterization of necking behavior beyond the ultimate stress is considered.

Some useful techniques for material modeling are also presented in Servis et al. (2002), Simonsen and Ocakli (1999), Simonsen and Lauridsen (2000), and Simonsen and Tornqvist (2004). In this section, we now describe a new technique.

We compare three types of material models, denoted by models I, II, and III. In traditional finite-element simulations, the true stress–strain curve is defined by using the engineering stress–strain curve up to the ultimate tensile stress only, as shown in Figure 8.7. This modeling is termed "material model I" hereafter. Material model I is often not appropriate because it overestimates the strain-hardening characteristics and does not account for the necking or softening behavior beyond the ultimate tensile stress, as shown in Figure 8.7.

To account for the necking behavior beyond the ultimate tensile stress, the entire set of data points defining the engineering stress–strain curve, that is, until the rupture takes place, are now used to get the true stress–strain curve, as shown in Figure 8.8. This modeling is termed material model II. As shown in Figure 8.7, however, material model II overestimates the strain-hardening characteristics although the necking behavior is more properly accounted for than model I.

Material model III represents both strain-hardening and necking behavior correctly, as shown in Figure 8.9. In this model, the following equation is used to

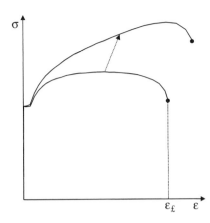

Figure 8.8. A schematic of material model II – the modified approach to transform the engineering stress–strain curve to the true stress–strain curve.

transform the engineering stress-versus-strain relationship to the true stress-versus-strain relationship [for symbols, unless specified below, see Eq. (8.10)]:

$$\sigma_t \approx f(\varepsilon_e)\,\sigma_e(1+\varepsilon_e), \quad \varepsilon_t = \ln(1+\varepsilon_e), \tag{8.12}$$

where $f(\varepsilon_e) =$ knock-down factor as a function of engineering strain.

It turns out that material model III is often more appropriate to model the stress-versus-strain characteristics of material because it reflects quite accurately the necking behavior as well as the strain-hardening characteristics. The knock-down factor function $f(\varepsilon_e)$ can be approximately based on existing tensile coupon-test data. $f(\varepsilon_e)$ may be derived approximately as follows:

$$f(\varepsilon_e) = \begin{cases} \frac{C_1-1}{\ln(1+\varepsilon_u)}\{\ln(1+\varepsilon_e)\}+1, & \text{for } 0 < \varepsilon_e \le \varepsilon_u \\ \frac{C_2-C_1}{\ln(1+\varepsilon_f)-\ln(1+\varepsilon_u)}\{\ln(1+\varepsilon_e)\}+C_1 - \frac{(C_2-C_1)\{\ln(1+\varepsilon_u)\}}{\ln(1+\varepsilon_f)-\ln(1+\varepsilon_u)}, & \text{for } \varepsilon_u < \varepsilon_e \le \varepsilon_f, \end{cases}$$
$$\tag{8.13}$$

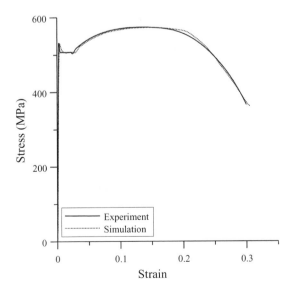

Figure 8.9. A schematic of material model III – to accurately represent both strain-hardening and necking behavior.

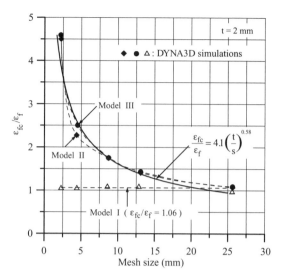

Figure 8.10. Variation of the critical fracture strain used for LS-DYNA finite-element simulations as a function of mesh size, at a quasistatic loading condition.

where ε_f = fracture strain obtained by the material tensile coupon test; ε_u = strain at the ultimate tensile stress obtained by the material tensile coupon test; and C_1, C_2 = test constants of material. The constants C_1 and C_2 are generally affected by plate thickness as well as material types. However, it is interesting to note that these coefficients may be approximated as constant values with $C_1 = 0.9$ and $C_2 = 0.85$ for mild- and high-tensile steel.

The accuracy in the (critical) strain, at which fracture is said to take place, is affected by various factors including mesh size and strain rate. It is also affected by material models of the finite-element simulation in association with the stress-versus-strain relationships. Figure 8.10 shows variations of the predicted critical fracture strain used for LS-DYNA finite-element simulations at quasistatic loading condition as a function of the mesh size and material model type on a tensile coupon-test specimen with the plate thickness of 2mm. Figure 8.10 shows that the critical fracture strain, when using model I, is almost constant at $\varepsilon_{fc}/\varepsilon_f = 1.06 \approx 1.0$ regardless of mesh size.

However, the predicted critical fracture strain varies with the finite-element mesh size when material models II or III are applied. When the mesh size is infinitely small, the critical fracture strain ε_{fc} to be defined for finite-element modeling may be several times larger than the nominal fracture strain ε_f obtained from material tensile coupon-test results, and it approaches the same value of ε_f as finite-element size increases. The finite-element simulations with relatively coarse mesh and material models II or III require the critical fracture strain to be similar to the nominal fracture strain in the test. This is because the length of the coarse mesh elements roughly corresponds to the yielding area in the test.

Based on the insights of Figure 8.10, the following equation available for models II and III may be proposed as being useful to determine the critical fracture strain as a function of element size:

$$\frac{\varepsilon_{fc}}{\varepsilon_f} = 4.1 \left(\frac{t}{s}\right)^{0.58}, \tag{8.14}$$

Table 8.1. *Sample coefficients for the Cowper–Symonds constitutive equation*

Material	$C\,(s^{-1})$	q	Reference
Mild steel	40.4	5	Cowper and Symonds (1957)
	7.39	4.67	Schneider and Jones (2004)
	114	5.56	Hsu and Jones (2004a)
Higher-tensile steel	3,200	5	Paik and Chung (1999)
Aluminum alloy	6,500	4	Bodner and Symonds (1962)
	9.39×10^{10}	9.55	Hsu and Jones (2004b)
α-Titanium (Ti 50A)	120	9	Symonds and Chon (1974)
Stainless steel 304	100	10	Forrestal and Sagartz (1978)

where ε_{fc} = critical fracture strain to be used for finite-element simulations; ε_f = nominal fracture strain obtained by tensile coupon testing; s = mesh size (length); and t = plate thickness.

8.4.4 Dynamic Material Properties

Note that the nominal fracture strain ε_f is significantly dependent on the strain rate and structural deterioration, among other factors. This is because a material such as steel may tend to become relatively less ductile or more brittle as the strain rate increases. Also, stress concentration due to initial deterioration or defects will make the fracture occurrence much earlier. Therefore, ε_f (and subsequently ε_{fc}) used for finite-element simulations should take a much smaller value than that obtained from tensile coupon testing as the strain rate and/or deterioration increases.

The dynamic yield strength of the material may be expressed as follows (Karagiozova and Jones 1997; Jones 1989):

$$\frac{\sigma_{Yd}}{\sigma_Y} = f(\dot{\varepsilon})\,g(\varepsilon), \tag{8.15}$$

where σ_Y, σ_{Yd} = static and dynamic yield stresses; $f(\dot{\varepsilon})$ = a function of strain-rate sensitivity effect; $g(\varepsilon)$ = a material strain-hardening function; and $\dot{\varepsilon}$ = strain rate.

If the strain-hardening effect is negligible, one can take as $g(\varepsilon) = 1$. The strain-rate sensitivity function $f(\dot{\varepsilon})$ is often given using the Cowper–Symonds equation (Cowper and Symonds 1957) as follows:

$$\frac{\sigma_{Yd}}{\sigma_Y} = 1.0 + \left(\frac{\dot{\varepsilon}}{C}\right)^{1/q}, \tag{8.16a}$$

where C and q are coefficients to be determined based on test data; see Table 8.1. It is evident that the coefficients C and q are dependent on material types, among other factors.

Figure 8.11 plots the Cowper–Symonds equation together with the relevant coefficients for mild- or high-tensile steels when $g(\varepsilon) = 1$. As shown in Figure 8.11, Paik and Chung (1999) found that the higher-tensile steel is less sensitive to the strain rate than mild steel. This was later also confirmed by Jones (2001).

In some cases, strains are much larger than the yield strain and the dynamic yield stress can change with the strain magnitude, as well as the strain rate, so that the

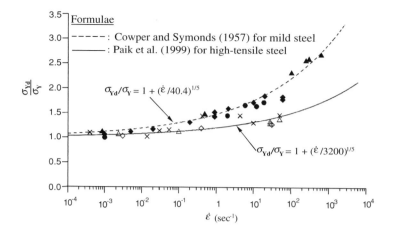

Figure 8.11. Dynamic yield strength σ_{Yd} (normalized by the static yield strength σ_Y) plotted versus strain rate $\dot{\varepsilon}$ for mild- and high-tensile steels, extracted from Paik and Chung (1999) and Paik and Thayamaballi (2003).

constants C and q in Eq. (8.16a) must be a function of the strain ε. To take into account this effect, Jones (1989) modified Eq. (8.16a) in the following form:

$$\frac{\sigma_{Yd}}{\sigma_Y} = 1.0 + \left(\frac{\dot{\varepsilon}}{D + E\varepsilon} \right)^{1/q},$$
(8.16b)

where D, E = test constants, which can be obtained from dynamic test results at the yield and ultimate tensile stresses, respectively.

Both crushing effects and yield strength can increase as the loading speed gets faster, although any fracture or tearing of steel (at the welded regions) of a structure tends to occur earlier for steel. The following approximate formula, which is the inverse of the Cowper–Symonds constitutive equation for the dynamic yield stress, is then useful for estimating the dynamic fracture strain as a function of the strain rate:

$$\frac{\varepsilon_{fd}}{\varepsilon_f} = \xi \left[1.0 + \left(\frac{\dot{\varepsilon}}{C} \right)^{1/q} \right]^{-1},$$
(8.17)

where ε_f, ε_{fd} = static or dynamic fracture strains; and ξ = ratio of the total energies to rupture for dynamic and static uniaxial loadings. The dynamic fracture strain ε_{fd} will then be used for the finite-element simulations in place of ε_f.

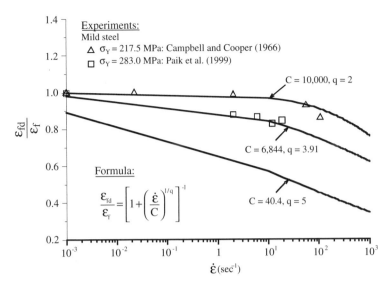

Figure 8.12. Dynamic fracture strain (normalized by the static fracture strain) versus strain rate for mild steels (Paik and Thayamballi 2003).

If the energy to failure is assumed to be invariant, that is, independent of the strain rate $\dot{\varepsilon}$, then it may be taken that $\xi = 1$. Figure 8.12 plots Eq. (8.17) with $\xi = 1$ for three sets of the coefficients as well as experimental results for mild- or high-tensile steels. The expression in Eq. (8.17) represents the decrease of the dynamic fracture strain with increase in the strain rate, but the coefficients for the dynamic fracture strain differ from those for the dynamic yield strength. It is again evident that the strain rate is a primary parameter affecting the impact mechanics and the structural crashworthiness. Also, it is seen from Figure 8.12 that Eq. (8.17), with $C = 40.4$ and $q = 5$ for mild steel, gives a too small value of the fracture strain. Rather, it is recommended to adopt C in the range of 7,000–10,000 and q in the range of 2 to 4.

It is, however, interesting to note that the rupture strain increases with strain rate for some materials such as higher strength steel (Peixinho et al. 2003a, 2003b) or aluminum alloys (Hsu and Jones 2004b). The analysis of dynamic behavior can be more complex for such materials that exhibit an increase of rupture strain in range of small strain rate but conversely also exhibit a decrease of rupture strain outside this range of strain rate, as observed by Kawata and colleagues (1968, 1977). Indeed, it is found that aluminum alloy 6063 T6 has a minimum rupture strain at a strain rate of $0.03\mathrm{s}^{-1}$, approximately (Hsu and Jones 2004b).

Inertia effects may sometimes need to be considered for impact-response simulations of thin-walled structures. Due to the inertia effects and stress wave propagation phenomena within the structures during impact actions, the strain distribution (or deformation pattern) at any moment in time would be nonhomogeneous (Paik and Thayamballi 2003). Although the inertia effect is usually neglected for the evaluation of ship collision consequence, it may become more important when the collision speed is very fast.

During the process of the collisions, the influence of friction would be normally large when a relative velocity exists between the striking and struck bodies. This

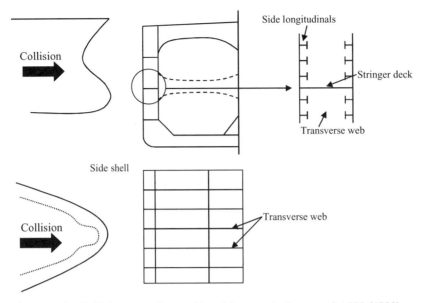

Figure 8.13. Collision scenario considered in a particular test of ASIS (1993).

situation is often seen at ship-grounding events when a ship with forward speed runs on to a rock pinnacle, for instance. The influence of friction may be ignored when the relative velocity between striking and struck bodies is comparatively small, although it is significant for a raking type of collision, such as on the side shell.

A very useful description of some recent developments in the dynamic inelastic behavior of structures, including dynamic material properties and dynamic energy absorption of structures, is presented by Jones (2006).

8.4.5 Illustrative Examples

The Association for Structural Improvement of the Shipbuilding Industry of Japan (ASIS 1993) has carried out several collision and grounding tests. One of the collision test models they used is a double-side structure model made of mild steel; the aim was to investigate the structural crashworthiness of double-side structures of tankers, as shown in Figure 8.13. The collision test model was prepared as illustrated in Figure 8.14. Table 8.2 indicates material properties of the test model plates. Figure 8.15 shows

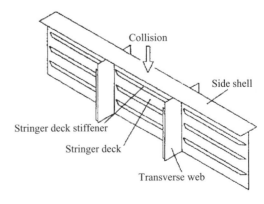

Figure 8.14. A double-skinned structure model considered in the collision test of ASIS (1993).

Table 8.2. *Material properties of the collision test model plate elements*

t (mm)	E (GPa)	σ_Y(MPa)	ε_f
7	210.2	310.2	0.4315
8	210.7	316.1	0.4401
10	210.7	282.7	0.4299

Notes: t = plate thickness; E = Young's modulus; σ_Y = material yield stress; and ε_f = nominal fracture strain at a quasistatic loading condition.

the modeling of the engineering stress–strain curves of the plates applying material model III that accounts for both strain-hardening and necking behavior of material, as described in the previous section.

The test structure was impacted from the outer side shell by an 8.4-ton weight (striking bow) freely fallen from a height of 4.8m. The weight struck the double-hull model at a speed of 9.7m/s. LS-DYNA is now used for finite-element simulations with the critical fracture strain determined following the technique described previously. The strain-rate sensitivity effect on the material yield stress of mild steel was accounted for using the Cowper–Symonds formula, Eq. (8.16a) with C = 40.4 and q = 5. The dynamic fracture strain or the critical fracture strain is estimated from Eq. (8.17) with C = 7,000, q = 4, and ξ = 1.0. The three different types of material models described herein are considered, and their resulting simulations are compared.

In finite-element modeling, the mesh size should be fine enough so that the pertinent deformation patterns can be properly captured in the analysis. It is usually desirable that the shape of the element is rectangular and that the aspect ratio of the element is close to 1.0. Although the deformation patterns of steel plates under axial compression at the ULS tend to have a sinusoidal shape, vessel collisions and grounding cause more complex deformation patterns involving folding and tearing as well as localized yielding.

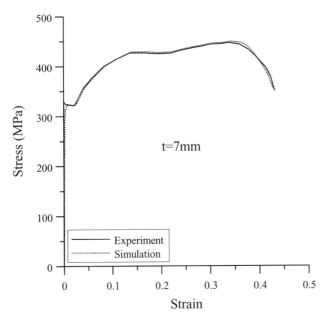

Figure 8.15. The engineering stress–strain curve of steel used for the collision test model 7mm-thick plate in a quasistatic loading condition.

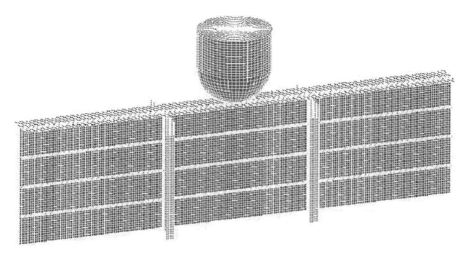

Figure 8.16. Finite-element mesh modeling for the collision test model.

The fold length of web plate between support members is predicted by Eq. (8.8). Because b = 2,000mm and t = 7mm for the test model, a half-length of one fold is given as H = 298.5mm. Therefore, it is recommended that the element size must be smaller than 298.5/8 = 37.3mm so that at least eleven elements are necessary between a stiffener spacing (i.e., 400mm). In the present benchmark example, thirteen elements were used for modeling of plating between stiffeners, which corresponds to the element size of 400/13 = 30.77mm. In terms of collision resistance of actual ship-type structures, shell plating deformation as well as web plate folding may also play a large role, and web plate folding will typically govern the selection of mesh size.

Figure 8.16 shows the finite-element mesh modeling so established. Figure 8.17 shows a typical shape of deformation after impact action is ended. Figures 8.18(a) and 8.18(b) compare the collision force versus penetration curves and the absorbed energy penetration curves, respectively. It is seen that the mesh size adopted is fine enough to capture the crushing mechanism. In this collision test example, the difference between material models is negligible because fracture behavior associated with necking is not a dominant failure mode, although material model III tends to more accurately simulate the collision behavior as well.

A realistic safety evaluation example is now shown when the side structure of a ship-shaped offshore unit with 46,000 deadweight tons in fully loaded condition is struck amidships by the bow of a trading tanker with the same size at its partially loaded condition, as shown in Figure 8.19(a). Figures 8.19(b) and 8.19(c) represent the LS-DYNA simulation results showing the damage extent of bow and side structures, respectively, when the collision velocity of the striking vessel is 10 knots. The analyses for failures of the striking bow and the struck side were performed separately by regarding the indentor as rigid body. The angle of encounter was presumed to be 90 degrees.

By varying collision parameters (e.g., collision velocity), the collision force penetration curves could be obtained. By integrating the areas below the collision force penetration curves, the collision energy penetration curves were calculated, as shown in Figure 8.20. The maximum energy absorption capability is determined as the

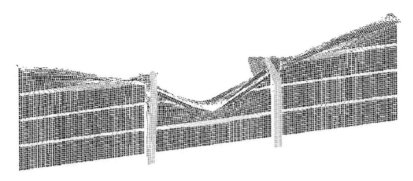

a

b

Figure 8.17. (a) Deformed shape of the collision test model after impact action is applied, as obtained from finite-element simulations. (b) Deformed shape of the collision test model after impact action is applied, as obtained from mechanical testing (courtesy of ASIS).

energy absorbed until the inner side shell is ruptured. Figure 8.21 shows design curves developed based on the computational results together with Eq. (8.6) as a function of collision velocity. It is evident from Figure 8.21 that the offshore unit or the environment can be safe as long as the absorbed energy capability is greater than the kinetic energy loss applied and vice versa.

8.5 Dropped Objects

8.5.1 Fundamentals

Impacts, arising from dropped objects and swinging load incidents involving cranes and lifting devices, can cause physical damage on piping systems and deck plates, usually as local denting (HSE 1996). Denting damage on a steel-plated structure of offshore units struck by an object depends on the shape and sharpness of the object as shown in Figure 8.22, and also on other factors such as mass and impact speed. Resulting impact damage may not only consist of localized dents together with global deformation, as shown in Figure 8.23, but also can extend to perforation or tearing (Muscat-Fenech and Atkins 1998).

The local and global deformation on the structure due to dropped object impacts can be quite accurately analyzed using nonlinear finite-element programs as described in the previous section. The safety of a dented structure may be judged from

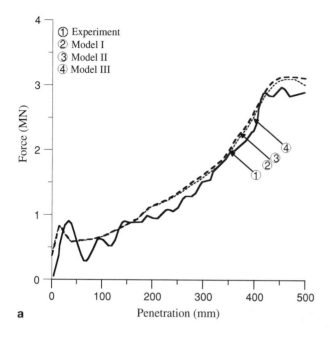

a

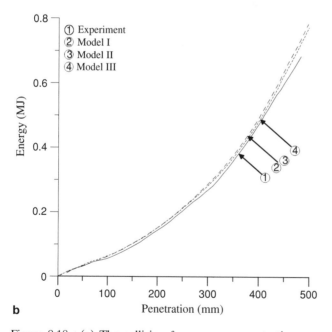

b

Figure 8.18. (a) The collision force versus penetration curves of the collision test model obtained from the test and finite-element simulations applying the three different material models. (b) The absorbed energy versus penetration curves of the collision test model obtained from the test and finite-element simulations applying the three different material models.

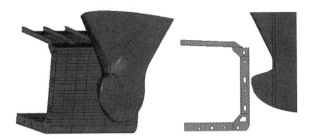

a

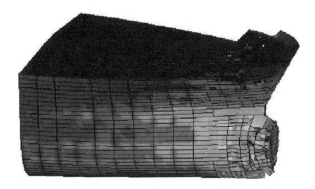

b

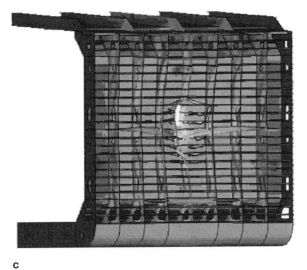

c

Figure 8.19. (a) A collision scenario for a fully loaded tanker struck by a partially loaded tanker. (b) Bow damage extent of striking vessel obtained by LS-DYNA simulations at a collision velocity of 10 knots. (c) Side damage extent of struck vessel obtained by LS-DYNA simulations at a collision velocity of 10 knots.

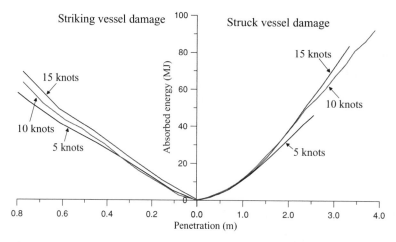

Figure 8.20. The absorbed energy versus penetration relations with varying collision velocity.

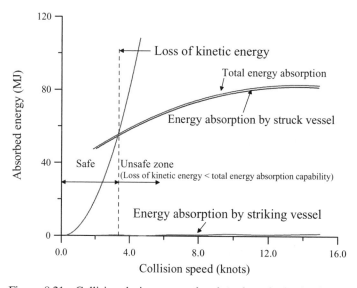

Figure 8.21. Collision design curves developed on the basis of numerical computations.

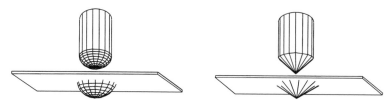

Figure 8.22. Typical shapes of impacting object (left: blunt or spherical object; right: sharp or conical object).

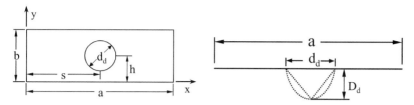

Figure 8.23. Nomenclature: geometric parameters of local denting due to dropped objects.

Eq. (8.1a) in terms of a critical deformation or strain until rupture occurs. The use of energy-based safety evaluation is also possible based on Eq. (8.1b). In the latter case, the energy absorption capability until the plating is ruptured by denting can be obtained by integrating the impact force versus dent damage curves obtained by numerical computations; however, the kinetic energy for the falling object in air is given by

$$E_k = \frac{1}{2}MV^2, \tag{8.18}$$

where E_k = kinetic energy of falling object; M = mass of falling object; $V = \sqrt{2gh}$ = velocity of falling object; h = traveled distance from drop position; and g = acceleration of gravity.

For dropped objects in water, such as on subsea structures, the velocity is reduced prior to impact because of drag associated with hydrodynamic resistance during the fall in water. Some simplified procedures to calculate the kinetic energy of falling objects in water may be found in NORSOK N004 (1998).

Although the effects of local denting are rarely critical to global integrity of the structure, dents can disrupt certain functions and also be of importance in estimating the residual strength of damaged structures because of many reasons. In such cases, it is also usually necessary to seek a rational life-cycle maintenance and repair strategy for keeping structural integrity without unacceptably large economic penalties (Smith and Dow 1981; Jones 1997; Paik et al. 2003b; Paik 2005).

8.5.2 Ultimate Strength Characteristics of Dented Plates

For residual strength assessment of dented structures, it is important to examine the effect of denting on the ultimate strength.

8.5.2.1 Under Axial Compressive Loads

To investigate the effect of denting on plate ultimate strength, nonlinear finite-element analyses were performed on simply supported rectangular plates under unaixial compressive loads in the plate length direction as shown in Figure 8.24, with

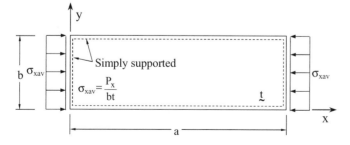

Figure 8.24. A simply supported rectangular plate under axial compressive loads (P_x = axial compressive load, σ_{xav} = average axial compressive stress).

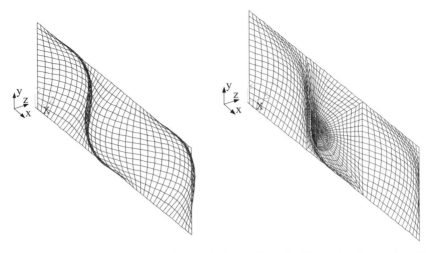

Figure 8.25. Deformed shape of a steel plate with and without denting at the ultimate limit state under axial compressive loads (a × b = 2,400 × 800 mm).

related nomenclature given in Figure 8.23 (Paik et al. 2003b). Figure 8.25 shows deformed shapes of undented and dented plates at the ULS under axial compressive loads.

Figure 8.26 shows the elastic/plastic large deflection behavior of plates with a blunt (spherical) or sharp (conical) shape of local denting and under axial compressive loads, varying the dent size (i.e., depth and diameter). It is seen from Figure 8.26 that the ultimate compressive strength decreases significantly as the depth and/or

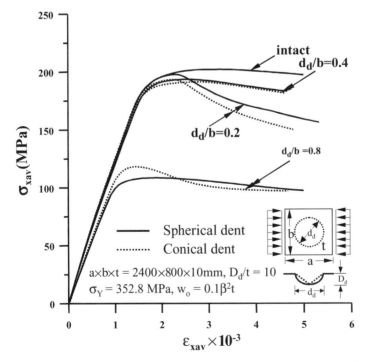

Figure 8.26. Average stress ε_{xav} versus average strain ε_{xav} curves for a dented steel plate under axial compressive loads, varying the dent depth and diameter, for $D_d/t = 10$ (w_o = initial, deflection $\beta = \frac{b}{t}\sqrt{\frac{\sigma_Y}{E}}$).

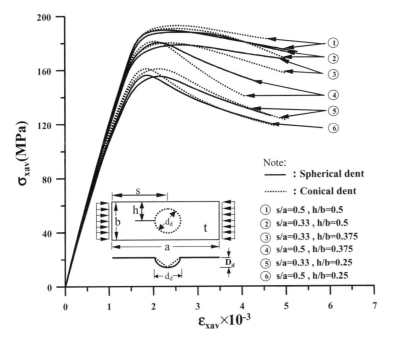

Figure 8.27. Average stress versus average strain curves for a dented steel plate under axial compressive loads, varying the dent location.

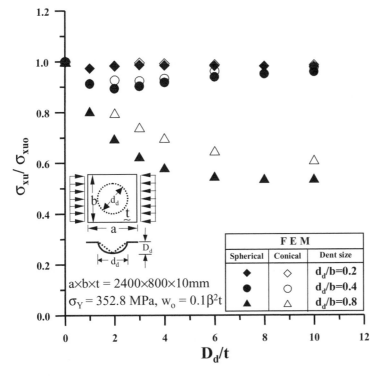

Figure 8.28. Variations of the ultimate compressive strength of a dented plate as a function of the D_d/t ratio.

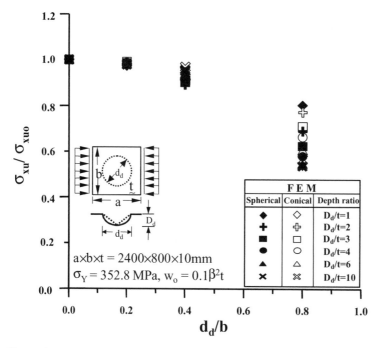

Figure 8.29. Variations of the ultimate compressive strength of a dented plate as a function of the d_d/b ratio.

diameter of local denting increases. Also, it is apparent that the collapse behavior for a blunt dent can be similar to that for sharp dent, but the blunt dent case is more likely to reduce the load-carrying capacity than the sharp dent as long as the depth and diameter of denting are the same. Figure 8.27 shows the effect of dent location on the collapse behavior of dented plates. Figure 8.27 shows that in the case considered, the plate falls in the worst situation in terms of the load-carrying capacity when the local denting is located at the plate center rather than at other places.

Figures 8.28–8.32 show the variations of the ultimate strength of a dented plate as a function of the D_d/t ratio, the d_d/b ratio, the dent location, the plate thickness, and the plate aspect ratio, respectively, with the nomenclature depicted in Figure 8.23, as those obtained by nonlinear finite-element analyses (Paik et al. 2003b). In these figures, σ_{xu} and σ_{xuo} are the ultimate compressive strengths for a dented or undented (intact) plate, respectively. It is evident that the size (depth, diameter) and location of local denting can, in many cases, be quite important to the normalized ultimate compressive strength, that is, $R_{xu} = \sigma_{xu}/\sigma_{xuo}$; however, the influence of the dent depth alone on the plate ultimate compressive strength is not significant as long as the dent diameter is small.

Because the expected plate collapse behavior for a blunt dent is similar to that for a sharp dent and the former can sometimes be slightly worse than the latter as long as the damage is not severe and does not involve rupture, the blunt dent may often be taken as representative of the local dent shape for the purpose of the plate ultimate strength prediction, regardless of the actual shape of denting in the types of denting situations studied. As the dent location becomes closer to the unloaded plate edges,

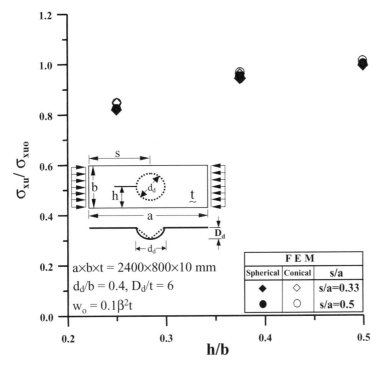

Figure 8.30. Variations of the ultimate compressive strength of a dented plate as a function of the dent location (σ_{xuc} = ultimate strength of the plate with local denting at center).

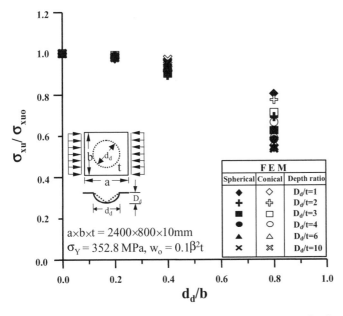

Figure 8.31. Variations of the ultimate compressive strength of a dented plate as a function of the plate thickness.

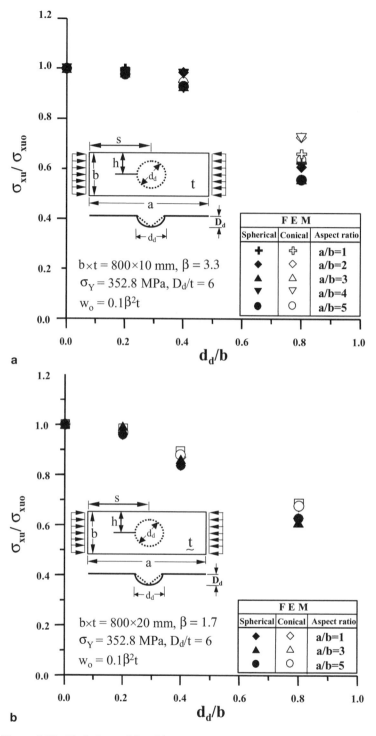

a

b

Figure 8.32. Variations of the ultimate compressive strength of a dented plate: (a) as a function of the plate aspect ratio, t = 10mm, as obtained by nonlinear finite-element analyses; (b) as a function of the plate aspect ratio, t = 20mm.

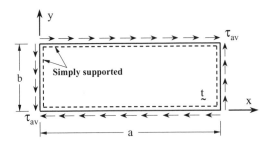

Figure 8.33. A simply supported rectangular plate under edge shear.

the ultimate strength is increased by 20 percent, compared to that of the dent located at the plate center. However, it is found that the plate thickness and the aspect ratio are not influential parameters for the strength reduction factor R_{xu}.

8.5.2.2 Under Edge Shear Loads

Similar numerical computations were performed on simply supported rectangular plates under edge shear, as shown in Figure 8.33 with nomenclature as in Figure 8.23 (Paik 2005). Figure 8.34(a) shows deformed shapes of undented or dented plates at ULS under edge shear. Figure 8.35 shows the elastic/plastic large deflection behavior of plates with blunt or sharp shape of local denting and under edge shear loads, varying the dent size (i.e., depth and diameter).

Figure 8.35 shows that the ultimate shear strength decreases significantly as the depth and/or diameter of local denting increases. Also, it is evident that the collapse behavior for a blunt dent can be similar to that for a sharp dent, although the former case is more likely to reduce the load-carrying capacity than the latter as long as the depth and diameter of denting are the same. Figure 8.36 shows the effect of dent location on the collapse behavior of dented plates. It is again found from Figure 8.36 that the plate falls in the worst situation in terms of the load-carrying capacity when the local denting is located at the plate center rather than other places in the cases considered to be similar to the dented plates under axial compressive loads.

A series of the nonlinear finite-element elastic/plastic large deflection analyses on dented steel plates under edge shear were also undertaken, varying the dent depth, the dent diameter, the dent location, the plate thickness, and the plate aspect ratio. Figures 8.37–8.41 show the variations of the ultimate shear strength of such a dented plate as a function of the D_d/t ratio, the d_d/b ratio, the dent location, the plate thickness, and the plate aspect ratio, respectively. In these figures, τ_u and τ_{uo} are the ultimate shear strengths for a dented or undented (intact) plate, respectively.

It is evident from the figures that the size (depth, diameter) and location of local denting are generally quite sensitive to the normalized ultimate shear strength, that is, $R_\tau = \tau_u/\tau_{uo}$, although the influence of the dent depth on the plate ultimate shear strength is not significant as long as the dent diameter is small.

The plate ultimate shear strength is not affected much by the dent damage as long as the dent diameter is small regardless of the dent depth. However, as the dent diameter increases, the plate ultimate shear strength decreases significantly. In this case, increasing dent depth serves to accelerate the strength reduction tendency

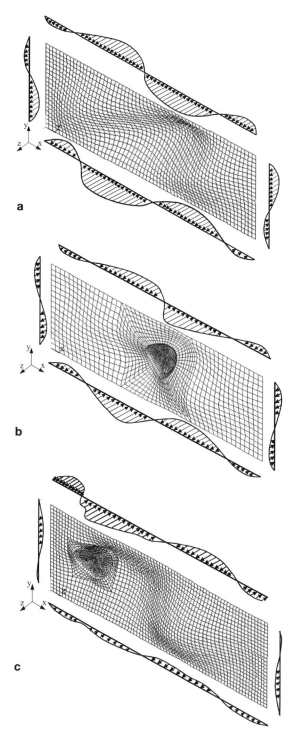

Figure 8.34. Deformed shape and membrane stress distribution of a steel plate, immediately after the ultimate limit state is reached under edge shear, for a × b = 2,400 × 800 mm: (a) without denting; (b) with denting located at the plate center; (c) with denting located at the plate side.

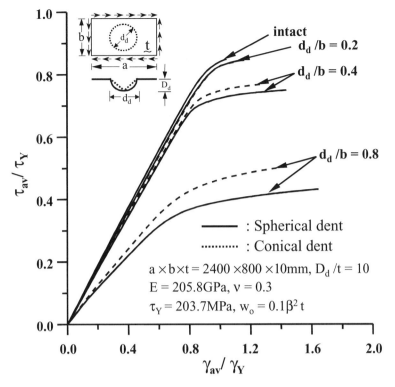

Figure 8.35.   Average shear stress τ_{av} versus average shear strain γ_{av} curves for a dented steel plate under edge shear, with varying the dent diameter, for $D_d/t = 10$ (τ_Y = shear yield stress, γ_Y = shear yield strain).

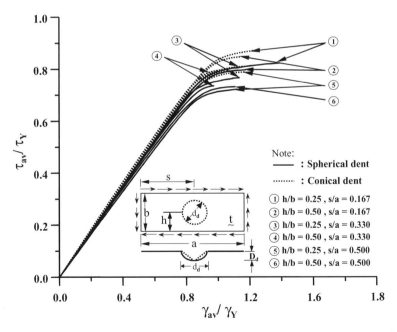

Figure 8.36.   Average shear stress versus average shear strain curves for a dented steel plate under edge shear, with varying the dent location.

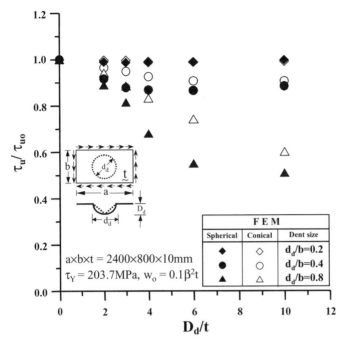

Figure 8.37. Variations of the ultimate shear strength of a dented plate as a function of the D_d/t ratio.

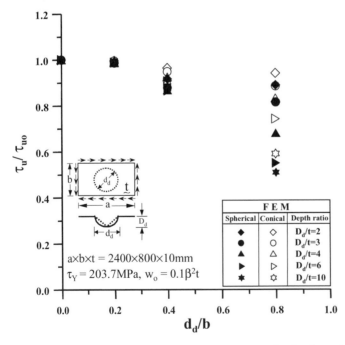

Figure 8.38. Variations of the ultimate shear strength of a dented plate as a function of the d_d/b ratio.

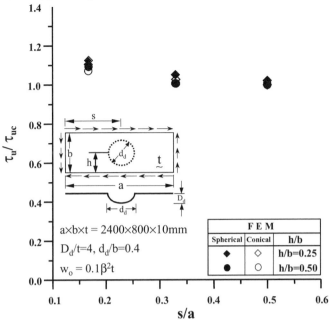

Figure 8.39. Variations of the ultimate shear strength of a dented plate as a function of the dent location (τ_{uc} = ultimate strength of the plate with local denting located at the plate center).

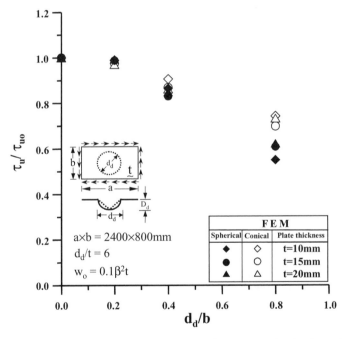

Figure 8.40. Variations of the ultimate shear strength of a dented plate as a function of the plate thickness.

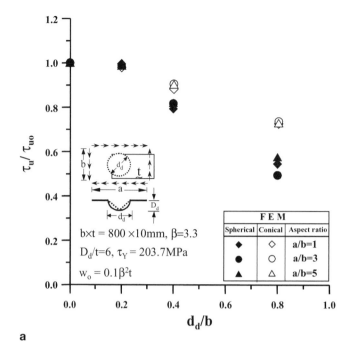

a

b

Figure 8.41. Variations of the ultimate shear strength of a dented plate as a function of the plate aspect ratio.

as well. The effect of the dent location on the plate shear strength is small, although the plate falls in an unfavorable situation in terms of the load-carrying capacity when the dent is located at the plate center under the boundary conditions considered.

This is in contrast to the dented plate under axial compressive loads described in the previous example, where the location change of localized dent in the longitudinal direction of the plate affects the ultimate compressive strength. The aspect ratio and thickness of the plate are not influential parameters to the normalized ultimate shear strength of dented plates. This implies that the two parameters (i.e., plate thickness and aspect ratio) primarily govern the collapse behavior of the plate in this case rather than local denting-induced behavior.

Because the plate collapse behavior for a blunt (spherical) dent is similar to that for a sharp (conical) dent and the former is slightly worse than the latter, the blunt (spherical) dent may be taken as representative of the local dent shape for the purpose of the plate ultimate strength prediction, regardless of the actual shape of denting. However, it is found that the plate thickness and the aspect ratio are not influential parameters for the strength reduction factor R_τ.

8.5.3 Closed-Form Expressions for Ultimate Strength of Dented Plates

8.5.3.1 Under Axial Compressive Loads

When axial compressive loads are applied, the plate ultimate strength reduction (or knock-down) factor due to local denting may pessimistically be expressible as a function of D_d/t, d_d/b, and h/b (with nomenclature in Figure 8.23), as follows (Paik et al. 2003b):

$$\frac{\sigma_{xu}}{\sigma_{xuo}} = C_3 \left[C_1 \ln \left(\frac{D_d}{t} \right) + C_2 \right],$$
(8.19)

where

$$C_1 = -0.042 \left(\frac{d_d}{b} \right)^2 - 0.105 \left(\frac{d_d}{b} \right) + 0.015;$$

$$C_2 = -0.138 \left(\frac{d_d}{b} \right)^2 - 0.302 \left(\frac{d_d}{b} \right) + 1.042;$$

$$C_3 = -1.44 \left(\frac{H}{b} \right)^2 + 1.74 \left(\frac{H}{b} \right) + 0.49;$$

σ_{xu}, $\sigma_{xuo} =$ ultimate compressive strengths of dented or undented (intact) plates, respectively; $\beta = \frac{b}{t}\sqrt{\frac{\sigma_Y}{E}} =$ plate slenderness ratio with $\sigma_Y =$ material yield stress and $E =$ elastic modulus; and $H = h$ for $h \leq \frac{b}{2}$ and $H = b - h$ for $h > \frac{b}{2}$.

Equation (8.19) is applicable for $a/b \geq 1$ and $d_d/b < 1$. σ_{xuo} can be obtained from the methods described in Section 6.5 of Chapter 6, with Eq. (6.5), for instance. Figure 8.42 shows the validity of Eq. (8.19). It is interesting to note that the ultimate compressive strength of a dented plate with large dent depth is close to that of a perforated plate, as shown in Figure 8.42(b).

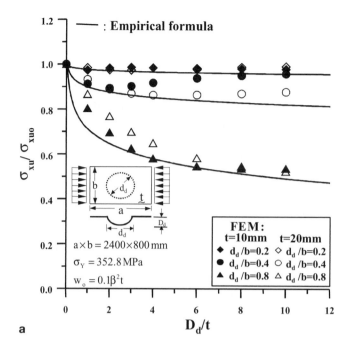

a

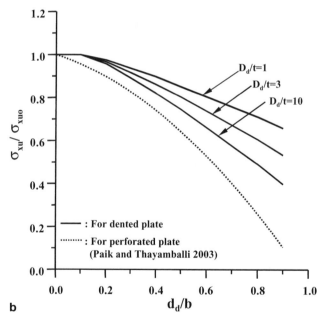

b

Figure 8.42. (a) Comparison of ultimate strength between the empirical formula, Eq. (8.19), and nonlinear finite-element solutions for dented plates under axial compression. (b) Variation of ultimate compressive strength for a dented plate as obtained by Eq. (8.19) and for a perforated plate (circular hole) as obtained by Paik and Thayamballi (2003).

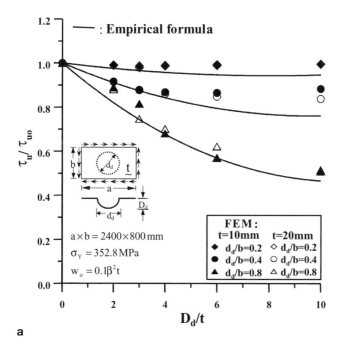

a

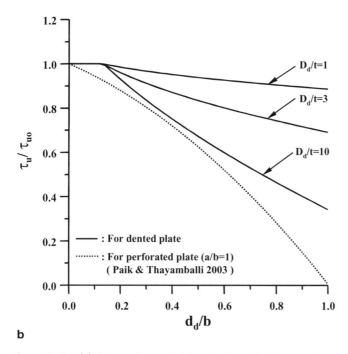

b

Figure 8.43. (a) Comparison of ultimate strength between the empirical formula, Eq. (8.20), and nonlinear finite-element solutions for dented plates under edge shear. (b) Variation of ultimate shear strength of dented plate as obtained by Eq. (8.20) and the perforated plate (circular hole) case as obtained by Paik and Thayamballi (2003).

8.5.3.2 Under Edge Shear Loads

Similarly, the plate ultimate shear strength reduction (or knock-down) factor due to local denting may pessimistically be expressible as a function of D_d/t and d_d/b, neglecting the influence of the dent location (Paik 2005), as follows:

$$\frac{\tau_u}{\tau_{uo}} = \begin{cases} C_1 \left(\frac{D_d}{t}\right)^2 - C_2 \left(\frac{D_d}{t}\right) + 1, & \text{for } 1 < \frac{D_d}{t} \leq 10, \\ 100C_1 - 10C_2 + 1, & \text{for } \frac{D_d}{t} > 10, \end{cases} \tag{8.20}$$

where

$$C_1 = 0.0129 \left(\frac{d_d}{b}\right)^{0.26} - 0.0076; \quad C_2 = 0.1888 \left(\frac{d_d}{b}\right)^{0.49} - 0.07;$$

and τ_u, τ_{uo} = ultimate shear strengths of dented or undented (intact) plates, respectively. τ_{uo} can be obtained the methods described in Section 6.5 of Chapter 6, with Eq. (6.7), for instance.

Figure 8.43 shows the validity of Eq. (8.20). Note again that the ultimate shear strength of a dented plate with large dent depth is close to that of a perforated plate as shown in Figure 8.43.

8.6 Fire

8.6.1 Fundamentals

In offshore oil and gas installations, fire sources are usually classified into two types: liquid and gas. Liquid fire may be classified into a few types (Skallerud and Amdahl 2002; Nolan 1996), as follows:

- Pool fire in the open air, which may take place when there is an ignition of a liquid spill, which is released on a horizontal solid surface in the open air, for example, on the ground
- Pool fire on the sea surface
- Pool fire in an enclosed area, when liquid fuel is released within an enclosed space, which may suffer from various degrees of air deficiency
- Fire ball, which results from boiling liquid expanding to vapor explosion, where an immediate ignition of the pressurized and liquefied fuel occurs
- Running liquid fire, when the liquid fuel is released on a surface that is not horizontal, for example, the sloping walls of a tank container, where the fuel burns as it flows down the surface
- Spray fire, when the liquid fuel is released under high pressure, and is subsequently dispersed into droplets

On the other hand, when a flammable gas is released into the atmosphere, somewhat different types of fires may take place according to the release mode and the degree of delayed ignition. Gas fire may be classified into the following types:

- Jet fire or flare fire, which results from a high-pressure leakage of a flammable gas: The jet fire is often said to be momentum-controlled because the momentum force prevails over the buoyancy force in large part of the flame plume. A jet fire is a pressurized stream of combustible gas (e.g., a high-pressure release from a gas pipe or wellhead blowout event) that is burning. Most fires involving gas in the oil and gas industry are associated with high pressure and regarded as jet fires.

- Flash fire or cloud fire, which results from a delayed ignition of a release of gas or vapor forming a cloud, which may disperse downwind: A flash fire is transient during a very short time period and its main hazard is thermal radiation to human beings. Flash fire is unlikely to cause any fatalities, but it will damage structures.
- Diffusive gas fire, which results from a diffusive release of a flammable gas through a relatively large opening: The diffusive gas fire is often said to be buoyancy-controlled because the buoyancy is dominant in the entire flame plume, in contrast to the jet fire.

Wind-tunnel testing, as described in Appendix 5, is highly desirable to analyze fire- and smoke-related ventilation problems on board a ship-shaped offshore unit (HSE 2000j). The wind-tunnel test may also be necessary to model emergency gas releases and fire scenarios and to identify the regions of poor ventilation.

8.6.2 Practices for Fire Assessment

A primary hazard due to fire is the temperature rise. Although the material properties of structural components exposed to fire can be changed, blast overpressure can also be induced so that the structure under fire can collapse. Therefore, the safety analysis must be carried out to estimate the transient thermal actions (e.g., temperature rise) and blast overpressures on the affected part of the structure exposed to fire, and also to identify critical structural components and to design them by avoiding global collapse of the structure. For this purpose, various structural collapse mechanisms for the selected fire scenarios must be considered as well.

The fire safety analysis methods may be classified into two parts: an external mechanics part and an internal mechanics part. The external mechanics part deals with transient temperature variations and fire action characteristics. For this purpose, empirical methods, phenomenological methods, field methods, and computational fluid dynamics (CFD) methods are relevant, although it is recognized that the CFD methods may give relatively more accurate results (Sinai et al. 1995; Holen et al. 1990a, 1990b; HSE 1999c). An elaborate description on heat-transfer analysis for fire is presented by Skallerud and Amdahl (2002).

The internal mechanics part of the analysis deals with progressive collapse response of the structure exposed to fire considering the external mechanics obtained from the former part. Nonlinear finite-element programs are again useful for this purpose. Fire tests are necessary to investigate and elaborate on the characteristics of fire consequences and also to verify the theoretical and numerical methodologies. A jet fire resistance test procedure and its development are described in HSE (1997).

For fire structural analysis, flux levels or history and temperature versus time relationships must be identified to represent the thermal states and actions during fire. The fire resistance can be evaluated in terms of the spreading measure of heat and smoke and load-bearing capacity. The spreading measure can be characterized by insulation, radiation, and smoke leakage, and the load-bearing capacity can be based on mechanical property changes of material and residual strength of structural

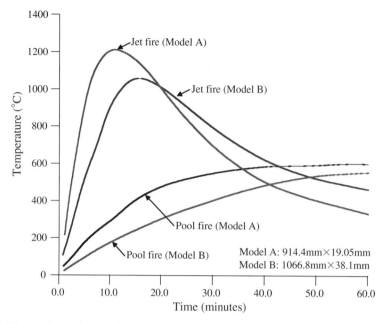

Figure 8.44. A sample temperature variation with time on unprotected tubular members under fire.

components and global structure. Both geometric and material nonlinearities must be taken into account for the progressive collapse analysis under fire actions.

Mechanical properties of structural steel will vary with temperature; therefore, the degradation of the material properties at the elevated temperatures must be identified and considered for fire-related structural analysis. Regarding material behavior, changes in temperature can cause the following effects:

- Elastic constants (e.g., Young modulus, Poisson ratio) can change.
- Strain can develop without mechanical loading.
- Material yield strength decreases with increase in temperature.
- Material can lose ductility with decrease in temperature.

In a fire event, the temperature rise is not generally uniform for heated structural components and is usually affected by the plate thickness of individual structural components, the heated perimeter, and the fire types. Figure 8.44 shows a sample of temperature profile for two unprotected tubular structural components with different cross-sectional dimensions for two types of fire, pool fire and jet fire, following the work of Hossain (2004). In his computations, the heat flux of 100 kW/m² for pool fire was kept constant, but for jet fire, the initial heat flux of 250 kW/m² was dropped off to 0 kW/m² after 30 minutes. Figure 8.44 shows that the temperature rise in a jet fire can be relatively quick although the temperature decays after peak. On the other hand, the temperature in a pool fire increases progressively until a peak is reached. It is also observed that the temperature variations are different depending on the structural geometry and dimensions as well.

For fire safety, the load-carrying capacity of structural components should be adequate to resist the applied loads with the fire present. Although the behavior of steel in a fire is affected by the heating rate, steel begins to lose strength at temperatures

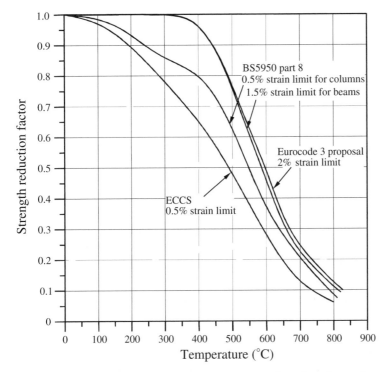

Figure 8.45. Strength reduction factor for structural steel at elevated temperatures (Lawson 1992).

above 300°C and reduces in strength at an approximately steady rate until around 800°C, as shown in Figure 8.45. The strength reduction factor in Figure 8.45 represents the yield strength of steel at a particular temperature relative to that at room temperature. To avoid excessive deformations due to fire, regulations recommend strain limits: ECCS (1982) uses a 0.5 percent limit for temperatures exceeding 400°C, BS 5950 (1985) uses a 1.5 percent limit for beams and 0.5 percent limit for columns, and Eurocode 3 (1992) uses a 2 percent limit. These strain limits can be useful to define the fire design criteria in specific cases.

Fire may also cause blast pressure actions as well as temperature rise, and thus their interacting effect must be considered for the collapse strength analysis. For clad (protected) types of structural components, the blast pressure versus time history needs to be applied as transient pressure actions to the nonlinear structural analysis. For unclad (unprotected) types of structural components, which may be slender in geometry, blast-wind velocity predicted from the hazard analysis can be used to calculate the transient blast-drag forces (Hossain et al. 2004).

8.7 Gas Explosion

8.7.1 Fundamentals

In offshore oil and gas installations, the risk of gas explosion is considered relatively high particularly for the topside modules. Impact-pressure actions arising from gas explosion can cause severe damage on the structure and also to various equipment; this in turn may lead to a threat to structural integrity, health, and the environment (Bjerketvedt et al. 1997).

The principle of ALS design for gas explosion is to reduce the explosion probability and the potential explosion forces (blast, impact pressure), as well as explosion consequences (e.g., damage). To reduce the probability of gas explosion, the following must be reduced:

- Potential for gas leaks
- Possible ignition sources
- Potential for gas clouds

Explosion risk assessment, therefore, must be performed to develop safety measures for risk mitigation (Czujko 2001; Korndorffer et al. 2004). In terms of structural layout, the following measures may help reduce explosion consequences:

- Prevention of high equipment congestion or blockage to reduce turbulence
- Installation of blast- and fire-resisting walls

The explosion safety analysis methods can again be classified into the external mechanics part and the internal mechanics part. The external mechanics part analyzes the profile of explosion actions (loads) in terms of impact pressure versus time history, and the internal mechanics part predicts the explosion consequences in terms of structural damage amount and the residual ultimate strength or resistance of the exploded structure.

For the analysis of the external mechanics part, empirical models, phenomenological models, or CFD models are relevant, although it is recognized that the CFD models are relatively more sophisticated and accurate (HSE 1999d). The analysis for the internal mechanics part can be made by the nonlinear finite-element program. A comprehensive description on the methods for the analysis of external and internal mechanics on gas explosion is given by Czujko (2001).

The Health and Safety Executive of UK has performed extensive work on gas explosions as well as other significant issues related to the safety of offshore oil and gas installations. For explosion impact pressure analysis, see HSE (2000e, 2000f, 2000g); for explosion resistance (strength) analysis, see HSE (2000c, 2000i); for gas explosion experiments, see HSE (1998, 1999e, 2000a, 2000b, 2000d, 2000e); and for blast wall design, see HSE (2000h).

The characteristics of impact actions arising from explosion are primarily affected by the following factors (Czujko 2001):

- Frequency, location, momentum, direction, and rate of gas leakage
- Wind and ventilation conditions
- Frequency, location, and source type of gas ignition
- Congestion and confinement
- Mitigation measures
- Characteristics of flammables

8.7.2 Practices for Gas Explosion Action Analysis

As previously described in Section 4.10.2 of Chapter 4, the problem of impact-pressure actions arising from gas explosion can be dealt with in the following three domains, depending on the duration time of impact-pressure pulse (NORSOK N004 1999):

- Quasistatic domain when $3T < \tau$
- Dynamic/impact domain when $0.3T < \tau \leq 3T$
- Impulsive domain, when $\tau \leq 0.3T$

where τ = duration of pressure pulse; and T = natural period of the structure exposed to explosion.

For the impulsive domain, the impact-pressure profile can be characterized by only two parameters, that is, the peak pressure and the pressure duration time. However, it must be represented by four parameters: rise time, peak pressure, pressure decay type, and duration time for the dynamic/impact domain. When the duration time is very long, the structural response can be analyzed quasistatically with peak pressure value only. The impact-pressure profile associated with gas explosion is more likely to be represented in the impulsive domain. It is also recognized that the nonlinear impact structural behavior in the dynamic domain can be analyzed approximately in the impulsive domain (Paik et al. 2004).

The distribution of impact pressure would normally not be uniform over the structure during explosion, and the related pressure differences can develop drag forces associated with normal and shear stresses over the surface of the structure exposed to explosion. Therefore, the effect of drag forces must be accounted for in the nonlinear structural analysis as necessary.

With the peak explosion pressure value and the pressure duration known, the structural damage can be calculated in the impulsive domain using the same methodology described in Section 5.6 of Chapter 5. Nonlinear finite-element methods can also be applied for analysis of explosion consequences and resistance using the computer programs noted in the previous section. The modeling techniques of finite-element mesh and material properties, similar to those used for analysis of side collisions, which can be applied, are described in Section 8.4.3.

It is desirable to apply risk-assessment tools for identifying actions arising from gas explosion and also for analyzing the consequences of the actions. An extensive description on probabilistic and risk-assessment methods for gas explosion analysis is presented in Czujko (2001).

Large-scale tests on gas explosions (HSE 2000a, 2000b, 2000c) have demonstrated that explosion actions (loads) are higher in cases of congestion. CFD methods are found to be capable of providing realistic values of explosion loads for design purposes, but the method choice must be based on more extensive analyses to get sufficiently accurate solutions. Such computations will require significant computing time.

8.7.2.1 Prescriptive Methods

In a closed vessel, gas explosion will generate impact pressure even with a slow combustion process – that is, regardless of a flame velocity – as long as there is no or very little relief (i.e., venting) of the explosion pressure.

The consequences of a gas explosion are affected by various parameters, including the following (Czujko 2001):

- Type of fuel and oxidizer
- Size and fuel concentration of the combustible cloud
- Location of ignition point
- Strength of ignition source
- Size, location, and type of explosion vent areas
- Location and size of structural elements and equipment
- Mitigation schemes
- Initial turbulence of the flow field

It is important to realize that impact actions arising from gas explosion are transient with variation in both time and space. Also, the related actions are composed of pressure actions on large surfaces and drag actions on slender objects. Extensive description of prescriptive analysis models for impact actions due to gas explosion is given in Czujko (2001).

8.7.2.2 Probabilistic Methods

Gas dispersion is a random event that happens at a chain of undesired events. Probabilities of those events vary from case to case. After gas leakage or fluid evaporation, a combustible gas cloud can be formed or not, depending on leakage rate, position of leakage with respect to vent system, and wind conditions. In this regard, gas dispersion by itself may not be a hazardous event, and it can still be detected and mitigated.

When the fuel concentration in a cloud passes the lower flammability limit, however, a gas explosion can happen if the cloud meets an ignition source. Various cases of gas explosion are usually considered, depending on the position of the gas cloud in the investigated space, degree of filling, and position of ignition source with respect to cloud and structure, each of which may, of course, result in different consequences.

Due to limitations in computer speed and capacity, simplified approaches have usually been previously applied by simulating a very high number of scenarios using simple models or by performing a low number of CFD simulations and extrapolating the results. However, it is recognized that these approaches are not often successful in terms of accuracy, and they may provide very conservative estimates for the risk and also lead to a wrong trend evaluation with regard to sensitivity analyses. On the other hand, probabilistic methods can take into account the effects of gas clouds with various sizes, locations, and concentrations more accurately. Various CFD simulations for gas explosions are presented in Bjerketvedt et al. (1997).

In the risk-based approaches, relevant scenarios for gas explosion analysis will be selected based on the probability of occurrence of parameters that affect the scenarios. Resulting probabilistic models of explosion loads and exceedance curves are then derived from probability of the scenarios considered. Generic probabilistic approaches that take advantage of numerical simulations may also be useful. Various probabilistic models for gas explosion analysis are presented Czujko (2001).

8.7.3 Practices for Gas Explosion Consequence Analysis

Single degree of freedom systems (SDOF) implemented into various forms of the Biggs method (Biggs 1964) or nonlinear dynamic finite-element methods as described in Section 8.4.3 can be used for the consequence analysis of gas explosion. As a consequence of gas explosion, structural damage or permanent set deflection will be computed, and it can be obtained using the pressure versus impulse relationship function.

The extent of consequences depends mainly on the integrity of walls and decks that can be subjected to considerable overpressure compared with ordinary design loads. It is often assumed that the interacting effect on gas explosion consequences between structural parts (e.g., walls, decks, beams) is small. This assumption makes it possible to deal with the consequence analysis of individual structural parts separately. Also, it is beneficial in terms of considerable reduction of the computational costs because a large number of probabilistic simulations do not need to be performed. This approach may be acceptable in part because the structural design criteria make use of the ductile behavior of material so that the system reliability is likely governed by local structural behavior rather than by stiffness distribution between different structural parts or by the statistical correlation between gas explosion loads imposed to different structural parts. However, it is a hypothesis and not a proven fact, and interactions may need to be better addressed in the future.

In spite of the simplifications stated above, any relatively rigorous reliability-based treatment of gas explosions is time-consuming mainly due to the need for nonlinear dynamic finite-element analysis to compute the structural damage for a given value of random pressure. For this problem, an SDOF system approach that replaces the complex finite-element model of a structural part (e.g., wall, deck, beam) has been employed using the Biggs method (Biggs 1964). The equivalent properties of the simple SDOF system are determined based on a few nonlinear dynamic response analyses performed by finite-element modeling and analysis for the SDOF system model, both under different values of peak pressure and impulse. A series of the dynamic response analyses with the SDOF systems so defined are then performed varying the pressure versus impulse combinations at relatively low computational costs.

8.7.4 Illustrative Examples

Figure 8.46 shows a sample result for the permanent set deflection contour for a wall exposed to gas explosion obtained by the SDOF system approach noted previously (Czujko 2001). In such a case, the design criterion of Eq. (8.1a) can be applied to limit the maximum permanent set deflection to an allowable value. In Figure 8.47, the response of the structure (wall) exposed to gas explosion is divided into two regions, that is, safe (acceptable) and unsafe (unacceptable) regions, in terms of basic pressure versus impulse random variables. Figure 8.48 shows sample results representing the probability of exceedance of the P (pressure) versus I (impulse) relationship for the steel wall exposed to gas explosion, when the maximum values of structural damage equivalent to allowable limits are assumed.

Figure 8.49(a) shows sample results of the impact pressure versus time history for steel plating under the impulsive type of pressure actions, obtained by nonlinear dynamic finite-element simulations. After the peak pressure is reached, the pressure

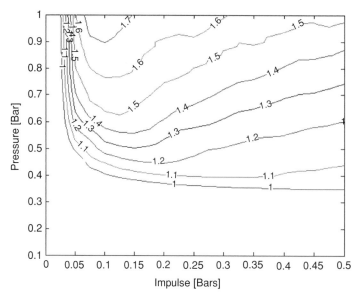

Figure 8.46. Sample results for the permanent set deflection contour of a steel wall exposed to gas explosion (deflection in meters) as obtained by the probabilistic SDOF method (Czujko 2001, 2005; courtesy of Czujko).

pulse oscillates. This is because no damping was accounted for in the numerical simulations. But it is apparent that the plate deflection converges to a value that is then regarded as permanent set deflection of the plating. Figure 8.49(b) shows the effect of peak pressure and duration time on the permanent set deflection of the

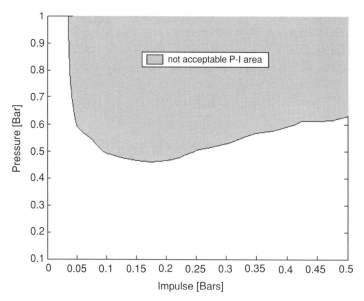

Figure 8.47. Sample results of the structure (wall) divided into acceptable and unacceptable regions in terms of pressure P versus impulse I, as obtained by the probabilistic SDOF method (Czujko 2001, 2005; courtesy of Czujko).

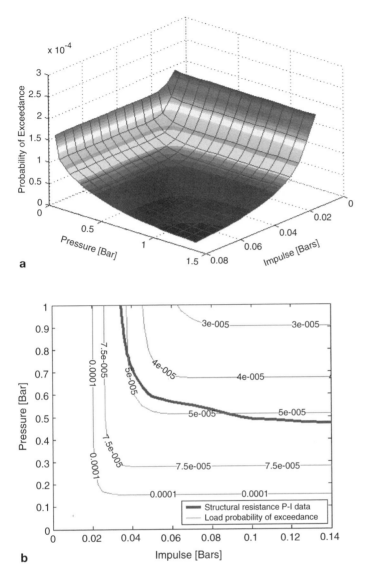

Figure 8.48. (a) Sample results of probability of exceedance of the impact pressure and impulse for the steel wall exposed to gas explosion (Czujko 2001, 2005; courtesy of Czujko). (b) Structural resistance P (pressure) versus I (impulse) relation for the steel wall exposed to gas explosion, comparing the load probability of exceedence with structural resistance (Czujko 2001, 2005; courtesy of Czujko).

plating. It is evident that the permanent set deflection is affected significantly by the impact-pressure duration time as well as peak pressure itself. It is also seen from Figure 8.49(b) that the permanent set deflection does not vary with impact-pressure duration time when the latter is longer than the natural period of vibration of the structure.

Figure 8.50 shows an example of nonlinear finite-element modeling for a stiffened plate structure (e.g., deck structure) used for obtaining the effects due to

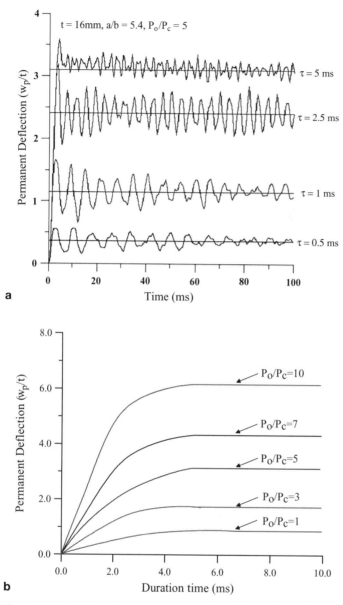

Figure 8.49. (a) Sample pressure versus time history for steel plating under impact-pressure action, as obtained by nonlinear dynamic finite-element simulations. (b) Effect of pressure pulse duration on the permanent deflection of steel plating, varying the peak pressure.

impact-pressure actions arising from gas explosion in the impulsive domain (Paik et al. 2004). One finite-element size is 47mm × 47.2mm. The four edges of the structure are clamped in the numerical computations, considering the characteristics of lateral deflection of the structure under pressure loading. Uniform impact pressure is applied and no drag force effect is considered in this example. Peak pressure value and duration time are varied in the analysis.

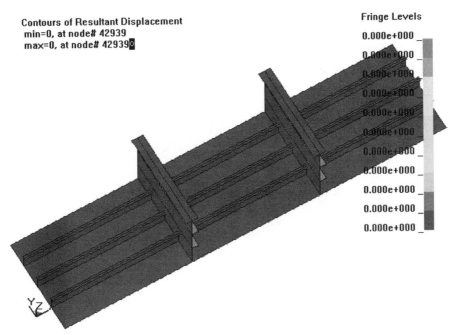

Contours of Resultant Displacement
min=0, at node# 42939
max=0, at node# 42939

Fringe Levels

0.000e+000
0.000e+000
0.000e+000
0.000e+000
0.000e+000
0.000e+000
0.000e+000
0.000e+000
0.000e+000
0.000e+000
0.000e+000
0.000e+000

Figure 8.50. Nonlinear dynamic finite-element modeling for the stiffened panel under explosion pressure actions.

Two types of explosion pressure loading direction were considered: compression in plate side (CIP) and compression in stiffener side (CIS), as shown in Figure 8.51. Figure 8.52 shows deformed shapes after the impact pulse ends. It is seen from Figure 8.52 that the permanent deflection in CIP is larger than that in CIS, when the longitudinal stiffeners are very strong, but the opposite trend is apparent when the longitudinal stiffeners become weaker as would be expected. Figure 8.53 presents the effect of pressure duration time on the permanent deflection of the structure exposed to gas explosion. It is evident that the structural damage due to gas explosion is significantly affected by the impact-pressure duration time as well as the peak pressure itself.

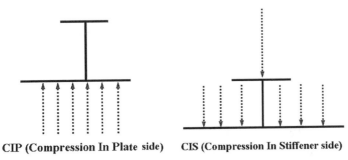

CIP (Compression In Plate side)   **CIS (Compression In Stiffener side)**

Figure 8.51. Two types of explosion pressure loading and directions considered.

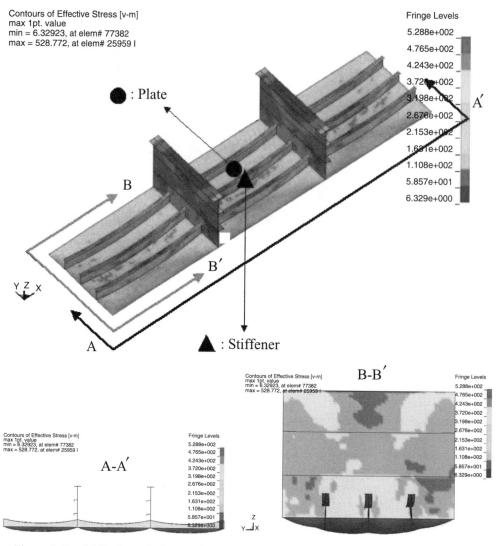

Figure 8.52. (a) Deformed shapes of the stiffened panel in the CIS mode, with $P_o/P_c = 10$ and $\tau = 2$ ms (P_c = quasistatic collapse pressure load, P_o = applied peak explosion pressure load, τ = explosion pressure duration time).

8.8 Progressive Collapse of Heeled Hulls with Accidental Flooding

In ALS considerations, it is in principle required to ensure that the integrity of the global structure remains at a sufficient level so as to avoid total loss even after structural damage and/or unintended flooding due to accidental actions, such as side collisions and, to a lesser extent, grounding. Figure 8.54 shows a schematic of a heeled vessel with structural damage due to accidental flooding arising from a collision.

The global structural integrity of a ship-shaped offshore unit in such a case can be checked by comparing the extreme hull girder bending moment with the ultimate hull girder bending moment for the selected scenarios involving heel angle and levels of presumed structural damage. The extreme hull girder bending moment can be

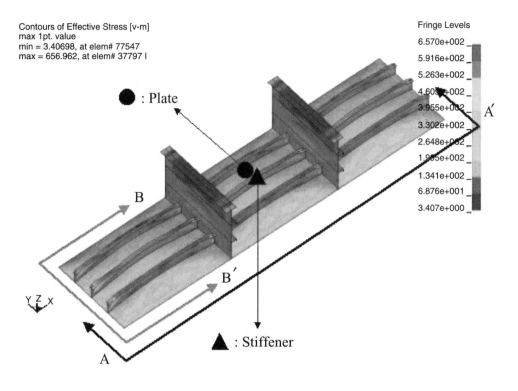

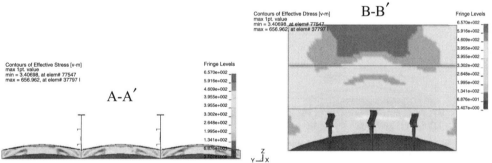

Figure 8.52. (*cont.*) (b) Deformed shapes of the stiffened panel in the CIP mode, with $P_o/P_c = 10$ and $\tau = 2$ ms (P_c = quasistatic collapse-pressure load, P_o = applied peak explosion pressure load, τ = explosion pressure duration time).

calculated as the sum of still-water bending moment and wave-induced bending moment, taking into account the effect of heel and cargo outflow or water ingress as pertinent. The ultimate hull girder strength needs to be calculated by the progressive hull girder collapse analysis methods, as described in Section 6.7.3 of Chapter 6, but taking into account the effects of heel and structural damage.

Figures 8.55–8.57 show some selected results obtained by the progressive hull girder collapse analysis for a ship-shaped offshore structure in a heeled condition together with and without structural damage. ALPS/HULL (2006), as described in Section 6.7.3 of Chapter 6, is used for the progressive collapse analysis. It is evident that the failure modes for structural components can be unsymmetric with

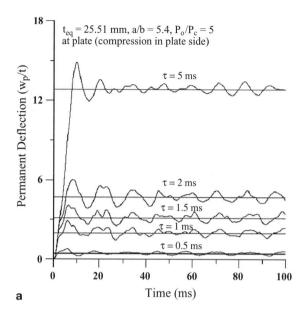

a

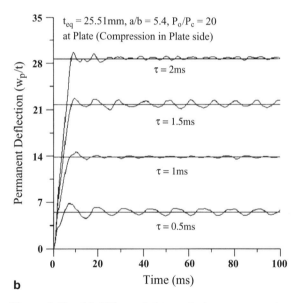

b

Figure 8.53. (a) Effect of the explosion pressure duration time on permanent deflection of the stiffened panel, when $P_o/P_c = 5$. (b) Effect of the explosion pressure duration time on permanent deflection of the stiffened panels, when $P_o/P_c = 20$. (a = length of plating between stiffeners, b = breadth of plating between stiffeners, and t_{eq} = equivalent plate thickness; for other symbols, see Figure 8.52).

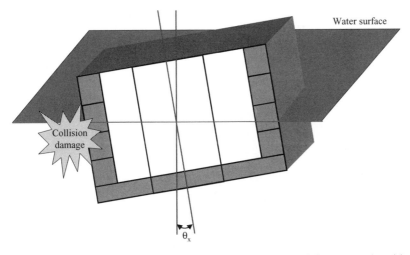

Figure 8.54. A schematic of a heeled vessel with structural damage and accidental flooding.

respect the to vessel center line because of heeled condition, in contrast to that of the upright vessel, even without any damage. It is also seen that, with a larger heeling angle, the ultimate hull girder vertical bending strength itself could become larger when compared to that of the upright case. In addition, the structural damage caused by accidental actions such as side collision can significantly reduce the ultimate hull girder strength.

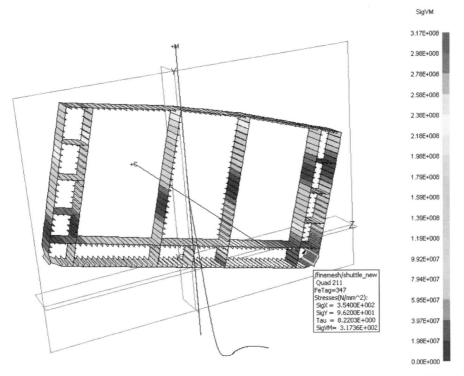

Figure 8.55. Von Mises stress distribution for a ship-shaped offshore structure with heel and damage at ULS under the vertical bending moment obtained by ALPS/HULL (2006).

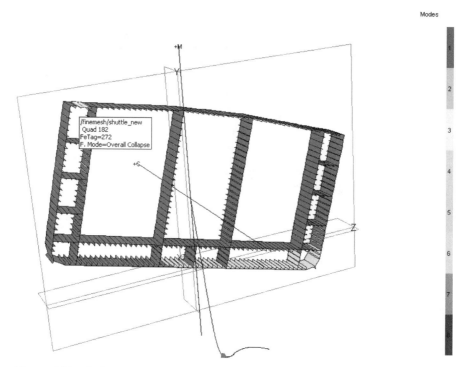

Figure 8.56. Collapse mode distribution for a ship-shaped offshore structure with heel and damage at ULS under the vertical bending moment obtained by ALPS/HULL (2006).

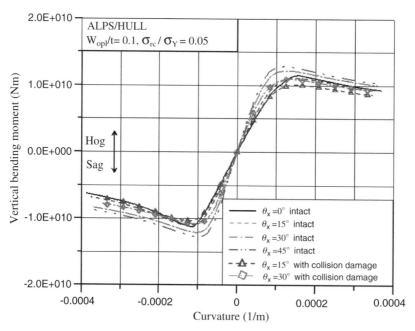

Figure 8.57. Vertical bending moment versus curvature curves for a ship-shaped offshore structure at ULS with varying heel angle and structural damage obtained by ALPS/HULL (2006).

In terms of structural integrity associated with vessel hull girder collapse in such cases, it is apparent that the heel angle would necessarily be considered an influential parameter, and also that the damage amount must play a dominant role.

8.9 Considerations for ALS Applications to Ship-Shaped Offshore Units

ALS applications as applied to ship-shaped offshore units are interwoven into their design process. In addition to initial design by simplified methods, subsequent checks and modifications are also made after so-called "safety case studies" later in design. Such safety case studies are carried out during detailed design with the help of specialists and typically consider various aspects, including the following:

- Gas dispersions analysis to confirm safe dispersion of gases
- Ventilation analysis to confirm that all intakes are from safe areas and that spaces will be adequately ventilated to ensure no hazardous situation or pockets can occur
- Drainage to ensure no collection of pools of liquids that can cause hazards or safety situation
- Fire risk and related studies for all spaces including machinery room, pump room, cargo area, galley, process areas, and any other areas that may have some fire risk
- Material handling and dropped-object study within and outside the vessel boundary; the study outside the vessel boundary is needed to include the subsea system
- Toxic gas release study
- Flare heat and radiation studies
- Traffic safety studies in vicinity of an installed unit

The results of these studies, which may include scenarios, procedures, and criteria, are then fed back into detailed design to the extent required. In addition to these, specifications for ship-shaped offshore units also contain various related and prescriptive provisions that affect ALS design.

These include, for example, requirements to meet International Maritime Organization (IMO) damage stability criteria for tankers (see Section 3.10 in Chapter 3); inclusion of a heeled case in structural analysis; additional strengthening of fendering areas; design of side and end(s) for specified minimum energy absorption levels during supply boat and shuttle tanker collisions; design of accommodation walls for a specified blast pressure (e.g., 0.5 bar); providing fire rated subdivisions and windows where necessary; and various other requirements, including the consideration of residual hull girder strength after collisions.

REFERENCES

Adhia, G., Ximenes, M., Kakimoto, M., and Ando, T. (2004). *Owner and shipyard perspective on new build FPSO contracting scheme, standards, and lessons.* Offshore Technology Conference, OTC 16706, Houston, May.
ALPS/HULL (2006). A computer program for progressive collapse analysis of ship hulls, Proteus Engineering, Stevensville, MD (http://www.proteusengineering.com).

ALPS/SCOL (2006). A computer program for analysis of collision and grounding mechanics, Proteus Engineering, Stevensville, MD (http://www.proteusengineering.com).

Amdahl, J., and Stornes, A. (2001). "Energy dissipation in aluminum high-speed vessels during collision and grounding." *Proceedings of the International Conference on Collision and Grounding of Ships*, Copenhagen, July, pp. 203–219.

ASIS (1993). *Prediction methodology of tanker structural failure and consequential oil spill*. Association of Structural Improvement of Shipbuilding Industry of Japan, III.39–III.47.

Biggs, J. M. (1964). *Introduction to structural dynamics*. New York: McGraw-Hill.

Bjerketvedt, D., Bakke, J. R., and von Wingerden, K. (1994). "Gas explosion handbook." *Journal of Hazardous Materials*, 52: 1–150.

Bodner, S. R., and Symonds, P. S. (1962). "Experimental and theoretical investigation of the plastic deformation of cantilever beams subjected to impulsive loading." *Journal of Applied Mechanics*, 29: 719–728.

BS 5950 (1985). *The structural use of steelwork in building*. (Part 8), British Standards Institution, London.

Cowper, G. R., and Symonds, P. S. (1957). *Strain-hardening and strain-rate effects in the impact loading of cantilever beams*. (Technical Report, No. 28), Division of Applied Mathematics, Brown University, September.

Czujko, J. (2001). *Design of offshore facilities to resist gas explosion hazard*. Oslo, Norway: CorrOcean ASA.

Czujko, J. (2005). Private communication. December.

ECCS (1982). *European recommendations for the fire safety of steel structures*. ECCS Technical Committee 3, European Convention for Constructional Steelwork.

Ellinas, C. P., Supple, W. J., and Walker, A. C. (1984). *Buckling of offshore structures: A state-of-the-art review*. Houston: Gulf Publishing Company.

Eurocode 3 (1992). *Design of steel structures: Part 1.2 Fire resistance*. British Standards Institution, London.

Forrestal, M. J., and Sagartz, M. J. (1978). "Elastic-plastic response of 304 stainless steel beams to impulse loads." *Journal of Applied Mechanics*, 45: 685–687.

Holen, J., Hekkelstrand, B., Evanger, T., Byggstoyl, S., and Magnussen, B. F. (1990a). *Finite difference calculation of pool fires and jet fires offshore*. (Eurotherm Seminar, No. 14), Louvain-la-Neuve, Belgium.

Holen, J., Brostrom, M., and Magnussen, B. F. (1990b). "Finite difference calculation of pool fires." *Proceedings of 23rd International Symposium on Combustion*, Orleans, France, pp. 1677–1683.

Hossain, A., Meek, A., and Warwick, A. (2004). *Integrated structural and safety engineering for fire and blast design – Case studies and lessons learned*. Offshore Technology Conference, OTC 16424, Houston, May.

HSE (1996). *An examination of the number and frequency of serious dropped object and swinging load incidents involving cranes and lifting devices on offshore installations for the period 1981–1992*. (Offshore Technology Report, OTO 1995/959), Health and Safety Executive, UK.

HSE (1997). *Validation of the jet fire resistance test procedure*. (Offshore Technology Report, OTO 1996/056), Health and Safety Executive, UK.

HSE (1998). *AutoReaGas modeling of test condition variations in full scale gas explosions*. (Offshore Technology Report, OTO 1998/048), Health and Safety Executive, UK.

HSE (1999a). *Effective collision risk management for offshore installations*. (Offshore Technology Report, OTO 1999/052), Health and Safety Executive, UK.

HSE (1999b). *Review of approaches to blast, fire, and accidental loads*. (Offshore Technology Report, OTO 1999/028), Health and Safety Executive, UK.

HSE (1999c). *CFD calculation of impinging gas jet flames*. (Offshore Technology Report, OTO 1999/011), Health and Safety Executive, UK.

HSE (1999d). *Review of analysis of explosion response*. (Offshore Technology Report, OTO 1998/174), Health and Safety Executive, UK.

HSE (1999e). *The repeatability of large scale explosion experiments.* (Offshore Technology Report, OTO 1999/042), Health and Safety Executive, UK.

HSE (2000a). *Structural response measurements during gas explosions in a test rig representing an offshore module.* (Offshore Technology Report, OTO 2000/009), Health and Safety Executive, UK.

HSE (2000b). *Measurement of the structural response of a large scale explosion test rig.* (Offshore Technology Report, OTO 2000/010), Health and Safety Executive, UK.

HSE (2000c). *Experimental and analytical studies of the structural response of stiffened plates to explosions.* (Offshore Technology Report, OTO 2000/087), Health and Safety Executive, UK.

HSE (2000d). *Explosions in full scale offshore module geometries – Main report.* (Offshore Technology Report, OTO 1999/043), Health and Safety Executive, UK.

HSE (2000e). *Explosion loading on topsides equipment: Part 1. Treatment of explosion loads, response analysis, and design.* (Offshore Technology Report, OTO 1999/046), Health and Safety Executive, UK.

HSE (2000f). *Explosion loading on topsides equipment: Part 2. Determination of explosion loading on offshore equipment using FLACS.* (Offshore Technology Report, OTO 1999/047), Health and Safety Executive, UK.

HSE (2000g). *Explosion pressures evaluation in early project phase.* (Offshore Technology Report, OTO 1999/048), Health and Safety Executive, UK.

HSE (2000h). *Blast wall design review.* (Offshore Technology Report, OTO 1999/078), Health and Safety Executive, UK.

HSE (2000i). *Explosion resistance of floating offshore installations.* (Offshore Technology Report, OTO 1999/093), Health and Safety Executive, UK.

HSE (2000j). *Review of model testing requirements for FPSO's.* (Offshore Technology Report, OTO 2000/123), Health and Safety Executive, UK.

HSE (2001). *Stability.* (Offshore Technology Report, OTO 2001/049), Health and Safety Executive, UK.

HSE (2004). *Collision resistance of ship-shaped structures to side impact.* Health and Safety Executive, UK.

Hsu, S. S., and Jones, N. (2004a). "Quasi-static and dynamic axial crushing of thin-walled circular stainless steel, mild steel and aluminum alloy tubes." *International Journal of Crashworthiness*, 9(2): 195–217.

Hsu, S. S., and Jones, N. (2004b). "Dynamic axial crushing of aluminum alloy 6063-T6 circular tubes." *Latin American Journal of Solids and Structures*, 1(3): 277–296.

Jones, N. (1997). "Dynamic plastic behaviour of ship and ocean structures." *RINA Transactions*, 139: 65–97.

Jones, N. (1989). *Structural impact.* Cambridge, UK: Cambridge University Press.

Jones, N. (2001). "Dynamic material properties and inelastic failure in structural crashworthiness." *International Journal of Crashworthiness*, 1(1): 7–18.

Jones, N. (2006). "Some recent developments in the dynamic inelastic behaviour of structures." *Ships and Offshore Structures*, 1(1): 37–44.

Karagiozova, D., and Jones, N. (1997). "Strain-rate effects in the dynamic buckling of a simple elastic-plastic model." *Journal of Applied Mechanics*, 64: 193–200.

Kawata, K., Fukui, S., and Seino, J. (1968). "Some analytical and experimental investigation on high velocity elongation of sheet materials by tensile shock." *Proceedings of IUTAM, Behavior of Dense Media Under High Dynamic Pressure*, Dunod, Paris, pp. 313–323.

Kawata, K., Hashimoto, S., and Kurokawa, K. (1977). "Analyses of high velocity tension of bars of finite length of BCC and FCC metals with their own constitutive equations." In *High velocity deformation of solids*, K. Kawata and J. Shioiri, eds. New York: Springer-Verlag, pp. 1–15.

Korndorffer, W., Schaap, D., van der Heijden, A. M. A., and Versloot, N. H. A. (2004). *An integrated approach for gas dispersion, gas explosion, and structural impact analysis for an offshore production platform on the Dutch Continental Shelf.* Offshore Technology Conference, OTC 16058, Houston, May.

Lawson, R. M. (1992). "Fire resistance and protection of structural steelwork." In *Constructional steel design: An international guide*. London: Elsevier Applied Science.

Lehmann, E., and Yu, X. (1998). "Inner dynamics of bow collision to bridge piers." In *Ship collision analysis*. Rotterdam, The Netherlands: Balkema, pp. 61–71.

Lewis, E. V. (Ed.) (1988). *Principles of naval architecture: Volume I. Stability and strength*. Jersey City, NJ: The Society of Naval Architects and Marine Engineers.

Muscat-Fenech, C. M., and Atkins, A. G. (1998). "Denting and fracture of sheet steel by blunt and sharp obstacles in glancing collisions." *International Journal of Impact Engineering*, 21(7): 499–519.

Nolan, D. P. (1996). *Handbook of fire and explosion protection engineering principles for oil, gas, chemical, and related facilities*. Westwood, NJ: Noyes Publications.

NORSOK N-004 (1999). *Design of steel structures*. NORSOK Standard, Norway.

Paik, J. K. (2005). "Ultimate strength of dented steel plates under edge shear loads." *Thin-Walled Structures*, 43: 1475–1492.

Paik, J. K., Amdahl, J., Barltrop, N., Donner, E. R., Gu, Y., Ito, H., Ludolphy, H., Pedersen, P. T., Rohr, U., and Wang, G. (2003a). "Collision and grounding." *International Ships and Offshore Structures Congress, Specialist Committee V.3*. Vol. 2, pp. 71–107, San Diego, CA.

Paik, J. K., and Chung, J. Y. (1999). "A basic study on static and dynamic crushing behavior of a stiffened tube." *KSAE Transactions*, 7(1): 219–238 (in Korean).

Paik, J. K., Chung, J. Y., and Chun, M. S. (1996). "On quasi-static crushing of a stiffened square tube." *Journal of Ship Research*, 40(3): 51–60.

Paik, J. K., Lee, J. M., and Lee, D. H. (2003b). "Ultimate strength of dented steel plates under axial compressive loads." *International Journal of Mechanical Sciences*, 45: 433–448.

Paik, J. K., Lee, J. M., Shin, Y. S., and Wang, G. (2004). "Design principles and criteria for ship structures under impact pressure actions arising from sloshing, slamming and green seas." *SNAME Transactions*, 112: 292–313.

Paik, J. K., and Pedersen, P. T. (1996). "Modelling of the internal mechanics in ship collision." *Ocean Engineering*, 23(2): 107–142.

Paik, J. K., and Thayamballi, A. K. (2003). *Ultimate limit state design of steel-plated structures*. Chichester, UK: John Wiley & Sons.

Paik, J. K., and Wierzbicki, T. (1997). "A benchmark study on crushing and cutting of plated structures." *Journal of Ship Research*, 41(2): 147–160.

Paik, J. Y., Choe, I. H., Thayamballi, A. K., Pedersen, P. T., and Wang, G. (1999). "On rational design of double hull tanker structures against collision." *SNAME Transactions*, 107: 323–363.

Peixinho, N., Jones, N., and Pinho, A. (2003a). "Experimental and numerical study in axial crushing of thin-walled sections made of high-strength steels." *Journal of Physics IV*, 110: 717–722.

Peixinho, N., Jones, N., and Pinho, A. (2003b). "Determination of crash-relevant properties of dual-phase and trip steels." *Proceedings of the 8th International Symposium on Plasticity and Impact Mechanics (IMPLAST 2003)*, pp. 343–353.

Ramberg, W., and Osgood, W. R. (1941). *Determination of stress-strain curves by three parameters*. (Technical Notes, No. 503), National Advisory Committee on Aeronautics (NACA), US.

Saitoh, T., Yoshikawa, T., and Yao, H. (1995). "Estimation of deflection of steel panel under impulsive loading." *The Japanese Society of Mechanical Engineers*, 61: 2241–2246 (in Japanese).

Schneider, F. D., and Jones, N. (2004). "Impact of thin-walled high-strength steel structural sections." *Proceedings of the Institution of Mechanical Engineers*, 218 (Part D): 131–158.

Servis, D., Samuelides, M., Louka, T., and Voudouris, G. (2002). "The implementation of finite-element codes for the simulation of ship-ship collisions." *Journal of Ship Research*, 46(4): 239–247.

Simonsen, B. C., and Lauridsen, L. P. (2000). "Energy absorption and ductile fracture in metal sheets under lateral indentation by a sphere." *International Journal of Impact Engineering*, 24: 1017–1039.

Simonsen, B. C., and Ocakli, H. (1999). "Experiments and theory on deck and girder crushing." *Thin-Walled Structures*, 34: 195–216.

Simonsen, B. C., and Tornqvist, R. (2004). "Experimental and numerical modeling of ductile crack propagation in large-scale shell structures." *Marine Structures*, 17: 1–27.

Sinai, Y. L., Owens, M. P., and Smith, P. (1995). "Advances in CFD assessment of fire and smoke movement. Offshore Mechanics and Arctic Engineering." *Safety and Reliability, The American Society of Mechanical Engineers*, 2: 379–386.

Smith, C. S., and Dow, R. S. (1981). "Residual strength of damaged steel ships and offshore structures." *Journal of Constructional Steel Research*, 1(4): 1–15.

Skallerud, B., and Amdahl, J. (2002). *Nonlinear analysis of offshore structures*. England: Hertfordshire Research Studies Press Ltd.

Symonds, P. S., and Chon, C. T. (1974). "Approximation techniques for impulsive loading of structures of time-dependent plastic behaviour with finite-deflections." *Mechanical Properties of Materials at High Strain Rates, Institute of Physics Conference Series*, 21: 299–316.

Vinnem, J. E., Hokstad, P., Dammen, T., Saele, H., Chen, H., Haver, S., Kieran, O., Kleppestoe, H., Thomas, J. J., and Toennessen, L. I. (2003). *Operational safety analysis of FPSO – shuttle tanker collision risk reveals areas of improvement*. Offshore Technology Conference, OTC 15317, Houston, May.

Wang, G., Jiang, D., and Shin, Y. (2003). *Consideration of collision and contact damage risks in FPSO structural designs*. Offshore Technology Conference, OTC 15316, Houston, May.

Wang, G., Davydov, I., Kujala, P., Lee, S. G., Marino, A., Pedersen, P. T., Sirkar, J., Suzuki, K., Vredeveldt, A., and Wang, Z. L. (2006). *Collision and grounding*. International Ships and Offshore Structures Congress, Specialist Committee V.1. Southampton, UK: Southampton University Press.

Wierzbicki, T., and Abramowicz, W. (1983). "On the crushing mechanics of thin-walled structures." *Journal of Applied Mechanics*, 50: 727–734.

Zhu, L., and Atkins, A. G. (1998). "Failure criteria for ship collision and grounding." *Proceedings of International Conference on Practical Design of Ships and other Floating Structures*, The Hague, pp. 141–148.

CHAPTER 9

Topsides, Mooring, and Export Facilities Design

9.1 Introduction

As described in Chapter 1, the general arrangement and layout of ship-shaped offshore units designed for oil and gas operations may be grouped into several major parts: hull structures including storage tanks, topsides (processing facilities), export facilities, mooring facilities, accommodations, machinery space, subsea systems, and flowlines. All of these various parts are equally important to achieve successful operation, with due consideration of safety, health, environment, and costs versus benefits.

This chapter focuses on topsides, moorings, and export facilities. The material presented herein is aimed at the nonspecialist introductory reader. It is consistent with the content of this book and is included, primarily, to complete the coverage of the various aspects relating to ship-shaped offshore units.

Topsides consist of processing facilities that are typically located as elevated modules that are several meters (say, 3m or more) above the main deck of the vessel hull, but related piping systems may be located on the main deck of the vessel hull. Depending on the vessel size and topsides layout, the topsides modules may have multiple decks that contain the oil-, water-, and gas-processing facilities; utility systems; and similar functions. The preferred configuration, however, may be that to the extent possible, the topsides facilities would be incorporated as single-layer "pancake" units. The single-layer unit arrangement requires a larger main deck area for a given set of needs.

Offloading or export facilities are used to unload cargo to shuttle tankers that will be appropriately moored, say, at buoys, and may also possibly use dynamic positioning systems to keep the tanker in place during offloading. For ship-shaped offshore installation itself, it is important that the installations have its own mooring facilities to achieve the required station-keeping of the vessel both during production and offloading. Either a weathervaning turret system or a spread-mooring system will be primarily employed for this purpose.

Generally, the design and construction of topsides modules and mooring systems are carried out by independent specialized contractors in parallel with the hull design and construction, although a single engineering procurement and construction (EPC) contractor may manage the entire project processes until commissioning. The topsides modules are typically prepared as preassembled units (PAU) that are then mounted onto the deck of the floating offshore unit using heavy-lift cranes. The various complexities essentially mean that related structural design issues pertaining to the interface between hull and topsides modules, including modules supports, must be considered in a great detail during design, construction, and installation. For example, there are a number of design teams involved for the various disciplines, and their designs are interdependent in reality. Therefore, such interdependent design data must be transferred back and forth between interfaces, and among disciplines such as topsides and mooring, at appropriate points in time. Related data accuracy, consistency, and timeliness must be maintained, and the sequence and progress work of the various disciplines involved must be appropriately monitored and controlled.

This chapter presents some design considerations for topsides, mooring, and export facilities in terms of functional requirements and also considering their structural interfaces to the vessel hull.

9.2 Topsides Facilities

The topsides facilities may consist of several types of modules, skids, units, or facilities, each addressing one of the following functions:

- Oil and water separation
- Gas compression
- Water injection
- Cargo handling and offloading
- Utility and support
- Safeguards

Figure 9.1 shows example design graphics of various topsides modules in an FPSO. For the design of topsides facilities, as well as other structures or facilities of ship-shaped offshore installations, risk management and control in terms of safety, health, the environment, and financial expenditures are necessary. Active and passive fire protection measures must be installed, including fire and blast walls. Strategies and means for escape, evacuation, and rescue must be considered, developed, and implemented.

In most jurisdictions, an environmental impact study is invariably required to ensure that the impacts associated with the operation of the offshore installation are minimal in terms of the interactions between the offshore installation (including the export system and subsea infrastructure) and the native ecology of the field (e.g., fish breeding grounds, cetacean migration routes, seabird activity, and seabed flora/fauna such as corals).

In the following sections, a general description of the functions, components, and selected design considerations for important items of topsides facilities in an offshore unit meant for oil and gas production, storage, and offloading are discussed. Note that the following discussion is not meant to be comprehensive. The systems involved are not unique, as many variations can be found in practice. Related and more comprehensive discussions as well as design guidelines for topsides facilities are presented in Lapidaire and de Leeuw (1996) and UKOOA (2002).

9.2.1 Oil and Water Separation Facilities

Crude oil is stabilized and dewatered by the oil and water separation facilities. Typically, multiple stages are required for the processing of segregation. The stabilized

Figure 9.1. (a) A bird's eye view of an FPSO with an external turret-mooring system (courtesy of Samsung Heavy Industries). (b) A computer graphic of the topsides facilities including flare tower (courtesy of Samsung Heavy Industries).

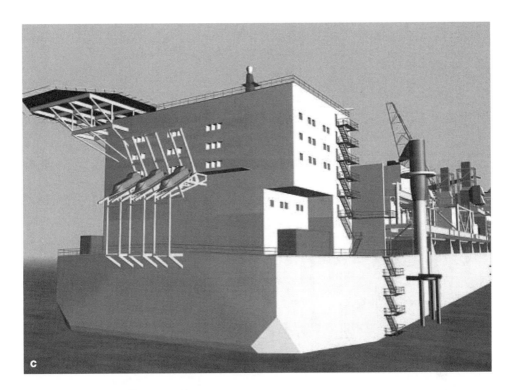

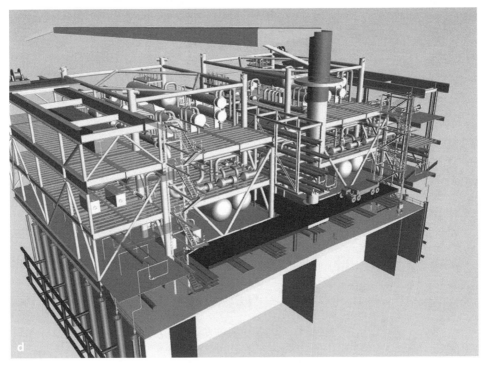

Figure 9.1. (*cont.*) (c) A computer graphic sea-level view of the accommodation area at stern with fire water pumps at both sides and free-fall lifeboats (courtesy of Samsung Heavy Industries). (d) A computer graphic view of crude separation modules equipped with maintenance lanes for removal of equipment (courtesy of Samsung Heavy Industries).

Figure 9.1. (*cont.*) (e) A computer graphic view of central pipe-rack (courtesy of Samsung Heavy Industries). (f) A computer graphic view of ladders for escape from topsides level to lower main deck level (courtesy of Samsung Heavy Industries).

crude oil may typically flow through crude coolers to a wet crude reception tank, which is usually a center cargo tank of the vessel. In there, the oil is further separated by gravity settling to basic sediment and water. The stabilized dry crude oil that comes out of the reception tank is then sent to the cargo oil tanks for temporary storage before export to shore by shuttle tankers.

The produced water may be disposed overboard after deoiling in hydrocyclones and degassing in a produced water flash drum. The reject streams developing in hydrocyclones will be sent to the low-pressure flare drum. Gas coming out of the produced water flash drum will also be sent to the low-pressure flare system.

For possible discharge overboard, the oil-in-water content must be sufficiently low to prevent marine pollution, say 40 parts per million (ppm) (DTI 1971). Recent trends show that the produced water is more often reinjected into the reservoir not only to provide pressure support but also to prevent potential marine pollution possibly arising from continuous overboard disposal.

With the reinjection option, the produced water may be injected on its own or with treated seawater. The seawater treatment is commonly carried out by the use of batteries of hydrocyclones; such operations can sometimes result in low levels of changes to the ambient hull motions.

Consideration must be given to handling, storage, and disposal of the produced water in emergency cases when the injection pumps may fail to work. For outages of short duration, the produced water may be stored in a separate tank. In the event that the pump is not available for a longer period, the produced water may be disposed overboard, provided it meets the necessary oil content specification, say 40 ppm.

9.2.2 Gas Compression Facilities

The gas compression facilities that include gas coolers and suction scrubbers compress the gas by electrical motor-controlled compressors; these must usually operate at a constant speed with constant discharge pressure. Typically, multiple processing stages of gas compression are necessary to keep the pressure above a specified level. In the early years of field production, the reservoir gas pressure may be relatively high. In this case, gas coming out of the processing stages during degassing may sometimes be mixed with the gas from the high-pressure gas compressor.

In the later years of production, the reservoir gas pressure will usually have decreased compared to the earlier years. In this case, gas from the separation stages may be sent to the inlet of the high-pressure gas compressor for pressure increase by additional compression as required.

The gas is dehydrated in the glycol contactor and then compressed to the export pressure in the export gas compressor. The glycol contactor is an absorption column where the gas is contacted with tri-ethylene glycol to remove water vapor from the gas. The water-rich tri-ethylene glycol is then sent from the contactor to be regenerated in the tri-ethylene glycol regeneration package. The gas will be exported through a gas line or other means with the appropriate fiscal metering provided by a gas-metering package.

9.2.3 Water Injection Facilities

For reservoir support during production in the field, high-pressure water usually needs to be injected to the reservoir. The water to be injected will be filtered or deaerated seawater or produced water or perhaps a mixture of them. Some additional treatment may also be carried out by including addition of corrosion inhibitors, scale inhibitors, or biocides to the water.

Oxygen will be removed from seawater by vacuum deaeration or by nitrogen stripping. Seawater filtration will be performed by coarse filters first and then by fine filters. The filter unit selection will consider line-cleaning requirements.

The water injection will be carried out via high-pressure centrifugal pumps. Only a single high-pressure pump may need to be used in some cases when the required water volume and the water injection criticality for reservoir support are limited, but multiple pumps may be required in the more demanding cases. In the design of high-pressure injection pipe work, the possibility of high-surge pressures arising, for example, from the rapid closure of valves on the subsea injection flow lines will need to be considered.

9.2.4 Cargo Handling Systems

A well-thought-out cargo management philosophy is necessary for the efficient and safe handling of the oil cargo. Some of the design issues that will need to be addressed may relate to the following:

- Offloading to shuttle tankers
- Reception tanks
- Transfer between tanks and associated valving
- Inerting and venting of tanks
- Drainage
- Tank inspection and isolation for entry
- Tank washing
- Cargo pump type and maintenance of cargo pumps
- Protection against over- and under-pressure in tanks
- Manifold configuration and isolation
- Product heating and cooling
- Backups in case of equipment breakdown
- Safety and accidents (e.g., leaks, pollution, fire, explosions)

Offloading of cargo to shuttle tankers is considered in Section 9.5. A reception tank is needed on board to store the stabilized crude oil that will be further dehydrated by gravity, settling to a specified basic sediment and water content, as described in Section 9.2.1. Transfer of oil to a shuttle tanker is carried out via a transfer hose as described in Section 9.5, and the safety of valve systems must be assured to reduce the risk of tank overpressure, for example.

For the safe handling and discharging of cargoes or other fluids, inerting of the spaces involved is to be carried out. Two inert gas systems are usually pertinent for ship-shaped offshore installations, one for blanketing and purging of production equipment and the other for blanketing of main cargo tanks. The system for blanketing and purging of production equipment may use nitrogen and the system for blanketing of main cargo tanks may use carbon dioxide. The nitrogen may be produced from compressed air in a membrane-exchange unit. The carbon dioxide may be generated from the combustion of diesel oil in an inert gas generator. Careful handling is required to prevent leakage and spillage of nitrogen because the very low temperature of the escaping nitrogen impinging on a surface of steel components can cause fracture. Also, unlike trading tankers, ship-shaped offshore installations would not share a common inert gas system for the inerting of the slop and cargo tanks.

In the control of overpressure and fast-flow velocities, vent systems are deployed. For cargo handling in tanks, a larger venting system is needed, although a smaller venting system will be used for hydrocarbon production facilities. On a new-build offshore unit, the use of a single venting system that can work both for cargo handling and hydrocarbon production may be considered, but two separate systems are often used for a tanker conversion.

There are three types of drain systems associated with production: closed drains, hazardous open drains, and nonhazardous drains. The closed-drain system gathers the drains from all piped hydrocarbon drain points into a slop tank collection system that can be located below the topsides deck level to achieve the drainage by gravity. In deck drains, the possibility of spillages and leaks must be dealt with. The hazardous open-drain systems may use a trap system to store small spillages that can then flow into a hazardous drains collection tank. In there, the drainage then may be decomposed into oil and water. Recovered oil may be returned to the process, and the oily water may be directed to the slop tank; the separated water may be disposed of overboard providing the right conditions are met. Traps must be cleaned and inspected on a regular basis to avoid any debris accumulation.

The nonhazardous drains system gathers seawater, rainwater, washdown water, and deluge into the slop tank collection system, also through traps. Large volumes of water overflowing the traps can usually be readily disposed of overboard because of its nonhazardous quantities.

Provisions must be made to offer full flexibility for tank inspections and personnel entry into tanks with the highest levels of safety assured. Considerations for heating, cooling, airflow, and other types of systems for the purpose are described in Section 9.2.5.

9.2.5 Utility and Support Systems

The utility and support systems available in the topsides facilities may include the following:

- Fuel gas system
- Heating and cooling medium systems
- Starting air/atomizing air for gas turbines
- Seawater treatment and injection system
- Chemical injection and storage systems
- Flare system
- Nitrogen generation system

In ship-shaped offshore installations, many facilities and systems require diesel fuel to operate, and these include main power generation units (turbines or reciprocating engines), gas turbine drivers for large compressors and pumps, emergency generators, fire pumps, and an inert gas generator for cargo tank blanketing. To prevent problems in the fuel gas system, the diesel must be filtered and centrifuged to remove moisture, sulphur, and particulates. The risk of fuel gas leaks in the utilities areas and machinery spaces must be considered in design.

Appropriate design of the heating medium system may help in optimizing the operation of the installation. Waste heat, for example, coming from the exhaust of power generation turbines, may be captured by waste heat recovery units. This can result in meaningful savings because the turbines usually work continuously, thus providing a steady source of waste energy. The cooling medium system is required to cool down the heated fluids; the coolant employed is usually seawater.

9.2.6 Safeguard Systems

Various systems are required to control and safeguard the operational integrity of the overall installation. These include, for example, the emergency shutdown system (ESD) and the fire and gas control system. The safety systems are controlled as for monitoring, normal operation, startup, and shutdown from a centralized control room usually located in the accommodations unit.

The design of safety systems must comply with various high standards of marine practice and detailed, and often prescriptive, classification and regulatory requirements. Design considerations related to safety systems include the following:

- Provision for smoke and gas detection
- Prevention of smoke ingress into accommodation and control areas
- Smoke clearance, especially in accommodation and control areas and along designated escape routes
- Interactions between main fire and gas system
- Ventilation of the temporary refuge and control areas in emergencies
- Provision of required life support during upset conditions
- Implications of hazardous area classification
- Dispersion of gas in areas where gas escapes may occur
- Fire protection and rating of dampers and ductwork
- Controls, control stations, and control system configuration and complexity
- Position of inlets and exhaust relative to hazardous areas
- Pressurization of spaces and airlocks
- Equipment redundancy and spares

For fire protection, a combination of active and passive protection measures is usually used for ship-shaped offshore installations. Active protection measures are generally provided by fire water (e.g., deluge, sprinklers, monitors, hose reels), foam, CO_2 blanketing, and portable extinguishers. Passive protection measures are provided by fire-rated partitions and fire-resistant coatings applied to primary structural components, bulkheads, decks, frames, equipment foundations, and pressure vessel shells/saddles/skirts. Fire protection should also be provided to flare tower and turbine exhaust supports. The type and extent of the fire protection methods can be affected by the results of fire, heat radiation, and other safety-related studies.

When selecting appropriate materials and features used for passive fire protection, their properties and abilities related to the following need to be considered:

- Withstanding the impact of firewater deluge and fire hose jets
- Remaining intact after an explosion
- Withstanding jet fires
- Being noncombustible
- Accommodating the movement of structural components and equipment affected by the flexing of the hull
- Imperviousness to seawater, firewater, hydrocarbons, and corrosive chemicals
- Chemical inertness to prevent deterioration over time
- Absorbing impacts from tools without spalling

The fire and gas system will contain many detection devices for gas, flame, and smoke. Fire detectors may be supplemented by thermal imaging cameras or closed-circuit television. It is important to ensure that the design of the fire and gas system and the monitoring and control hardware must be made by vendors with a proven track record in design, manufacturing, testing, delivery, installation, and commissioning of such systems offshore.

9.3 Structural Design and Fabrication Considerations for Topsides and Their Interfaces with the Hull

The topsides modules are fitted at an elevated deck level a few meters (3–4m) above the upper deck of the vessel hull. In between the hull upper main deck and the topsides modules, foundation structures (module support structures), as shown in Figure 9.2, are present to support the topsides modules. Their adjoining spaces are also used for the required pipe work.

As necessary, the topsides modules may have a single deck or multiple decks to arrange the process equipment and utility systems employed. The actual elevation between the hull upper deck and the topsides flooring may be determined based on the required space and height underneath the topsides modules that is necessary to fulfill various safety, functionality, and other requirements, in addition to the space required for the module support arrangements themselves.

As described in Section 2.19 of Chapter 2, careful consideration is required for marinization, motions, specifications, layout, and integration with the hull. Also, it is important to realize that some of the standard topsides equipment used for processing on an FPSO may have been developed for fixed-type offshore platforms or onshore applications.

9.3.1 Types of Topsides Supports

The primary functions of the foundation structures (module support structures) and the spaces adjoining them are as follows (Krekel and Kaminski 2002):

- Provide support to the topsides modules on the hull upper deck
- Provide space for deck piping and hull equipment

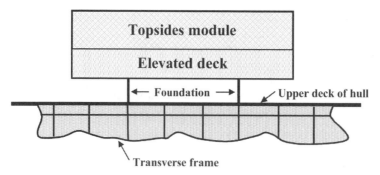

Figure 9.2. A schematic of the foundation structures supporting a topsides module on the hull upper deck.

- Provide space for safeguard and utility systems on the hull upper deck
- Allow for sufficient natural ventilation on the hull upper deck; for example, so as to prevent explosive gaseous mixtures
- Help in creation of a fire barrier between the topsides and the hull upper deck
- Help in the creation of a hazardous-area subdivision for equipment (e.g., electrical)

As the size of the topsides modules increases, the influence of the module interactions with the vessel hull becomes greater. The associated relative deflection and other effects between topsides modules and the vessel hull need, in any event, to be adequately considered and addressed in the design of the structural interfaces, the module support structure, interconnecting pipe work between modules, and the lower parts of the modules structures.

A number of types of design solutions can be relevant for supporting the topsides modules on a vessel deck. Three types of support arrangements are common: multipoint support columns, sliding/flexible support stools, and transverse web girder supports.

9.3.1.1 Multipoint Support Columns

In this type of arrangement, the topsides structure is supported on an array of vertical columns that are welded directly to the vessel structure, usually at typical intersections between longitudinal and transverse deck strength members. The column sections may comprise tubular or square hollow sections, or prismatic I sections.

The columns may need to be more slender with minimal diagonal bracing in the longitudinal direction in order to adequately "absorb" the vessel deflections. The diagonal bracing may be located at the center of the structure to act as a point of "fixity" and avoid excessive deformation of the outermost columns. Often, this type of arrangement may be considered for small to medium sized "pancake"-type topsides structures.

The advantage of this type of arrangement is that it is relatively simple to fabricate and reasonably well suited to integrate with the vessel structure, which is an important consideration in tanker conversions. Because the supports can be numerous, the topsides deadweight can be maximized and fairly well distributed over the vessel deck. Strengthening under deck may be minimized and could in ideal cases be limited to reinforcement of the deck girders in the form of web stiffeners. With regard

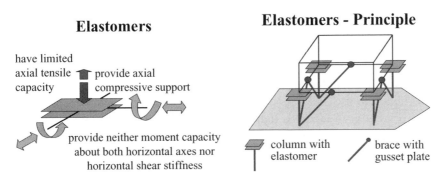

Figure 9.3. PAU supported on elastometric pads (Krekel and Kaminski 2002; courtesy of Society of Petroleum Engineers/Offshore Technology Conference).

to safety, the design exhibits a high level of redundancy where the failure of any one individual support would not necessarily cause collapse of the whole topsides structure.

The disadvantage of this type of arrangement is that the numbers of supports involved could make the accurate prediction of support reaction loads difficult without a very detailed consideration and analysis of numerous load cases. The flexibility of the deck, the flexibility of the topsides structure, the construction tolerances, and the existing deck levels are all factors that can affect the magnitude and distribution of the support reaction loads. Additionally, these same factors need to be accounted for when the topsides module is a PAU and integrated with the support structure as a single unit. It is to prevent the eventuality that the topsides end up resting on a few points, which will require extensive corrective work later. If the vessel is to be spread moored, a complicated sheaving arrangement may become necessary in some cases for anchor-line installation in order to avoid interference with the support columns. For pipe routing and access for maintenance, congestion may also be an issue.

9.3.1.2 Sliding/Flexible Support Stools

In comparison to the multipoint design, this arrangement is more capable of being composed of fewer support points; typically, around four to six per module. Each individual support will typically include a stool structure located on the vessel deck, which also aligns with a matching structure on the topsides module.

To better compensate for vessel flexure, two or more of the supports will be designed to provide a laterally flexible connection. This usually makes use of low-friction coating and/or composite elastomer material applied between the stool and module mating surfaces. For example, the PAU may be supported on elastometric pads as shown in Figure 9.3. The elastometric pads are stiff in compression but flexible in shear. With a suitable arrangement of stoppers and retainers, the PAU can be decoupled from the hull's deck deformations including elongation. Generally, this type of supports needs to be considered for large-span PAU modules, common in large new builds.

The use of standard bridge bearings has also been previously proposed for this application. Here in such a case is an array of roller and sliding joints fitted between the hull strong points and the PAU will serve to isolate the PAU from the hull deformations. The support locations will also be arranged with limiting devices for resisting uplift and significant transverse loading as necessary.

The advantage of this class of supports (i.e., of the sliding/flexible type) is that the clearance between topsides and vessel deck can be greatly increased with resulting benefits for access and maintenance primarily because the supports can be fewer in number. The ability to use larger modules can also significantly reduce the amount of hook-up work between modules that has to be performed typically toward the end of the project. Concerning the design itself, the reaction loads can be relatively accurately predicted, and the associated installation onto the vessel can also be relatively more easily managed and controlled.

The disadvantage of this type of supports is that the distribution of weight of topsides structure onto the hull may be less uniform, and some of the resulting support reactions can be relatively high because the topsides module needs to be self-supporting between the limited number of stool points. In turn, significant strengthening may be required below the vessel deck. For many new builds, this is not generally a big problem. At times, however, for a conversion project, the amount of strengthening so needed may in fact dictate the feasibility of using or not using these types of supports and, consequently, large PAUs. For the design of the flexible supports, accurate prediction and compensation for all load effects are also a necessity.

9.3.1.3 Transverse Girder Supports

This type of support generally comprises rows of deep web T-sections mounted transversely across the width of the vessel in line with the transverse deck girders and with the flange sections at the uppermost point, as shown in Figure 9.4. A typical example of the transverse frame spacing in the hull may be about 5m, but this will depend on the size of the vessel. The topsides modules sit astride these supports with their weight transmitted through the section web. Depending on the size of the modules, some form of flexible joint between the module and the top flange may be required to better accommodate the vessel flexure effects.

The advantage of this type of support system is that it may allow for greater flexibility in the design, layout, and arrangement of the topsides modules. This system is relatively less reliant on the layout of the strength members below the vessel deck. The T-sections and transverse girders can form near integral strength members through which the topsides loads can be transmitted and, perhaps, be more evenly distributed. Compared to the multipoint design, installation of the modules can be more easily controlled by this type of support as well.

The disadvantage of this type of support is that access fore and aft on the vessel deck is limited by the size of access holes that can be safely incorporated in the section web. Piping arrangements may also be affected given the limited space available. Adjacent modules sitting on a common support that are staggered longitudinally can give rise to cross-link effects that may result in related shear and fatigue problems in the support structure unless adequately analyzed and designed for.

9.3.2 Types of Topsides Flooring

The layout of the hull upper deck and the topsides modules together with foundation structures, pipe work, equipment, escape/access routes (see Figure 9.1(f)), and mechanical handling headers must be defined as clearly and as early as possible.

Figure 9.4. A photograph of topside foundation supports consisting of transverse girders, in a ship-shaped offshore unit (Krekel and Kaminski 2002; courtesy of Society of Petroleum Engineers/Offshore Technology Conference).

When the fabricators for hull structures and topsides are different, full communication between them as to the various interface matters is also a necessity.

Another issue is the selection of the type of flooring for the topsides; such selection needs to maximize both safety and functionality. Two flooring types are usually relevant: plated type and grated type.

Advantages of a steel-plated floor include the possibility that it may work as a fire barrier between the hull upper deck and the topsides. Also, leaks of gas and spillages of oil from the topsides to the hull upper deck can be prevented by the plated floor, although relevant drains systems are in this case required to deal with the drainage of the spills and leaks with rainwater, wash-down water, firewater deluge, and seawater, as previously described in Section 9.2.

A disadvantage of a plated topsides deck is that the pipe-work areas on the hull upper deck are now even more completely covered by the topsides; this can disturb natural ventilation of gas or gaseous mixtures, thereby increasing the risk of fire and explosion. Furthermore, the blast pressure in explosions involving a plating-type floor is likely to be greater than with a grating-type floor.

9.3.3 Types of Topsides Fabrication

Two types of topsides module fabrication are relevant: built-in grillage deck and preassembled unit (PAU).

Figure 9.5. An assembly of large PAUs by heavy lift; for accommodations (left) and topsides facilities (right) (courtesy of Samsung Heavy Industries).

9.3.3.1 Built-In Grillage Deck

In a built-in type construction, one builds the topsides deck at an elevation above the hull upper deck. Facilities and equipment are at most prepackaged onto skids and then are lifted on the built-in grillage deck. Piping, electrical, and instrumentation systems are interconnected after that. This type of fabrication will lessen issues associated with heavy-lift operations, but it may not meet the possibility of precommissioning of topsides facilities equally well.

9.3.3.2 Preassembled Units

In the PAU method, single or multiple assemblies of topsides modules containing the necessary facilities and equipment with as many preconnected piping, electrical, and instrumentation systems as possible are fabricated. The PAUs are then lifted onto the hull upper deck by relatively expensive heavy-lift operations, as shown in Figure 9.5.

Requirements must be met to ensure that the supports of the foundation structures are fully welded to the hull upper deck without pregap, as shown in Figure 9.6, and also that they are sufficiently designed to prevent fatigue cracking. Note that Figure 9.6 shows this type, but it is only an example. Better strength characteristics may be achieved by even partial penetration welding when compared to a fillet weld. In any event, fabrication with PAUs is very useful for fast-track project schedules.

9.3.4 Structural Analysis of Topsides Modules and Interfaces

From the structural design point of view, three major aspects need to be considered for topsides modules and their interfaces with the hull, as follows:

- Strength of topsides modules considering vessel motions and accelerations
- Minimizing the deformations of the topsides supports
- Fatigue life of the topsides supports

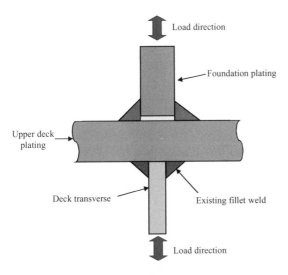

Figure 9.6. A schematic of unsuccessful fillet welding due to a large gap between the foot of the foundation structures and the hull upper deck.

Vessel motions, in terms of roll, pitch, and heave, and also aspects of weathervaning that are associated with wind, waves, and currents can significantly affect the behavior of topsides modules. Trim or heel effects also need to be considered under normal operating and offloading conditions. Vessel motion analyses must be undertaken to identify the pertinent effects of motions and accelerations on the design of topsides modules and interfaces, both under extreme weather conditions and under normal operating conditions.

For onsite cases, site-specific environmental data will be applied for the motion analyses, as described in Chapter 7. The transit conditions to or from the field must also be similarly considered. Figure 9.7 shows an example of part of a finite-element model involving topsides, foundation structures, and hull-deck structures. In the finite-element modeling shown, topsides modules are modeled as space-framed structures (which is often typical), although hull and foundation structures are modeled as plated structures.

In practice, there will be limits placed on the roll, pitch, and heave motions in order that the normal operations of topsides facilities are not restricted. Due to interacting hull girder bending action, the foundation structures will deform as shown, for example, in Figure 9.8. A primary concern, in terms of design, is that the maximum elongation of the foundation structure supports specifically in free support frames must not exceed a specified limit. This design criterion can be important to prevent failure of pipe work and electrical cable runs mounted on the hull upper deck. Typically, the permissible elongation of the foundation structure supports in such cases is about 0.1 percent of the transverse frame spacing.

Fatigue limit-state design requirements (see Chapter 7) must also be met along the interfaces between vessel hull and topsides modules; for example, the supports of foundation structures.

9.3.5 Interface Management and Other Lessons Learned

Although the split-design execution of hull structures and topsides modules by various parties is not the ideal situation to be recommended, it is common in reality

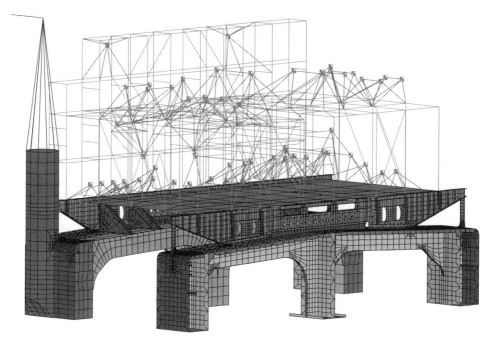

Figure 9.7. An example of the finite-element modeling involving topsides modules, foundation structures, and hull upper deck structures (Krekel and Kaminski 2002; courtesy of Society of Petroleum Engineers/Offshore Technology Conference).

and, therefore, the identification and detailed consideration of issues of the interfaces between such systems and their contractors invariably will arise (Keolanui et al. 1998).

The technical agreement between the hull contractor and the topsides modules contractor must be kept in terms of design and fabrication parameters from the

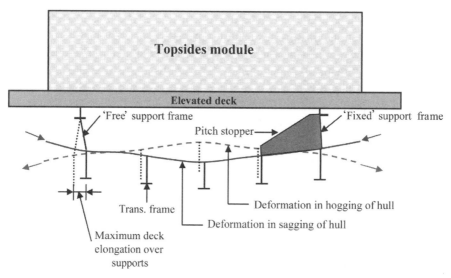

Figure 9.8. A schematic of the distribution of foundation structure deformations affected by hull girder bending.

earliest design stage to avoid expensive and complex troubleshooting later. If not, it is possible, for example, that the hull, topsides, and their interfaces may need to be strengthened at a later design stage or even at the stage of fabrication. Consistency in design data such as loads is particularly important in this case.

In practice, when adequate design is not achieved for the supports of foundation structures against some of the loads eventually predicted to occur – for instance, those associated with the transportation to site and fatigue conditions – troubleshooting possibilities may necessitate the following changes because it will be too late in the process to make any essential modifications to the conceptual design:

- Increase the size of support columns.
- Add haunch details at some of the joint connections.
- Tie-in the vertical cross bracings to decks located at the upper levels.
- Grind welds, chamfer bracket toes.

Such measures may become necessary due to the fact that the contractor may not wish to consider changes to the conceptual arrangement to avoid any potential changes to the process equipment layout. In one project, during a meeting with the hull contractor to discuss fatigue strength criteria for the design of the hull deck, the topsides contractor may casually remark that a high proportion of the topsides structure fatigue life has been estimated to be used up during transportation of the installation to site; that may be the signal for the start of a series of troubleshooting operations.

When reviewing the proposed designed against the design criteria required by the specifications, the company then expresses a concern that the casual remark may imply that either the design or calculation process is in error. Further meetings and a more detailed review of the design process may conclude that the hull contractor, who is responsible for deriving vessel motions and managing the interface, had possibly forwarded incorrect design information to the topsides contractor.

A further review of the design calculation documents issued a few months later may show that although some of the original mistakes are now rectified, the maximum accelerations used in the design are still significantly lower than those expected for the governing conditions required by the company specification. Further investigation may show again that not all the correct design information has been forwarded to the topsides contractor. Accordingly, more detailed and refined analyses and further design modifications are required to ensure that the structure conforms adequately to the transportation and in-service requirements of the specification, particularly with regard to fatigue strength. Because material has been ordered, the layout and arrangement of the structure has been generally finalized, and fabrication has already begun, various difficulties may now lie ahead in meeting the project schedule.

The topsides contractor would, of course, have investigated some options for changing the layout of the support structure, particularly with respect to a more efficient and less complicated bracing arrangement. However, the resulting redistribution of support reactions may still not be supportable by the deck, which has already been designed without any significant reserve strength capacity.

In summary, the type of structure eventually fabricated may not be optimal in design, is inefficient in its use of material, and in some instances may barely conform to the requirements of the specification. Due to the minimal amount of bracing, very heavy column sections may turn out to be required to maintain adequate flexural strength. As a result, in order to maintain adequate fatigue strength, complicated reinforced joint connections could be required that can subsequently prove a challenge to fabricate. The decision to tie-in the cross braces at the second- and third-deck levels may result in very complicated and unusual details that prove relatively difficult to align and fabricate and that consequently require very detailed analysis to ensure compliance with the design criteria.

In another example case, the limiting features including limited deck area may effectively dictate a topsides structure with a multitude of support columns. This in turn results in a congested space into which the pipe work is then to be fitted.

In this instance, it may have been that the hull contractor also acted as the interface manager and that there may have been a tendency for the sake of convenience, for the hull contractor to dictate the layout and design of the topsides. Little attempt may have been made to develop a more optimal solution for the unit as a whole, particularly with respect to access space and piping arrangements between the hull and process decks. In effect, the solution eventually derived is more akin to that expected for a conversion project rather than a new-build design.

Both hull and topsides contractors tend to optimize their respective designs at an early stage in the design process, which may turn out to be based on preliminary and perhaps incorrect design information. This may minimize allowances for mistakes or the possibility of implementing more advantageous design changes. To offset this effect and to ensure that the design does not fall below certain standards, it may be advisable to specify a limiting set of particular loading criteria that should always be satisfied prior to confirmation and verification for the project-specific design conditions.

Key elements of interface data passed between the contractors should be continually verified and updated. The design review should focus on more than only the limited documents and design information periodically issued by the respective contractors to the company. The imposition and use of a formal interface management system, which includes a register of interface design and construction data that can be reviewed and updated by all parties on an ongoing basis, can help improve the design process.

9.4 Mooring Facilities

9.4.1 Types of Moorings

For floating offshore structures, the mooring is of vital importance for station-keeping. The mooring systems used in related projects can be classified into the following two main groups: spread mooring and single-point mooring (SPM).

Additional means such as thrusters and dynamic positioning (DP) may be used to assist, particularly as part of the single-point mooring systems, to improve station-keeping. The use of DP as the sole means of station-keeping, although much less frequent, is also possible. For good examples of the related mooring systems and

Figure 9.9. A spread-moored FPSO (courtesy of Samsung Heavy Industries).

variations thereof, see Baltrop (1998) and the websites of important offshore contractors such as SBM, www.singlebuoy.com.

9.4.1.1 Spread Moorings

As shown in Figure 9.9, a spread mooring is a combination of chain, wire, or synthetic cable (rope) fixed in groups at strategic points on the floating vessel hull so it maintains station, usually on a fixed position and heading. Because moorings are in this case independent of production risers (unlike most turret moorings), risers can be brought on board at convenient points on the hull and tied into the process train or to a header.

This system may be suitable for certain types of offshore applications where designing for the wave direction effects concerned is not onerous and remains cost effective. However, in a purely technical sense, it is usually best to allow a vessel to weather-vane around the mooring so that mooring forces and vessel roll motions can be kept to a minimum.

An advantage of spread moorings is that it is usually cheaper than turret moorings. The main disadvantage of spread moorings is likely lower availability for offloading operations than turret moorings. If the environment has strong directionality, the FPSO can be aligned with the environment, and shuttle tankers may be able to moor for much of time. However, if there is no strong directionality, shuttle tankers could have difficulty approaching and mooring in tandem to the FPSO under certain conditions. In such cases, it is necessary to have an export terminal, usually a buoy, in addition to the FPSO. This of course can reduce or remove any lost advantages of a spread mooring.

Figure 9.10. A photograph of an FPSO moored at a fixed-tower mooring system (courtesy of SBM Offshore NV).

In addition to traditional spread moorings, there is also the concept of a semi-weathervaning spread mooring; for example, the DICAS system described in Section 9.4.3.

9.4.1.2 Single-Point Moorings

SPMs of various types usually located at the vessel's bow can provide the weather-vaning characteristic in a mooring design. In this regard, many ship-shaped offshore units adopt a single-point mooring system of some type. There are many types of SPM, including the following (Barltrop 1998):

- Fixed tower
- Catenary anchor leg mooring (CALM) buoy
- Single-anchor leg mooring (SALM) buoy
- Articulated loading platform (ALP)
- Single point and reservoir (SPAR)
- Single-anchor loading (SAL)
- Turret mooring

(1) Fixed Tower. Figure 9.10 shows a picture of a fixed-tower mooring that is a rela-tively simple method of mooring a floating vessel while allowing weathervaning. Floating hoses are typically used between the tower and the vessel. The tower incor-porates a bearing to allow the section that is normally connected to the hoses and mooring hawser/chain to rotate as the vessel weathervanes around the tower. In such cases, the possible risk of collision by the vessel itself is a consideration, and perhaps most important, a fixed tower is not normally the preferred choice in deep waters.

(2) CALM. In a CALM system, a catenary moored buoy is used instead of a fixed tower, together with floating flexible hoses to transfer the oil, as shown in Figure 9.11. This system can be applied to deeper waters, and there may be lower

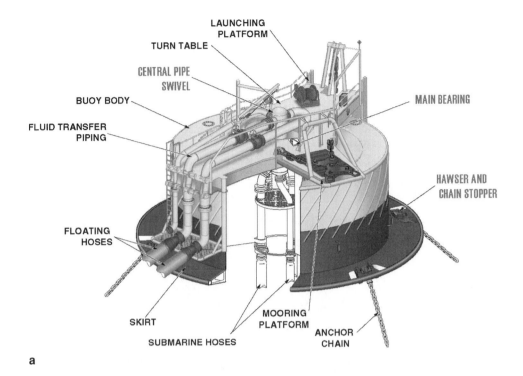

a

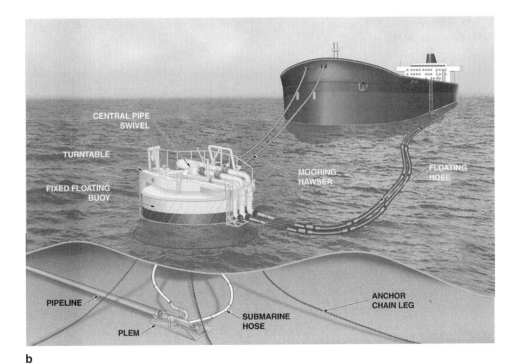

b

Figure 9.11. (a) A schematic of a CALM system (courtesy of SBM Offshore NV). (b) An artist's impression of a CALM arrangement with an FPSO (courtesy of SBM Offshore NV).

c

Figure 9.11. (*cont.*) (c) A photograph of a CALM arrangement with an FPSO (courtesy of SBM Offshore NV).

expected collision consequences than the fixed tower. Although shuttle tankers do occasionally collide with or "kiss" buoys, serious damage is somewhat rare. However, there is still a significant risk of damage to the CALM due to a vessel impact that may require costly maintenance and repair to the mooring and fluid swivels on the catenary-moored buoy.

(3) SALM. In a SALM, a single vertical tensioned chain mooring leg is installed as shown in Figure 9.12, instead of multiple catenary anchor lines that are used for the CALM. The fluid swivels can often be placed on the seabed in shallow water, whereas they can be attached part way up the mooring leg in deep water. That the swivels are placed at seabed is a disadvantage because the maintenance of underwater swivels is costly and not easy. It is normally desirable to keep mechanical components above water, where they can be inspected. A SALM uses buoyancy to generate the restoring force when the vessel moves off station. This contrasts with a CALM, where the restoring force is generated by imbalance in weight of the mooring legs from which they are lifted or laid down on the seabed.

(4) ALP. The ALP system as shown in Figure 9.13 is somewhat of a cross between the fixed tower and the SALM. The swivels in this case are located above the water, with a related risk of collision present. A helideck can be provided on an ALP. The bending moments of ALP can become large due to dynamic equilibrium needs of the system in deep water. The system of this type generally may not be

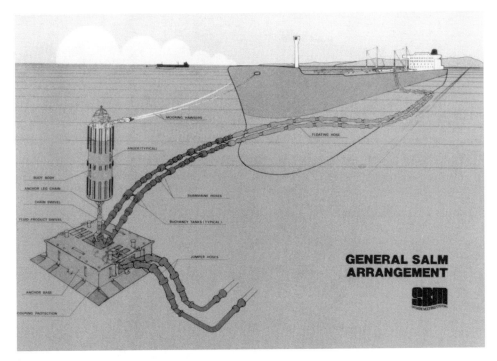

a

b

Figure 9.12. (a) An artist's impression of a SALM arrangement with an FPSO (courtesy of SBM Offshore NV). (b) A photograph of a SALM installation (courtesy of SBM Offshore NV).

Figure 9.13. A vessel moored at an ALP ready for loading (courtesy of SBM Offshore NV).

the easiest nor the most attractive in terms of the required maintenance. The ALP incorporates a rotating mooring head, similar to a fixed tower. A roller bearing is located at the top of the cylindrical structure, and the mooring head above is free to rotate under the action of the howser and jumper hose as the vessel weathervanes in response to changes in environmental loading.

(5) SPAR. The SPAR (which is not the production platform described in Section 1.3.2 in Chapter 1) may often be part of a very large CALM buoy, and it is usually employed to support process equipment rather than for storage or simply as marker buoys. However, it very rarely consists of a cylindrical shape tank to be used for storage, as shown in Figure 9.14. This type of arrangement may be useful in some cases because considerable quantities of oil can be stored in the SPAR when the ship-shaped offshore unit does not have sufficient storage.

(6) SAL. The SAL system, as shown in Figure 9.15, may be useful in situations with less demanding operational requirements. It will typically have an upper operational limit dependent on water depth and vessel size type. The weakness of the SAL system is that all mechanical parts are submerged (and are difficult to inspect and maintain), but it can be cheaper than a CALM buoy in benign conditions.

(7) Turret Mooring. A turret is a device directly built into the moored vessel, incorporating a bearing arrangement for purposes of weathervaning, and attached to the seabed usually by catenary anchor lines. The moored part of the turret is fixed relative

Figure 9.14. A vessel moored at a SPAR (courtesy of SBM Offshore NV).

Figure 9.15. A computer graphic of a SAL system (courtesy of APL).

Figure 9.16. An FPSO with an external turret-mooring system (courtesy of Samsung Heavy Industries).

to the sea bottom; flexible risers are suspended from this fixed part and are connected to Pipe Line End Manifold (PLEM) arrangements or directly to wellheads. A swivel connects the flexible risers to the fixed piping mounted on the vessel.

Turrets can be grouped into two main types: permanent and disconnectable. The former is permanently built in the floating vessel, but a part of the latter is made capable of being disconnected in the event that certain design environmental conditions and associated limits are exceeded; for example, in case of severe hurricane warnings. Three types of permanent turrets may be relevant (de Boom 1989; d'Hautefeuille 1991): external bow/stern turret (Figure 9.16), internal turret (Figure 9.17), and clamped-riser turret.

An external turret-mooring system can be used in moderate to severe environments. Figure 9.16 shows a typical external turret-mooring system. For harsh environments, consideration must be given to protection of risers from wave damage and this could be a limiting factor. The typical limit on the number of risers could be about twenty.

An internal turret-mooring system has the turret inside the vessel, as shown in Figure 9.17. This system can also be used in moderate or harsh environments. The typical limit on the number of risers in such cases may be about 100. Integration of the turret into the hull is an important consideration in the structural design of the associated vessel hull. It is also important, in such cases, to ensure that hull girder strength is not reduced by the internal turret arrangement. Also, because of the internal turret,

Figure 9.17. (a) An FPSO with an internal disconnectable turret-mooring system and its schematic (courtesy of SBM Offshore NV).

available deck area for the topside footprint may be reduced. Although internal turrets can be designed to accommodate even 100 risers, this requires the use of a split bearing or a bogie system because single-piece roller bearings of sufficient size are not available.

A clamped-riser turret system has been developed for the purpose of protecting flexible risers in splash zone areas; the turret structure mounted at the vessel bow consists of a cylinder supported by two bearing assemblies, one at vessel deck level and the other just above the vessel keel level. The chain-table is then mounted at the bottom of the turret structure and usually remains submerged. This arrangement provides for not only the protection of flowlines but also lessens potential interference between chains and the vessel hull.

A disconnectable turret arrangement is designed to deal with disconnection in high-sea states and to allow reconnection, as shown in Figure 9.18. It is usually used under harsh environmental conditions or for other special reasons. Typically, the anchor legs and submarine hoses will remain on a floating or submerged buoy to ease the reconnection in such cases. Disconnectable turrets are sometimes classified into the following two types (de Boom 1989; d'Hautefeuille 1991): riser turret mooring (RTM) and buoyant turret mooring (BTM).

Figure 9.17. (*cont.*) (b) A photograph of an FPSO with an internal turret-mooring system (courtesy of Samsung Heavy Industries).

An RTM is made of a long slender tubular riser, suspended from a rigid mooring arm mounted at the vessel bow, as shown in Figure 9.19. The mooring arm houses the main weathervaning bearing from which a long mooring riser is suspended via an articulated joint. The caternary lines and the fluid risers are attached to the structural riser. To disconnect the vessel, a mechanical connector just below the articulated joint is released, thus dropping the riser, which should then provide sufficient buoyancy to stay afloat with the attached mooring chains and fluid risers. The disconnection may be made in case of shutdown of the field or disconnection of fluid lines. The risers may remain free-floating at the sea surface when disconnected and, as such, they must resist potentially harsh environmental conditions. For reconnection, a cable is pulled up by a winch arrangement mounted at bow.

A BTM system intends to improve the efficiency of disconnection and reconnection of the RTM system. For a BTM, the buoy is always submerged when disconnected, and it has to be pulled into the FPSO moonpool for reconnection, so it must be located below keel level when disconnected. A relatively small-sized buoy that remains on site after disconnection is used in contrast to the RTM riser, which can be a very long cylinder (say, 75–100m) and with a fairly large displacement (typically 2,000 tons). For a more detailed description of permanent and

Figure 9.18. A photograph showing the installation of an external disconnectable turret during the connection operation (courtesy of SBM Offshore NV).

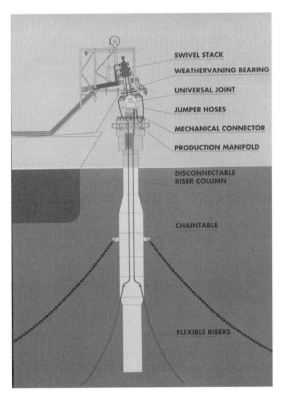

SWIVEL STACK
WEATHERVANING BEARING
UNIVERSAL JOINT
JUMPER HOSES
MECHANICAL CONNECTOR
PRODUCTION MANIFOLD
DISCONNECTABLE RISER COLUMN
CHAINTABLE
FLEXIBLE RISERS

Figure 9.19. A schematic of the RTM (courtesy of SBM Offshore NV).

Table 9.1. *A survey of FPSO mooring systems (Ronalds and Lim 1999)*

Mooring	Number
Spread mooring	9
Catenary anchor leg mooring	12
Single-anchor leg mooring	11
External turret	7
Internal turret	22
Disconnectable riser turret	4
Disconnectable buoyant turret	2
Dynamic positioning	2
Total	69

disconnectable turret configurations, see de Boom (1989) and d'Hautefeuille (1991).

A turret allows the vessel to weather-vane while providing a geostationary hull area to bring in flow and mooring lines. An efficient bearing system is required for this purpose. Interfacing bearing systems with the turret and vessel structure requires extensive design and analysis to assure that their relative deformation will not affect the longevity of the bearing system (Pollack et al. 1997).

9.4.2 Mooring System Selection for an FPSO in Deep Water

The selection of the appropriate mooring system may vary from one project to another and be influenced by many factors (Hewlett 1992). In addition to station-keeping, other factors such as off-take frequency, environmental conditions, export tanker size, and risk of collision are also important considerations in this regard.

For mooring of a ship-shaped offshore unit in deep water, spread-mooring and turret-mooring systems are the most relevant (d'Hautefeuille 1991; Adhia et al. 2004). As noted in Section 9.4.1, a spread-mooring system uses a number of cables and charms to moor the vessel. This system is usually adopted for the installations operated in a moderately benign environment. There is little practical limit on the number of risers in such cases. It may be typical practice to use a spread-mooring system together with a CALM buoy for export operations of a shuttle tanker. In recent projects in benign environmental areas, such as West Africa and Indonesia, a spread-mooring system has been used quite often.

Internal turrets are becoming more common in many FPSO applications; see Table 9.1, which indicates the results of a related survey performed by Ronalds and Lim (1999).

In the selection of the mooring system in any particular case, a variety of features and related cost-versus-benefit tradeoffs must be considered, including the following (d'Hautefeuille 1991):

- Environmental conditions
- Water depth
- Weather thresholds for disconnection and reconnection
- Installation complexity
- Operation and maintenance
- Safety and reliability from operational and survival points of view
- Design and fabrication schedules
- Capital costs
- Operating costs
- Downtime
- Emergency repairs

9.4.3 Design Considerations for Mooring Systems

As previously described in Section 2.20 in Chapter 2, the mooring system design becomes increasingly more specialized and perhaps complex as one moves to deeper waters and harsh environmental areas with tropical cyclones. The new-build FPSOs usually have a large water plane area and thus the vessel motions are usually large in both wave-energy spectral period (3–25 seconds) and slowly varying period range (> 50 seconds). Also, weathervaning is applied to minimize the mooring forces.

The critical issues involved in the design of mooring systems include the following (Nestegard and Krokstad 1999; Huang 2000):

- Maximum excursions of the vessel that can be accommodated by the risers.
- Line dynamics due to 6-degrees-of-freedom wave-frequency vessel motions, that is, in surge, sway, heave, roll, pitch, and yaw.
- Low-frequency vessel motions in surge, sway, and yaw.
- Effects of noncollinear environments of winds, currents, and waves on the responses of the vessel and its mooring system, which require motion and structural response analysis of coupled deep-water systems under dynamic actions due to winds, currents, and waves.
- Because of longer exposed risers and moorings, current loading and viscous forces become stronger, which implies that flow-induced vibration phenomena may increase.

Barltrop (1998) and Huang (2000) present excellent descriptions of mooring-system design, including the mooring system dynamics involved. Nestegard and Krokstad (1999) review various computer tools for the analysis of coupled solution of mooring dynamics and vessel motions. Useful feedback from operational experience and inspections of FPSO moorings are provided by Mastrangelo (2000) and Brown et al. (2005).

Mooring lines are continually exposed to corrosion attack, impact, abrasion, fatigue, bending, and tension, and subsequently the systems can degrade over time. The impact of mooring failure is of significance, potentially leading to damage or rupture of the risers, and eventually shutdown of the offshore unit. The design premise of the majority of ship-shaped offshore units is that they should be able to withstand a single-mooring-line failure without causing damage to the risers. However, if a single-mooring-line failure is undetected over time and/or if severe degradation exists, multiple-line failure can take place, particularly in harsh weather conditions.

Many theoretical studies on mooring system designs including risers and anchor designs have been undertaken in the literature. The methods for analysis and design of mooring and riser systems are presented by McClure et al. (1989), Huang and Judge (1996), Connaire et al. (1999), Barusco (1999), Ward et al. (1999), Baar et al. (2000), and Duggal et al. (2000). Portella and Mendes (2002) provide an overview of the concept of a differential compliance anchoring system (DICAS) mooring system and its related benefits. Mooring and anchoring systems designed for operation in deep water are also studied by Barusco (1999) and Ehlers et al. (2004). Regulatory considerations for FPSO moorings are presented by Lee (1997).

Various important features for the design and operation of mooring systems, such as causes of mooring system degradation, consequences of mooring line failure fatigue, implications of friction-induced bending, options for in-water inspection, importance of connector design, and methods to detect mooring line failure, are discussed by Brown et al. (2005). Properties of fiber rope deep-water mooring lines are studied by Flory et al. (2004).

Descriptions of many studies involving model testing of mooring systems are also available in the literature. Yashima and Miyamoto (1989) conducted a large-scale model test of turret-mooring system for an FPSO. Model test results for the dynamic response characteristics of a SALM system considering the effects of noncollinear environments are presented by Ho (1991), where a floating storage tanker moored to a SALM via a rigid yoke is considered.

Considering the relative differences in motions due to heave, roll, and pitch, the selection and design of risers for ship-shaped offshore units can be more complicated than those for other types of floating offshore such as semisubmersibles, spars, and tension leg platforms. Indeed, the risers that are feasible on the latter types of systems may not always work on the former. In particular, for riser systems used in conjunction with ship-shaped offshore units in harsher environments, more compliant and decoupled configurations compared to simple caternary or top-tensioned vertical risers may be required. Petruska et al. (2002) carried out an assessment of technical feasibility and commercial viability on several riser options for a 2-million barrel new-build FSO for possible deployment in approximately 1,370m (water depth) in the Gulf of Mexico. They studied three types of riser systems in detail: the steel lazy-wave riser (SLWR), the single-line hybrid riser (SLHR), and the tension leg riser (TLR).

9.5 Export Facilities

9.5.1 Methods of Export

The following two methods are relevant to the export the separated oil or gas from ship-shaped offshore installations to the shore: shuttle tankers and high-pressure pipelines feeding into a larger pipeline gathering system onshore.

The export using shuttle tankers generally requires high-capacity cargo transfer pumps for offloading of cargo from storage tanks within the turnaround times required. Both methods will require the use of riser technology appropriate to the depth of water, flow rates, and pressures involved.

The cargo pumps in new builds may be deep-well hydraulically powered units. In some cases, when the ship-shaped offshore units have been converted from trading tankers, existing pumps typically located in machinery spaces may also be used. Design considerations must be given to dismantling and removal of the pumps for maintenance purposes.

The oil cargo would be directed to a shuttle tanker via a transfer hose, which is generally deployed at the stern of the FPSO. The necessary hose diameter is determined based on the flow rates. During the period when offloading is not taking place, the transfer hose may need to be stowed appropriately. Design consideration must be given to the proximity of the transfer hose location to living quarters, temporary refuge, escape routes, and lifeboat embarkation stations in order to consider the possibility of accidental leakage and spillage during offloading. Also, procedures and support devices for the removal and repair of damaged hose sections need to be considered.

The presence of the hydraulic power systems involved requires one to consider minimizing the effects of vibration and noise in hull-machinery spaces and on deck, specifically in the vicinity of living quarters. Other than causing human discomfort, vibration may also result in the fatigue cracking at stress-concentration areas due, for example, to its transmission via the structure, pipe work, and piping supports.

Due to environmental concerns and by related regulatory requirements, it is increasingly not allowed to flare or vent gas surplus in the air except in an emergency. The separated gas may, therefore, need to be reinjected into the reservoir or adjacent suitable geological formations for storage, and may or may not be recovered in the future. Where a pipeline infrastructure exists, and assuming economic feasibility, the gas can also be exported.

9.5.2 Types of Shuttle Tanker Export

For export operation of an FPSO by a shuttle tanker, the following three types are relevant (Adhia et al. 2004):

- Tandem (Figure 9.20)
- Side-by-side (Figure 9.21)
- CALM buoy, usually located at a distance from an FPSO (Figure 9.22)

The selection of the type of export system to be employed may depend on various factors. A backup export arrangement is also highly desirable in case of failure of the primary export means. Most fields use the export process as the means to make oil production profitable. Therefore, the offloading system should be able to safely offload the oil or gas from the installation and also should be able to accurately measure the quantity and quality of the export operation. The oil or gas must be exported at sufficient rate to avoid incurring demurrage of the tankers involved. For example, a million-barrel oil parcel may need to be offloaded in less than 36 hours measured from arrival of the shuttle tanker, or when the shuttle tanker indicates its readiness to berth at the terminal, to its departure after all the paperwork has been completed.

The offloading rate selection should allow for connection time, slow-down during start and finish (topping-up) operations, and paperwork and disconnection times. Typically, an offloading system may then need to provide the capability for offloading a full parcel within 24–26 hours with the remaining time being taken up by other related activities like connection, start, topping-up, and paperwork. For a larger parcel size, one may consider allowing for longer periods, for example, 72 hours for a 2-million barrel parcel size.

The use of dynamic positioning systems attached to ship-shaped offshore units and/or shuttle tankers can help secure the required station-keeping accuracy, disconnection limits, and prevention of production downtime due to weather during

Figure 9.20. A photograph of a tandem export from an LPG FPSO (Adhia et al. 2004; courtesy of American Society of Mechanical Engineers).

production and/or offloading. In most dynamic positioning systems, thrusters are used to control the vessel motions. Comprehensive descriptions related to the design of dynamic positioning systems are presented by Barltrop (1998).

9.5.3 Design Considerations for Export Systems

It is important to design the offloading system and to optimize the related operations considering the need to maintain operations in relatively harsh environments so as to minimize downtime (see Section 2.21 in Chapter 2). In this regard, Morandini et al. (2002) present some guidelines and specific criteria for design of offloading operations, where the following types of offloading systems were considered:

- CALM buoy + FSO + shuttle tanker
- Turret-moored FPSO + shuttle tanker
- Spread-moored FPSO + shuttle tanker

Daughdrill and Clark (2002) present some considerations for reducing risks faced by FPSOs and shuttle tankers during offloading. Viennem et al. (2003) study the likelihood of collision between FPSO and shuttle tanker during offloading, which is based on data from incidents and near-misses as well as from expert opinion.

Figure 9.21. A computer graphic of a side-by-side export arrangement for an FPSO (Adhia et al. 2004; courtesy of American Society of Mechanical Engineers).

Figure 9.22. A photograph of an export arrangement with a CALM buoy-moored shuttle tanker located at a distance from an FPSO (Adhia et al. 2004; courtesy of American Society of Mechanical Engineers).

Section 13.6.1 of Chapter 13 further discusses the consideration of collision risks for an FPSO.

Although the design, building, and operation of ship-shaped offshore units for production, storage, and offloading are focused on in this book, consideration of the same aspects for the case of shuttle tankers is also important as well. In this regard, Williams et al. (1999) discuss the operational risk management of shuttle tankers, with the design and construction issues, in terms of ship size/capacity, operating thresholds, transit speed, and cargo threshold.

REFERENCES

Adhia, G. J., Pellegrino, S., and Ximenes, M. C. (2004). "Practical considerations in the design and construction of FPSO's." *Proceedings of OMAE–FPSO 2004, OMAE Specialty Symposium on FPSO Integrity, OMAE–FPSO'04-0090*, Houston, August 30–September 2.

Baar, J. J. M., Heyl, C. N., and Rodenbusch, G. (2000). *Extreme responses of turret moored tankers.* Offshore Technology Conference, OTC 12147, Houston, May.

Barltrop, N. D. P. (1998). *Floating structures: A guide for design and analysis.* The Centre for Marine and Petroleum Technology (CMPT), Herefordshire, England: Oilfield Publications Ltd.

Barusco, P. (1999). *Mooring and anchoring systems developed in Marlin field.* Offshore Technology Conference, OTC 10720, Houston, May.

Brown, M. G., Hall, T. D., Marr, D. G., English, M., and Snell, R. O. (2005). *Floating production mooring integrity JIP – Key findings.* Offshore Technology Conference, OTC 17499, Houston, May.

Connaire, A., Kavanagh, K., Ahilan, R. V., and Goodwin, P. (1999). *Integrated mooring and riser design: Analysis methodology.* Offshore Technology Conference, OTC 10810, Houston, May.

Daughdrill, W. H., and Clark, T. A. (2002). *Considerations in reducing risks in FPSO and shuttle vessel lightering operations.* Offshore Technology Conference, OTC 14000, Houston, May.

de Boom, W. C. (1989). *The development of turret mooring systems for floating production units.* Offshore Technology Conference, OTC 5978, Houston, May.

d'Hautefeuille, B. B. (1991). *Floating production storage and offloading: Disconnectable or not?* SPE Asia-Pacific Conference, Society of Petroleum Engineers, SPE 22987, Perth, Australia, November.

DTI (1971). *Prevention of Pollution Act 1971.* Department of Trade and Industry, UK.

Duggal, A. S., Heyl, C. N., and Vance, G. P. (2000). *Global analysis of the Terra Nova FPSO turret mooring system.* Offshore Technology Conference, OTC 11914, Houston, May.

Ehlers, C., Young, A. G., and Chen, J. H. (2004). *Technology assessment of deepwater anchors.* Offshore Technology Conference, OTC 16840, Houston, May.

Flory, J. F., Banfield, S. P., and Petruska, D. J. (2004). *Defining, measuring, and calculating the properties of fiber rope deepwater mooring lines.* Offshore Technology Conference, OTC 16151, Houston, May.

Hewlett, C. W. (1992). *FPSO Ocean Producer: A unique design for shallow water marginal fields.* Offshore Technology Conference, OTC 7051, Houston, May.

Ho, R. T. (1991). *Effects of noncolinear environments on SALM-yoke loads.* Offshore Technology Conference, OTC 6607, Houston, May.

Huang, K. (2000). *Mooring system design considerations for FPSOs.* American Bureau of Shipping, Houston (http://www.eagle.org/news/TECH/offshore/Mooring.pdf), accessed June 2006.

Huang, K., and Judge, S. (1996). *Turret mooring system design and analysis for harsh environments.* Offshore Technology Conference, OTC 8260, Houston, May.

Keolanui, G., Lunde, P., and Jeannin, O. (1998). *Modular or turnkey FPSO: A world of interfaces.* Offshore Technology Conference, OTC 8810, Houston, May.

Krekel, M. H., and Kaminski, M. L. (2002). *FPSOs: Design considerations for the structural interface hull and topsides.* Offshore Technology Conference, OTC 13996, Houston, May.

Lapidaire, P. J. M., and de Leeuw, P. J. (1996). *The effect of ship motions on FPSO topsides design.* Offshore Technology Conference, OTC 8075, Houston, May.

Lee, M. Y. (1997). FPSO mooring: design and regulatory considerations. Offshore Technology Conference, OTC 8390, Houston, May.

Mastrangelo, C. F. (2000). *One company's experience on ship-based production system.* Offshore Technology Conference, OTC 12053, Houston, May.

McClure, B., Gay, T. A., and Slagsvold, L. (1989). *Design of a turret-moored production system (TUMOPS).* Offshore Technology Conference, OTC 5979, Houston, May.

Morandini, C., Legerstee, F., and Mombaerts, J. (2000). *Criteria for analysis of offloading operation.* Offshore Technology Conference, OTC 14311, Houston, May.

Nestegard, A., and Krokstad, J. R. (1999). *JIP-DEEPER: Deepwater analysis tools.* Offshore Technology Conference, OTC 10811, Houston, May.

Petruska, D. J., Zimmermann, C. A., Krafft, K. M., Thurmond, B. F., and Duggal, A. (2002). *Riser system selection and design for a deepwater FSO in the Gulf of Mexico.* Offshore Technology Conference, OTC 14154, Houston, May.

Pollack, J., Pabers, R. F., and Lunde, P. A. (1997). *Latest breakthrough in turret moorings for FPSO systems: The forgiving tanker/turret interface.* Offshore Technology Conference, OTC 8442, Houston, May.

Portella, R. B., and Mendes, C. (2002). *DICAS (differentiated complacent anchoring system) mooring system: Practical design experience to demystify the concept.* Offshore Technology Conference, OTC 14309, Houston, May.

Ronalds, B. F., and Lim, E. F. H. (1999). *FPSO trends.* SPE Annual Technical Conference, Society of Petroleum Engineers, SPE 56708, Houston, October.

UKOOA (2002). *FPSO design guidance notes for UKCS service.* Offshore Operators Association, UK.

Vinnem, J. E., Hokstad, P., Dammen, T., Saele, H., Chen, H., Haver, S., Kieran, O., Kleppestoe, H., Thomas, J. J., and Toennessen, L. I. (2003). *Operational safety analysis of FPSO – Shuttle tanker collision risk reveals areas of improvement.* Offshore Technology Conference, OTC 15317, Houston, May.

Ward, E. G., Haring, R. E., and Devlin, P. V. (1999). *Deepwater mooring and riser analysis for depths to 10,000 feet.* Offshore Technology Conference, OTC 10808, Houston, May.

Williams, J. S., Brown, D. A., Shaw, M. L., and Howard, A. R. (1999). *Tanker loading export systems for harsh environments: A risk-management challenge.* Offshore Technology Conference, OTC 10905, Houston, May.

Yashima, N., and Miyamoto, A. (1989). *A large-scale model test of turret mooring system for floating production storage offloading (FPSO).* Offshore Technology Conference, OTC 5980, Houston, May.

CHAPTER 10

Corrosion Assessment and Management

10.1 Introduction

While in service, most structural systems such as ships, offshore structures, bridges, industrial plants, land-based structures, and other infrastructure will be subject to age-related deterioration that can potentially cause significant issues in terms of safety, health, the environment, and financial expenditures. Indeed, such age-related deterioration has reportedly been involved in many of the known failures of ships and offshore structures, including total losses. Although the loss of a total system typically causes great concern, repair and maintenance of damaged structures is also very costly to society, in general, and important to the economic viability of the enterprises involved, in particular. It is thus of great importance to develop advanced technologies, which can allow for the proper management and control of such age-related deterioration.

One of the most important factors in the safety and integrity of ship-shaped offshore units at sea is that they are affected by corrosion. Corrosion for such structures becomes a problem particularly where surfaces are unprotected, and this can be an issue for internal surfaces in cargo tanks where conditions of high humidity, and often inspection and access difficulties, also exist.

Periodic surveys can detect corrosion problems and help estimate the remaining thickness of structural components prior to their needing replacement. For high-quality asset management and for the development of optimal maintenance programs, refined strategies for corrosion management and control, including corrosion corrective or preventive measures, are required. One of the financial issues for corrosion management is that severe degradation can lead to significant downtime costs through lost production, steel renewals, and related high direct and indirect costs.

In addition to anticorrosion measures, two approaches may be relevant in terms of structural design for corrosion: (1) timely renewal of heavily corroded structural components, and/or (2) prior addition of corrosion margins to newly built structures. These approaches are based on being able to define the amount of corrosion permitted before structural component renewal is required.

Adding a corrosion margin to completely avoid steel renewals in service can, in theory, be even more attractive when the renewal of corroded areas is not straightforward because of costs and/or greater difficulties related to dry-docking, repair, and maintenance in offshore installations. But, even in these cases, the approach may often not be feasible for economic reasons. In any event, the prediction of likely future corrosion loss is then important for the structures as it will enable one to maintain the necessary margins or allowances as the structure ages.

In a corrosion risk-management scheme for a structure, both the structural safety and the optimal operational serviceability are the matters of prime concern. The use of an appropriate corrosion risk-management scheme will necessitate the specification of an optimal corrosion protection scheme. Experience shows that the areas of corrosion in trading tankers that we are relatively more concerned with are inside cargo holds and ballast tanks rather than outer surfaces of vessel hulls, which are well coated; and it also helps that the tanker dry-docks at 5-year intervals.

In ship-shaped offshore units, on the other hand, such concern also must extend to the underwater hull. Further, coatings begin to fail after 10 years or so, and minimizing related effects requires very good quality control at their initial application, plus the regular renewal of backup anodes. Also, the important parts of cargo tanks, such as the deck head and the bottom or inner bottom, must often be coated to start with in contrast to typical practice in trading tankers. The external hull is usually protected by coatings and an induced current corrosion protection (ICCP) system, although the primary structure has been protected by anodes rather than coatings in some offshore structures.

This chapter presents current practices and recent advances on relevant corrosion mechanisms, corrosion wastage models, design considerations (design corrosion margin values), and preventive measures (coatings, ICCP, or anodes) with emphasis on ship-shaped offshore units. Although a great deal of knowledge on corrosion science and technology has been obtained during the last decade, it is also true that there are still a large number of problem areas that must be resolved specifically in a marine environment because corrosion is a very complex issue involving a variety of influential parameters.

More detailed and useful information about corrosion mechanisms on steel and other types of materials can be found in corrosion handbooks (Schumacher 1979; AISI 1981; Uhlig and Revie 1985; Battelle 1986; Fontana and Greene 1986; Korb 1989; Borenstein 1994; Craig and Anderson 1995; Roberge 1999; Revie 2000; and DECHEMA 2003). MacMillan et al. (2004) address key corrosion protection design and fabrication issues for new-build FPSOs and their corresponding impact on inspection, maintenance, and repair during operation. Paik et al. (2006) provide a recent review of state of the art regarding corrosion of trading ship structures. The effect of corrosion wastage on the ultimate strength of steel plates and FPSO hulls is discussed in Chapter 11 in this book.

10.2 Marine Corrosion Mechanisms

10.2.1 Fundamentals

Marine corrosion is the degradation of metals in the marine environment. Offshore engineers must have a good understanding of corrosion principles and the methods to mitigate or prevent corrosion effects so that inadequate designs and subsequent structural failures and high maintenance and renewal costs can be avoided. Initiation and progression of marine corrosion may be related to the following actions:

- Electrochemical actions
- Galvanic actions
- Intergranular actions
- Crevice actions
- Erosion actions
- Microbial actions

Electrochemical actions can cause corrosion in a metal (anode) when an electrolyte exists between the anode and a cathode, where ions are transferred and electrons flow from the anode to the cathode. It is the anode that preferentially corrodes. In so-called cathodic protection, the potential of the part to be protected is preferentially driven below the corrosion potential with reference to a noncorroding anode or emitter plate by an induced current system. This is often called an induced current corrosion protection (ICCP) system.

Galvanic corrosion occurs when two dissimilar metals, which are immersed in sea-water or other conductive liquid acting as an electrolyte, are connected directly or by a metallic path. *Intergranular corrosion* is a microscopic form of corrosion caused by a potential difference between grain bodies of the metal and grain boundaries. When the grain body is anodic to the grain boundaries, corrosion occurs along the boundaries, for example, along weld zones. *Crevice corrosion* may occur in a relatively confined space such as in slightly open joints (crevices), under nuts, bolt heads, and washers. *Erosive corrosion* may occur under high-velocity seawater flows in bends and elbows of pipes and at strike plates. Corrosion due to cavitation can also be erosive although the initiation mechanism involves more than the flow velocity alone.

Corrosion gives rise to rust. It may be quite hard and, therefore, protective or it may be friable and nonprotective. Friable rust is likely to occur on internal surfaces that do not have a protective coating. Under certain conditions, such as with abrasion or excessive metal flexing, the rust scale can break off, exposing fresh metal to corrosive attack. Rust can also dissolve in some types of liquids, including some petroleum products. In any event, the loss of metal thickness due to corrosion cannot always be judged visually, and ultrasonic inspection techniques may need to be used. Failure to remove mill scale from the steel surface during construction of the vessel has been known to cause accelerated corrosion in service. Severe corrosion can lead to significant steel loss and/or the possibility of leaks of oil or gas.

10.2.2 Types of Corrosion

There are several different types of corrosion wastage possible for mild- and low-alloy steels in marine applications, as follows (Figure 10.1):

- Uniform (general) corrosion
- Pitting
- Grooving
- Weld metal corrosion

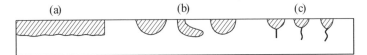

Figure 10.1. Typical types of corrosion wastage: (a) general corrosion; (b) localized corrosion (pitting); and (c) cracks initiated from localized corrosion.

10.2.2.1 General Corrosion

Corrosion wastage, which is formed almost uniformly on the surface as shown in Figure 10.1(a), is called "general corrosion" (or uniform corrosion). Figure 10.2 represents one often important mechanism of general corrosion on the back of the upper-deck steel plates of oil tankers and FPSOs, primarily when such a structure is uncoated or the coating has failed (SRAJ 2002). In cargo oil tanks, there are various chemical elements such as H_2O, O_2, CO_2, H_2, S, and SO_x, which originate from the crude oil together with the inert gas. Also, dew may occur on the back of the deck plate due to day to night temperature fluctuations. Solid sulphur (S) may be extracted onto the plate surface by catalysis of iron oxide (FeOOH).

The stratified corrosion product consisting of a layer of elemental S and iron rust is formed. The characteristics of corrosion progress may depend on the density of H_2S, among other factors. The reaction of S precipitation on FeOOH surface has no relationship to the reaction of steel corrosion, as shown in Figure 10.2. Although the flaking-off of the corrosion product occurs repeatedly from this layer of elemental S, chemical analysis appears to confirm that any acceleration of the corrosion rate in such cases may not be due to that phenomenon alone (SRAJ 2002).

The additional corrosion margin generally is meant to guard against this and other types of quasiuniform corrosion loss over relatively large but still local areas (to the extent of, say, a few plate panels between stiffeners and web frames). Uniform

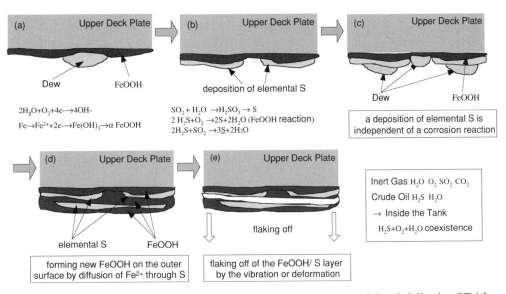

Figure 10.2. Mechanism of general corrosion in steel plates at a deck head, following SRAJ (2002).

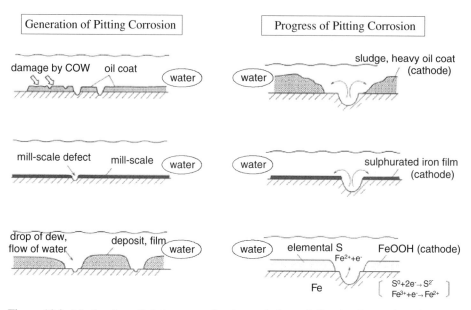

Figure 10.3. Mechanism of pitting corrosion in steel plates, following SRAJ (2002).

(general) corrosion is important for structural strength considerations. In assessing corroded strength in such cases, the thickness is considered to be equal to the original thickness minus a uniform reduction or corrosion wastage.

In principle, the allowable uniform corrosion reduces with increase in the amount of structural area affected. For example, in trading tanker structures, it is often suggested that a plate panel may be permitted by a 20 percent reduction of the thickness, although the entire deck may be permitted at best by a 10 percent reduction, and so on.

10.2.2.2 Pitting Corrosion

Pitting is a localized form of corrosion, as shown in Figure 10.1(b), that typically occurs on bottom plating, on other horizontal surfaces, and at structural locations that may trap water, for example, in the aft-most bays of tanks assuming that the vessel trim is by aft, for example, for reasons of drainage.

Figure 10.3 shows a mechanism of pitting corrosion on the tank top plate in cargo oil tanks (SRAJ 2002). Pitting initiates from a defect in the protective covering which may be an oil coat, mill-scale, or corrosion product. The initial damage of the protective covering may be caused by the flow of crude oil, local moisture, crude oil washing (COW), or seawater washing. A type of corrosion that resembles pits is generated in the defective areas of paint coating. It has been determined that pitting corrosion is more likely to occur with paint coating, but corrosion in cargo hold areas is more likely to be general corrosion when coating protection is not provided (Nakai et al. 2005).

Pit surface sizes, depth, and density (i.e., number of pits per unit of area) can vary. Pit shape through the depth can also vary, sometimes in a staircase fashion. As the pit density increases, say, beyond 30 percent, the situation may become more akin to uniform corrosion for strength assessment purposes. Pitting in low densities is primarily a leakage concern.

Pit repairs and plate renewals are usually the only viable options in such cases; the addition of specific pit-related corrosion margins up front in design being economically difficult to justify in most cases because the possibility of the occurrence of pitting, and its likely density, usually cannot be well predicted, the phenomenon involved is not well understood.

Pit growth rates can be many times that for general corrosion. This is usually thought to be associated with an acidic environment at the pit bottom, although in coated surfaces, an "area effect," such as a local coating breakdown, has also been cited as a reason for the rate magnification. Sulphate-reducing bacteria (SRB) have also been implicated in pitting in some cases.

10.2.2.3 Grooving

Grooving is another poorly understood and, therefore, generally difficult to predict phenomenon. It is usually manifested as localized line corrosion, which occurs at structural intersections where water or heavy moisture collects or flows. A groove is eventually formed, usually along welding-heat-affected zones. The effect may be exacerbated by structural flexibility and result in loss in scale. In coated surfaces, the coating must somehow fail in order for the corrosion and grooving to initiate.

This type of corrosion is sometimes referred to as "in-line pitting attack" and is often observed on longitudinal bulkheads in trading tankers. A specific margin is rarely added for pitting or grooving because such addition is not economically justifiable.

10.2.2.4 Weld Metal Corrosion

Weld metal corrosion is defined as preferential corrosion of the weld deposit. It is not a well-understood form of corrosion either. The most likely reason for this mode of corrosion is galvanic action with the base metal that may initially lead to pitting and, perhaps, grooving. This more often is determined to occur in hand welds as opposed to machine welds. Although a general corrosion margin may be applied to the base material, it is cautioned that an adequate amount of corrosion margin may also need to be present in the welds in order to minimize the effects of weld corrosion in general.

10.2.3 Factors Affecting Corrosion

Localized corrosion and general corrosion must be considered for corrosion management and control because localized corrosion can cause oil or gas leaks, and the generalized corrosion is more likely to lead to structural strength problems (see Section 2.22 in Chapter 2). Factors affecting marine corrosion in an enclosed or open space include the following:

- Types of structural material (e.g., steel, aluminum alloy)
- Corrosion protection scheme (e.g., coating, anodes, impressed current cathodic protection)
- Types of cargo or stored material (e.g., oil, seawater, wax content, oxygen content, salinity, reactivity)

- Dry–wet cycles related to loading/unloading of cargo or stored material
- Humidity
- Temperature
- Oxygen
- Water velocity

In ship-shaped offshore units, five types of cargo and ballast tank spaces may be considered for the areas of corrosion concern:

(1) Segregated ballast spaces
(2) Cargo only spaces
(3) Cargo and/or clean ballast spaces
(4) Cargo and/or dirty ballast spaces
(5) Cargo and/or storm ballast spaces

Note that in new builds today, whether in ship-shaped offshore units or trading tankers, the last three categories for tanks need not occur as long as today's IMO MARPOL regulations (see Appendix 6 in this book) are being met by design in operation. However, even today, the need for consideration of (past) corrosion in combined use spaces can arise when older tankers are being converted.

The frequency of filling ballast tanks in trading tankers is decided by economic factors and the characteristics of the trade route and weather conditions. In storage and offloading of cargo in ship-shaped offshore units, such as FPSOs, empty cargo holds may sometimes be partly filled with ballast water to adjust freeboard or trim, which increases the possibility of wet–dry cycles. The preferred situation, however, would be that such draft and trim control occur mainly by changes to dedicated ballast spaces, including peak spaces, not empty cargo holds.

The type of cargo affects corrosion rates. Some types of oil lead to higher corrosion rates. For example, a sour crude is more likely to cause corrosion than a sweet one, and cargo that is higher in oxygen content, such as gasoline, can lead to higher corrosion rates. However, the deposition of wax from high-wax-content crude might ameliorate corrosion effects.

In trading tankers and ship-shaped offshore units, the spaces prone to corrosion with some certainty are the ballast tanks, due to their repeated exposure to seawater, humidity, a saline atmosphere when empty, and increases in temperature when deck and sides are exposed to sunlight. Combined ballast and cargo tanks, where present, are usually relatively less prone to corrosion, even though they are exposed to water washing that can destroy the protective nature of oil films that tend to remain on surfaces, thereby increasing the probability of exposing unprotected steel. At the sides and the top of cargo-only tanks, the oil and a good inert gas blanket normally provide some protection, but the bottom area usually is exposed to the highly acidic water that separates from the oil. However, significant deck-head corrosion has also been known to occur, in some cases, by mechanisms discussed in Section 10.2 2.1.

Similar to trading tankers, corrosion rates in an offshore unit depend on the location and orientation of structural components and, of course, the type of corrosion-protection measures employed. In ballast tanks, which are normally coated, corrosion will start at locations of coating breakdown, in high-stress zones at structural details, and at free edges of cutouts. Significant corrosion of structural components in ballast tanks adjacent to heated cargo tanks or tanks with consumables is also possible. An increased degree of local structural flexibility has been recognized to increase corrosion rates as time progresses. This is apparently because of serial increases in scale loss and structural flexibility. Locations of necking and grooving also can be disproportionately affected.

It is important to note that the corrosion rates for ship-shaped offshore structures can be different from those of "at-sea" stationary immersion corrosion, such as in a steel pile, because of differences in circumstances. Temperature inside ballast or cargo tanks can be warmer than that of the sea. In storage and offloading of cargo, ballasting and deballasting are used to adjust freeboard or trim. Such ballast cycles may accelerate corrosion process because the steel surface becomes repeatedly dry or wet by seawater.

Where coatings are present, the progress of corrosion will normally depend on the degradation characteristics of such anticorrosion coatings. Adequate levels of appropriate maintenance to ensure the effectiveness of corrosion protection systems is strongly recommended.

By virtue of extensive investigations based on marine corrosion science principles, useful insights have been developed in recent years for corrosion loss, including pit depth, by Melchers (1997, 2002, 2003a, 2003b, 2004a). Melchers also studied the effects of environmental parameters, including the effects of water velocity, depth, dissolved oxygen, surface finish, and water pollution (Melchers 2003c, 2004b, 2005; Melchers and Jeffrey 2004, 2005).

It has been shown that temperature (including water temperature) has a significant effect on corrosion wastage (Melchers 2002). In-situ (field) observations of water-velocity effects extending over 2 years have also been made (Melchers 2004b). Corrosion of steel in real seawater occurs in a relatively harsh, complex biotic environment that does not occur during laboratory testing. Bacterial action plays an important role from the very beginning of exposure, and laboratory observations do not normally replicate this. Regarding temperature effects, other data indicate that a doubling of corrosion rate for every $10°C$ increase in temperature is possible.

In real seawaters, macrofouling and bacterial biofilms begin to form on the steel surface immediately on exposure, causing considerable localized changes in pH. Localized values as low as 2 pH have been recorded under the biofilms at the steel surface (Edyvean and Videla 1992). These changes also affect dissolved oxygen. Under these conditions, the properties of the bulk seawater may be of little interest. The reason why these very aggressive pH conditions do not lead to extremely high-corrosion losses is that the nature of the corrosion environment at the corroding surface can change as corrosion progresses. In contrast, the situation in the leading depth of a pit would be different.

Melchers (2003d) has suggested phenomenological models to represent marine corrosion in seawater considering various parameters of influence, as illustrated in Figure 10.4. Corrosion in the marine environment can be affected by pH; it can also be affected by the presence of sulphate-reducing bacteria (SRB). In surface corrosion affected by SRB, there is a short period of very high corrosion (i.e., phase 0 of the model), during which corrosion products (rust layers) begin to build up. This both influences and ameliorates the effect of pH.

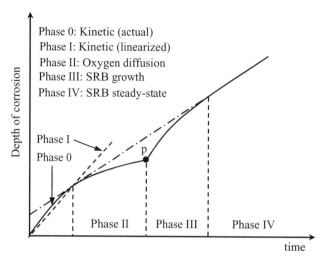

Figure 10.4. A phenomenological representation of corrosion progress by the Melchers model (SRB = sulfate-reducing bacteria).

In the next phase, nutrient and oxygen transport to any corrosion-affecting bacteria present becomes more difficult, and within days to weeks (depending largely on average water temperature), the oxygen diffusion from the bulk water controls the rate of corrosion in phase I of the Melchers model. The pH in the normal ranges encountered "at-sea" is then no longer relevant in this phase.

The important role of realistic bulk water variations in pH lies in its (long-term) effect on the carbonate solubility of the seawater, particularly in brackish situations. Eventually, such effects are involved in controlling the rate of diffusion of oxygen to the corroding surface and, hence, the instantaneous corrosion rate. This is part of the phase II of the Melchers model. The importance of hardness and pH in the case of fresh water has been known for a time, but the situation for brackish waters has only recently been clarified following an extensive review of experimental data (Melchers 2005). In any case, the typical variations in pH for actual brackish seawaters are small (±1), meaning that pH may have a minor influence on immersion corrosion in this phase overall.

The possible effect of steel composition on corrosion was also noted by Melchers. The effect may be negligible for short-term corrosion with low-alloy carbon steels typically used for ship-shaped offshore structures as well as trading tankers, but where somewhat higher levels of alloying are involved, the longer term corrosion effects can be significant (Melchers 2003e, 2004c).

10.3 Mathematical Models for Corrosion Wastage Prediction

The corrosion mechanism, including initiation and progression in ship-shaped off-shore structures, at first, may be similar to that in trading tanker structures as long as the corrosion environment is similar. The Tanker Structure Co-operative Forum (TSCF) has devoted significant effort to understanding and control of corrosion, both in ballast tanks and combined-use tanks (TSCF 1992, 1997). Guidelines for inspection of ship hulls and recommendations for repair of corroded areas have been developed

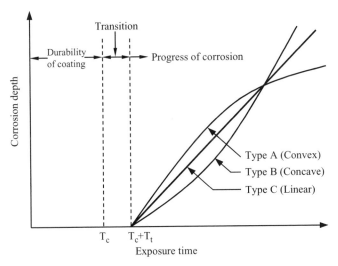

Figure 10.5. A mechanical representation of corrosion progress by the Paik corrosion model.

by TSCF (1995). Classification societies have in turn developed nominal design corrosion values (NDCV) for the different parts of a ship structure and for various vessel types (e.g., ABS 2000). Guidelines for coating systems for corrosion prevention of ship structures have also been developed (e.g., IMO 1995; IACS 1998; DNV 1998, 1999).

At a more theoretical level, several aspects for condition assessment of corroded ships and ship-shaped offshore units have been considered by various other investigators. For example, Gardiner and Melchers (2001a, 2001b, 2002a, 2002b, 2003) studied corrosion processes in enclosed coal and iron ore cargo spaces. Gudze et al. (2001, 2004) studied corrosion in naval vessel ballast tanks and the influence of environmental parameters. Most recently, a series of the time-variant corrosion wastage models for oil tankers, bulk carriers, and ship-shaped offshore units were developed by Paik and his colleagues (Paik et al. 1998, 2003a, 2003b, 2004; Paik 2004).

10.3.1 Overall Behavior of Corrosion

Generally, corrosion wastage increases with time and it is the rate of progression that is of major interest. However, in coated surfaces, corrosion does not commence until there has been a breakdown in protective coating; similar logic would also extend to other types of corrosion protection such as the relatively rare extensive use of anodes as the primary means of protection. These features can be represented schematically as shown in Figure 10.5. The figure incorporates features related to (a) the durability or life of the coating, (b) a transition period, and (c) the progression of corrosion (Paik and Thayamballi 2003). This discussion generally assumes that the primary corrosion protection is by coatings; the anode, or ICCP system, is secondary.

Coating breakdown is estimated, usually visually, by the percentage of surface that has blistered, cracked, finely corroded, or more coarsely corroded. Although various factors may be involved in deterioration of coating, including electrochemical (current, voltage) and/or mechanical (strain) reactions (Matsuoka et al. 1985; Martin

et al. 1996), it is widely observed in practice that coating breakdown occurs gradu-
ally and, subsequently, corrosion increases progressively over the surface as a result
(Melchers and Jiang 2006). Some associated features are described in more detail in
Section 10.5.

Nevertheless, it is common practice to define a *coating life* by measuring the time
until a predefined degree of corrosion has occurred on the average over the surface,
within a few percentage points. The coating life, so defined, depends on the type of
coating systems used, details of its application (e.g., surface preparation, stripe coats,
film thickness, humidity, salt control during application), and relevant maintenance,
among other factors. The coating life has been assumed to follow either the normal or log-normal probability distribution toward a predefined state of breakdown
(Yamamoto and Ikegami 1998).

After the nominal effectiveness (i.e., "life") of a coating is lost locally, some transition time may be considered to exist before the corrosion "initiates" over a large
enough and measurable area. This transition time is often considered an exponentially distributed random variable. As an example, the mean value of the transition
time for transverse bulkhead structures of oceangoing bulk carriers was said to be
3 years for deep-tank bulkheads, 2 years for watertight bulkheads, and 1.5 years for
stool regions (Yamamoto and Ikegami 1998). When the transition time is assumed to
be zero, it implies that corrosion will start immediately after the coating effectiveness
has been lost locally.

In Figure 10.5, three possibilities (types A–C) are indicated for the corrosion loss
curve. The convex curve of type A shows the corrosion rate (i.e., the curve gradient)
decreasing as corrosion progresses. This type of behavior is common in many environments, particularly for stationary immersion, and is brought about by the gradual
build-up of protective rust layers. It is said to have been observed, for example, in
the upper parts of relatively rigid cargo holds.

The concave curve of type B is representative of accelerating corrosion with time.
This is characteristic of situations where there is increasing structural surface strain
from flexure of dynamically loaded structures together with significant thinning of
structural components. This is often said to be seen in the very advanced stages of
corrosion with accelerated degradation in terms of the "domino effect" (Herring and
Titcomb 1981; Ohyagi 1987; Contraros 2003).

The linear curve of type C may be characteristic of situations where the rust layers
are continually removed but dynamic flexing is not involved. This is said to be typical
of the lower parts of cargo holds of vessels used for aggressive cargoes, but with
relatively rigid structure. The same linear modeling, that is, curve C, is often also used
as a first (and sometimes only) approximation to the other two types of corrosion
behavior (i.e., curves A and B) in the absence of better information.

10.3.2 Mechanical Models

10.3.2.1 Corrosion Depth Formulations

The plate thickness loss due to corrosion may be expressed as a function of the time
(year) after the commencement of corrosion,

$$t_r = C_1 T_e^{C_2}, \tag{10.1}$$

where t_r = corrosion depth in mm; T_e = exposure time in years, after breakdown of the coating, taken as $T_e = T - T_c - T_t$, with T = exposure time in years; T_c = life of coating in years; T_t = duration of transition in years; and C_1, C_2 = coefficients to be determined by statistical analysis of plate thickness measurements.

The coefficient C_1 is the initial corrosion rate, obtained from Eq. (10.1) at $T = 0$. It also allows for the derivation of a nominal or "annualized" corrosion rate r_r, which is obtained by differentiating Eq. (10.1) with respect to T as follows:

$$r_r = C_1 C_2 T_e^{C_2-1}. \tag{10.2}$$

It is important to note that much effort is needed in specific cases to obtain the coefficients C_1 and C_2 from actual corrosion loss data for the appropriate environments, structure, and other important conditions. In principle, they can be simultaneously determined from carefully collected statistical corrosion data, although this may not be always practical. Another possibility is to determine the coefficient C_1 for a constant and preselected value of C_2. This is mathematically convenient, but it does not negate any of the limitations arising due to usual methods of data collection in surveys including the (usual) lack of continuous monitoring at the same spots over time.

When $C_2 < 1$, it means that the (instantaneous or tangent) corrosion rate decreases or stabilizes over time (type A in Figure 10.5). On the other hand, for dynamically loaded structures, it is possible for corrosion to be accelerated because of structural surface straining or flexing effects and because of the possibility of fresh steel being exposed to corrosion; such behavior will have a value of $C_2 > 1$ (type B in Figure 10.5). The trends for corrosion progression thus vary with C_2 as would be expected. For practical design purposes, it has often been assumed that $C_2 = 1$ (type C in Figure 10.5), implying that the corrosion wastage is linearized over a conveniently small timeframe.

Once the coefficient C_2 is selected and set to a constant value over any convenient time period, the next step is to determine the coefficient C_1, which corresponds to the annualized corrosion rate over that period, and its statistics (mean, variance, and type of probability density distribution). It is evident that the characteristics of the coefficient C_1 must be evaluated in a probabilistic form because of the uncertainties and related scatter in the data involved. From Eq. (10.1), the coefficient C_1 can be given for a sampling point, when the transition time is taken as $T_t = 0$, as follows:

$$C_1 = \frac{t_r}{(T - T_c)^{C_2}}, \tag{10.3}$$

where t_r = depth of corrosion loss as obtained by measurements.

10.3.2.2 Data Collection of Corrosion Measurements

Given a set of available statistical corrosion data, the coefficient C_1 can be calculated from Eq. (10.3), and the relative frequency of the coefficient C_1 can be evaluated. It is apparent from Eq. (10.3) that the coating life may significantly affect the coefficient C_1 and, therefore, the annualized corrosion rate, which, obviously, is also quite uncertain in nature.

Because the statistical corrosion data is usually very scattered due to many uncontrolled and sometimes unknown factors, it may also be of interest to investigate the

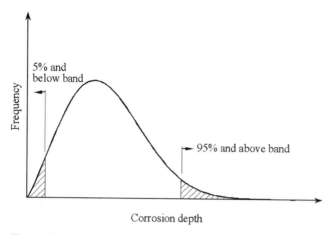

Figure 10.6. A schematic of corrosion depth distribution at a certain exposure time.

upper-bound probabilistic characteristics of the coefficient C_1, which represents the 95 percentile and the *above band* as illustrated in Figure 10.6. The characteristics determined by all of the corrosion data in a particular case will, of course, represent an average corrosion rate, while upper-bound data will lead to a more severe corrosion model.

Even today, the worldwide "fleet" of aged ship-shaped offshore units is relatively small in number, although the number of trading tankers is quite large. The temperature, humidity, structural flexibility, and wet–dry cycle effects in corrosion characteristics of ship-shaped offshore units may be different from those of trading tankers; for example, loading and unloading of ship-shaped offshore units are a lot more frequent, sometimes every week, and this may accelerate the corrosion progress. They may also be placed in long-term environments of high temperature and humidity, although in certain geographical areas, such as the polar regions, the reverse can also be true regarding temperatures, at least.

Also, ship-shaped offshore units typically can be at a specific site with zero forward speed, and this aspect may likely mitigate the dynamic flexing in some areas (e.g., under benign environments) such as the bottom-slamming region, keeping the corrosive scale therein relatively static, compared to that of oceangoing tankers. In any event, for various reasons, and depending on the specifics of the offshore unit, there can be differences in corrosion experience expected in a ship-shaped offshore unit versus the corrosion experience expected in a trading tanker.

The required onsite corrosion margin in ship-shaped offshore units is often based on data obtained for trading tankers, with some adjustments for significant differences to be made largely by judgment. This applies to both new builds and conversions. The implicit assumption, in such an approach, is that the corrosion models available for trading tankers may be applicable to corrosion prediction of ship-shaped offshore units as long as the corrosion environment is similar. The very few FSO/FPSO specific corrosion studies thus far available in the literature appear to support such an approach but additional studies, including the collection and analyses of related data, are certainly needed.

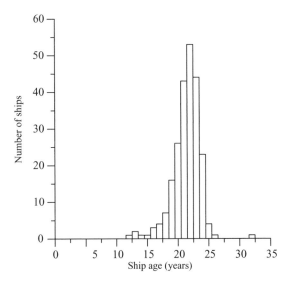

Figure 10.7. Age distribution for more than 200 trading tankers surveyed for corrosion in the study of Paik et al. (2003a).

Measured data for the corrosion loss in structural components for a total of 230 trading tankers carrying crude oil or products have been collected and analyzed in a study conducted with participation of Pusan National University and others (Paik et al. 2003a). Figure 10.7 shows the distribution of the ship age in that study. The average age of vessels involved appears to be more than 20 years, meaning that some vessels were built before IMO MARPOL segregated ballast-related changes, so that others may follow post-MARPOL expected tank-usage patterns. We have included some useful information from the study in this section. The discussion herein also serves to illustrate some of the real-world difficulties encountered in such a study.

All of the trading tankers surveyed by Paik et al. (2003a) have single-hull structures. The trading routes of the vessels expand worldwide, including Korean waters. Although the corrosion wastage data for some of the vessel structures might have been affected by tank heating, no specific differentiation was found in the study.

Effects of renewal of steel in corroded areas over time are also implicit in the data, although normally the related amount of steel renewals can be expected to be small when compared to the overall hull steel weight. The remaining thickness or corrosion loss was mostly measured by the technique of ultrasonic thickness measurements. This usually implies that the measurements were made at several points within a single plating, and a representative value (e.g., average) of the measured corrosion loss is then determined to be the depth of corrosion. It is possible that the corrosion data and model presented herein may conceivably include some localized pit corrosion data as well, in some cases; for example, in the case of cargo tank bottom structures.

A total of 33,820 measurements for thirty-four different member groups that include fourteen categories of plate parts, eleven categories of stiffener webs, and nine categories of stiffener flanges were obtained, as indicated in Table 10.1 and Figure 10.8. The member groups were defined by locations and categories of structural components. As shown in Figure 10.8, the member groups also represent differing corrosion propensities and environment. For instance, A/B-H indicates a "horizontal" (H) member group located in between "air" (A) and "ballast water" (B), and

Table 10.1. *Number of gathered data from corrosion loss measurements for thirty-four primary structural member groups in trading tankers*

No.	Member type	Example	Gathered data number
1	B/S-H	Bottom shell plating (segregated ballast tank)	148
2	A/B-H	Deck plating (segregated ballast tank)	1,410
3	A/B-V	Side shell plating above draft line (segregated ballast tank)	33
4	B/S-V	Side shell plating below draft line (segregated ballast tank)	274
5	BLGB	Bilge plating (segregated ballast tank)	164
6	O/B-V	Longitudinal bulkhead plating (segregated ballast tank)	361
7	B/B-H	Stringer plating (segregated ballast tank)	19
8	O/S-H	Bottom shell plating (cargo oil tank)	849
9	A/O-H	Deck plating (cargo oil tank)	5,557
10	A/O-V	Side shell plating above draft line (cargo oil tank)	86
11	O/S-V	Side shell plating below draft line (cargo oil tank)	692
12	BLGC	Bilge plating (cargo oil tank)	348
13	O/O-V	Longitudinal bulkhead plating (cargo oil tank)	1,082
14	O/O-H	Stringer plating (cargo oil tank)	42
15	BSLB(W)	Bottom shell longitudinals in ballast tank – Web	672
16	BSLB(F)	Bottom shell longitudinals in ballast tank – Flange	678
17	DLB(W)	Deck longitudinals in ballast tank – Web	975
18	SSLB(W)	Side shell longitudinals in ballast tank – Web	913
19	SSLB(F)	Side shell longitudinals in ballast tank – Flange	913
20	LBLB(W)	Longitudinal bulkhead longitudinals in ballast tank – Web	1,024
21	LBLB(F)	Longitudinal bulkhead longitudinals in ballast tank – Flange	973
22	BSLC(W)	Bottom shell longitudinals in cargo oil tank – Web	2,030
23	BSLC(F)	Bottom shell longitudinals in cargo oil tank – Flange	2,205
24	DLC(W)	Deck longitudinals in cargo oil tank – Web	2,215
25	DLC(F)	Deck longitudinals in cargo oil tank – Flange	34
26	SSLC(W)	Side shell longitudinals in cargo oil tank – Web	2,187
27	SSLC(F)	Side shell longitudinals in cargo oil tank – Flange	2,091
28	LBLC(W)	Longitudinal bulkhead longitudinals in cargo oil tank – Web	2,850
29	LBLC(F)	Longitudinal bulkhead longitudinals in cargo oil tank – Flange	2,634
30	BGLC(W)	Bottom girder longitudinals in cargo oil tank – Web	154
31	BGLC(F)	Bottom girder longitudinals in cargo oil tank – Flange	42
32	DGLC(W)	Deck girder longitudinals in cargo oil tank – Web	94
33	DGLC(F)	Deck girder longitudinals in cargo oil tank – Flange	36
34	SSTLC(W)	Side stringer longitudinals in cargo oil tank – Web	35
		Total	33,820

B/S-V represents a "vertical" (V) member group located in between "ballast water" (B) and "seawater" (S). An example of A/B-H is deck plating on ballast tanks and an example of B/S-V is side shell plating below draft line. For further description, see Table 10.1. These member groups can then potentially be applied to comparable corrosion areas and situations in a ship-shaped offshore unit, as illustrated in Figure 10.9.

In some member groups, there is very little data behind the numbers, as indicated in Table 10.1. Although the corrosion models developed for these member groups may still be illustrative, it is necessary to refine the related estimates in the future as more related corrosion measurement data becomes available.

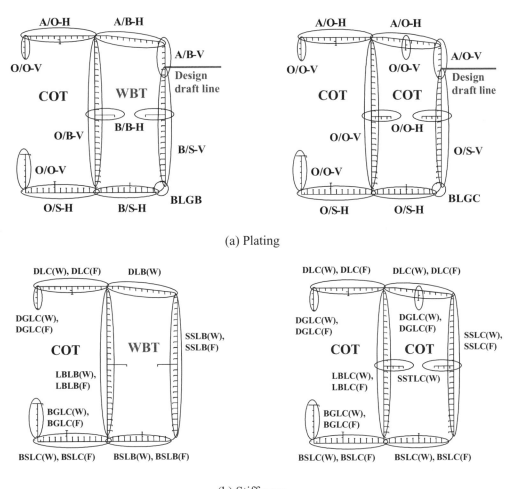

(a) Plating

(b) Stiffeners

Figure 10.8. The thirty-four member groups defined by location, category, and corrosion environment of trading tankers (A = air; B = ballast water; O = Oil; S = seawater; H = horizontal member; V = vertical member; W = web; and F = flange).

10.3.2.3 Characteristics of Observed Corrosion Wastage

Figures 10.10(a)–(c) show the frequency distribution of corrosion depth (thickness loss) for three example member (location/category) groups as a function of the ship age. It is seen from the figures that the data for corrosion wastage is very scattered. Although the sources of such uncertainty involved are various as mentioned in the previous sections, the coating life is also a factor in such uncertainties. Note that some of the data used may pertain to uncoated spaces, especially in the cargo tanks; however, specific information about coating, or the lack of it in such cases, was not available.

In Eq. (10.1), the coefficients C_1 and C_2 are closely correlated. Figure 10.11 shows a sample best-fit formulation of Eq. (10.1) using the least-squares method, varying C_2. Five cases were considered, varying the value of C_2 in the range of 0.5 to 1.5.

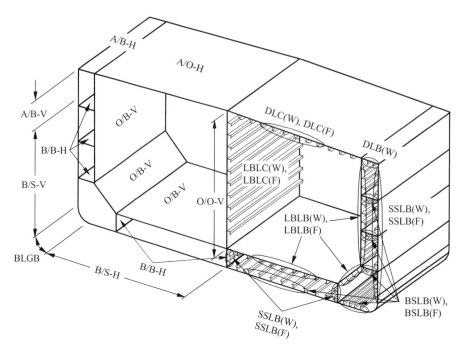

Figure 10.9. Application of the various corrosion member groups to a ship-shaped offshore unit based on possible similarity of corrosion environment to that of a trading tanker, assuming a double-skinned-hull structure.

When C_2 is less than 1.0 (but larger than 0.0), the corrosion rates apparently decrease or stabilize over time, showing the convex curve (type A) in Figure 10.5 or the curves denoted by 1 and 2 in Figure 10.11. As previously noted, this corrosion behavior may typically be plausible for statically loaded structures. However, for dynamically loaded structures such as ship-shaped offs structures subjected to wave loading in which corrosion scale is continually being lost and new material is being exposed to corrosion environment because of structural flexing, such values of C_2 may not always be appropriate. The situation may be better represented by the concave curve (type B) in Figure 10.5 or the curves denoted by 4 and 5 in Figure 10.11 with C_2 having a value greater than 1.0.

Clearly, the coefficient C_2 affects the implied trend of the corrosion progress. Considering the scatter of corrosion progress characteristics and also for the purpose of practical design, however, $C_2 = 1$ can be used, with the corrosion behavior perhaps linearized over convenient and small enough extents of time as depicted by the linear curve (type C in Figure 10.5) or the curve denoted by 3 in Figure 10.11. Considering these discussions, Eq. (10.1) is now simplified to the following:

$$t_r = C_1 (T - T_c). \tag{10.4}$$

The only coefficient now left to be determined is C_1. T_c is treated as a constant parameter. The mean and coefficient of variation (COV) of C_1 are then determined by the statistical analysis of corrosion measurement data. In corrosion loss measurements, information on the coating life is normally unclear. In fact, a 5-year coating life is considered to represent an undesirable situation, and a coating life of 10 years or

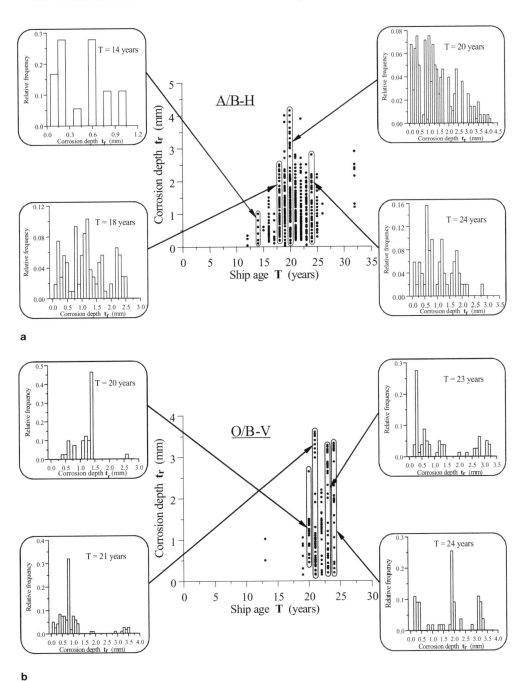

Figure 10.10. (a) The corrosion depth versus the vessel age for the measurements of deck plating in ballast tanks. (b) The corrosion depth versus the vessel age for the thickness measurements of the O/B-V group.

longer will be representative of a relatively desirable state of affairs. In this regard, for any given set of corrosion data with several unknown or uncontrolled factors, a parametric approach may be used by varying the coating breakdown time to, say, 5, 7.5, and 10 years.

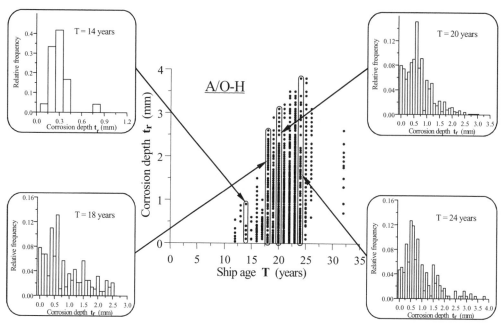

Figure 10.10. (*cont.*) (c) The corrosion depth versus the vessel age for the thickness measurements of the A/O-H group.

10.3.2.4 Annualized Corrosion Rates

With t_r and T_c known (t_r is from the corrosion measurement data, and T_c is assumed), C_1 can be readily given from Eq. (10.3) for a sampling point. For a given set of available statistical corrosion data, therefore, the statistical characteristics of the coefficients C_1 can be analyzed. The statistical distribution of the coefficient C_1

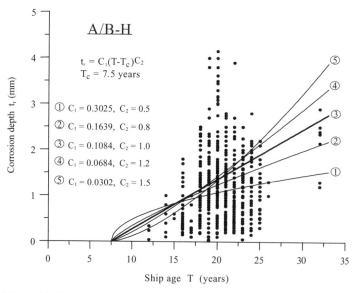

Figure 10.11. Sample best fit formulations of the corrosion depth measurements for the A/B-H group as a function of vessel age, varying the coefficient C_2.

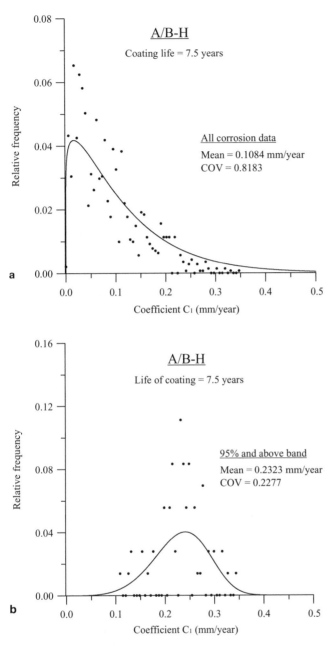

Figure 10.12. (a) The best fit Weibull distribution for mean corrosion coefficient C_1 for the A/B-H group assuming a coating life of 7.5 years. (b) The best fit Weibull distribution for severe corrosion coefficient C_1 for the A/B-H group assuming a coating life of 7.5 years.

corresponding to the most probable (average) trend follows a Weibull function as shown in Figure 10.12(a), but that at the upper-bound trend more likely follows the normal function, as shown in Figure 10.12(b).

Tables 10.2(a) and 10.2(b) summarize the computed results for the mean value and COV of the coefficient C_1 for the thirty-four different member location/category groups. Using the results of these tables, the corrosion behavior (thickness loss,

Table 10.2(a). *Mean and COV of the annualized corrosion rate (coefficient* C_1*) for plating*

ID No.	Member group	Coating life (years)	Average corrosion Mean (mm/year)	COV	Severe corrosion Mean (mm/year)	COV
1	B/S-H	5	0.0518	0.8439	0.1483	0.2387
		7.5	0.0597	0.9901	0.1717	0.2290
		10	0.0704	0.9894	0.2159	0.1974
2	A/B-H	5	0.0824	0.9039	0.1908	0.2498
		7.5	0.1084	0.8183	0.2323	0.2277
		10	0.1208	0.8922	0.3012	0.1942
3	A/B-V	5	0.0552	1.1258	0.1582	0.3227
		7.5	0.0661	1.1341	0.1897	0.3227
		10	0.0762	1.1147	0.2436	0.3207
4	B/S-V	5	0.0545	1.0033	0.1566	0.2387
		7.5	0.0622	1.0030	0.1823	0.2185
		10	0.0731	1.0020	0.2382	0.1942
5	BLGB	5	0.0539	0.9134	0.1525	0.3008
		7.5	0.0619	0.8821	0.1805	0.2167
		10	0.0728	0.8559	0.2371	0.2387
6	O/B-V	5	0.0792	0.8162	0.1616	0.2498
		7.5	0.1012	0.7994	0.1919	0.2277
		10	0.1184	0.8369	0.2483	0.1866
7	B/B-H	5	0.1111	0.2290	0.2206	0.0000
		7.5	0.1408	0.2704	0.2586	0.0000
		10	0.1790	0.2708	0.3125	0.0000
8	O/S-H	5	0.0526	0.8439	0.1503	0.2601
		7.5	0.0607	0.8248	0.1777	0.2167
		10	0.0709	0.7793	0.2217	0.2080
9	A/O-H	5	0.0489	0.8430	0.1434	0.2495
		7.5	0.0581	0.8262	0.1689	0.2290
		10	0.0682	0.8240	0.2113	0.1942
10	A/O-V	5	0.0444	1.0023	0.1339	0.2601
		7.5	0.0523	1.0111	0.1529	0.2167
		10	0.0633	0.9993	0.1928	0.1942
11	O/S-V	5	0.0346	0.9134	0.1318	0.2387
		7.5	0.0423	0.7601	0.1497	0.2290
		10	0.0532	0.7563	0.1841	0.1827
12	BLGC	5	0.0340	1.0010	0.1290	0.2704
		7.5	0.0414	1.0033	0.1446	0.2167
		10	0.0513	0.9993	0.1776	0.1827
13	O/O-V	5	0.0475	0.8108	0.1406	0.2498
		7.5	0.0577	0.8162	0.1621	0.2185
		10	0.0671	0.8170	0.2014	0.2055
14	O/O-H	5	0.0330	1.1979	0.1251	0.2495
		7.5	0.0405	1.1341	0.1423	0.2277
		10	0.0509	1.1258	0.1727	0.2080

annualized corrosion rate) of exposure areas over time can be predicted from Eq. (10.4) when ship age (T) and coating life (T_c) are known. However, we caution that the coating life T_c used for derivation of the present corrosion model is not real but simply assumed.

Table 10.2(b). *Mean and COV of the annualized corrosion rate (coefficient C_1) for longitudinal stiffeners*

ID No.	Member group	Coating life (years)	Average corrosion Mean (mm/year)	COV	Severe corrosion Mean (mm/year)	COV
15	BSLB(W)	5	0.1184	0.8922	0.2126	0.2495
		7.5	0.1367	0.7802	0.2461	0.2290
		10	0.1613	0.9325	0.3052	0.1942
16	BSLB(F)	5	0.0976	1.1147	0.2024	0.2704
		7.5	0.1127	1.0121	0.2343	0.1827
		10	0.1330	1.1433	0.2905	0.1942
17	DLB(W)	5	0.2081	1.0020	0.3667	0.2498
		7.5	0.2403	0.9165	0.4244	0.1942
		10	0.2836	1.0139	0.5263	0.1974
18	SSLB(W)	5	0.1224	0.8559	0.2242	0.2601
		7.5	0.1413	1.0097	0.2595	0.1942
		10	0.1667	0.9153	0.3218	0.2387
19	SSLB(F)	5	0.0764	0.9134	0.1408	0.2495
		7.5	0.0882	0.8966	0.1630	0.2167
		10	0.1041	1.0283	0.2021	0.1866
20	LBLB(W)	5	0.1697	0.7793	0.3318	0.2387
		7.5	0.1960	0.9993	0.3840	0.1827
		10	0.2313	0.7955	0.4762	0.1866
21	LBLB(F)	5	0.1543	0.9894	0.2985	0.2498
		7.5	0.1782	0.9941	0.3455	0.2055
		10	0.2103	1.0394	0.4284	0.1974
22	BSLC(W)	5	0.0404	0.8240	0.0767	0.3227
		7.5	0.0466	1.1156	0.0888	0.2387
		10	0.0550	0.9062	0.1101	0.1942
23	BSLC(F)	5	0.0378	0.9993	0.0723	0.2387
		7.5	0.0437	1.1341	0.0837	0.1866
		10	0.0516	1.0238	0.1038	0.1942
24	DLC(W)	5	0.0620	0.7563	0.1082	0.3008
		7.5	0.0716	0.8902	0.1252	0.2167
		10	0.0845	0.8263	0.1552	0.1827
25	DLC(F)	5	0.0509	0.9993	0.0916	0.2601
		7.5	0.0588	1.0032	0.1060	0.1866
		10	0.0694	1.0211	0.1314	0.2055
26	SSLC(W)	5	0.0364	1.0258	0.0700	0.2387
		7.5	0.0420	1.0517	0.0810	0.2185
		10	0.0496	1.1224	0.1004	0.2080
27	SSLC(F)	5	0.0344	1.0507	0.0683	0.3008
		7.5	0.0397	0.8551	0.0790	0.1866
		10	0.0468	1.1350	0.0980	0.2055
28	LBLC(W)	5	0.0476	0.9003	0.0814	0.2498
		7.5	0.0550	0.8129	0.0942	0.1758
		10	0.0649	0.9859	0.1168	0.2080
29	LBLC(F)	5	0.0440	1.1341	0.0796	0.2601
		7.5	0.0508	1.0012	0.0921	0.2167
		10	0.0599	1.1944	0.1142	0.1942
30	BGLC(W)	5	0.0326	1.0030	0.0617	0.2495
		7.5	0.0377	0.9824	0.0714	0.2395
		10	0.0445	1.1079	0.0885	0.1866
31	BGLC(F)	5	0.0276	0.8821	0.0499	0.2387
		7.5	0.0319	0.8439	0.0578	0.2290
		10	0.0376	0.9039	0.0717	0.2055
32	DGLC(W)	5	0.0413	0.9432	0.0778	0.3008
		7.5	0.0477	1.0818	0.0900	0.2277
		10	0.0563	1.0071	0.1116	0.2080
33	DGLC(F)	5	0.0389	0.8248	0.0745	0.2601
		7.5	0.0449	0.9533	0.0862	0.1974
		10	0.0530	0.8972	0.1069	0.1942
34	SSTLC(W)	5	0.0226	1.0111	0.0378	0.2495
		7.5	0.0261	1.0926	0.0437	0.1827
		10	0.0308	1.1255	0.0542	0.1974

377

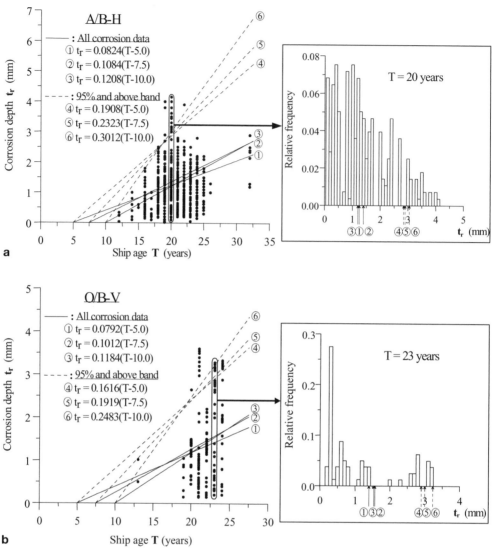

Figure 10.13. (a) Comparison of the time-dependent corrosion wastage models together with the measured corrosion data for the A/B-H member group. (b) Comparison of the time-dependent corrosion wastage models together with the measured corrosion data for the O/B-V member group.

Figure 10.13 shows some selected plots of the corrosion modeling so obtained together with the underlying corrosion wastage measurements themselves. Such corrosion modeling can be used to represent the relevant average or severe corrosion behavior of various corrosion areas. A comparison of the present corrosion rates with the TSCF corrosion rates (TSCF 1997) is made in Table 10.3. Both models (i.e., the present ones and those from TSCF) may correlate well, and a benefit of the present model is to provide corrosion rates for a larger variety of member groups. Figure 10.14 is a related summary of average and severe corrosion rates of various areas in ship-shaped offshore structures when the coating life is assumed to be 7.5 years.

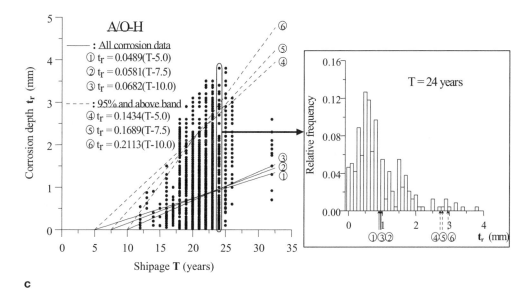

Figure 10.13. (*cont.*) (c) Comparison of the time-dependent corrosion wastage models together with the measured corrosion data for the A/O-H member group.

10.3.3 Phenomenological Models

The empirical models described in the previous section are essentially calibrated to measured data for different parts of marine structures. It is clear that such empirical models are limited by the data used. For large aggregations of information from many different structures operating under very different conditions, the models may, in fact, give little actual insight into what really happens and, furthermore, what might happen when conditions change. This lack of predictive power is a major problem for the empirical models.

In practice, the optimal management of individual marine structure assets is of primary interest, not that of undefined large aggregations of unknown vessels operating under ill-defined conditions. It is possible, in principle, to "drill-down" into the database to uncover more specific data, but in so doing the number of supporting data and, hence, the confidence in the information will reduce. This is the inevitable outcome of a reliance only on recorded data without discrimination of the specific environmental (and other) and influences on corrosion.

All of these factors, and others, are important for the practical description of the progression of corrosion with time. However, in common with other areas of endeavor, there can be no progress in reducing the very large uncertainties associated with the empirical models unless a better level of understanding of the issues involved is reached.

Early phenomenological corrosion models based on ion transport limitations are described by Evans (1966) and Tomashev (1966) and noted by Melchers (2003d). Much more recent fundamental corrosion research has shown that metal ion transportation is very unlikely to be a corrosion rate-controlling process. A model that corrosion was limited by oxygen diffusion to the corroding surface was proposed by

Table 10.3. *A comparison of the corrosion-rate modeling for trading tanker structures by Paik et al. (2003a) and TSCF (1997)*

ID No.	Member type	Average corrosion rate by the Paik model (mm/year)	Severe corrosion rate by the Paik model (mm/year)	Corrosion rate by TSCF (mm/year)
1	B/S-H	0.0597	0.1717	0.04–0.10
2	A/B-H	0.1084	0.2323	0.10–0.50
3	A/B-V	0.0661	0.1897	0.06–0.10
4	B/S-V	0.0622	0.1823	0.06–0.10
5	BLGB	0.0619	0.1805	–
6	O/B-V	0.1012	0.1919	0.10–0.30
7	B/B-H	0.1408	0.2586	–
8	O/S-H	0.0607	0.1777	0.04–0.10
9	A/O-H	0.0581	0.1689	0.03–0.10
10	A/O-V	0.0523	0.1529	0.03
11	O/S-V	0.0423	0.1497	0.03
12	BLGC	0.0414	0.1446	–
13	O/O-V	0.0577	0.1621	0.03
14	O/O-H	0.0405	0.1423	–
15	BSLB(W)	0.1367	0.2461	–
16	BSLB(F)	0.1127	0.2343	–
17	DLB(W)	0.2403	0.4244	0.25–1.00
18	SSLB(W)	0.1413	0.2595	0.10–0.25
19	SSLB(F)	0.0882	0.1630	–
20	LBLB(W)	0.1960	0.3840	0.20–1.20
21	LBLB(F)	0.1782	0.3455	0.20–0.60
22	BSLC(W)	0.0466	0.0888	0.03
23	BSLC(F)	0.0437	0.0837	–
24	DLC(W)	0.0716	0.1252	0.03–0.10
25	DLC(F)	0.0588	0.1060	–
26	SSLC(W)	0.0420	0.0810	0.03
27	SSLC(F)	0.0397	0.0790	–
28	LBLC(W)	0.0550	0.0942	0.03
29	LBLC(F)	0.0508	0.0921	–
30	BGLC(W)	0.0377	0.0714	–
31	BGLC(F)	0.0319	0.0578	–
32	DGLC(W)	0.0477	0.0900	–
33	DGLC(F)	0.0449	0.0862	–
34	SSTLC(W)	0.0261	0.0437	–

Chernov (1990) and Chernov and Ponomarenko (1991). It included semi-empirical factors to correct for seawater temperature, velocity, and salinity effects. These factors had not been considered earlier, although Reinhart and Jenkins (1972) had proposed an entirely empirical corrosion model that incorporated a linear functional relationship for both water temperature and seawater oxygen content. For corrosion in tropical waters, Southwell and Alexander (1970) observed that for longer term immersion of steel, corrosion was strongly influenced by SRB and proposed a simple bilinear (but, again, entirely empirical) model.

Building on these various efforts and attempting to achieve better consistency among the various observations, Melchers (1997, 2003d) proposed a more refined

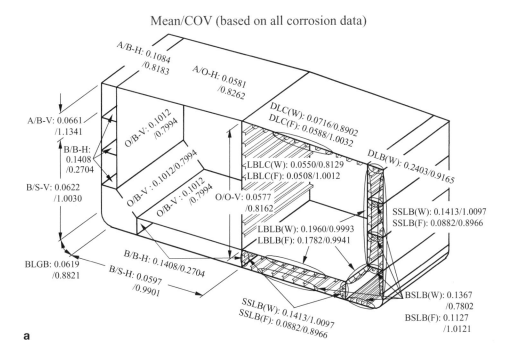

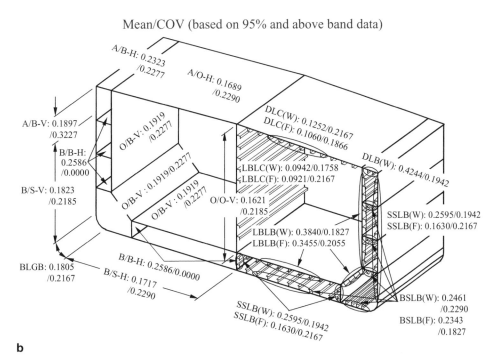

Figure 10.14. (a) Mean and COV of the average (most probable) corrosion rate (coefficient C_1) for the thirty-four member location/category groups assuming a coating life of 7.5 years, following Paik et al. (2003a). (b) Mean and COV of the severe (upper bound) corrosion rate (coefficient C_1) for the thirty-four member location/category groups assuming a coating life of 7.5 years, following Paik et al. (2003a).

physio-chemical model incorporating aspects of earlier models. Its main innovations include the following:

- Oxygen diffusion through the rust layers cannot be the controlling mechanism at the very start of immersion when there is no rust.
- The processes involved change with continued corrosion. Therefore, each consecutive "phase" requires a different mathematical model.
- The corrosion process changes, eventually, from oxidation to a process involving anaerobic bacterial activity.
- The action of SRB, when present, must be part of the long-term corrosion process.
- The notion that nutrient supply controlled the rate of activity of SRB.

The typical schematic function presented by the Melchers model for corrosion loss as a function of time is illustrated in Figure 10.4, together with the parameters employed. The model and its phases are now calibrated to a very wide range of field observations for a range of seawater temperatures. Note that for high-temperature tropical waters, phases I and II become very short (e.g., shorter than 1 year) and thus would not be detected in corrosion-loss observations that have the first observation point one or more years after first exposure, as is common in many experimental field-observation programs. This is one reason for the idealized simple linear model that was perhaps first proposed, entirely empirically, by Southwell and Alexander (1970) on the basis of very extensive field observations extending over 16 years in the Panama Canal zone:

$$
\begin{aligned}
c &\le a & T &= 0, \\
c &= a + bT & T &> 0,
\end{aligned}
\tag{10.5}
$$

where c = mean depth of corrosion (obtained from corrosion coupon weight-loss measurements); a, b = constants; and T = exposure time. The effect of seawater temperature was not considered explicitly as it varied little at the test locations. This behavior is more consistent with phase IV of the Melchers model illustrated in Figure 10.4.

Although such phenomenological models are more sophisticated than the purely empirical models to predict the corrosion loss, it is evident that there are a lot of problem areas yet to be resolved, and basic understanding has yet to be refined. In this regard, the empirical corrosion models presented in the previous section may be applied for practical purposes of design and condition assessment.

10.4 Options for Corrosion Management

Various options for corrosion management are relevant, as follows:

- Corrosion margin addition
- Coating
- Cathodic protection
- Ballast water deoxygenation
- Chemical inhibitors

Table 10.4(a). *Example of average corrosion margin values for various areas assuming a coating life of 7.5 years*

| ID No. | Member category | C_1 (mm/year) | Average corrosion rate case | |
			$C_1 \times 17.5$ years (mm)	Corrosion margin value (mm)
1	B/S-H	0.0597	1.0448	1.0
2	A/B-H	0.1084	1.8970	2.0
3	A/B-V	0.0661	1.1568	1.0
4	B/S-V	0.0622	1.0885	1.0
5	BLGB	0.0619	1.0833	1.0
6	O/B-V	0.1012	1.7710	2.0
7	B/B-H	0.1408	2.4640	2.5
8	O/S-H	0.0607	1.0623	1.0
9	A/O-H	0.0581	1.0168	1.0
10	A/O-V	0.0523	0.9153	1.0
11	O/S-V	0.0423	0.7403	1.0
12	BLGC	0.0414	0.7245	1.0
13	O/O-V	0.0577	1.0098	1.0
14	O/O-H	0.0405	0.7088	1.0
15	BSLB(W)	0.1367	2.3923	2.5
16	BSLB(F)	0.1127	1.9723	2.0
17	DLB(W)	0.2403	4.2053	4.0
18	SSLB(W)	0.1413	2.4728	2.5
19	SSLB(F)	0.0882	1.5435	1.5
20	LBLB(W)	0.1960	3.4300	3.5
21	LBLB(F)	0.1782	3.1185	3.0
22	BSLC(W)	0.0466	0.8155	1.0
23	BSLC(F)	0.0437	0.7648	1.0
24	DLC(W)	0.0716	1.2530	1.5
25	DLC(F)	0.0588	1.0290	1.0
26	SSLC(W)	0.0420	0.7350	1.0
27	SSLC(F)	0.0397	0.6948	1.0
28	LBLC(W)	0.0550	0.9625	1.0
29	LBLC(F)	0.0508	0.8890	1.0
30	BGLC(W)	0.0377	0.6598	1.0
31	BGLC(F)	0.0319	0.5583	0.5
32	DGLC(W)	0.0477	0.8348	1.0
33	DGLC(F)	0.0449	0.7858	1.0
34	SSTLC(W)	0.0261	0.4568	0.5

10.4.1 Corrosion Margin Addition

To reduce the costs associated with downtime for repairs and renewal of corroded structural components, prior addition of corrosion margins to new-build structures is necessary in ship-shaped offshore structures like in trading ships. In some types of offshore structures, such as jacket platforms, the margins added, together with the corrosion protection schemes employed, may even virtually eliminate the need for steel renewals for the life of the platform, except as usual in the topsides.

A design corrosion margin often needs to be determined so that a representative maximum predicted thickness loss for the entire life of the offshore unit is added to the "net" structure, which has been designed for the relevant design demands

Table 10.4(b). *Example of severe corrosion margin values for various areas assuming a coating life of 7.5 years*

ID No.	Member category	C_1 (mm/year)	Average corrosion rate case	
			$C_1 \times 17.5$ years (mm)	Corrosion margin value (mm)
1	B/S-H	0.1717	3.0048	3.0
2	A/B-H	0.2323	4.0653	4.0
3	A/B-V	0.1897	3.3198	3.5
4	B/S-V	0.1823	3.1903	3.0
5	BLGB	0.1805	3.1588	3.0
6	O/B-V	0.1919	3.3583	3.5
7	B/B-H	0.2586	4.5255	4.5
8	O/S-H	0.1777	3.1098	3.0
9	A/O-H	0.1689	2.9558	3.0
10	A/O-V	0.1529	2.6758	3.0
11	O/S-V	0.1497	2.6198	3.0
12	BLGC	0.1446	2.5305	2.5
13	O/O-V	0.1621	2.8368	3.0
14	O/O-H	0.1423	2.4903	2.5
15	BSLB(W)	0.2461	4.3068	4.5
16	BSLB(F)	0.2343	4.1003	4.0
17	DLB(W)	0.4244	7.4270	7.5
18	SSLB(W)	0.2595	4.5413	4.5
19	SSLB(F)	0.1630	2.8525	3.0
20	LBLB(W)	0.3840	6.7200	7.0
21	LBLB(F)	0.3455	6.0463	6.0
22	BSLC(W)	0.0888	1.5540	1.5
23	BSLC(F)	0.0837	1.4648	1.5
24	DLC(W)	0.1252	2.1910	2.0
25	DLC(F)	0.1060	1.8550	2.0
26	SSLC(W)	0.0810	1.4175	1.5
27	SSLC(F)	0.0790	1.3825	1.5
28	LBLC(W)	0.0942	1.6485	1.5
29	LBLC(F)	0.0921	1.6118	1.5
30	BGLC(W)	0.0714	1.2495	1.5
31	BGLC(F)	0.0578	1.0115	1.0
32	DGLC(W)	0.0900	1.5750	1.5
33	DGLC(F)	0.0862	1.5085	1.5
34	SSTLC(W)	0.0437	0.7648	1.0

associated with actions (loads) alone. The value of corrosion wastage (depth) with time can then be predicted from Eq. (10.4).

For example, Table 10.4(a) and Figure 10.15(a) indicate the corrosion margin values for each primary member location/category group based on the average corrosion wastage models presented in Section 10.3.2. Here, the design service life is assumed to be T = 25 years, and the coating life is considered to be 7.5 years; thus, the associated corrosion exposure time becomes $25 - 7.5 = 17.5$ years. A similar derivation of corrosion margin values is possible, under the same circumstances, using the severe (or upper bound) corrosion model; related results are indicated in Table 10.4(b) and Figure 10.15(b).

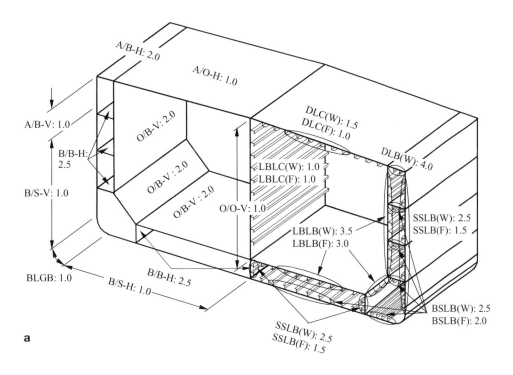

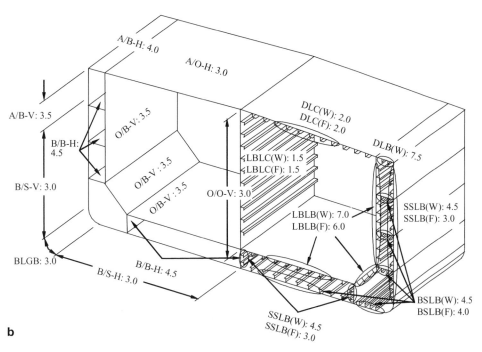

Figure 10.15. (a) Example of average corrosion margin values (mm) for primary member location/category groups, following Paik et al. (2003a). (b) Example of severe corrosion margin values (mm) for primary member location/category groups, following Paik et al. (2003a).

10.4.2 Coating

Protection of metal surfaces by coating – for example, painting – is the most common option for corrosion control used for ship-shaped offshore structures as well as trading tanker structures. The life of coating in places can possibly be as low as a few years and, thus, the repair of worn, scratched, or chipped spots should be routinely undertaken in order to prolong the coating condition over the required time period.

Note that paints with a longer life have been developed in recent years, and recommended paints, surface preparation, painting, and maintenance protocols for average coating lives of 15 or 20 years can be found in the industry literature (Paik et al. 2006). More rarely are they to be found in classification society guidelines because, to date, most classification societies have invariably opted to stay out of having to monitor coating application and related quality control, whether in a new-build stage or for maintenance. Classification societies do monitor the general visual condition of coatings for ship-shaped offshore structures classed with them, however. This situation will probably change somewhat in the next few years for tankers at least, in light of ongoing work related to the development of ballast tank coating performance standards; see IMO (2004).

10.4.2.1 Surface Preparation

The durability of coatings is significantly affected by the quality of surface preparation. Details as to what may be required will vary somewhat with the source of the guidelines. Devanney (2006) presents a readable and useful guidance for surface preparation. All structural steel must be blasted before coating per ISO 8501-1 (Sa3) and immediately shop-primed with inorganic zinc primer having a minimum dry film thickness of 20 microns. Sharp edges resulting from steel cutting – for example, with plasma arc torches – must be treated, for example, by grinding to convert them into a reasonably smooth arc with a minimum radius of at least 2mm. Otherwise, paint can stick to these edges, and surface tension forces may pull the coating away from the corner due to a phenomenon called "pull-back," which is inversely proportional to the radius of the corner. Secondary surface preparation could also be required at block and erection stages. For example, ballast tanks, slop tanks, and the bottom half-meter of cargo tanks to be coated may need to be reblasted per ISO 8501-1 (Sa3) and to a surface profile of 75–125 microns per NACE RP 0287.

Blasting methods for surface preparation include the following:

- Dry (open) grit blasting
- Water-enclosed grit blasting
- Slurry blasting
- Ultrahigh-pressure water jetting
- Other systems (e.g., jet systems)

The dry grit blasting is the most common method of surface preparation for large surface areas in today's shipbuilding industry. Although this method is relatively fast and produces good surface profiles for subsequent coating, it may not be as efficient or effective for the removal of contaminants (e.g., salts) from weathered-steel surfaces. Additional treatments such as washing and drying may be required in such cases although that would increases the costs and also time involved. Also, open blasting is a noisy and dusty operation that may not be allowed or possible in some areas.

Water-enclosed grit blasting is similar to open blasting, but a water shroud is added to reduce the amount of dust generated. Also, the use of water in this method more effectively removes salt contaminants from the substrate, producing a cleaner surface than with dry blasting. Run-off water together with used grit must be collected, separated, and treated before the disposal.

In the slurry blasting method, water and grit are mixed together at the blast pot and this mixture is used as the cleaning media. The ultrahigh-pressure water jetting method uses the water of ultrahigh pressure greater than 1,500 bar. This method is very effective in removing salts from contaminated surfaces, but it is noisy although no dust is generated. Compared with dry grit blasting, this method is also slower. There are also "jet" systems that increase the mechanical impact of cleaning particles on a surface during dry blasting. For example, abrasive from a blast pot initially traveling at a normal velocity may be accelerated several times before impact. This can provide a higher cleaning efficiency.

Appropriate surface preparation standards will be specified. For example, it may be required that all steel plates, profiles and fittings be grit-blasted to give a finish equal to ISO 8501-1 (Sa2.5) and coated with shop primer soon after. The minimum water washing pressure will be specified. Measurement requirements for the salt that can remain will be included as this is a critical component in lasting coating adhesion. Water from the washing process will usually need to be removed physically and from the spaces completely and not allowed to evaporate.

Appropriate edge preparation will also be required. For example, all edges of structures of tanks, void spaces, and the exterior hull will be required to be coated, to be ground to a certain radius, and to be stripe-coated before starting on each full coat.

10.4.2.2 Types of Coating

In practice, vessel specifications will require that appropriate coating be applied to most surfaces including all ballast tank surfaces, cargo tank top, and bottom at least, and antifouling be applied to the external hull, in addition to the corrosion margins that are added in design. In new builds, at least, permanent means of access suitable for inspection and maintenance of the corrosion protection measures, such as walkways and enlarged stringers, will need to be employed.

Normally, ship-shaped offshore unit hulls are coated with anticorrosive paints and antifouling paints on top of the underwater hull surfaces, similar to trading ships. The purpose of the antifouling coating on the offshore units, however, is not to keep the underwater hull surfaces smooth but rather to delay the corrosion by reducing the sea growth. Table 10.5 shows the characteristics of selected types of coatings for marine corrosion protection (Randall 1997).

With paints containing biocide compounds not now preferred and tin-based paints off the market, it might be difficult to get an antifouling coating that would last more than perhaps 5 years with adequate effectiveness. Silicon-based "foul release" type coatings have for some time now been used on naval ships, LNG carriers, and fast ferries. These are somewhat like teflon coatings and prevent fouling growth due to a nonsticky surface. Even though some marine growth may occur, this can be cleaned more easily.

This type of coating is environmentally ideal as there is no biocide compound in the coating material. Unlike ordinary antifouling coatings that are dissolved in water

Table 10.5. *Features of generic types of coatings for marine corrosion protection*

Type of coating	Feature
Epoxy paints	Coal tar–epoxy paints offer good resistance to water soil and inorganic acids, but no longer preferred; polyamide-hardened more resistance to moisture but less resistance to acids; amine-hardened more resistance to chemicals; epoxy-ester easier to apply but less corrosion resistance
Silicon paints	Excellent water repellent; maximum temperature of $650°C$; poor chemical resistance
Zinc paints	Used for galvanic protection; organic requires less surface preparation; inorganic easier to topcoat; used effectively in neutral and slightly alkaline solutions; inorganic more heat resistance
Oil-based paints	Easy application; relatively inexpensive; permeable; recommended for mild atmospheric conditions
Alkyd paints	Must be baked to dry; better corrosion resistance than oil-based paints; not suitable for resistance to chemicals
Urethane paints	Good resistance to abrasion; corrosion resistance approaches vinyl and epoxy paints
Vinyl paints	Better corrosion resistance than oil and alkyd paints; adherence and wetting often poor; good resistance to aqueous acids and alkalines; maximum temperature of $65°C$

and consumed with time, these may not require recoating, except for local damage repairs. For new builds, however, most commercial shipyards are reluctant to apply a silicon-based coating system, despite the few examples noted herein for naval and some special ships.

As several ships are under construction at the same time in a large shipyard, any overspray of silicon-based paints from the ship-shaped offshore unit blocks or hull could contaminate the coating of the conventional ship blocks in the neighborhood. Perhaps being in a somewhat different situation as to such contamination, repair yards may generally be more amenable to applying this type of coating to commercial ships during periodic overhauls or conversions.

Even in the case of a new-build unit, silicon-based foul-release coatings may be possible if the final silicon coating is done at a different dry-dock, for instance. One possible consideration in applying silicon coatings to ship-shaped offshore unit hulls is that less than the entire area is available or appropriate for silicon coating on the bottom because of closely spaced support blocks (relatively more blocks for supporting heavier hull weights with topsides facilities).

To assure integrity, ballast and cargo tanks also need to be adequately coated to prevent rapid corrosion and pitting. To help inspection, the coating ideally needs to be of a light color. Although it is always recommended to also consider alternatives such as thicker scantlings or secondary cathodic protection, good coatings are still invariably needed because such alternatives will not always efficiently solve the problem of fast progress of localized corrosion (e.g., pitting, grooving) and subsequent failures at welded joints.

No coating system will be 100 percent fail-proof and last through the expected long service life of the offshore units. Hence, as a backup, it is normal to provide an ICCP for the underwater hull and sacrificial zinc anodes in the ballast tanks. Design of the

ICCP system for the offshore unit hull provides several challenges. One consideration is that the ICCP system of trading ships is serviced every few years when the ships are dry-docked, but the ICCP system for ship-shaped offshore units has to be designed to last through the service life of the offshore units without dry-docking.

The additional reliability can usually be achieved by spare anodes and reference electrodes that can be activated when the working units fail. Another possible complication is that the ship-shaped offshore unit is normally continuously grounded through the mooring chains and, hence, the related stray current needs to be analyzed and designed to ensure that the system will protect the hull and will not make some other parts into anodes causing current drain through the chains.

The antifouling paint in ship-shaped offshore units has a limited life in water. Therefore, a dry-docking for hull coating may sometimes need to be considered after a relatively long onsite service period. In any event, a good, properly applied underwater coating is critical for achieving high onsite lives.

The coating system for a new-build hull will generally be selected and specified based on the required service life and associated maintenance and upkeep philosophy. The coating system for the cargo tanks, slop tanks, reception tanks, off-spec tanks, and the like will be specified to be compatible with the expected temperatures of the cargo and liquids therein. For the external hull and ballast spaces, abrasion-resistant epoxies may be preferred. Soft coating systems are not now used for hull structures. Skid-resistant coatings will be applied to selected locations, such as walkways and stairways around equipment and along important access routes.

The specifications will cover the intended service, the performance standards required of the coating, the surface preparation to be achieved, and also various environmental and human safety matters, such as avoiding tin-free paints and coal-tar epoxies. The coating system specification will include the required dry film thicknesses and number of coats for various areas, and also specify the allowable variability in design of dry film thicknesses and the measurement and control of the same.

The coating application will consider all manufacturer recommendations. Freshwater washing will normally be used to remove salt residue from surfaces before any surface preparation or coating application.

It may also be required that the coating system be tested according to various applicable standards (e.g., ISO, ASTM, NORSOK; see Appendix 6) for aspects such as thickness, adhesion, brittleness, and osmosis. These various factors pertain to the overall effectiveness of the coating system. The coating system as applied will usually be required to be warranted for a specified period (e.g., 5 years).

10.4.2.3 Selection Criteria of Coating Material

The selection of coating material is based on several considerations and related criteria. As an example, the following main considerations may apply in the case of epoxy coatings (Devanney 2006):

- Adhesive strength
- Permeability
- Glass transition temperature
- No solvent
- Right filler

Adhesion is a measure of the coating's ability to resist being lifted off the surface by corrosion or bubbling. The bigger the adhesion, the better the adhesive strength. A coating with an adhesion of at least 140 bar is required per ASTM D4541. Permeability is a measure of the ease with which water can work its way through the coating. The smaller the permeability, the better the coating durability. Most epoxy coatings when heated may become semiplastic at a critical temperature, 50–60°C, glass transition temperature (GTT). The temperature in the top of ballast tanks may often be 55–65°C in tropical areas; this is, in part, the reason why the coating breakdown sometimes starts from the top of ballast tanks. Because it is important to keep the top of the tank cool if possible, coatings with a high GTT, at least 55°C, must be considered. With modern catalyst-activated coating technology, solventless or 100 percent solids coatings are feasible and desirable. A hydroscopic filler, such as clay or calcium carbonate, is not normally recommended because it may attract water and weaken the coating. The use of aluminum or aluminum oxide filler is thought to be more desirable.

10.4.2.4 Methodologies for Coating-Life Prediction

For corrosion management involving coatings, and also for strength assessment of corroding structures, it is essential to identify the expected life of protective coatings. In practice, useful guidelines for the application of protective coatings as a function of the life expected are available in the maritime industry; for example, DNV (1992, 1998, 1999), TSCF (2002), and IACS (2002). Some of these guidelines may also provide for rather general estimates of the expected life for various types of coatings as a function of the application protocol.

The durability of coatings is affected by various parameters, including the applied coating system itself, but is generally uncertain. In practice, therefore, a pragmatic approach is often used when in-situ coatings are periodically inspected to estimate the coating condition, including the remaining life of the coating, and also the time before remedial action is likely to be required. Certainly, reliability-based approaches are more desirable to apply for probabilistic estimates of coating life (Martin et al. 1996; Faber and Melchers 2001), although data suitable for such purposes is currently very lacking in the literature (Melchers and Jiang 2006). However, it may be expected that reliability-based approaches will become increasingly more useful as more data becomes available, particularly from results of the enhanced survey program for tankers (e.g., IACS 2002), which has now been in place for many years; and the associated more detailed recording of the results of periodic coating inspections (Paik et al. 2006).

Factors affecting the coating life in marine structures (Melchers and Jiang 2006) include the following:

- Surface preparation, including the condition of the surface immediately prior to coating application; salt content on the surface is an important consideration.
- Achieved dry film thickness, including its variability over the surface, and considering possible "holidays," "peel-backs," and other shortcomings including at plate edges, holes, and discontinuities and at welds.
- Severity of the (local) environment, including temperature range; relative humidity, both during the coating application process and during the coating life; and chemical exposure.

As described in Section 10.4.2.1, surface preparation is a primary parameter of influence on the durability of coating (Staff 1996; Flores and Morcillo 1999). Also, coating thickness has an important bearing on the permeability of moisture into the protective coating. Relatively thicker coatings tend to have longer effective lives, provided they are applied in multiple layers; but not so thick as to lead to coating cracking (Lambourne and Strivens 1999; Friar 2001).

Johnson (1999) indicates that most coating breakdown may be due to inadequate surface preparation or its application error. The local corrosive environment may also be a factor to consider. For instance, horizontal stiffeners are typically more prone to corrosion than vertical surfaces because horizontal stiffeners are more likely to trap moisture than vertical stiffeners.

Breakdown of coatings may be considered to start with small defects that may increase in size and severity with time. The level of coating deterioration is often measured by the percentage of the deteriorated surface. Fatigue cracking can also be a possible source of coating deterioration initiation (Perera 1995; Eliasson 2003).

In practice, the condition of coating is usually evaluated in a qualitative manner, for example, as "good," "fair," and "poor," based on the percentage of deteriorated surface area (IACS 2002). Typically, when 1–2 percent coating deterioration is measured, maintenance measures, particularly recoating, are triggered although such a modest level is not necessarily considered to be the criterion "for failure" of coatings for backup anode design.

Relevant data on coating life for reliability modeling purposes may be available from direct field observations (e.g., Rolli 1995) and from laboratory testing (e.g., Scully and Hensley 1994). Martin et al. (1996) reviewed several prescriptive and reliability-based methods for predicting the coating life. Emi et al. (1994) developed a procedure for coating life prediction based on a knowledge-based system. Progress of coating breakdown is time-variant in nature. Some investigators (Pirogov et al. 1993; Sakhnenko 1997; Yamamoto and Ikegami 1998) proposed mathematical models for coating breakdown considering that the final coating life obtained can be taken as a normal random variable.

Melchers and Jiang (2006) performed some very interesting work including a survey of the coating breakdown times for trading ships based on the opinions of vessel users, coating contractors and suppliers, and an independent expert using a prepared questionnaire. Figures 10.16 and 10.17 show selected results from that survey, representing the coating-breakdown functions measured by percentage of the surface area of coating deterioration as a function of exposure time. It is interesting to note that the users tended to be perhaps more pessimistic and the contractors and suppliers tended to be perhaps more optimistic in their estimates compared with those of the independent expert. It was also said that the coating life as measured by the percentage surface area of coating breakdown is approximately normal distributed with increase in both mean and variance of coating deterioration over time.

10.4.3 Cathodic Protection

For large areas of submerged steel, cathodic protection is the most common form of corrosion protection. Typically, the cathodic protection is applied together with

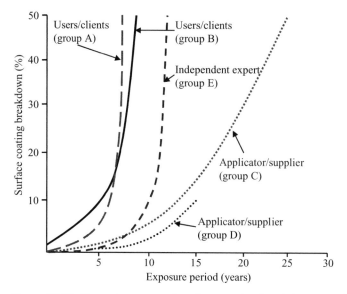

Figure 10.16. Percentage area of surface coating breakdown as a function of exposure time; estimated from the questionnaire survey from users, coating contractors, suppliers, and independent experts, following Melchers and Jiang (2006).

coatings and acts as backup in case of coating breakdown to a specified level. Two types of cathodic protection are relevant: impressed current system and galvanic system. The former type is a more reliable long-term protection system, although it requires the use of continuous external electrical power.

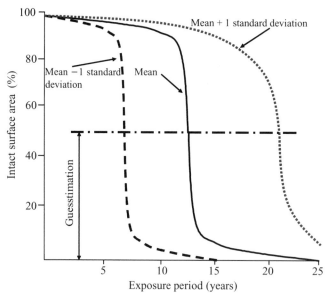

Figure 10.17. Deterioration function showing remaining intact surface area as a function of exposure time; estimated from the questionnaire survey from users, coating contractors, suppliers, and independent experts, following Melchers and Jiang (2006).

The latter type uses aluminum, magnesium, or zinc anodes that are attached to the steel material exposed to corrosion in seawater. The steel material then becomes the cathode that can no longer corrode as before. For offshore applications such as offshore platforms, pipelines, and mooring chains, aluminum or aluminum-zinc anodes are often employed.

Aluminum anodes can be more durable than zinc anodes since aluminum anodes have a greater current-to-weight ratio than zinc anodes. Therefore, aluminum anodes may corrode less in brackish water over time, and zinc is more sacrificial. When the anodes deteriorate, they need to be replaced. Also, with certain aluminum anodes, there can be sparking concerns to be dealt with in case the anode detaches and strikes another metallic surface. Outer surfaces of vessel hulls and submerged parts of offshore units are usually protected by a cathodic method.

The following is a typical process for the design of a cathodic protection system (Randall 1997):

- Determine design current density based on geographical location.
- Calculate surface areas to be protected.
- Calculate total anode material (number of anodes) required for selected life of the system (structure).
- Determine anode geometry and initial current density assuming adequate driving potential; for example, 0.45 volts between steel and aluminum alloy anodes.
- Determine the durability (life) of anodes for polarized material; for example, 0.25 volts potential for polarized steel.
- Distribute the anodes evenly over the areas to be protected.

Vessel specifications will specify the various attributes required of the cathodic protection system and typically will also require that such systems be designed in accordance with certain accepted guidelines (e.g., DNV-RP-B401); see Appendix 6 in this book. A self-regulating ICCP system including multiple anodes and multiple reference electrodes will usually be specified for the underwater surfaces of the hull, turret, hull parts adjacent to the mooring system, and risers. The cathodic potential will be required to be within certain limits, for example, –800 mV to –1,000 mV (Ag/AgCl). The system specified for the hull will also be required to be compatible with the corresponding systems to be used for the mooring and subsea arrangements.

The design current densities for the hull will usually be based on full-load draft and a specified level of coating damage. Anode life required will be specified for the external hull; for example, the service life. In cargo and ballast tanks, replaceable zinc sacrificial anodes may be installed consistent with the maintenance philosophy to be adopted. The distribution and number of anodes will be appropriately designed. The ICCP system and sacrificial anodes will need to be maintainable in service to the greatest extent possible, including those attached to the external hull.

10.4.4 Ballast Water Deoxygenation

In ballast water tanks, oxygen is an important factor that promotes corrosion. Therefore, an alternative to prevent or mitigate the corrosion wastage is to remove

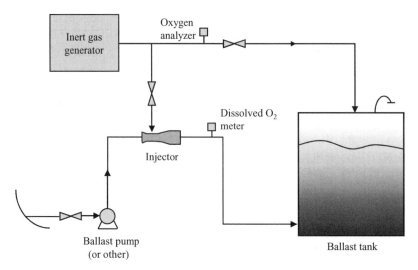

Figure 10.18. A schematic representation of a ballast water deoxygenation system (Tamburri and Ruitz 2005).

oxygen from the ballast water. It has been observed, based on the laboratory tests, that the corrosion rate may be reduced by as much as 90 percent when oxygen levels are reduced and maintained below 0.5 percent (Tamburri et al. 2003).

Various methods to deoxygenate from the ballast water have been considered; for example, by vacuum, biological process, or with the help of an inert type of gas (Devanney 2006). The introduction of inert gas into the ballast water is, perhaps, the most effective way to accomplish the stated aim (Tamburri and Ruiz 2005).

For example, nitrogen gas may be pumped into the ballast water through pipes installed into the ballast tank (Matsuda et al. 1999). In this method, the ballast water tank needs a seal at the deck to permit nitrogen purging of the headspace. This method achieves deoxygenation through the contact of the nitrogen bubbles with the water to an extent, but it also very much relies on preventing the diffusion of oxygen through the water surface. This method takes time to deoxygenate from the ballast water, say, a day or two per tank, and it can also be expensive because large volumes of gas would need to be pumped into the tanks.

A more efficient and cost-effective method for deoxygenation may be to introduce microfine bubbles of an inert gas as the water is being pumped into the tanks (Tamburri et al. 2003; Tamburri and Ruitz 2005). In this method, the ballast water is basically deoxygenated using specially designed equipment before entering the tanks, as shown in Figure 10.18. For crew safety, the oxygen level in ballast water tanks can be measured and controlled to meet tank entry requirements when needed. The efficiency of this system is said to increase with smaller bubbles because the gas-to-water contact surface area increases. Deoxygenated water itself is relatively benign when discharged, and it may also be reoxygenated and mixed with water received in harbors prior to the disposal as necessary.

10.4.5 Chemical Inhibitors

The corrosion protection of closed systems such as engines or boilers usually uses chemical inhibitors. Five broad types of inhibitors may be considered: absorption, hydrogen evolution "poisons," scavengers, oxidizers, and vapor phase inhibitors. Absorption inhibitors may affect anodic and cathodic reactions. Scavengers can remove the oxygen needed for the cathodic reaction. The first four methods can be used for the corrosion protection of metal. For example, corrosion of engines and radiators is often reduced by introducing rust inhibitors.

REFERENCES

ABS (2000). *Rules for building and classing steel vessels.* American Bureau of Shipping, Houston.

AISI (1981). *Handbook of corrosion protection for steel structures in marine environment.* American Iron and Steel Institute, Washington, DC.

Battelle (1986). *Seawater corrosion of metals and alloys.* Columbus, Ohio: Battelle Laboratories.

Borenstein, S. W. (1994). *Microbiologically influenced corrosion handbook.* New York: Industrial Press.

Chernov, B. B. (1990). "Predicting the corrosion of steels in seawater from its physiochemical characteristics." *Protection of Metals*, 26(2): 238–241.

Chernov, B. B., and Ponomarenko, S. A. (1991). "Physiochemical modeling of metal corrosion in seawater." *Protection of Metals.* 27(5): 612–615.

Contraros, P. D. (2003). "The Domino effect: Coating breakdown – corrosion – structural failure leading to possible design ramifications." *Proceedings of Lloyd's Register Conference on the Prevention and Management of Marine Corrosion*, London, April 2–3.

Craig, B. D., and Anderson, D. S. (1995). *Handbook of corrosion data* (materials data series). 2nd ed. Materials Park, OH: ASM International.

DECHEMA (2003). *Corrosion handbook.* Elsevier Science (CD-ROM edition).

Devanney, J. (2006). *The tankship tromedy: The impending disasters in tankers.* Tavernier, FL: Center for Tankship Excellence (CTX) Press.

DNV (1992). *Guidelines for corrosion protection of ships.* (Paper No. 92-P001), Det Norske Veritas, Oslo.

DNV (1998). *Type approval programme for protective coating systems.* (Paper No. 1-602.1), Det Norske Veritas, Oslo.

DNV (1999). *Corrosion prevention of tanks and holds.* (Classification Notes, No. 33.1), Det Norske Veritas, Oslo.

Edyvean, R. G. J., and Videla, H. A. (1992). "Biofouling and MIC interactions in the marine environment: An overview. In *Microbial corrosion*, C. A. C. Sequeira and A. K. Tiller, eds. London: The Institute of Metals, pp. 18–32.

Eliasson, J. (2003). "Economics of coatings / corrosion protection of ships." *Proceedings of Lloyd's Register Conference on the Prevention and Management of Marine Corrosion*, London, April 2–3.

Emi, H., Matoba, M., Arima, T., and Umino, M. (1994). "Corrosion protection system for long ships' water ballast tanks. *Proceedings of International Conference on Marine Corrosion Prevention* (Paper No. 15), The Royal Institution of Naval Architects, London, October 11–12.

Evans, U. R. (1966). *The corrosion and oxidation of metals: Scientific principles and practical applications.* London: Edward Arnold Ltd.

Faber, M. H., and Melchers, R. E. (2001). "Aspects of safety in design and assessment of deteriorating structures safety." In *Safety, risk, and reliability – Trends in engineering.* Malta: IABSE, pp. 161–166.

Flores, S., and Morcillo, M. (1999). "Anticipated levels of soluble salts remaining on rusty steel prior to painting." *Surface Coatings International*, 82(1): 19–25.

Fontana, M., and Greene, N. (1986). *Corrosion engineering*. New York: McGraw-Hill.

Friar, D. E. (2001). "A new concept in corrosion protection for ships hulls." *Proceedings of International Conference on Marine Corrosion Prevention* (Paper No. 8), The Royal Institution of Naval Architects, London, October 11–12.

Gardiner, C., and Melchers, R. E. (1998). "Aspects of bulk carrier hold corrosion." *Proceedings of International Conference on the Design and Operation of Bulk Carriers* (Paper No. 7), The Royal Institution of Naval Architects, London.

Gardiner, C. P., and Melchers, R. E. (1999). "Enclosed atmospheric corrosion in an unloaded bulk carrier cargo hold." *Proceedings of Australian Corrosion Congress* (Paper No. 45), Sydney, Australia, November.

Gardiner, C. P., and Melchers, R. E. (2001a). "Bulk carrier corrosion modeling." *Proceedings of the 11th International Offshore and Polar Engineering Conference*, Stavanger, Norway, June 17–22, pp. 609–615.

Gardiner, C. P., and Melchers, R. E. (2001b). "Enclosed atmospheric corrosion within ship structures." *British Corrosion Journal*, 36(4): 272–276.

Gardiner, C. P., and Melchers, R. E. (2002a). "Corrosion of mild steel in porous media." *Corrosion Science*, 44: 2459–2478.

Gardiner, C. P., and Melchers, R. E. (2002b). "Corrosion of mild steel by coal and iron ore." *Corrosion Science*, 44: 2665–2673.

Gardiner, C. P., and Melchers, R. E. (2003). "Corrosion analysis of bulk carriers: Part 1 Operational parameters influencing corrosion rates." *Marine Structures*, 16(8): 547–566.

Gudze, M., Cannon, S., and Melchers, R. E. (2001). "Structural deterioration modelling issues for reliability based management." *Proceedings of International Maritime Conference* (Pacific 2002), pp. 142–151.

Gudze, M., Cannon, S., and Melchers, R. E. (2004). "Ballast tank corrosion using naval ship operational profiles." *Proceedings of International Maritime Conference* (Pacific 2004), Sydney, February 3–5, Vol. 1, pp. 280–289.

Herring, L. D., and Titcomb, A. N. (1981). *Investigation of internal corrosion and corrosion control alternatives in commercial tankers*. Ship Structure Committee, SSC-312, Washington, DC.

IACS (1998). *Guidelines for acceptance, application, and survey of semi-hard coatings on ballast tanks*. (IACS Recommendation, No. 54), International Association of Classification Societies, London, September.

IACS (2002). *Bulk carriers: Guidelines for surveys, assessments, and repair of hull structures*. 2nd ed. International Association of Classification Societies, London.

IMO (1995). *Guidelines for the selection, application, and maintenance of corrosion prevention systems of dedicated seawater ballast tanks*. (Resolution A. 798(19)), International Maritime Organization, London.

IMO (2004). *Safety of life at sea (SOLAS): Consolidated edition: International convention for the safety of life at sea*. International Maritime Organization, London.

Johnson, J. R. (1999). "A primary cause of coating failure." *Materials Performance*, 38(6): 48–49.

Korb, L. J. (1989). *ASM handbook: Corrosion*. Materials Park, OH: ASM International.

Lambourne, R., and Strivens, T. A., eds. (1999). *Paint and surface coatings: Theory and practice*. Norwich, NY: William Andrew Publishing.

LaQue, F. L. (1975). *Marine corrosion*. New York: Wiley-Interscience.

MacMillan, A., Fischer, K. P., Carlsen, H., and Goksoyr, O. (2004). *Newbuild FPSO corrosion protection – A design and operation planning guidelines*. Offshore Technology Conference, OTC 16048, Houston, May.

Martin, J. W., Saunders, S. C., Floyd, F. L., and Wineburg, J. P. (1996). *Methodologies for predicting the service lives of coating systems*. Blue Bell, PA: Federation of Societies for Coatings Technology.

Matsuda, M., Kobayashi, S., Miyuki, H., and Yoshida, S. (1999). *An anticorrosion method for ballast tanks using nitrogen gas*. (Technical Report), Ship and Ocean Foundation, Tokyo, Japan.

Matsuoka, K., Arita, M., and Ohnaga, K. (1985). "Early stage deterioration of anti-corrosive epoxy coating for offshore structure use." *Journal of the Society of Naval Architects of Japan*, 157: 498–506 (in Japanese).

Melchers, R. E. (1997). "Modeling of marine corrosion of steel specimens." In *Corrosion testing in natural waters*. Vol. 2. R. M. Kain and W. T. Young, eds. Philadelphia: American Society for Testing and Materials, pp. 20–33.

Melchers, R. E. (2002). "Effect of temperature on the marine immersion corrosion of carbon steels." *Corrosion*, 58(9): 768–782.

Melchers, R. E. (2003a). "Modeling of marine immersion corrosion for mild and low alloy steels: Part 1. Phenomenological model." *Corrosion*, 59(4): 319–334.

Melchers, R. E. (2003b). "Probabilistic models for corrosion in structural reliability assessment: Part 1. Empirical models." *Journal of Offshore Mechanics and Arctic Engineering*, 125(4): 264–271.

Melchers, R. E. (2003c). "Probabilistic models for corrosion in structural reliability assessment: Part 2. Models based on mechanics." *Journal of Offshore Mechanics and Arctic Engineering*, 125(4): 272–280.

Melchers, R. E. (2003d). "Mathematical modelling of the diffusion controlled phase in marine immersion corrosion of mild steel." *Corrosion Science*, 45(5): 923–940.

Melchers, R. E. (2003e). "Effect on marine immersion corrosion of carbon content of low alloy steels." *Corrosion Science*, 45(11): 2609–2625.

Melchers, R. E. (2004a). "Pitting corrosion of mild steel in marine immersion environment: Part 1. Maximum pit depth." *Corrosion*, 60(9): 824–836.

Melchers, R. E. (2004b). "Mathematical modelling of the effect of water velocity on the marine immersion corrosion of mild steel coupons." *Corrosion*, 60(5): 471–478.

Melchers, R. E.(2004c). "Effect of small compositional changes on marine immersion corrosion of low alloy steel." *Corrosion Science*, 46(7): 1669–1691.

Melchers, R. E. (2005). "Effect of nutrient-based water pollution on the corrosion of mild steel in marine immersion conditions." *Corrosion*, 61(3): 237–245.

Melchers, R. E., and Jeffrey, R. (2004). "Surface roughness effect on marine immersion corrosion of mild steel." *Corrosion*, 60(7): 697–703.

Melchers, R. E., and Jeffrey, R. (2005). "Early corrosion of mild steel in seawater." *Corrosion Science*, 47(7): 1678–1693.

Melchers, R. E., and Jiang, X. (2006). "Estimation of models for durability of epoxy coatings in water ballast tanks." *Ships and Offshore Structures*, 1(1): 61–70.

Nakai, T., Matsushita, N., and Yamamoto, N. (2005). "Pitting corrosion and its influences on local strength of hull structural members." *Proceedings of the 24th Offshore Mechanics and Arctic Engineering Conference*, OMAE2005–67025, Halkidiki, Greece, June 12–17.

Ohyagi, M. (1987). *Statistical survey on wear of ship's structural members*. (NK Technical Bulletin), Nippon Kaiji Kyokai, Tokyo, pp. 75–85.

Paik, J. K. (2004). "Corrosion analysis of seawater ballast tanks." *International Journal of Maritime Engineering*, 146(Part A1): 1–12.

Paik, J. K., Brennan, F., Carlsen, C. A., Daley, C., Garbatov, Y., Ivanov, L., Rizzo, C. M., Simonsen, B. C., Yamamoto, N., and Zhuang, H. Z. (2006). *Condition assessment of aged ships. Report of the ISSC Specialist Committee V.6, International Ship and Offshore Structures Congress*. Southampton, UK: Southampton University Press.

Paik, J. K., Kim, S. K., and Lee, S. K. (1998). "Probabilistic corrosion rate estimation model for longitudinal strength members of bulk carriers." *Ocean Engineering*, 25: 837–860.

Paik, J. K., Lee, J. M., Hwang, J. S., and Park, Y. I. (2003a). "A time-dependent corrosion wastage model for the structures of single- and double-hull tankers and FSOs and FPSOs." *Marine Technology*, 40(3): 201–217.

Paik, J. K., and Thayamballi, A. K. (2003). *Ultimate limit state design of steel-plated structures.* Chichester, UK: John Wiley & Sons.

Paik, J. K., Thayamballi, A. K., Park, Y. I., and Hwang, J. S. (2003b). "A time-dependent corrosion wastage model for bulk carrier structures." *International Journal of Maritime Engineering*, 145(Part A2): 61–87.

Paik, J. K., Thayamballi, A. K., Park, Y. I., and Hwang, J. S. (2004). "A time-dependent corrosion wastage model for seawater ballast tank structures of ships." *Corrosion Science*, 46: 471–486.

Perera, D. Y. (1995). "Stress phenomena in organic coatings." In *Paint and coating testing manual. Gardner–Sward Handbook.* Philadelphia: American Society for Testing and Materials.

Pirogov, V. D., Lyublinskii, E. Y., Samsonov, A. L., Shevchuk, P. R., Galapats, B. P., and Shevchuk, V. A. (1993). "Analytical method for calculating the life of ship paint coatings." *Protection of Metals*, 29(2): 291–296.

Randall, R. E. (1997). *Elements of ocean engineering.* Jersey City, NJ: The Society of Naval Architects and Marine Engineers.

Reinhart, F. M., and Jenkins, J. F. (1972). *Corrosion of materials in surface seawater after 12 and 18 months of exposure.* (Technical Note, No. 1213), Naval Civil Engineering Laboratory, Port Hueneme, CA.

Revie, R. W. (Ed.) (2000). *Uhlig's corrosion handbook.* New York: Wiley-Interscience.

Roberge, P. R. (1999). *Handbook of corrosion engineering.* New York: McGraw-Hill Professional Publishing.

Rolli, M. S. (1995). "A reliable method for predicting long-term coatings performance." *Corrosion Management*, 4(1): 4–7.

Sakhnenko, N. D. (1997). "Imitation model form predicting the lifetime of paint coatings." *Protection of Metals*, 33(4): 393–397.

Schumacher, M. (1979). *Seawater corrosion handbook.* Park Ridge, NJ: Noyes Data Corporation.

Scully, J. R., and Hensley, S. T. (1994). "Lifetime prediction for organic coatings on steel and a magnesium alloy using electrochemical impedance methods." *Corrosion*, 50(9): 705–716.

Southwell, C. R., and Alexander, A. L. (1970). "Corrosion of metals in tropical waters – Structural ferrous metals." *Materials Protection*, 9(1): 179–183.

SRAJ (2002). *Study on cargo oil tank corrosion of oil tanker.* (Report of Ship Research Panel 242), The Shipbuilding Research Association of Japan, Tokyo.

Staff, C. M. (1996). "Surface preparation – The key to coating performance on steel." *Corrosion Management*, 5(1): 23–32.

Tamburri, M. N., Little, B. J., Ruiz, G. M., Lee, J. S., and McNulty, P. D. (2003). "Evaluations of Venturi Oxygen Stripping™ as a ballast water treatment to prevent aquatic invasions and ship corrosion." *Proceedings of the 2nd International Ballast Water Treatment R&D Symposium*, International Maritime Organization, London, July 21–23.

Tamburri, M. N., and Ruiz, G. M. (2005). *Evaluation of a ballast water treatment to stop invasive species and tank corrosion.* Presented at the 2005 SNAME Annual Meeting, The Society of Naval Architects and Marine Engineers, October, Houston.

Tomashev, N. D. (1966). *Theory of corrosion and protection of metals.* New York: The MacMillan Co.

TSCF (1992). *Condition evaluation and maintenance of tanker structures.* Tanker Structures Co-operative Forum, London: Witherby & Co. Ltd.

TSCF (1995). *Guidance for inspection and maintenance of double hull tanker structures.* Tanker Structures Co-operative Forum, London: Witherby & Co. Ltd.

TSCF (1997). *Guidance manual for tanker structures.* Tanker Structures Co-operative Forum, London: Witherby & Co. Ltd.

TSCF (2002). *Guidelines for ballast tank coatings systems and surface preparation.* Tanker Structures Co-operative Forum, London: Witherby & Co. Ltd.

Uhlig, H. H., and Revie, R. W. (1985). *Corrosion and corrosion control.* New York: Wiley-Interscience.

Yamamoto, N., and Ikegami, K. (1998). "A study on the degradation of coating and corrosion of ship's hull based on the probabilistic approach." *Journal of Offshore Mechanics and Arctic Engineering*, 120: 121–128.

CHAPTER 11

Inspection and Maintenance

11.1 Introduction

Inspection may be defined as an activity performed during the service life of a functioning structural unit in order to help detect and evaluate deterioration in the structural components or equipment by visual, electronic, or other means. *Maintenance* indicates the total set of activities (not including inspections, for the sake of a correct definition) undertaken to enable the installation to remain fit-for-service, including repairs, replacements, adjustments, and modifications.

Inspection and maintenance play a significant role in the operation of ship-shaped offshore units as they do in other types of structures. The methods, frequencies, and acceptance criteria used for inspection and maintenance can significantly affect the structural integrity of the units.

The inspection and maintenance technologies for ship-shaped offshore units have been based on those of trading ships (HSE 1998), but with certain modifications to suit the particular mission of the offshore units involved. Traditionally, the inspection frequencies adopted for the marine and offshore industries have thus been determined largely by prescriptive practices usually at specified time intervals based on the age of the unit concerned, although more flexible risk-based approaches are now being considered, and sometimes employed, for the same purposes. The traditional practice developed on the basis of operational experience with generic classes of structures usually follows what regulatory requirements and classification societies guidelines presumably can be deemed adequate when applied to structures of the same class and circumstances.

However, such a practice may not take into account the likelihood (frequency) of structural failure nor the failure impacts (consequences) in an explicit manner. Further, experience with ship-shaped offshore installations under particular circumstances may not always be plentiful. In some cases, this may mean that the inspection activity could be considered excessive, although it may be said that it is not enough in other cases. Furthermore, there is usually a lack of specific measures to ensure that the inspection activities performed are commensurate with any safety and reliability improvements implemented over and above minimum requirements; hence, the corresponding cost benefit may not be fully realized.

In this regard, a systematic approach to inspection for marine structures that can achieve a balance between risk and cost benefit is necessary. Risk-based inspection techniques can provide an excellent tool to evaluate the likelihood and consequences

400

of structural failures and lead to an optimum inspection process that effectively reduces the associated risk of failure, while keeping the inspection-influenced life-cycle costs at the appropriate level (ABS 2003; LR 2003; Ku et al. 2004).

The same may be said for maintenance activities, where it is desirable to use a systematic approach applying planned maintenance methods based on a predefined time schedule together with predictive maintenance technologies and application of condition monitoring (DNV 2003). A risk-based maintenance (including reliability-centered maintenance) scheme is a systematic maintenance approach applying risk-management and risk-control techniques to meet maintenance needs in a rational manner (Conachey and Montgomery 2003; ABS 2004; Lanquetin 2005).

This chapter presents recent advances and practices in inspection and maintenance schemes together with current practices applicable to ship-shaped offshore units. The current practices for condition assessment of trading tankers that can be useful for ship-shaped offshore units are reviewed. Principles and practices for risk-based inspections and maintenance are addressed. Time-variant reliability of ship-shaped offshore units affected by corrosion and fatigue cracking is also addressed and illustrated together with repair strategies for aging structures.

For a recent state-of-the-art review on condition assessment of aging marine structures involving inspection and maintenance schemes, see Paik et al. (2006).

11.2 Types of Age-Related Deterioration

Typical types of deterioration involved in marine structures while in service, including those due to age, are as follows:

- Corrosion
- Fatigue cracks
- Dropped objects and impact damage
- Inadequate fabrication such as out-of-tolerance misalignments
- Coating breakdown

Corrosion can be of various types, such as general corrosion and localized corrosion as described in Section 10.2 in Chapter 10. Cracking may be caused by fatigue due to the dynamic actions arising from environmental phenomena, operation, and other causes, such as high local stresses and hard spots. Low-temperature exposure may make the material brittle resulting in brittle fracture given the appropriate high stress–strain rates (Drouin 2006); although brittle fracture is much less common today with the attention now being paid to materials and their selection for use (Sumpter and Kent 2004).

Deck plates of offshore structures may be subjected to impacts due to objects dropped from cranes. Such mechanical damage can result in denting, cracking, and residual stresses or strains due to plastic deformation or coating damage leading to pit corrosion; see Figure 11.1. Inadequate fabrication can lead to significant initial defects and misalignments that may also increase the probability of fatigue cracking. As described in Section 10.4.2.4 in Chapter 10, the durability of protective coating is affected by various parameters and is also of interest in the present context.

a   b

Figure 11.1. Examples of mechanical damage: (a) local denting; (b) corrosion due to coating damage by local denting.

11.3 Methods for Damage Examination

For the inspection of aging structures, various tools are available for detecting and measuring age-related deterioration as well as other types of deterioration and damage. Nondestructive examination (NDE) methods are typically applied for the detection and measurements of defects and deterioration in marine structures (Halmshaw 1997; Porter 1992; Bøving 1989); however, their actual application or use may depend on a vessel's type and condition, a surveyor's experience and motivation, and the environment around the structure (Demsetz et al. 1996). Table 11.1 shows various methods for detection and measurements of age-related and other defects and deterioration. In the following sections, the methods for detecting and measuring each type of deterioration are presented (Paik et al. 2006).

11.3.1 Corrosion Wastage Examination

In the evaluation of corrosion wastage, a primary decision arises as to which parameter must be detected and measured: average remaining thickness, minimum thickness, and maximum pit depth or pit intensity (as a percentage of the plate surface). In current practices, average remaining thickness and maximum pit depth are considered to be primary parameters of corrosion in terms of repair criteria, but the trend is now toward a more quantitative definition of corrosion intensity.

Visual or close-up detection is a primary method to detect corrosion wastage, although it is largely affected by the detector's skill, experience, and local conditions. In design, it is desirable to use light color paints so that any coating breakdown and small rust spots can be more easily detected. In all types of visual inspection, photographic records will be very useful for a postinspection defect and damage assessment. Computer-aided digital-imaging methods using modern digital cameras, instead of direct visual detection, may also be considered to inspect for corrosion wastage, sometimes without a person having to enter a tank.

To detect and measure corrosion wastage, the method using ultrasonic sensors is widely employed, although it is time-consuming because point-by-point examination, preparation of surface, and preparation of coupling medium is necessary. For

Table 11.1. *Methods for examining defects and deterioration*

| Method of examination | Type of defect / deterioration | | | Remarks |
	Corrosion	Cracks	Mechanical damage	
Visual detection Close-up detection	✓	✓	✓	Small equipment such as hammer, flash, caliper, and measuring tape are needed
Digital imaging	✓	✓	✓	Automatic processing is usually required
Leak or pressure tests	✓	✓		Pit corrosion and small cracks can be detected
Dye penetrants, chemical sensors		✓		Affected by cleanliness
Ultrasonic tests	✓	✓		Time consuming and requires operator skill like all other methods
Magnetic particle		✓		Only for magnetic materials; only (sub)surface defects are detected
Strain gauges	✓	✓		Reduction of stiffness due to damage can be detected
Electro-magnetic field techniques		✓		Surface and subsurface cracks at weld seams, heat-treatment variations, steel thickness, coating thickness, crack depth
Radiometry (X-ray)		✓		Danger of radiation; specialized expertise needed
Acoustic emission or natural frequencies	✓	✓		For preliminary assessments; specialized firms are needed
Thermal imaging	✓			Limited to specific materials or situations
Moirè contours			✓	Deformation patterns of dents; an emerging technique
Replica		✓		Simple, cost effective, records surface defects
Test coupons	✓			Preliminary calibration is needed

structures with many corrosion pits, it is not always easy to remove the heavy rust and correct the thickness measurement because the surface is uneven after the rust is removed.

Specialized NDE wastage assessment technology is an alternative when coating breakdown is not significant (Bøving 1989; Saidarasamoot et al. 2003). Some advanced methods are also available (Agarwala and Ahmad 2000); the *acoustic emission* and *natural frequency measurement methods* are cheap and reliable in terms of detecting significant changes in structural responses and can also be tailored for the detection of both general and pitting corrosion.

Radiographic methods can detect the variations in the thickness of metallic components. *Thermal-imaging methods* may be useful for detecting hidden corrosion. The *weight-loss coupon method* periodically monitors loss of weight in a coupon exposed to corrosion. *Galvanic thin film microsensors* may be employed for in-situ monitoring of coating durability and hidden corrosion. *Electrochemical impedance spectroscopy* can be used to measure the early-stage deterioration of coating and substrate corrosion underneath a (paint) coating, although electrochemical techniques are affected by temperature and pH, among other factors. *Eddy current arrays* can provide a high-resolution readout with fast response, although eddy current arrays may not always be easy to apply to the large and geometrically complex structures. *Hydrogen measurement probes* can be used in cases where corrosion proceeds with measurable evolution of hydrogen.

The use of chemical sensors of certain types, particularly those relying on fluorescence and color change adopted for dye-penetrant testing, has not proven very practical because corrosion is typically widespread. The methods using strain gauges are also not practical because they need a calibration with the noncorroded elements and are generally affected by the corrosive environment as the strain gauges need to be bonded to the structure in large quantities. In *magnetic flux measurement*, a sensor is immerged to sense the current flow between anodic and cathodic areas; then, by measuring the metal loss, the corrosion wastage distribution can be obtained by computer-controlled data processing.

11.3.2 Fatigue and Other Crack Examination

In practice, fatigue cracking is repeatedly seen at geometrically similar locations. Therefore, it will be wise to know critical areas prone to fatigue cracking beforehand. This may be easier for standard details but more difficult in new types of structures (Ma et al. 1999) and can, in any event, be achieved by appropriate detailed stress- and fatigue-analysis.

A visual inspection is a primary method to detect cracks where it is needed to determine the type of crack in situ and examine whether cracks are likely to propagate. Dye penetrant and magnetic particle testing may follow after visual detection so that surface crack lengths can be approximately measured; it is usually difficult to measure the crack depth without the removal of the material affected.

Various NDE methods are available for detection and measurement of fatigue cracking. Tiku and Pussegoda (2003) compare the applicability of such methods (see Table 11.2). In addition, more advanced NDE techniques, such as acoustic emission, infrared thermography, laser shearography, potential drop test, alternating current field measurement, crack propagation gauges, and automated ball indentation, are also available.

Eddy current, ultrasonic, and potential drop tests can characterize the crack dimensions and locations with different accuracies, but these tests are generally better than visual inspection (Ditchburn et al. 1996). Vanlanduit et al. (2003) present a new method for in-service monitoring of fatigue cracking applying ultrasonic surface-guided waves where dynamic actions are allowed so that open cracks can be detected. Talei-Faz et al. (2004) present a digital photogrammetric technique that allows for three-dimensional measurements in real time for the cracking and deformation occurrence in local areas.

Table 11.2. *Comparison of nondestructive examination (NDE) methods for cracks (Tiku and Pussegoda 2003)*

Item	Ultrasonics	X-ray	Eddy current	Magnetic particle	Liquid penetrant
Capital cost	Medium to high	High	Low to medium	Medium	Low
Consumable cost	Very low	High	Low	Medium	Medium
Time of results	Immediate	Delayed	Immediate	Short delay	Short delay
Effect of geometry	Important	Important	Important	Less important	Less important
Access problems	Important	Important	Important	Important	Important
Type of defect	Internal	Most	External	External	Surface breaking
Relative sensitivity	High	Medium	High	Low	Low
Formal record	Extensive	Standard	Extensive	Unusual	Unusual
Operator skill	High	High	Medium	Low	Low
Operator training	Important	Important	Important	Important	Important
Training needs	High	High	Medium	Low	Low
Portability of equipment	High	Low	High to medium	High to medium	High
Dependent on material composition	Very	Very	Very	Magnetic only	Little
Ability to automate	Good	Fair	Good	Fair	Fair
Capabilities	Thickness gauging	Thickness gauging	Thickness gauging	Defects only	Defects only

11.3.3 Mechanical Damage Examination

A close-up visual inspection is typically considered for the detection of mechanical damage (e.g., local denting) as long as deformations are within specified limits in terms of extension and depth of the dent so that the inspection can be made safely. However, it is important to realize that such mechanical damage is usually accompanied by other types of deterioration, for example, cracking and coating damage, and these usually would need to be looked for as well.

Song et al. (2003) studied a guided wave technique for measuring mechanical damage where an approximate image of damage in terms of a discontinuity locus map is constructed using guided waves reflected from the damage. Measurement of dent geometry and stresses, studied by Babbar et al. (2005), uses the magnetic flux leakage inspection technique, which provides information on both geometry and stress effects.

11.3.4 Probability of Detection and Sizing

Uncertainties associated with the detection and measurement of deterioration originate from many sources, such as geometry, material properties, location of structural components, life of coating, type of cargo, operational conditions, loading cycles, sea water, internal temperature, humidity, environment, measuring sensors, lighting, and access.

Although some manuals, guidelines, and standards for NDE techniques address the uncertainties of the methods and measuring sensors (e.g., Halmshaw 1997; Porter 1992; Berens 1989; Bøving 1989), it is noted that a major source of data scatter is attributable to operator skill and practical difficulties rather than measuring equipment. For example, gauging for remaining thickness measurements may have some errors mainly due to errors inherent in measuring a sensor's location (Ma et al. 1999), which are not easy to quantify at the post-inspection stages of damage evaluation.

The uncertainties associated with damage detection and measurements are then characterized by statistical distributions in terms of probability of detection (POD) or probability of sizing (POS) (Rummel et al. 1989; Rudlin and Wolstenholme 1992; HSE 1997, 2000a). Figure 11.2 shows examples of the POD curves for fatigue cracking in ship structures (Fujimoto et al. 1996) and ship-shaped offshore structures (HSE 1997) as a function of crack size. The larger the crack size, the higher the probability of crack detection. Also, for cracks at structural details that are hard to detect, POD will be low. Moan et al. (2000) also discuss POD curves for cracks in offshore structures. Li et al. (2003) present a Bayesian updating method for POD curves in the case of corrosion inspections. Ivanov and Wang (2004) discuss the statistical distributions of uncertainties regarding corrosion measurements.

Demsetz et al. (1996) and Demsetz and Cabrera (1999) review the factors influencing the likelihood that the inspector can find the existing damages or failures. Goyet et al. (2004) report the lessons learned from the experience of a particular class society regarding inspection uncertainties and discuss its experience with the implementation of a risk-based inspection planning program that highlights that skill and experience are still very necessary and important in the inspection process. For a good discussion on the development and use of POD and POS curves in the inspection of offshore structures, see Dover et al. (2003).

11.4 Recommended Practices for Trading Tankers

The practices and guidelines available for condition assessment of trading tankers will usually be very useful for developing inspection and maintenance plans for ship-shaped offshore units, although certain differences will necessarily need to be accounted for. One reason is that the characteristics of loading and the environment affecting the deterioration in ship-shaped offshore units can be different from those in trading tankers; also, loading patterns and frequencies may be different.

The current procedures for condition assessment of trading tankers are based largely on experience gained over the years on the age-related deterioration processes as applicable to the trading ships concerned, together with some very costly lessons learned from required repairs and even catastrophic accidents involving age-related factors. The recent episodes related to the tanker accidents of the *Erika* and the *Prestige* and the related subsequent statutory developments is an example of lessons learned (Devanney 2006).

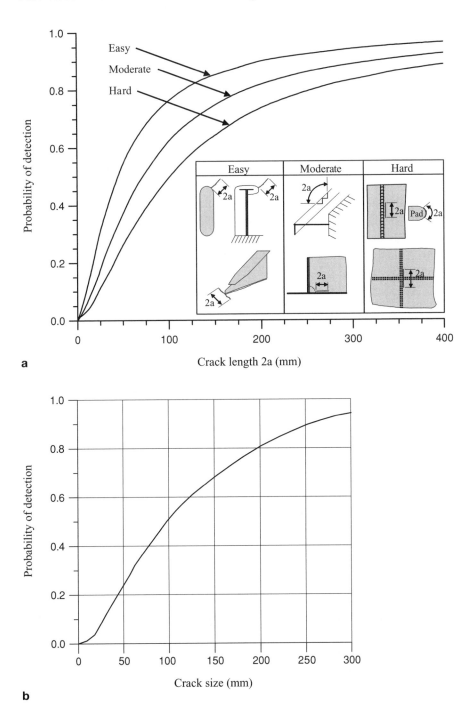

Figure 11.2. (a) An example of POD by visual inspection as a function of the crack length (size) in ship structures, depending on the complexity of structural details (Fujimoto et al. 1996). (b) An example of POD by visual inspection as a function of crack size in ship-shaped offshore structures (HSE 1997; courtesy of HSE).

It is interesting to note that since the early 1970s, trading tankers have been increasingly designed and built using the "minimum requirement approaches," that is, optimum structural scantlings plus minimum coating requirements (e.g., by painting perhaps only in bottom of cargo tanks), together with extensive use of high-tensile steel. Because the specifications of such designs were based on those specifications used for smaller ships built before and/or for ships in mild steel or lower-tensile-strength steels, they did not always cope well with subsequent age-related deterioration such as corrosion and fatigue cracking.

This was realized only over time as the tanker fleet started to age in the late 1980s, and it caused some concern among all the stakeholders interested in safety and quality of the ships – for example, ship owners, flag states, International Maritime Organization (IMO), classification societies, charterers, and insurance companies – each one involved in its own way to improve the situation. As a result, today there are more refined practices for inspection and maintenance of trading tankers available, as described in Sections 11.4.1–11.4.4.

11.4.1 Condition Assessment Scheme

The Condition Assessment Scheme (CAS) was originally started by classification societies as an optional response to the commercial need of charterers for greater information regarding vessel condition over and above minimum class. It still remains largely optional but with certain emerging exceptions discussed in this section. The Condition Assessment Programme (CAP) is a special, voluntary scheme of specified surveys, which leads to a rating of vessel condition from "excellent" to "poor." For ships over 15 years of age, the CAP rating often plays a part in chartering decisions. In the case of selection of vessels for conversions to floating offshore units, condition assessments are invariably necessary and provide important information that typical classification society records alone do not provide in general. For these purposes, however, typical class condition-assessment "packages" must invariably be supplemented as to the various inspections, gauging, and analyses required.

More recently, IMO Regulation 13G of MARPOL Annex I (refer to Appendix 6) has been amended following the impact of the *Erika* accident in 1999 and the *Prestige* accident in 2002 to bring the phasing-out schedule of single-skin tankers forward. The amended IMO regulation allows for continued operation of some of the affected vessels provided that a CAS is undertaken in a prescribed manner. In this regard, the applicable IMO CAS requirements (IMO Res. A.744(18)) are similar to the International Association of Classification Societies (IACS) Enhanced Survey Programme (ESP) requirements (IACS UR Z10.1), although there are some differences regarding survey planning, reporting, and flag state involvement.

Certain flag and coastal states may apply additional regional requirements. For instance, the Oil Pollution Act of 1990 (OPA 90) regulations of the United States mandates a high quality for all tankers operating in U.S. waters. Port cities including those in the United States that are increasingly concerned with the condition of ships passing their waters and entering their ports may also require operators to meet certain more stringent regulations.

Table 11.3. *Required frequency of inspection in ballast water tanks (IACS 2004)*

Year	5	6	7	8	9	10	11	12	13	14	15	16	17	18	19	20
Good	S			I		S			I		S			I		S
Fair or poor	S	A	A	I	A	S	A	A	I	A	S	A	A	I	A	S

Note: S = special survey; I = intermediate survey; and A = annual survey.

11.4.2 Enhanced Survey Programme

The ESP developed by IACS (2004, 2005a) is today the standard inspection and survey scheme required by classification societies and IMO for trading tankers. ESP procedures contain detailed and prescribed close-up inspections and thickness measurements to identify ongoing deterioration processes; the related requirements increase in scope as the vessel ages. ESP requirements have been improved over time, typically in the aftermath of major incidents such as those involving the *Erika* and the *Prestige* and also sometimes proactively.

As always, the corresponding classification society considers its various prescribed requirements including ESP to be a minimum regime, compliance with which is checked by classification societies and matters are rectified as needed only at certain points in time. As the ESP now specifically notes, the owner or operator is finally, and continually, responsible for the development and implementation of proper inspection and maintenance plans including ones additional to ESP. All observations of damage and defects by the operator must now be reported to the classification society in order to identify critical areas for future inspections and for adequate follow-up. The ship owner is also responsible for arranging safe access facilities for close-up surveys.

The special periodical survey (SPS) required by classification societies (and ESP) is usually every 5 years, with a couple of weeks often required for preparation of close-up access to critical structures. These surveys generally occur in dry docks in the case of tankers; and, as noted, their requirements are greater the older the vessel. The special surveys are supplemented by both intermediate surveys (IS) and annual surveys (AS). What is required during intermediate surveys can vary depending on the age of the vessel and other circumstances, including specific past experience. Annual surveys are typically carried out to determine the general condition of the vessel, and their scope may be altered depending on individual circumstances. For instance, in ballast water tanks, AS may be needed when the protective coating is in less than "good" condition, in cases where the "substantial corrosion" may exist, or if coating in ballast tanks was not applied to start with. Table 11.3 indicates the required frequency of inspection in ballast water tanks.

The coating condition is usually "measured" by three levels, that is, good, fair, and poor according to the guidance given in IACS Recommendation 87 (IACS 2004) for tankers; see also related guidelines by TSCF (1997). *Substantial corrosion* is currently typically said to have occurred when the corrosion wastage exceeds 75 percent of corrosion margins, although this particular definition could change after the implementation of IACS Common Structural Rules (CSR) for tankers (IACS 2005b). Information as to substantial corrosion must be recorded and reported to the

Table 11.4. *Structural parts for close-up surveys of trading tankers, following TSCF (1997)*

Class	Ship age ≤ 5 years	5 years < ship age ≤ 10 years	10 years < ship age ≤ 15 years	Ship age > 15 years
A	One web frame in a ballast wing tank, if any, or a cargo wing tank used primarily for water ballast	All web frame rings in ballast wing tanks, if any, or a cargo wing tank used for primarily water ballast	All web frame wings in all ballast tanks, a cargo wing tank, and each remaining cargo wing tank	As complete periodical survey as the case of the ship in 10 years < ship age ≤ 15 years; additional transverse including as deemed necessary by the classification society
B	One deck transverse in a cargo oil tank	One deck transverse in each remaining ballast tank, if any, or a cargo wing tank, and two cargo center tanks		
C		Both transverse bulkheads in a wing ballast tank, if any, or a cargo wing tank used primarily for water ballast	All transverse bulkheads in all cargo and ballast tanks	
D	One transverse bulkhead in a ballast tank, a cargo oil wing tank, and a cargo oil center tank	One transverse bulkhead in each remaining ballast tank, a cargo oil wing tank, and two cargo center tanks		
E			One deck and bottom transverse in each cargo center tank	
F			As considered necessary by the surveyor	

Notes: Class A = complete transverse web frame ring including adjacent structural members; Class B = deck transverse including adjacent deck structural members; Class C = complete transverse bulkhead including girder system and adjacent members; Class D = transverse bulkhead lower part including girder system and adjacent members; Class E = deck and bottom transverse including adjacent structural members; and Class F = additional complete transverse web frame ring.

classification society so that it will be an explicit part of future vessel inspection and maintenance. Such information could also be essential input data for the development of risk-based inspection and maintenance schemes by the owner or operator.

Table 11.4 indicates recommended structural parts of trading tankers for which close-up surveys are required according to TSCF guidelines (TSCF 1997). The quality of the close-up survey is in turn reflected in the confidence one may place on the results of thickness measurements (gauging). A special certification scheme has now been implemented as part of the ESP to cover the competence of thickness measurement companies. Guidelines for ultrasonic thickness measurements are given by IACS (2005a) in general terms and by the individual classification societies at a detailed level (e.g., DNV 2004a). Requirements of ESP are also employed, by default, to floating offshore structures in many cases, either because it is considered good practice, or because it is mandated by the classification society involved, or for a combination of reasons. However, in the case of such structures, operators do nowadays increasingly replace the prescriptive requirements of ESP by individually developed and tailored risk-based schemes of inspection and maintenance.

11.4.3 Emergency Response Services

The emergency response services (ERS) provided by recognized bodies, usually the for-profit parts of or outgrowths of classification societies, are special schemes to assist ship owners to handle emergency events such as collision, grounding, fire, explosion, and heavy weather damage. The aim of the ERS is to safeguard the ship, the crew, and the environment. This is generally an optional scheme although in some cases, jurisdictions may mandate the same. In the case of floating offshore structures, most major operators do have or have contracted for ERS capabilities as standard procedure.

The ERS may apply standard naval architectural calculations and also more sophisticated numerical simulations following accidental limit-state-based methods in the damaged condition as well as serviceability and ultimate limit-states-based methods in damaged and recovery conditions as may be necessary. Considerations may include damaged vessel stability, hull girder collapse due to accidental flooding together with structural damage, emergency ballasting, floating off of grounded vessels, temporary repairs, and rescue operations.

11.4.4 Ship Inspection Report Programme

The Oil Companies International Marine Forum (OCIMF)/Ship Inspection Report Programme (SIRE) started in 1993 with the view to establish a transparent database of ship condition in order to aid the vessel vetting process (see, e.g., INTERTANKO 2003). A detailed questionnaire reflecting the ship condition including its crew, quality, and safety systems is filled out by specially trained inspectors and is made available as part of the vetting process to oil majors and others through a centralized computer database in London.

Regarding the ship's hull condition, the questionnaire addresses the general condition, the repair history, and may also refer to the classification society's ESP results, class, statutory certificates, and status. An original aim of the program was to reduce multiple inspections that were undertaken on the part of various vetting interests. In addition to usual trading tankers, the OCIMF/SIRE scheme may also be used as part of the vetting and clearing process related to shuttle tankers berthing and offloading from ship-shaped offshore units.

11.5 Risk-Based Inspection

Inspection is required to detect deterioration before a possibly catastrophic, polluting and/or expensive failure can result. Inspections in themselves may not affect the likelihood of structural deterioration. But any excessive deterioration can be found by relevant inspections and the subsequent actions such as repairs, replacements, or adjustments can then be taken. If such potential problems can be identified in

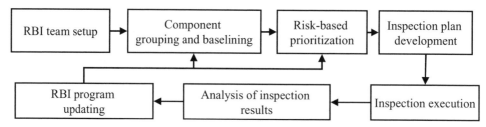

Figure 11.3. Main steps for developing risk-based inspection program, following ABS (2003).

a timely manner, proper actions with risk-corrective/preventive measures can be applied to reduce the likelihood of structural failure. Although risk in any event does not reduce to zero, it can be managed and controlled under an acceptable level, discussed in Chapter 13.

The aim of risk-based inspection (RBI) is to develop an inspection plan that can help prevent or greatly reduce failures of the offshore unit compromising its safety, the environmental or economic viability, by applying risk assessment and mitigation technology. Effective inspection programs including identification of required inspection frequency and practices are to be specifically provided. In the RBI scheme, the application of risk assessment is an essential component and it considers both the likelihood and the consequences of structural failures initiated by various types of structural deterioration and also preexisting defects and conditions.

Figure 11.3 shows the main steps involved in developing an RBI program (ABS 2003). The first step is to set up an RBI team who will establish the goals of the RBI program and also decide on the overall RBI approach needed to arrive at an inspection plan that successfully achieves the desired goals. The second step is about the component grouping and baselining, where the components that will be subjected to the RBI program are identified and grouped. In-service data as well as design features of such components are sought and will also be collected and examined for this purpose. In the third step, a risk-based prioritization is performed after a risk assessment so that the components involved are ranked on the basis of risk, from the highest risk to the lowest risk.

The fourth step is then to develop an inspection plan based on the results of the risk prioritization so that the risk of failure is kept under an acceptable level. In the fifth step, the next required inspection is executed and the inspection results are evaluated. This information is required by the RBI plan for the identification and implementation of measures for successful continued operation of the offshore unit until the next inspection. In the final conceptual step, the RBI plan may be updated for the future, on the basis of the inspection results, observed deterioration mechanisms, and other prior experience.

In Sections 11.5.1–11.5.7, the principles and practices of RBI plan developments are summarized; see ABS (2003) for a more detailed description.

11.5.1 RBI Team Setup

The composition of an RBI team will depend on the complexity of the project, installation, the scope of the RBI program, and also on any applicable regulatory requirements that may need to be satisfied. Experts who are familiar with risk assessment including the identification of potential causes of failure, their likelihood, and the determination of consequences of failure must be involved. Typically, the RBI team will consist of individuals who have the expertise in the following disciplines:

- Risk assessment
- Maintenance and inspection
- Structural deterioration and related failure mechanisms
- Structural integrity and reliability
- Operations and related hazards
- Production processes and related hazards
- Health and safety
- Materials and their selection and application

All of the RBI team members will usually need to stay involved in all tasks until the RBI plan has been developed. Brainstorming is an essential process that will be used to identify various types of items pertinent to the exercise.

11.5.2 Component Grouping and Baselining

To execute this step of the relevant RBI scheme, various forms of the information are required. These may include original design and construction data, subsequent structural modifications, inspection and maintenance records, and operational histories. When information is lacking or the level of its accuracy is low, some conservatism may result depending on the assumptions made.

The baseline data may be obtained from an initial baseline condition survey and also specific measurement data and/or previous inspection records for the structure concerned and, to a lesser extent, for similar structures. An indepth review of such data is then made to identify and formulate the various component groups that make up the offshore structure in question. The likelihood and consequences of hazards associated with deterioration in these groups must also be identified so that risk assessment can be subsequently carried out.

Based on the information, certain logical groupings of components will be defined as inspectable units. Note that an inspectable unit must be large enough to have significant consequences of deterioration, but at the same time it should be small enough to have similar load effect and deterioration-mechanism exposures.

Examples of possible inspectable higher-level units for offshore installations are as follows:

- Cargo tanks
- Ballast tanks
- Void spaces
- Watertight compartments
- Spaces with through-hull connections
- Pump rooms

11.5.3 Risk-Based Prioritization

A risk assessment is required to make the risk prioritization. Once the risk assessment is completed, the component groups to be subject to RBI inspection can be ranked on a risk basis, from those with the highest risk to those with the lowest risk. Such prioritization may be influenced by additional factors such as anomalies, repairs, or scheduled shutdown programs. Some details of related risk assessment and risk-based ranking methodology are presented in Chapter 13; see also the original document by ABS (2003). For a risk-based underwater inspection prioritization, see DeFranco et al. (1999).

11.5.4 Inspection Plan Development

With a risk-prioritized list of inspectable units developed, appropriate inspection strategies are then selected to consider the damage detection methods, scope (e.g., sample size, location, and extent of inspection), and frequency.

11.5.4.1 Inspection Strategy

The inspection strategy must address the following aspects:

- Which items are susceptible to deterioration and where are they located.
- What inspection methods must be adopted to deliver the required inspection results.
- How effective the selected inspection methods are at detecting the possible deterioration mechanisms.
- How much inspection is required to ensure the target inspection effectiveness.
- What frequency of inspection is required for each inspectable unit or component.

11.5.4.2 Scope of Inspection

An inspection plan must address where to inspect and how much to inspect in terms of sample size (number of test points), location, and extent of the inspection for purposes of measuring the level of activity of the deterioration process. Note that risk will generally increase as inspectable units deteriorate with time. As the number of components affected by the same deterioration mechanism increases, the associated likelihood of loss of integrity can increase as well. Most types of deterioration are time-variant and, therefore, risks are higher for older and more intensely used units.

The sample size (number of test points) must be large enough to be collectively representative of the entire deterioration mechanism. For example, uniform corrosion wastage characteristics in an inspectable unit must be measured at a sufficient number of points spread appropriately over the unit when the corrosion rate is of concern. For localized deterioration mechanisms (e.g., pitting, cracking), a much greater number of points must typically be inspected, but prioritization among them is possible if certain types of structural analyses results and specific past experience are available.

Inspection locations should be defined and selected so that common features susceptible to deterioration mechanisms can be considered within each inspectable unit. These may include the following:

- Weld seams and heat affected zones
- Hard spots and also complex connections involving structural components
- Heat-affected zones from welds on component surfaces (e.g., welded pipe supports)

- Vapor spaces in the deck head
- Process internals, phase boundaries
- Difficult-to-inspect internal structural components
- Areas subject to impingement of water

The extent of inspection within a component or inspectable unit may be decided by consideration of the component size and the likely uniformity of the deterioration environment. For a component or a larger unit, appropriately selected areas will need to be inspected considering efficiency and economy; the whole unit, or component, may be more readily inspected if it is small enough. However, the areas must be so selected that they, when taken together, adequately represent the deterioration behavior of the entire component, the unit, and, by extension, the structure as a whole during the service life.

11.5.4.3 Frequency of Inspection

The frequency of inspection is related to the time interval between planned inspections, which is to be determined considering the expected deterioration rate and the overall condition identified for a structural unit at an inspection. The earlier inspections usually are more conservative. After actual characteristics of deterioration and the rate of deterioration become sufficiently recognizable, for example, through the first few inspections, the inspection frequency can then be optimized.

11.5.5 Inspection Execution

For a successful RBI, the inspection plan developed must be executed correctly. This is because the results of each inspection have a significant impact on both the perceived integrity of the installation and the accuracy of the subsequent RBI program updates. Also, the inspections are the primary sources for gathering deterioration data.

The following are some of the prerequisites that will need to be considered for a successful RBI execution:

- Prior definition of clear and concise inspection work scope, including inspection control procedure
- Reporting format standardization
- Qualified inspectors
- Precision of equipment used for inspection
- Clear anomaly criteria and reporting process
- Clear management process for any possible change in the inspection procedure, allowing flexibility to respond to findings on a real-time basis
- Clear safety guidelines and policies

11.5.6 Analysis of Inspection Results

After inspection activities are completed, the inspection results must be analyzed so that important information for developing future inspection plans can be obtained. In some cases, anomalous data falling outside the normal operational boundaries or the acceptable risk level may be observed and some remedial actions may be required as a matter of urgency. The following are possible actions to resolve such issues:

- Reinspection to resolve data capture, measurement, or input errors
- Additional inspections including broader coverage and possibly more invasive techniques to refine the extent of the anomalous condition
- Technical analysis of the installation, unit, and its components to determine their fitness for purpose for continued service; for example, corrosion predictions using more accurate corrosion wastage models, refined fatigue analysis, fatigue-crack-growth analysis, and fracture mechanics analysis
- Development of repairs and modifications to restore the structure or its components to a state that is suitable for safe operation
- Modification of the RBI plan to increase and/or modify the inspection scope and frequency

Another aim of the analysis of inspection results is to obtain trending information related to the deterioration mechanisms. It is important to identify whether the current deterioration trends are comparable to the anticipated trends established from data and previous inspections, and whether the trends are still suitable or need to be modified.

11.5.7 RBI Program Updating

By a continuous feedback with and analysis of inspection data, the RBI program can be updated to improve its effectiveness. Because most types of deterioration in offshore units are time-variant, the RBI program by necessity must be modified periodically and at important stages during the unit's service life. The RBI program will be updated to include many of the same steps that were required for the development of the original RBI program, such as risk assessment, risk prioritization, and inspection plan. This is real-time data, and it is pertinent to the deterioration mechanisms that are present because the data is gathered from the previous inspections. It will help improve the future accuracy of the program.

The RBI program updating may result in revisions to the following:

- Risk ranking of components
- Inspection methods
- Inspection frequency and/or scope

11.6 Risk-Based Maintenance

A risk-based maintenance (RBM) scheme is essentially a systematic maintenance process in which risk management and control techniques are applied with the aim of maintaining, to appropriate levels, the reliability of components that govern the system reliability. Such schemes are developed, for example, for various systems involving machinery on board. In the following sections, concepts pertinent to a RBM scheme and its development are summarized; see ABS (2004) for greater detail.

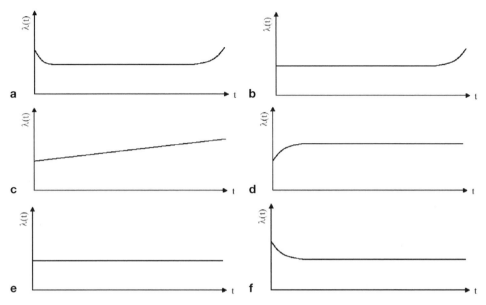

Figure 11.4. Various patterns of conditional failure rate: (a) pattern A: bathtub; (b) pattern B: traditional wear-out; (c) pattern C: gradual rise with no distinctive wear-out zone; (d) pattern D: initial increase with a leveling off; (e) pattern E: random failure; and (f) pattern F: infant mortality.

11.6.1 Time-Variant Failure Mechanisms

To develop an RBM scheme, the time-variant failure characteristics and mechanisms for the system components in question must be suitably identified. The failure distribution over time follows the Weibull function.

A useful measure to characterize the time-variant failure frequency is the conditional failure rate denoted by $\lambda(t)$ as a function of time t, which relates to the probability of a failure occurring during the next instant of time, given that the failure has not already occurred. This measure often gives useful information about the survival life and the expected future failure experience patterns. Six patterns of conditional failure rate are shown in Figure 11.4 (Smith 1993; Moubray 1997).

For decision making related to maintenance, failure characteristics representing the entire possible failure history in terms of wear-in failure, random failure, and wear-out failure must be known. Pattern A of the conditional failure rate is suitable to illustrate the failure characteristics where the three stages of failure behavior are identified as shown in Figure 11.5.

With the failure pattern identified, an appropriate maintenance strategy can be sought. For example, if the failure stage is characterized by wear-out failure, then replacement or rebuilding must be taken as the applicable strategy. However, this same maintenance strategy may not be the right one to use if the expected failure is due to infant mortality or wear-in. Furthermore, if the expected failure rate is high or failure will occur frequently, redesign may be a better strategy if possible when compared to more frequent maintenance. Associated costs versus benefits must be considered in such decisions.

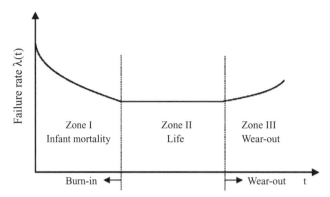

Figure 11.5. A schematic of conditional failure rate by pattern A (ABS 2003).

In an RBM scheme, failure corrective/preventive measures can be determined through the following actions:

- Proactive maintenance
- Redesign or modification
- Operational condition changes

The basic idea behind proactive maintenance actions is to prevent the failures before they occur or to detect the onset of failures in a sufficient enough time before they occur so that they and their effects can be managed and controlled. When the failure rates are too high or proactive maintenance actions are not enough, redesign or operational condition changes need to be applied. The proactive maintenance actions may be divided into the following four categories:

(1) Planned maintenance
(2) Condition-monitoring
(3) Combined planned maintenance and condition-monitoring
(4) Failure-finding

Regarding maintenance in general, the overall objectives of the maintenance strategy selected will be various. These will include maintaining the process plant and equipment such that safety and asset integrity is at the appropriate high levels, the uptime is maximized, and the letter and spirit of all applicable states and regulations are met. Maintenance is particularly important to equipment that is critical to safety and availability. These may include items that are important to the safeguarding of process or offloading equipment on the FPSO, the wells, subsea flow lines, manifolds, and risers. Therefore, a related criticality assessment is essential. The approach to the maintenance will also need to consider both vendor recommendation and associated data reliability wherever available. Other factors to consider may include the following:

- Potential for application of noninvasive inspection techniques and remote diagnostics
- Appropriate levels of operator training and maintenance of related skills over time
- Equipment layout and access

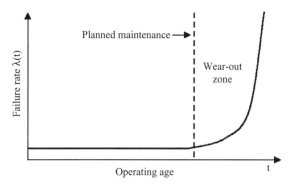

Figure 11.6. A schematic of wear-out failure mode showing a clear failure life (ABS 2003).

In Sections 11.6.2–11.6.5, we summarize each of the proactive maintenance-related actions.

11.6.2 Planned Maintenance

At a specified interval, planned maintenance (also called "preventive maintenance") is carried out regardless of the actual conditions of the components in question. When the failure type is of the wear-out type, the time interval for planned maintenance must necessarily be adequately short. Planned maintenance may be further subdivided into two categories: "restoration" action and "discard" action. The restoration action is a scheduled maintenance task performed before the end of a specified interval by restoring the capability of components. The discard action is a scheduled maintenance task performed before the end of a specified interval by replacing components in question.

The failure patterns A and B illustrated in Figure 11.4 exhibit a clear failure life, but other cases may not be so distinctive in that regard. For planned maintenance action to be effective in managing and controlling the failure, the failure patterns must be characterized by pattern A or B, as shown in Figure 11.6. After a planned maintenance action is performed, the failure rate may be reset, as illustrated in Figure 11.7.

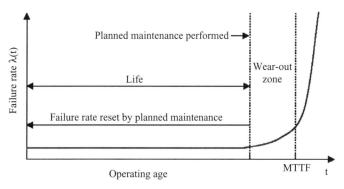

Figure 11.7. A schematic of failure rate reset by a planned maintenance (MTTF = mean time to failure) (ABS 2003).

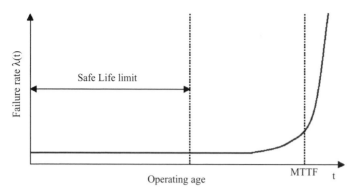

Figure 11.8. A schematic of safe life-limit concept (MTTF = mean time to failure) (ABS 2003).

The interval of planned maintenance action can be determined using the following data or information:

- Information on fabrication and construction
- Expert opinion
- Published data
- Regulatory requirements
- Statistical analysis of failure history including data for mean time to failure

Two concepts are relevant in the determination of the planned maintenance interval: the safe life limit and the economic life limit. Figures 11.8 and 11.9 illustrate these concepts.

The safe life-limit concept is applied when severe safety or environmental impacts or the highest risk events are anticipated. The action interval is set to ensure that there is little chance of failures occurring before the planned maintenance is carried out. In general, the interval is set well before the mean time to failure.

The economic life-limit concept is applied for all other failure modes. The action interval in such cases is determined on the basis of the economical points of the maintenance action and also the expected life of components in question. In some cases, the action interval may be set to be even beyond the point of the mean time to failure.

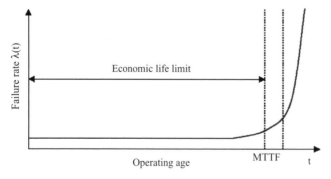

Figure 11.9. A schematic of economic life-limit concept (MTTF = mean time to failure) (ABS 2003).

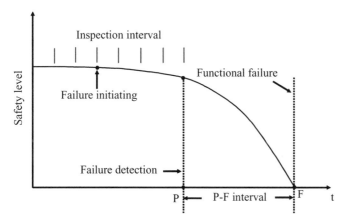

Figure 11.10. A schematic of potential failure diagram (P-F diagram) until functional failure (ABS 2003).

11.6.3 Condition Monitoring

The *condition monitoring action* (which may also be sometimes called "predictive maintenance") is a scheduled maintenance task used to detect the onset of a failure before it occurs so that appropriate maintenance action can be taken to prevent failure. For this purpose, theoretical and/or numerical simulations can also be performed to identify the time-variant condition of components associated with various failures.

Figure 11.10 shows a schematic of potential-failure diagram (called the P-F diagram) that represents the condition variation of components as a function of time until functional failure occurs. The time interval between point P and point F in Figure 11.10 is called the P-F interval, which implies the warning period, that is from the detectable time of failure onset to the time of occurrence of functional failure.

The available P-F interval, which indicates the actual time between discovery of the potential failure and the occurrence of functional failure, is usually shorter than the P-F interval. For example, when the inspection interval is once per year and the P-F interval is 3 years, then the available P-F interval is 2 years. The available P-F interval should, of course, be longer than the time required to take action before the functional failure. Methods similar to those used to determine the planned maintenance interval could also be applied for the determination of the condition-monitoring interval.

11.6.4 Combination of Planned Maintenance and Condition Monitoring

When the planned maintenance or the condition monitoring by itself is not sufficient to reduce the risk under an acceptable level, a combination of both actions may be taken. The decision as to whether to adopt a combined action approach or not can be based on the risk-assessment results obtained in association with failures of components in question.

11.6.5 Failure Finding

Failure finding is a scheduled maintenance task used to detect failures that may lie hidden and may not usually be detected during normal crew operations or regular

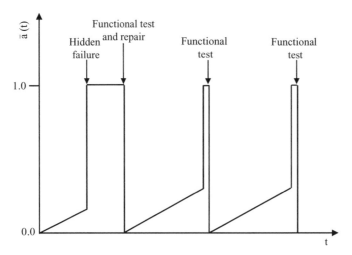

Figure 11.11. A schematic of functional test of a safety component for reducing the potential unavailability level using failure-finding action (ABS 2003).

inspections and when planned maintenance or condition monitoring is not applied. This is usually a scheduled function check to ensure that the required function of a component is performed. Most components considered for failure-finding action are either standby or protective equipment such as safety valves on a boiler or the power-up button on an emergency electrical generator.

Figure 11.11 shows a schematic of functional test as failure-finding action that is performed to reduce the unavailability level or increase the availability level of a safety component. The failure-finding action interval may be determined by mathematical considerations together with related risk acceptance criteria. The frequency of occurrence of the resulting failure event can be expressed as

$$F_m = F_i(1 - a) = F_i\bar{a}, \tag{11.1}$$

where F_m = frequency of occurrence of the multiple failure; F_i = frequency of occurrence of the initiating event due to the hidden failure event; a = availability of the safety component or backup component; and $\bar{a} = 1 - a$ = unavailability of the safety component or backup component.

When the acceptable frequency of the resulting failure occurrence is F_{mc}, the unavailability of the safety component can be obtained from Eq. (11.1) by replacing F_m with F_{mc}, as follows:

$$\bar{a} = \frac{F_{mc}}{F_i}. \tag{11.2}$$

The failure-finding interval may then be given by

$$T_m = 2\bar{a}T_f, \tag{11.3}$$

where T_m = failure-finding interval; T_f = mean time to failure of the component with hidden failure; and $\bar{a}$ = as defined in Eq. (11.2).

Table 11.5. *Causes of structural damages in FPSO structures, following LR (2003)*

Damage	Cause
Bow damage	Inadequate structural design and inadequate consideration of environmental loadings
Caisson damage	Improper material selection
Flare damage	Inadequate structural design and inadequate consideration of environmental loadings
Tank damage	Inadequate consideration of environmental loadings or errors in design process; unsatisfactory construction techniques; site-specific loadings not anticipated in design process
Breakdown of coating systems	Poor surface preparation, application, and/or selection
Swivel damage	Use of new technology

11.7 Recommended Practices for Ship-Shaped Offshore Units

11.7.1 Inspection Practices

Almost all of the significant and expensive failures on ship-shaped offshore units can be attributed to various mostly detectable and addressable causes. Table 11.5 indicates some examples of damage causes (LR 2003). As presented in Section 11.3, various methods are available for examining damages. Table 11.6 indicates practical applications of inspection techniques for FPSOs in terms of observed percentages as provided by (LR 2003) as part of a study undertaken for the UK Offshore Operators Association (UKOOA).

Special inspections for ship-shaped offshore units, used for storing (and offloading crude oil cargo (and similar to trading tankers) will also include a significant amount of close-up visual inspection that is usually carried out on site instead of in a dry dock (HSE 2000b).

Inspection for fatigue cracks is usually performed in the first instance by visual inspection. Nondestructive testing methods (DNV 2004a) using ultrasonic inspection or magnetic particle technique are useful for the focused examination of selected high-stressed and fatigue-prone areas and to better assess the size of defects. Whatever the method of inspection used, it is important to realize that the probability of detection or sizing may depend on the crack size, general visibility, access, inspector training, surface condition (e.g., painted, corroded, oil covered, or slime covered), and various other parameters.

Inspection for corrosion including pitting is generally visually undertaken, followed by thickness measurements in selected areas using ultrasonic thickness gauging. In theory, destructive methods such as drilling and cutting may be applied to get more accurate measurements, but this is not convenient except when studying material that has already been removed from the structure, for example, in the case of investigations of accidents.

The concepts for inspection planning may be divided into either deterministic or probabilistic approaches. In both approaches, one determines the inspection

Table 11.6. *Application of damage examination techniques to FPSOs, following LR (2003)*

System	Component		Average inspection interval (months)	General examination (%)	General visual (%)	Close visual (%)	Ultrasonic examination (%)	Radiography (%)	Magnetic particle inspection (%)	ROV examination (%)	Coating examination (%)	Anode inspection (%)	Vibr monitoring (%)	System test (%)	Rocking test (%)	Grease analysis (%)
Ballast water system			60													
	Pipework	GRP	48			67										
		Cunifer	48			17										
		Carbon Steel	54			100	83		33					17		
	Tanks	WB Tanks	54	17		83	100		83		100	100		17		
		WB Tanks	120			17										
		Forepeak	54	17		83	100		83		100	100		17		
		Forepeak	120			17										
		Afterpeak	54	17		83	100		83		100	100		17		
		Afterpeak	120			17										
	Pumps		25			100							100			
	Control systems		34			100								100		
Oil storage system	Pipework	Carbon steel	54			100	83	50								
	Tanks	Cargo tanks	53	17		83	83		83		100					
		Cargo tanks	120			17										
		Slops tanks	43	17		83	83		83		100					
		Slops tanks	120			17										
	Pumps		25			83							83			
	Control systems		27			100								100		
Hull	Tanks and above water	Continuous survey hull	72													
		Subsea	30													
	External	Sea chests	33			83	83			100	100	100				
	Internal	Sea chests	45													
	External	Turret	27			100				83	100	50				
		Cathodic protection	30							83				83		
		Wind and water area	25			100	17				100					
Caissons			52			33	33		33	17	33	17				
Deck structures	Pallets		33			100				83				17		
	Walkways		33			100								17		
	Deck plating		38			100	100							17		
Tank venting system			19													
	Pipework		50			100	83		83							
	P/V valves		34			100								17		
	Seals		26			100								17		
Cranes			42			100					33		33			
	Grease sampling		3													100
	Rocking test		6												100	
Thrusters			33			67					83	17	17	33		17
			21							67						
Swivels and drag chains	Swivels–leak recuperation		1		33	33										33
	Swivels–instrumentation		54		33									33		
	Swivel stack (mechanical)		60		33	33										
	Chains / stoppers / anchors		33		33	33					33					

interval so that the next inspection must be undertaken before the largest undetected defect reaches a "critical" size. In the probabilistic approach, explicit limitation of consequences to an allowable risk level is a part of the analysis.

In the deterministic approach, lower-bound capacity and upper-bound demand parameters are used together with a deterministic safety factor to accomplish a similar aim, although usually more pessimistically because of the nature of the assumptions that are made. The explicit consideration of variability in parameters including the probability of detection is what makes probabilistic approaches more powerful but flexible enough to be better tailored to the particular circumstances at hand. Both approaches will typically include crack growth calculations and fracture mechanics analysis, for example, as related to critical crack sizes. Different inspection intervals may be set for different areas or components in question through such approaches.

Risk-based inspection methods, which include reliability-based methods, therefore, allow one to better determine cost-effective inspection options while keeping the risk under an acceptable level. Risk-based inspection is highly desirable but a difficulty arises from the fact that the output of the risk-based approach is sensitive to the risk-assessment values that are quite subjective; however, the traditionally rule-based inspection approach is inflexible. In this situation, an intermediate strategy (LR 2003) may be offered as follows:

- Initial examination and response to the developments of inspection schemes can be largely driven by generic recommendations, pooling experience, and learning from a large fleet, bringing in relevant practice from outside areas, for example, from trading tankers, FPSOs under varying circumstances, and in various regions.
- Subsequent particularization of tactics and strategies would be driven more directly by vessel-specific experience.
- Modeling and analysis would be aimed at identifying on a rational basis (not necessarily a risk basis) the needed inspection and maintenance.

11.7.2 Maintenance Practices

For maintenance and repair actions of ship-shaped offshore units to be considered effective, the following factors must be achieved to the requisite degrees of success:

- Repair in situ, that is, without going off the field or dry-docking
- Repairs ideally affecting only the repair area, without functional stoppage or interruption including the production storage areas and offloading in other areas
- Repair, ideally without hot work such as cutting or welding
- Fast track and cost-effective repair
- Repair by easy-to-apply and readily or even locally available technologies and personnel
- Reliable repair methods backed up by a large amount of experience

LR (2003) has surveyed several operators' experiences regarding repairs and modifications on FPSOs operating in the UK Continental Shelf (UKCS), as summarized in Table 11.7. In addition to remedial actions for age-related deterioration, such as fatigue cracks and corrosion, it is also seen that a number of modifications required to improve the serviceability and operability of the units possibly arose because the original design may not have been adequate. In this table (and in similar experience compilations), we caution that all "causes" noted are generally speculative by nature.

Table 11.7. *Selected experience related to repairs and modifications for FPSOs, following LR (2003)*

Damage or inadequacy	Remedial actions
Fatigue cracks in water ballast tank frames	Fatigue cracks detected in lower flume openings after 2–3 years of operation as a converted FPSO after operation of about 15 years as a trading tanker. The cracks were drilled and ground. Modifications using rope access were made. These are now subject to annual monitoring.
Defects in cargo oil tanks	Defects found in two starboard cargo tanks in way of transverse lower support brackets. Repaired using additional brackets and new insert plates. A high level of nondestructive examination and strict welding control is required.
Breakdown of paint coating	Breakdown of paint coating in various areas of vessel hull structures was found. The cause is perhaps inadequate selection and application of coating. Recoating is necessary.
Corrosion in caissons	Extensive corrosion of seawater and firewater caissons in the water ballast tanks mainly caused by coating breakdown. Repairs by means of external plugs and recoating were partially successful. In some severe cases, repairs were attempted by recoating and by grouting a larger annular sleeve, but they were not successful. The cement leaked into and blocked base of caisson to a depth of 1–2 m.
Bow damage	Heavy weather damage to plating and internals of vessel's bow was found. Plating variously indented between stiffeners with internal brackets sprung. Repaired on location using heavier section bulb bar and larger brackets with strict welding control. Tears in way of inner deck were faired and rewelded.
Green water impacts	Green-water-impact effects were observed. Additional green-water protection added to protect the process equipment pallets aft of the forecastle.
Deformation on main deck foundations and supports	Process module main deck foundations and supports were found to be inadequate large after a structural motion analysis showing accelerations and forces attributable to the vessel movement to be in excess of the original design limits. Modifications would require substantial strengthening.
Excessive roll motions	Bilge keels added to alleviate the excessive roll of the vessel during heavy swells.

Unlike trading tankers, dry-docking of ship-shaped offshore units is usually not planned during the entire production period in the field, possibly more than 10 years to even 20 years. The repair work of offshore units in situ using welding or flame cutting that is common for traditional repairs of trading tankers may be concerns for high-fire and explosion risk. Large parts of an offshore structure may need to be isolated and/or closed during welding, which is also very costly. Some have argued that FPSO hot work repairs, when undertaken, cannot completely follow standard industry guidelines for safety during hot work, such as those from ICS/OCIMF/IAPH (1996), and must be dealt with considering particular circumstances and using risk-based approaches to identify the safety measures needed.

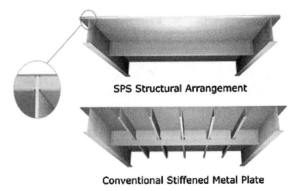

SPS Structural Arrangement

Conventional Stiffened Metal Plate

Figure 11.12. A schematic of the proprietary Sandwich Plate System (SPS 2005).

When extensive areas are repaired by welding, the offshore installation may expect production shutdown, transit to repair yard, dry-docking for repair, transit to field, and recommissioning. This may take several months. Repairs of small areas by welding may usually be accomplished on site following waiting for mild weather, limited production shutdown, repair, and production restart. Even small repairs may take several weeks. It is then to be emphasized that the most ideal way to minimize the possibility of expensive onsite repairs is by building in additional structural design safety margins to start with, and that these margins need to be greater to the extent possible when compared to trading tankers for which dry-docking every 5 years is the norm.

An alternative repair method to traditional welding is the use of adhesive bonded patches, for which no concern regarding fire hazards during repair exists because no hot work is undertaken. Composite (fiber–reinforced plastic) patches can be bonded or laminated over the structure to bridge and reinforce corroded or cracked areas.

There are potential difficulties as well, including the difficulty of restoring any lost strength and the possibility of cargo and gas accumulation within imperfectly bonded parts. In contrast, the prevention of leakage from corrosion is more readily possible. The use of repair methods with composite patches is said to have been successful in naval ships, bridges, and some infrastructure repairs (*Oil and Gas News* 2001); it may be expected that such methods will also be increasingly considered for offshore structures in the future.

Another new technology for repair is the Sandwich Plate System (SPS), which consists of two metal plates bonded to a compact elastomer core, as shown in Figure 11.12. The elastomer provides continuous support to the plates, stops local plate buckling, and is said to eliminate the need for stiffeners. SPS is meant also to possibly replace conventional stiffened metal plates in maritime, offshore, and civil-engineering structures. It is used for new construction and also as SPS overlay for repair and conversion. SPS overlay bonds a new top plate to the existing structure in a process that is said to be quick and economical (SPS 2005). The SPS technology is certified by several classification societies and beginning to be more widely used, for example, for new construction of barges and repair of damaged hull parts in the parallel mid-body, although some difficulties may still remain

Figure 11.13. A sample of a pitting
intensity diagram (degree of pit intensity = 20 percent).

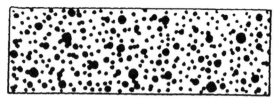

Figure 11.13. A sample of a pitting
intensity diagram (degree of pit intensity = 20 percent).

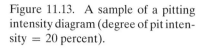

regarding obtaining required levels of shear strength and fatigue performance in
routine practice.

11.8 Effect of Corrosion Wastage on Plate Ultimate Strength

Corrosion wastage of plate elements of ship-shaped offshore units can reduce their
ultimate strength. Two types of corrosion damage are usually considered: general
(or uniform) corrosion and localized corrosion. General corrosion reduces the plate
thickness uniformly, but localized corrosion such as pitting appears nonuniformly in
selected regions, for example, the vessel bottom in crude oil cargo tanks. The ultimate
strength of a steel member with general corrosion can be easily predicted, that is, by
excluding the plate-thickness loss due to corrosion.

Based on a series of experimental and numerical studies on steel-plated structures,
however, we know that the plate ultimate strength reduction characteristics due to
general corrosion are quite different from those due to pit corrosion. The equivalent
plate thickness reduction approach, which represents a pitted plate with a plate of an
"equivalent thickness," is not always sufficient for accurately predicting the plate's
ultimate strength.

Figure 11.13 shows a sample of pitted plates. Figures 11.14 and 11.15 represent
examples of the structural models used for the experiment and nonlinear finite element computations respectively, undertaken by Paik et al. (2003).

To assess the magnitude of breakdown due to pit corrosion, a parameter denoted
by degree of pit corrosion intensity (DOP) is often used. DOP may be defined by a
volumetric basis:

$$\text{DOP} = \alpha = \frac{1}{abt} \sum_{i=1}^{n} V_{pi} \times 100 \ (\%), \tag{11.4a}$$

where n = number of pits; V_{pi} = volume of the ith pit; a = plate length; b = plate
breadth; and t = plate thickness. For more pessimistic strength measure, DOP is
often defined by a surface area basis:

$$\text{DOP} = \alpha = \frac{1}{ab} \sum_{i=1}^{n} A_{pi} \times 100 \ (\%), \tag{11.4b}$$

where A_{pi} = surface area of the ith pit, which may be calculated as $A_{pi} = \pi d_{ri}^2 / 4$
with d_{ri} = diameter of the ith pit, for a circular type of pit corrosion.

A series of experimental and numerical studies for steel-plated structures with
pits and under axial compressive loads or edge shear was performed by varying the
DOP, the depth of pit, the regularity of pit, the plate thickness, and the plate aspect

a

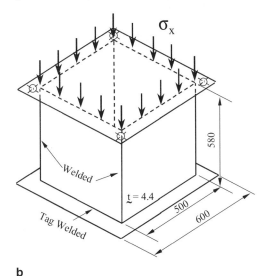

b

Figure 11.14. (a) Collapse test set-up for a box column type of a plated structure with idealized pits. (b) A schematic view of the test structure.

ratio (Paik et al. 2003). It is found from the experimental and numerical studies that the ultimate strength of a plate with pit corrosion can be estimated using a strength knock-down factor that can be calculated using the following approximations in the cases studied:

$$R_{xr} = \frac{\sigma_{xu}}{\sigma_{xuo}} = \left(\frac{A_o - A_r}{A_o}\right)^{0.73}, \quad \text{for axial compressive loads,} \quad (11.5)$$

where R_{xr} = a factor of ultimate compressive strength reduction due to pit corrosion; σ_{xu} = ultimate compressive strength for a member with pit corrosion; σ_{xuo} = ultimate compressive strength for an intact (uncorroded) member, which can be given by the methods described in Section 6.5.2 or by Eq. (6.5), for instance;

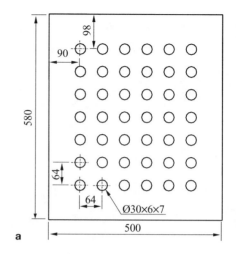

a

Figure 11.15. (a) Idealization of pit size and location. (b) Finite-element modeling of a plate with regular pit corrosion (s: symmetric boundary condition). (c) Finite-element modeling of a plate with random pits $t_r = t$ (through thickness pitting is assumed), $a \times b \times t = 800 \times 800 \times 15$ mm, $DOP = 11.3$ percent with $(A_o - A_r)/A_o = 0.56$ (the digits inside the figure represent the diameters of pits in mm).

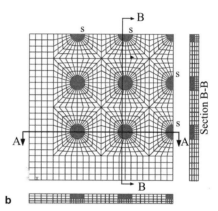

b

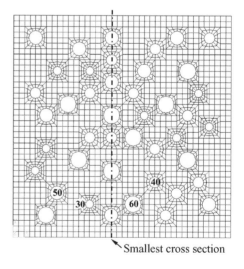

c

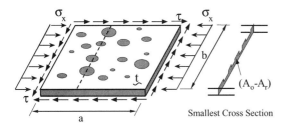

Figure 11.16. A schematic of localized pit corrosion and definition of the smallest cross-sectional area.

A_o = original cross-sectional area of the intact member; and A_r = cross-sectional area involved by pit corrosion at the smallest cross section; see Figure 11.16.

$$R_{\tau r} = \frac{\tau_u}{\tau_{uo}} = \begin{cases} 1.0 \text{ for } \alpha \leq 1.0 \\ -0.18 \ln \alpha + 1.0 \text{ for } \quad \alpha > 1.0 \end{cases}, \qquad \text{for edge shear}, \qquad (11.6)$$

where $R_{\tau r}$ = a factor of ultimate shear strength reduction due to pit corrosion; τ_u = ultimate shear strength for a pitted plate; $\alpha = $ DOP as defined in Eq. (11.4); and τ_{uo} = ultimate shear strength for an intact plate that can be obtained by the methods described in Section 6.5.2 in Chapter 6, or by Eq. (6.7), for instance.

Figure 11.17 compares Eq. (11.5) with the related numerical and experimental results. Figure 11.18 compares Eq. (11.6) with the nonlinear finite-element method computations. It is evident that the proposed strength knock-down factor approach can be useful for predicting the ultimate compressive or shear strength of pitted plates.

11.9 Effect of Fatigue Cracking on Plate Ultimate Strength

Under the action of repeated loading, fatigue cracks may be initiated in the stress concentration areas of the structure. Initial defects or cracks may also be formed in the structure by inappropriate fabrication procedure and may conceivably remain undetected over time. In addition to propagation of cracks by repeated cyclic loading, cracks may also grow in an unstable way under monotonically increasing extreme loads, a circumstance that eventually can in some cases lead to catastrophic failure

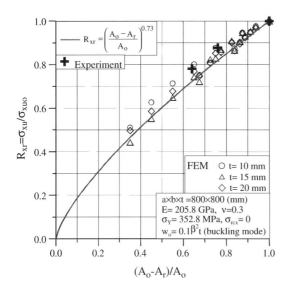

Figure 11.17. The ultimate compressive strength reduction factor as a function of the smallest cross-sectional area for a plate with pit corrosion.

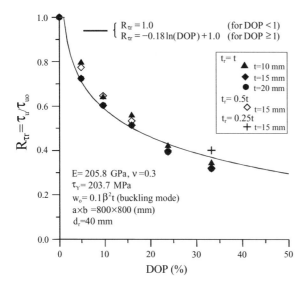

Figure 11.18. The ultimate strength versus the DOP ratio for a steel plate with pit corrosion under edge shear (symbols: nonlinear finite-element calculations).

of the structure. This possibility is usually tempered by the ductility of the material, and also by the presence of reduced stress intensity regions in a complex structure that may serve as crack arresters even in an otherwise monolithic structure.

For residual strength assessment of aging steel structures under extreme loads as well as under fluctuating loads, it is often necessary to evaluate the effects of a known or premised crack as a parameter of influence.

Figure 11.19 shows a schematic of a stiffened plate structure with three types of crack locations and orientations and under axial compression or edge shear. Strictly speaking, the ultimate strength behavior of panels depends on the types of crack orientations, among other factors. Figure 11.20 shows a sample finite-element

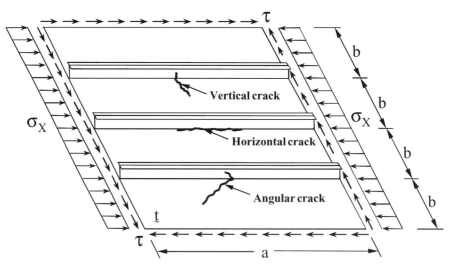

Figure 11.19. A schematic of a stiffened steel-panel component with three types of crack locations and orientations and under axial compression or edge shear.

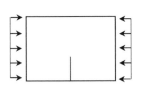 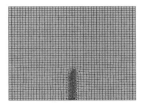

Figure 11.20. A sample finite-element mesh for a plate with one edge crack and under axial compression.

modeling for a steel plate with edge crack at one side and under axial compressive loads. Based on results of the experiment and nonlinear finite-element computations, the ultimate strength of cracked plates may be calculated by applying the strength knock-down factor approach (Paik et al. 2003.), as follows:

$$R_{xc} = \frac{\sigma_{xu}}{\sigma_{xuo}} = \frac{A_o - A_c}{A_o}, \quad \text{for axial tensile or compressive loading,} \qquad (11.7)$$

where R_{xc} = a factor of the ultimate tensile or compressive strength reduction due to cracking damage; σ_{xu} = ultimate axial strength of cracked plating; σ_{xuo} = ultimate axial strength of uncracked plating, which may be taken as $\sigma_{xuo} = \sigma_Y$ for axial tensile load and $\sigma_{xuo} = \sigma_u$ for axial compressive loads; σ_u = as described in Section 6.5.2 of Chapter 6, or by Eq. (6.5), for instance; A_o = cross-sectional area of uncracked (original) plating; and A_c = cross-sectional area involved by cracking damage.

$$R_{\tau c} = \frac{\tau_u}{\tau_{uo}} = \frac{A_o - A_c}{A_o}, \quad \text{for edge shear,} \qquad (11.8)$$

where $R_{\tau c}$ = a factor the ultimate shear strength reduction due to cracking damage; τ_u = ultimate shear strength for a plate with premised cracks; and τ_{uo} = as described in Section 6.5.2 of Chapter 6, or by Eq. (6.7), for instance.

Figure 11.21 compares Eq. (11.7) with the experimental results and nonlinear finite-element analyses. Figure 11.22 compares Eq. (11.8) with the numerical computations. It is seen that Eqs. (11.7) or (11.8) provide reasonable predictions of the ultimate strength of cracked plates.

11.10 Effect of Time-Variant Age-Related Deterioration on FPSO Hull Ultimate Strength Reliability: An Academic Example

The effect of time-variant age-related deterioration, for example, corrosion and fatigue cracking, on the ultimate strength reliability of a hypothetical FPSO hull is now illustrated and discussed using a simplified example mainly for educational purposes. A hypothetical double-hulled FPSO with a storage capacity of 113,000 dwt as shown in Figure 11.23 is selected for this purpose. Further description of this example can be found in Paik and Thayamballi (2005).

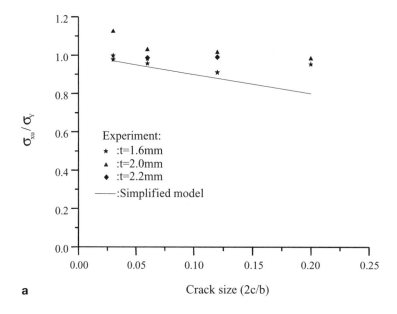

a

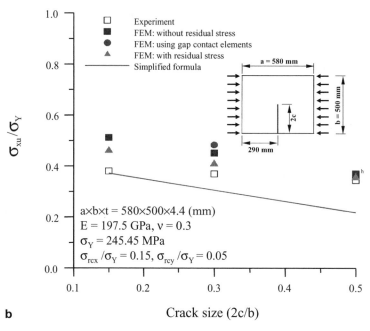

b

Figure 11.21. (a) Variation of the normalized ultimate tensile strength of a steel plate with a single center crack as a function of the crack size. (b) Variation of the normalized ultimate compressive strength of a steel plate with a single edge crack as a function of the crack size.

11.10.1 Scenario for Sea States and Operational Conditions

As described in Chapter 4 of this book, hull girder actions of ship-shaped offshore units during tow, or once on site, depend on various site-specific environments and

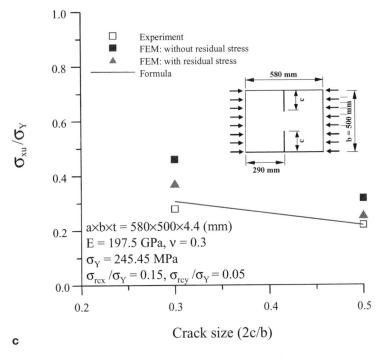

Figure 11.21. (*cont.*) (c) Variation of the normalized ultimate compressive strength of a steel plate (with multiple cracks) as a function of the crack size.

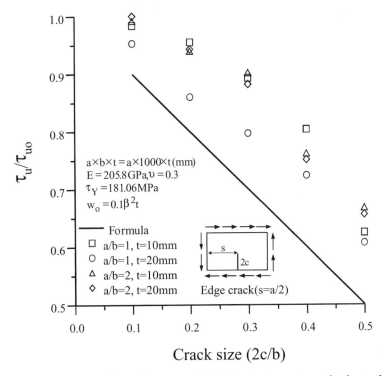

Figure 11.22. Variation of the normalized ultimate shear strength of a steel plate as a function of the crack size, varying the plate thickness and aspect ratio (symbols: nonlinear finite-element computation results).

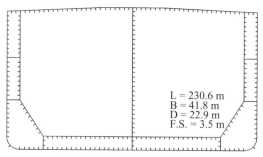

Figure 11.23. Midship section of a hypothetical FPSO (L = vessel length, B
= vessel beam, D = vessel depth, and
FS = transverse frame spacing).

L = 230.6 m
B = 41.8 m
D = 22.9 m
F.S. = 3.5 m

operational conditions. The still-water bending moments of the vessel can be obtained
as described in Section 6.3 of Chapter 6, considering that in a short-term sense the
vessel is loaded in the most onerous permitted still-water conditions; however, where
long-term reliability analysis is planned, such as in a design context, the entire range
of still-water bending moments possible, and the effects of different types of hull
behavior (hogging, sagging) must be correctly accounted for.

As purely a matter of convenience, however, the still-water bending moments in
this illustrative example are estimated from the empirical formula of trading tankers,
Eq. (3.25) of Chapter 3. On the other hand, the wave-induced bending moments are
obtained by the short-term response analysis that involves the operational conditions and sea states, rather than by the empirical formulae such as Eq. (3.28) in
Chapter 3 applicable for trading tankers. For wave-load prediction purposes, the
FPSO is assumed to have an equivalent operational speed of 10 knots in waves;
however, the FPSO usually remains at a specific location once installed.

Note that the present scenarios associated with the operational conditions and sea
states are adopted for illustrative purposes. The results of such reliability analyses
are then indicative – that is, only notional probabilities of failure – conditional on
the specific storm conditions and on the vessel being loaded in that storm in the very
onerous way noted.

11.10.2 Scenario for Time-Variant Corrosion Wastage

Age-related structural degradation and its effects need to be dealt with as a function
of a vessel's age. Specifically, the vessel is considered to be under the most onerous
still-water condition and in a given, reasonably severe short-term storm. The vessel
may be of varying age, and subject to certain generic patterns of corrosion and certain
idealized crack scenarios.

In Chapter 10 of this book, corrosion wastage models are presented for different
structural member groups by type and location, considering plating, stiffener webs,
and flanges. These models can be used to predict the corrosion depth in primary
members as the vessel ages. In the present reliability assessment, a most probable
(average) level of corrosion wastage is considered. Although it is assumed that corrosion starts immediately after the breakdown of coating, the coating life for all
structural members in the vessel structure is assumed to be 7.5 years. Figure 11.24
shows the progress of corrosion depth for selected members as the vessel ages. The

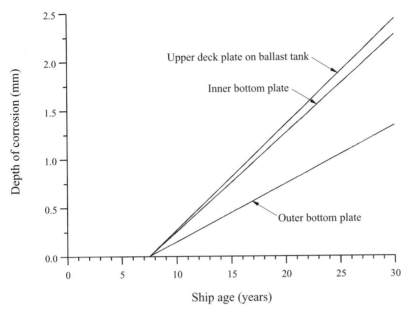

Figure 11.24. Presumed progress of corrosion diminution for selected members in the object FPSO structure.

figure neglects any effect of steel renewal after inspections and surveys; repair of heavily corroded members is considered in Section 11.10.5.

As described in Chapter 10, several types of corrosion are possible for mild steel and low-alloy steels used in marine applications. Although the general (or uniform) corrosion, which reduces the member thickness over large areas, is normally regarded as an idealized type of corrosion in today's vessels, localized corrosion such as pitting is more likely to be observed in marine structures. Also, the ultimate strength behavior of ship structures with pitting corrosion is different from that of general corrosion. For more realistic assessment of the reliability, therefore, it is important to take into account the effects of general and pitting corrosion.

In the present reliability assessment, it is assumed that the most heavily pitted cross section of any structural member extends over the plate breadth. This may provide a somewhat pessimistic evaluation of residual strength but is convenient for the reliability assessment. The mean and coefficient of variation (COV) values of annualized corrosion rates for individual structural component groups are defined in Figure 10.14 in Chapter 10. Note that, in reality, pits will be repaired once they reach certain depths and extents, regardless of the related strength criteria, although this is not accounted for in the present illustrative calculations.

11.10.3 Scenario for Time-Variant Fatigue Cracking

In Section 7.11 of Chapter 7, a time-dependent crack propagation model is presented. The crack length of any critical area is predicted by a closed-form formula as a function of vessel age. In the present application example, it is assumed that cracking initiates in all stiffeners and plating when the vessel is 5 years of age. The initial crack size is considered to be 1.0mm. These are simply assumptions for illustrative purposes. Cracks, of course, normally start as surface cracks and then progress through the thickness of the plating. They certainly do not simultaneously occur at all stiffeners and plating. Also, it has been assumed that the structure had been designed based on the fatigue limit state, using crack initiation technology to start with, and that they start at a specific time (5 years in this case). Even if the constants of the Paris–Erdogan equation may usually be considered the same at all joints, the crack growth characteristics may be different because the stress ranges affecting the stress intensity factors at individual joints vary for reasons of geometry, location of crack, and any differences in load effects that may apply.

Fatigue loading characteristics are random in nature and their sequence is normally unknown, although the long-term distribution of the fatigue loading is commonly known. Therefore, some postulated methodologies are used to generate a random loading sequence. With the fatigue loading sequence and the amplitude known, the dynamic stress range $\Delta\sigma_i$ at the ith joint may then be given by

$$\Delta\sigma_i = 2 \times \sigma_{xi} \times \mathrm{SCF}_i \times k_f = 2 \times \sigma_{xi}^* \times \mathrm{SCF}_i, \qquad (11.9)$$

where σ_{xi} = cyclic "peak" stress amplitude acting on the ith structural element, which may be given by $M_w z / I$; M_w = wave-induced bending moment; I = time-dependent moment of inertia due to time-variant damage; and z = distance from the time-dependent neutral axis to the point of the stress calculation: $\sigma_x^* = k_f \times \sigma_x$.

The SCF_i in Eq. (11.9) is the stress concentration factor at the ith critical joint. In the present illustrative examples, it is assumed that the SCF at all joints between plating and stiffeners (or support members) is 2.1. Note that more refined calculations can be conducted to determine the SCF values for different joints. The SCF values used here are simply illustrative assumptions based on related design guidance by some classification societies; for example, DNV-RP-C203 (DNV 2005). k_f is a knockdown factor accounting for the dynamic stress cycles and assumed to be 0.25, for present illustrative purposes.

At a given age of the vessel, the ultimate strength of structural members with known (or assumed) fatigue cracking damage can be predicted by the strength knockdown factor approach, as noted in Section 11.9. Fracture takes place if the crack size (length) of the member reaches the critical crack size that may be assumed to be the smaller of the plate breadth and the stiffener web height.

Table 11.8 indicates example probabilistic characteristics (i.e., mean, COV, distribution) of the random variables used for the present illustrative purposes. It is important to realize that the probabilistic characteristics of random variables will normally be different for different types of ship-shaped structures, operating scenarios, and applications.

11.10.4 Time-Variant Ultimate Hull Strength Reliability Assessment

It is more convenient to use a closed-form expression for predicting the ultimate hull strength formula in the reliability assessment process. For this purpose, Eq. (6.38) in Chapter 6 can be used. When time-variant structural degradation (e.g., corrosion, fatigue, cracking) and local denting areas considered, the value of member thickness at any particular time is a function of random variables associated with age-related

Table 11.8. *Examples of the probabilistic characteristics for random variables at a given age of the vessel*

Parameter	Definition	Distribution function	Mean	COV
E	Elastic modulus	Normal	205.8GPa	0.03
σ_Y	Yield stress	Log-Normal	As for each member	0.10
t_p	Thickness of plating	Fixed	As for each member	—
t_w	Thickness of stiffener web	Fixed	As for each member	—
t_f	Thickness of stiffener flange	Fixed	As for each member	—
T	Vessel age	Fixed	As for each age	—
T_c	Coating life	Normal	5.0 years	0.40
			7.5 years	0.40
C_1	Corrosion rate	Weibull	As for each member	As for each member
a_o	Initial crack size	Normal	1.0mm	0.20
C	$\frac{da}{dN} = C(\Delta K)^m$	Log-Normal	6.94E-12	0.20
m	$\frac{da}{dN} = C(\Delta K)^m$	Fixed	3.07	—

Note: a = crack size; b = plate breadth; N = number of stress cycles; and ΔK = stress intensity factor.

deterioration as well as all relevant geometric and material properties. Thus we have (for symbols, see Table 11.8)

$$M_u = M_u \left(E_i, \sigma_{Yi}, t_{pi}, t_{wi}, t_{fi}, T, T_{ci}, C_{1i}, a_{oi}, C_i, m_i \right), \qquad (11.10)$$

where M_u = ultimate hull girder moments in sagging or hogging. The subscript i represents the *i*th member.

In a reliability assessment, all the parameters noted in Eq. (11.10) are treated as random variables, with the probabilistic characteristics (i.e., mean, COV, and distribution function), as defined in Table 11.5.

Figure 11.25 shows the effects of the above damage scenarios on the time-dependent characteristics of ultimate hull girder strength and reliability when no repair or renewal is made. As the vessel ages, the corrosion depth and cracking size (length) increase and thus the ultimate hull girder strength and reliability index decrease (or failure probabilities increase).

The reliability indices for the hypothetical FPSO unit against hull girder collapse in the intact condition are seen to be about 2.5. At the age of around 15 years, the safety and reliability reduces to less than 90 percent of the original (as-built) states. If maintenance is not properly carried out, the levels of reliability can decrease rapidly.

11.10.5 Considerations for Repair Strategies

To maintain the vessel's safety and reliability at a certain target level or higher, a proper, cost-effective scheme for maintenance must be established. In this regard, some considerations for repair strategies of structural members postulated to be heavily damaged by corrosion and fatigue cracking are now illustrated. The International Maritime Organization (IMO 2000) requires that one should keep

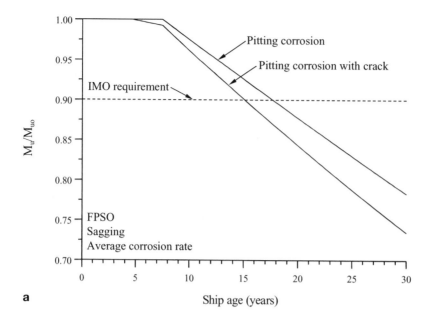

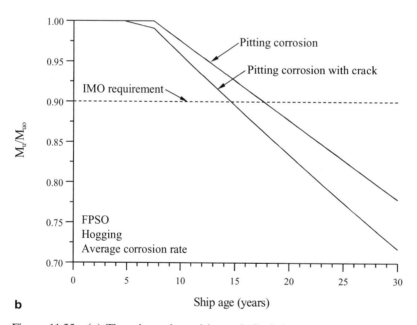

Figure 11.25. (a) Time-dependent ultimate hull girder strength of the hypothetical FPSO in sagging. (b) Time-dependent ultimate hull girder strength of the hypothetical FPSO in hogging.

the longitudinal strength of an aging ship at the level of at least 90 percent of the initial state. Although the IMO requirement is in fact based on the vessel's section modulus, the present illustrative examples are extended as a device to establish a more sophisticated maintenance scheme based on hull girder ultimate strength. The aim of the illustrated scheme is that the ultimate hull girder strength

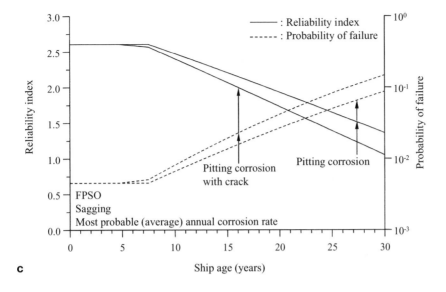

c

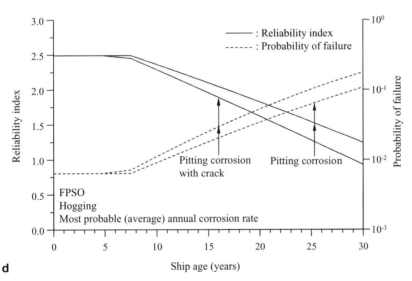

d

Figure 11.25. (*cont.*) (c) Time-dependent reliability of the hypothetical FPSO associated with hull-girder collapse in sagging. (d) Time-dependent reliability of the hypothetical FPSO associated with hull girder collapse in hogging.

of an aging vessel must always be at least 90 percent of the initial, as-built vessel value.

Figure 11.26 shows the time-dependent hull girder ultimate strength and reliability values after repair of postulated heavily damaged structural members so that the ultimate hull girder strength is always at least 90 percent of its original value. In these illustrations, the renewal criterion for any damaged member is based on the member's ultimate strength rather than member thickness as traditionally done. This is advantageous because member thickness-based renewal criteria cannot reveal the effects of pitting corrosion, fatigue cracking, or local dent damage adequately even

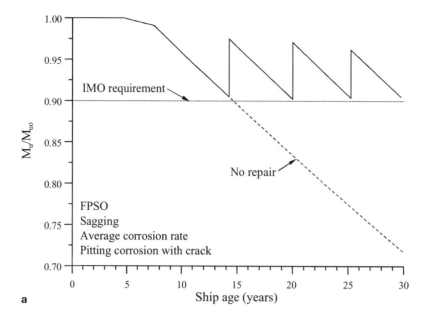

a

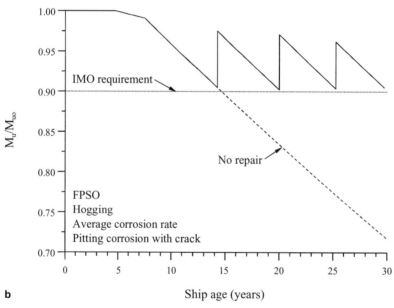

b

Figure 11.26. (a) Assumed repair and the resulting time-dependent ultimate hull girder strength of the hypothetical FPSO in sagging. (b) Assumed repair and the resulting time-dependent ultimate hull girder strength of the hypothetical FPSO in hogging.

though it may handle the thickness reduction effects of general corrosion reasonably well. On the other hand, member's ultimate strength-based renewal criteria are adequate and better equipped to deal with all types of structural damage.

As the illustrations imply, the more heavily damaged members need to be renewed (or repaired) to their as-built state, immediately before the ultimate longitudinal strength of an aging vessel reduces to a value less than 90 percent of the original

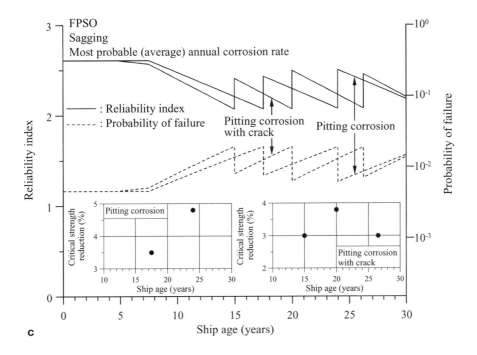

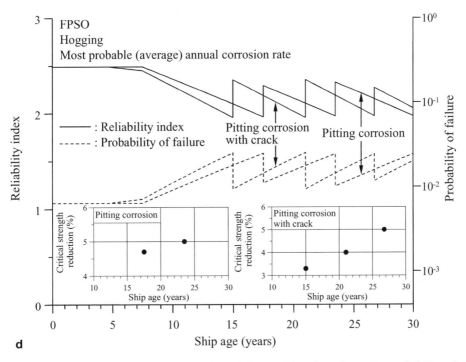

Figure 11.26. (*cont.*) (c) Assumed repair and the resulting time-dependent reliability of the hypothetical FPSO associated with hull girder collapse in sagging. (d) Assumed repair and the resulting time-dependent reliability of the hypothetical FPSO associated with hull girder collapse in hogging.

vessel. It is evident from Figure 11.26 that the structural safety and reliability of aging vessels can, of course, be controlled by proper maintenance strategies. It is also seen that the repair criterion based on member ultimate strength can provide a potential improvement to better control the age-dependent degradation of a vessel's longitudinal strength. The illustrations demonstrate that the percentage reduction in critical ultimate strength of structural members that need to be repaired is not constant as might be expected and is in the range of 2–7 percent of the as-built state.

REFERENCES

ABS (2003). *Guide for surveys using risk-based inspection for the offshore industry*. American Bureau of Shipping, Houston, December.

ABS (2004). *Guidance notes on reliability-centered maintenance*. American Bureau of Shipping, Houston, July.

Agarwala, V. S., and Ahmad, S. (2000) "Corrosion detection and monitoring a review." *Proceedings of Corrosion 2000* (Paper No. 271), NACE International, Orlando, March 26–31.

Babbar, V., Bryne, J., and Clapham, L. (2005). "Mechanical damage detection using magnetic flux leakage tools: The effect of dent geometry and stress." *NDT & E International*, 38: 471–477.

Berens, A. P. (1989). "NDE reliability data analysis." In *Metals handbook, Vol. 17: Nondestructive evaluation and quality control*. Materials Park, OH: ASM International.

Bøving, K. (1989). *NDE handbook: Nondestructive examination methods for condition monitoring*. Abington, UK: Woodhead Publishing Ltd.

Conachey, R. M., and Montgomery, E. L. (2003). *Application of reliability-centered maintenance techniques to the marine industry*. Presented at a meeting of the SNAME–Texas Section, The Society of Naval Architects and Marine Engineers, Houston, April 8.

DeFranco, S., O'Connor, P., Tallin, A., Roy, R., and Puskar, F. (1999). *Development of a risk based underwater inspection (RBUI) process for prioritizing inspections of large numbers of platforms*. Offshore Technology Conference, OTC 10846, Houston, May.

Demsetz, L., and Cabrera, J. (1999). *Detection probability assessment for visual inspection of ships*. Ship Structure Committee, SSC-408, Washington, DC.

Demsetz, L., Carlo, R., and Schulte-Strathaus, R. (1996). *Inspection of marine structures*. Ship Structures Committee, SSC-389, Washington, DC.

Devanney, J. (2006). *The tankship tromedy: The impending disasters in tankers*. Tavernier, FL: Center for Tankship Excellence (CTX) Press.

Ditchburn, R. J., Burke, S. K., and Scala, C. M. (1996). "NDT of welds: State of the art." *NDT & E International*, 29(2): 111–117.

DNV (2003). *Guidance for condition monitoring*. (Classification Notes, No. 10.2), Det Norske Veritas, Oslo.

DNV (2004a). *Guidelines for ultrasonic thickness measurements of ships classed with DNV*. Det Norske Veritas, Oslo.

DNV (2004b). *Non-destructive testing*. (Classification Notes, No. 7), Det Norske Veritas, Oslo.

DNV (2005). *Fatigue design of offshore steel structures*. (Recommended Practices, DNV-RP-C203), Det Norske Veritas, Oslo.

Dover, W. D., Brennan, F. P., Karé, R. F., and Stacey, A. (2003). "Inspection reliability for offshore structures." *Proceedings of the 22nd International Conference on Offshore Mechanics and Arctic Engineering* (OMAE 2003), Cancun, Mexico, June 8–13.

Drouin, P. (2006). "Brittle fracture in ships – A lingering problem." *Ships and Offshore Structures*, 1(3): 229–233.

Fujimoto, Y., Kim, S. C., Shintaku, E., and Ohtaka, K. (1996). "Study on fatigue reliability and inspection of ship structures based on an enquete information." *Journal of Society of Naval Architects of Japan*, 180: 601–609.

Goyet, J., Rouhan A., and Faber, M. H. (2004). "Industrial implementation of risk based inspection planning lessons learned from experience (Parts 1&2)." *Proceedings of the 23rd International Conference on Offshore Mechanics and Arctic Engineering* (OMAE 2004), (Paper Nos. 51572 and 51573), Vancouver, Canada, June 20–25.

Halmshaw, R. (1997). *Introduction to the non-destructive testing of welded joints.* 2nd ed. Abington, UK: Woodhead Publishing Ltd.

HSE (1997). *A review of monohull FSUs and FPSUs.* (Offshore Technology Report, OTO 1997/800), Health and Safety Executive, UK.

HSE (1998). *Review of classification society structural survey requirements for FPSO/FSU's.* (Offshore Technology Report, OTO 1998/163), Health and Safety Executive, UK.

HSE (2000a). *POD/POS curves for non-destructive examination.* (Offshore Technology Report, OTO 2000-018), Health and Safety Executive, UK.

HSE (2000b). *Review of current inspection practices for topsides structural components.* (Offshore Technology Report, OTO 2000/027), Health and Safety Executive, UK.

IACS (2004). *Recommendation 87: Guidelines for coating maintenance and repair for ballast tanks and combined cargo/ballast tanks of oil tankers.* International Association of Classification Societies, London.

IACS (2005a). *Unified Requirements, UR Z10.1.* (Rev. 12, 16 June 2005: Enhanced Survey Programme), International Association of Classification Societies, London.

IACS (2005b). *Common structural rules for double hull oil tankers.* International Association of Classification Societies, London.

IMO (2000). SOLAS /2 *Recommended longitudinal strength.* (MSC.108(73)), Maritime Safety Committee, International Maritime Organization, London.

INTERTANKO (2003). *A guide to the vetting process.* International Association of Independent Tanker Owners, UK.

ICS/OCIMF/IAPH (1996). *International safety guide for oil tankers and terminals (ISGOTT).* ICS (International Chamber of Shipping), OCIMF (Oil Companies International Marine Forum), and IAPH (International Association of Ports and Harbors), London: Witherby & Co. Ltd.

Ivanov, L. D., and Wang, G. (2004). "Uncertainties in assessing the corrosion wastage and its effect on ship structure scantlings." *Proceedings of 9th Symposium on Practical Design of Ships and Other Floating Structures* (PRADS 2004), Luebeck-Travemuende, Germany, September, 12–17, Vol. 2, pp. 586–593.

Ku, A., Serratella, C., Spong, R., Basu, R., Wang, G., and Angevine, D. (2004). "Structural reliability applications in developing risk-based inspection plans for a floating production installation." *Proceedings of 23rd International Conference on Offshore Mechanics and Arctic Engineering* (OMAE 2004), OMAE 2004–51119, Vancouver, Canada, June 20–25.

Lanquetin, B. (2005). *Floating units integrity management and life care enhancement.* Offshore Europe 2005, Society of Petroleum Engineers, SPE 96385, Aberdeen, Scotland, September.

Li, D., Zhang, S., and Tang, W. (2003). "Uncertainty and Bayesian updating considering inspection for ship structures subjected to corrosion deterioration." *Proceedings of the International Conference on Offshore Mechanics and Arctic Engineering* (OMAE 2003), OMAE 2003-37450, Cancun, Mexico, June 8–13.

LR (2003). *FPSO inspection, repair, and maintenance – Study into best practice.* (R20821-5-UKOOA), Lloyd's Register, London, May.

Ma, K., Orisamolu, I. R., and Bea, R. G. (1999). *Optimal strategies for inspections of ships for fatigue and corrosion damage.* Ship Structure Committee, SSC-407, Washington, DC.

Moan, T., Vårdal, O. T., Hellevig, N. C., and Skjoldli, K. (2000). "Initial crack depth and POD values inferred from in-service observations of cracks in North Sea jackets." *Journal of Offshore Mechanics and Arctic Engineering*, 122: 157–162.

Moubray, J. (1997). *Reliability-centered maintenance.* New York: Industrial Press Inc.

Oil and Gas News (2001). *Patch repair of FPSOs.* (No. 5), Det Norske Veritas, Oslo, December.

Paik, J. K., Brennan, F., Carlsen, C. A., Daley, C., Garbatov, Y., Ivanov, L., Rizzo, C. M., Simonsen, B. C., Yamamoto, N., and Zhuang, H. Z. (2006). *Condition assessment of aged ships. Report of ISSC Technical Committee V.6, International Ship and Offshore Structures Congress.* Southampton, UK: Southampton University Press.

Paik, J. K., Wang, G., Thayamballi, A. K., Lee, J. M., and Park, Y. I. (2003). "Time-variant risk assessment of aging ships accounting for general/pit corrosion, fatigue cracking, and local denting damage." *SNAME Transactions*, 111: 159–197.

Porter, R. (1992). "Non-destructive examination in shipbuilding." *Welding Review*, 11(1): 9–10.

Rudlin, J. R., and Wolstenholme, L. C. (1992). "Development of statistical probability of detection models using actual trial inspection data." *The British Journal of Non-Destructive Testing*, 34(12): 583–589.

Rummel, W. D., Hardy, G. L., and Cooper, T. D. (1989). "Applications of NDE reliability to systems." In *Metals handbook, Vol. 17: Nondestructive evaluation and quality control.* Materials Park, OH: ASM International.

Saidarasamoot, S., Olson, D. L., Mishra, B., Spencer, J. S., and Wang, G. (2003). "Assessment of the emerging technologies for the detection and measurement of corrosion wastage of coated marine structures." *Proceedings of International Offshore and Polar Engineering Conference (OMAE), OMAE 2003–37371*, Cancun, Mexico, June 8–13.

Smith, A. M. (1993). *Reliability-centered maintenance.* New York: McGraw-Hill.

Song, W. J, Rose, J. L., and Whitesel, H. (2003). "An ultrasonic guided wave technique for damage testing in a ship hull." *Material Evaluation*, 61(1): 94–98.

SPS (2005). *Sandwich plate system.* (www.ie-sps.com) accessed June 2006.

Sumpter, J. D. G., and Kent, J. S. (2004). "Prediction of ship brittle fracture casualty rates by a probabilistic method." *Marine Structures*, 17: 575–589.

Talei-Faz, B., Brennan, F. P., and Dover, W. D. (2004). "Residual static strength of high strength steel cracked tubular joints." *Marine Structures*, 17: 291–309.

Tiku, S., and Pussegoda, N. (2003). *In service nondestructive evaluation of fatigue and fracture properties for ship structures.* Ship Structure Committee, SSC-428, Washington, DC.

TSCF (1997). *Guidance manual for tankers.* Tanker Structures Cooperative Forum.

Vanlanduit, S., Guillaume, P., and van der Linden, G. (2003). "On-line monitoring of fatigue cracks using ultrasonic surface waves." *NDT & E International*, 36(8): 601–607.

CHAPTER 12

Tanker Conversion and Decommissioning

12.1 Introduction

For offshore oil and gas production, storage, and offloading, particularly in marginal fields, existing tankers are often converted to ship-shaped offshore units instead of using a new-build option. In fact, more than two thirds of all such ship-shaped offshore installations worldwide are currently thought to be built from converted tankers. More recent practice indicates that the application of converted offshore structures is more common in relatively benign environments, although a new-build installation may be more appropriate for special purposes in harsh conditions and/or for longer term use (e.g., more than 10–15 years). These are general statements, and operations using conversions have also been considered for fields with harsh environmental conditions as well. In marginal fields, it is often the relative cost advantage and better ability to fast track that makes a conversion more compelling than a new build. The number of vessels potentially available for conversions, however, have continued to dwindle over time.

In Section 1.6 of Chapter 1, the many advantages and disadvantages of the conversion option for ship-shaped offshore installations were discussed. The possible advantages include reduced capital costs, a less expensive and fast-track design and construction schedule, increased choices regarding construction facilities, and perhaps reduced overall project supervision requirements (Parker 1999). The disadvantages may include shorter design and remaining lives; greater site-specific environment limitations; perhaps increased operating costs because of difficulty in building in high safety factors; reduced or minimal resale and residual values; reduced reusability opportunities; and the relatively greater need for increased risk-mitigation measures related to regulatory compliance, which, as expected, have increased over time.

Therefore, some of the key issues that would need to be considered in the case of a conversion option arise from the need to minimize the disadvantages noted previously in specific cases. This chapter presents important aspects that must be considered when converting tankers to ship-shaped offshore units, particularly as related to suitable tanker selection, condition assessment of tanker hull structures, estimation of structural renewal and modification needs for strength and durability, reusability of existing equipment and systems, addition of new equipment and systems, conversion-yard appraisals, and the like.

The decommissioning of offshore installations has become an increasingly important activity for operators and also has had significant interests from external audiences particularly on environmental-, social-, and economic-impact issues. For both new builds and conversions, the decommissioning of offshore installations and related issues (including some design aspects such as avoidance of environmentally inappropriate materials) are also becoming increasingly important.

Consider that there are currently more than 7,000 offshore oil and gas platforms and production facilities of various types including ship-shaped offshore units in operation worldwide, and these are located in more than fifty countries and in a wide range of environmental conditions. These installations will eventually be decommissioned and, subsequently, the international oil and gas exploration and production industry and the offshore contractor community will at some time in the next two decades face the significant challenge of related decommissioning issues, which may include technical feasibility, cost effectiveness, environmental impact, health, safety, and alleviation of public concern.

In this regard, O'Neil (2001), the Secretary General of the International Maritime Organization (IMO), which is the United Nations agency whose goal is "safer shipping – cleaner oceans," says that a ship's demise should be prepared for even before its birth. The ship's design and construction must take into account how dismantling and recycling can be carried out. During the ship's operational phases, choices and decisions must not only take short-term results into account, but they must also have a long-term perspective that includes the possibility of safe and efficient recycling. Of course, these very same statements are also true for ship-shaped offshore structures. Their decommissioning, break up, or redeployment must ideally be prepared from the very beginning of the concept design phase, with adequate consideration of safety, health, and the environment.

Unlike other types of offshore platforms such as jackets, decommissioning of ship-shaped offshore installations is perhaps easier. Also, it is of large benefit that part or all of decommissioned ship-shaped offshore installations can potentially be reused. This chapter presents key issues that must be considered for safe decommissioning, dismantling, and disposal with an emphasis on ship-shaped offshore installations.

12.2 Tanker Conversion

It is important to understand that a ship-shaped offshore unit is quite different from a trading tanker in terms of many aspects associated with functional requirements as well as design, construction, and operation, as described in Section 1.5 of Chapter 1. Consequently, the selection of suitable tankers for conversion must bear in mind such differences; and the subsequent conversion engineering must also account for the differences in an adequate way.

In some cases, the service of ship-shaped offshore units can be more arduous than that of trading tankers, depending in part on the environment of operation. More severe service conditions in ship-shaped offshore installations can also occur because of higher tank temperature, more loading/offloading cycles, and more risk of hot produced water in tanks, among other factors. Deck structures of ship-shaped offshore units are relatively more heavily loaded than trading tankers, by topsides and process facilities, causing potential design concerns in terms of strength, freeboard,

Table 12.1. *Structural design trends for VLCCs built during the 1970s and 1980s*

	1970s		1980s	
Structure	Thickness (mm)	Material	Thickness (mm)	Material
Main deck	35	Mild steel	20	High-tensile steel
Side shell	23	Mild steel	17	High-tensile steel
Bottom shell	36	Mild steel	20	High-tensile steel

stability, and deck deflections. Also, ship-shaped offshore units are normally not planned for future dry-docking during their service life, although again this will not always be the case.

A better understanding of operational and other differences is essential to achieve successful conversion of trading tankers to ship-shaped offshore units. In addition to adequate condition assessment of a selected tanker, it is important to identify and evaluate what systems in the tanker can be reused or renewed and what systems must be newly installed into the converted vessel.

Useful information and practices in association with conversion of existing tankers to ship-shaped offshore units may be found in van Voorst et al. (1995), Johnson (1996), Park et al. (1998), da Costa Filho (1997), Assayag et al. (1997), Parker (1999), Neto and de Souza Lima (2001), Terpstra et al. (2001, 2004), Mones (2004), Lane et al. (2004), Terpstra et al. (2004), and Biasotto et al. (2005).

12.2.1 Selection of Suitable Tankers

In selecting a trading tanker to be converted, the following basic prerequisites and vessel-related factors may be considered, for example:

- Price of tanker
- Tank size or oil storage capacity
- Year of building (tanker age)
- Hull arrangements (e.g., single skin, double sides/single bottom or double sides/ double bottom)
- Condition of hull structures and systems
- Residual strength and fatigue lives

Most of the trading tankers with a Suezmax or very large crude oil carrier (VLCC) class are considered as candidates to be converted. In the case of the structure, an important consideration is the likelihood that the vessel can be converted to the new service with relatively modest amounts of steel additions, modifications, and repair. Thus far, conversion candidates built in the 1970s were preferred in this regard because of their relatively heavier scantlings and greater use of mild steel in their construction. Some trends of structural features for VLCCs built in the 1970s and 1980s are indicated in Table 12.1.

The condition of candidate tankers whether from the 1970s or later will vary regarding fatigue damage and corrosion wastage accumulated during tanker service life and also regarding quality of initial construction. The available tankers from the 1970s are now becoming increasingly fewer, but they do have thicker scantlings that are

Table 12.2. *Example of oil tanker average sale price, following Biasotto et al. (2005)*

	Average sale price (million US$)			
Year	Built in 1973–1985	Built in 1986–1995	Built in 1996–2000	Built in 2001–2004
2000	12	43	74	–
2001	13	35	79	83
2002	11	22	60	75
2003	13	33	52	65
2004	15	60	86	110
Number of tankers available	29	185	116	118
Average sale price	12	36	66	88

relatively good for expected future corrosion performance after conversion. However, their age is more than 30 years and this implies that residual fatigue lives must be accurately estimated. The tankers of the 1980s were built with thinner scantlings using relatively greater amounts of high-tensile steel. This causes more concern with fatigue cracking and corrosion wastage considering the long service life of ship-shaped offshore installations, for example, 10 years or more after conversion.

Before starting the construction work for conversion, the condition of candidate tankers must then be assessed carefully, and the amounts of steel renewal must be estimated; the conversion must be such that any steel renewals or repairs in situ over the onsite service life can be avoided to the greatest extent possible. All machinery must be suitable for the expected functions, service life, and maintenance philosophy to be employed.

Since the adoption of US OPA 90, tankers have been built with double-hull designs using varying amounts of high-tensile steel. MARPOL Annex I requirements (IMO 2003; also refer to Appendix 6) are mandatory for ship-shaped offshore units, but not the double-hull regulations, although some national or regional statutory bodies may require them. The use of double sides is preferred in ship-shaped offshore units because of collision concerns, but again are not usually required by regulation. In any event, it is mostly single-skin tankers that have been refurbished and converted; most available double-skin tankers are relatively new and are, therefore, more expensive to purchase. In the case of a single-hull tanker, the potential provision of double sides, when desired, can possibly be accomplished by additional structures such as sponsons.

Table 12.2 shows an example of the average sale price of existing tankers based on data from C. W. Kellock & Co., Ltd.; http://www.yachtworld.com/kellock(Biasotto et al. 2005). Table 12.2 shows that in recent years the sale price of existing tankers appears to have increased rapidly, in part, presumably because of the current trends of high oil demands worldwide and the consequent increasing demand for the transportation of oil by tankers.

12.2.2 Condition Assessment of Aged Tanker Hull Structures

12.2.2.1 Inspection and Maintenance

To avoid undesirable dry-docking or steel renewals during the service life of ship-shaped offshore units, the condition of aged tankers to be converted must be assessed carefully. It will be appreciated that it is based on such assessment that one decides the required levels of steel renewal and structural modifications for purposes of anticipated strength, corrosion, and fatigue performance on site. With such steel renewal and structural modifications, and the implementation of relevant maintenance schemes in service, the converted vessel will hopefully have been made fully suitable for the required service life. For structural modifications of cargo tanks with fatigue cracking, see Newport et al. (2004).

The primary aspects that must be considered during the condition assessment of the selected tanker hull structure at conversion are as follows:

- Original design parameters, assumptions, and specifications
- Original quality of shipyard construction
- Structural steel grade
- Trading history, route, and types of cargo carried
- Fatigue cracking and residual fatigue lives
- Corrosion wastage and remaining thickness of structural components
- Buckling and ultimate strength of primary structural components and the hull girder
- Damage to and the restoration/modification of coatings and the corrosion protection scheme in general
- Vessel records including gauging, dry-docking, and damage suffered

The historical trading data will be useful for the analysis of residual fatigue lives. The desired data may include voyage routes, loading conditions, operating speeds, and sea conditions, together with environmental data and time at sea and harbor.

Techniques of recommended practices described in Section 11.4 of Chapter 11 will be useful for condition assessment of candidate tankers. Close-up visual inspections and nondestructive examination must be carried out to detect and quantify fatigue cracking and corrosion wastage at conversion. The inspectors involved must understand that the tankers will be used for different functions after conversion. Although they may be more familiar with classification societies rule requirements and related inspection/survey procedures, for example, with a 5-year inspection interval, lessons learned from operations of ship-shaped offshore units converted from trading tankers show that the existing classification societies inspection procedures will need to be augmented and that a special survey and comprehensive inspections must be performed in the shipyard before the repair and refurbishment plan can be finalized and the refurbishment and conversion work started.

The structure will need, at some point, to be inspected in detail in a dry dock, for extensive measurements of loss of thickness/corrosion wastage and identifying construction defects and in-service damage such as dents and cracks. Close-up inspection and thickness measurement of the hull externals and internals will be carried out once the hull has been properly cleaned and made accessible for close-up inspection. Close-up visual inspection of cargo, slop, and ballast tanks may be carried out from scaffolding erected in the tanks or alternatively by rafting in some cases. Local pitting, grooving, and knife edging are of significant concern in view of the difficulties involved in repairs offshore and, therefore, must also be found to the maximum extent possible. Therefore, the inspections, gauging, and repair/refurbishment will tend to be comprehensive. At least the following elements must be included in the inspection process (Mones 2004):

- Visual inspection of vessel structure in all cargo, slop, ballast, fuel oil, forepeak, aft peak, and void spaces
- Close visual inspection (within 0.5m) of the toes of all transverse bottom webs and horizontal girders; where these toes connect directly to an oil tight bulkhead, they shall be preferrably tested by magnetic particle inspection (MPI)
- Close visual inspection of collar plates for all longitudinals protruding through watertight or oil-tight bulkheads; perhaps 20 percent of the welds will be tested by MPI
- Ultrasonic thickness measurements and close-up inspection of the entire main deck, entire bottom, and selected side shell strakes (at ballast and fully loaded waterline)
- Ultrasonic thickness measurements and close-up inspection of all horizontal stringers and the centerline girders
- Ultrasonic thickness measurements and close-up inspection of representative sections of selected web frames and all transverse bulkheads; in each cargo, slop, and ballast tank, one or more web frames should be inspected
- Ultrasonic thickness measurements and close-up inspection of one entire transverse girth belt in each tank (including the slop tank)
- Detailed measurements and mapping of all areas of significant pitting in cargo, slop, and ballast tanks

Heavily corroded areas must be renewed and enhanced to the required levels at conversion. The renewal plate thickness may be estimated as a sum of net plate thickness required by the new service demands plus a corrosion margin value. Although the net plate thickness of structural components must be determined based on the strength (stress and buckling) and also on fatigue criteria as necessary, corrosion-margin values can be based on appropriate corrosion wastage models as described in Chapter 10.

12.2.2.2 Renewal Scantlings for Tanker Conversion

General corrosion-related renewal scantlings need to be determined for purposes of conversion; these are the limits below which the member would be renewed. For overall corrosion of plating and stiffeners, renewal thickness at conversion may be defined such that "substantial corrosion" conditions will not be reached within the onsite life of the structure, taking into account anticipated corrosion losses during the service life (see Section 11.4 of Chapter 11).

The substantial corrosion margin is usually defined as 75 percent of the allowable corrosion loss to renewal. The allowable maximum corrosion loss to renewal will also need to be defined for this purpose; these will usually be in the 20–30 percent range depending on location of the member in the structure locally. The renewal thickness at conversion could then be found as follows:

$$\text{Renewal thickness} = \text{Required thickness} \,(1 - 0.75 \times \text{allowed corrosion percentage})$$
$$+ \,(\text{life in years} \times \text{yearly corrosion loss}).$$

This formula relates to general corrosion on a local structural level. The diminutions in plate thickness prior to renewal on allowable deck and bottom areas, and the hull girder section modulus and gross panel strength properties over the anticipated service life, must be defined and satisfied and are usually more stringent than the locally allowed percentages. These normally do not exceed 10 percent.

One needs to repair all structural damage, structural defects, and cracks and renew all steel wasted beyond acceptable limits. The structural condition as converted, and all related structural modifications and refurbishment, need to be validated by appropriate stress, buckling, and fatigue calculations using appropriate procedures and criteria.

12.2.2.3 Repair of Defects, Dents, Pitting, Grooving, and Cracks

The approach will be to repair and/or modify all structural defects, damage, and cracks, and all conditions that may lead to further structural degradation or damage during the service life of the unit. Existing structural damage should be repaired to levels acceptable to the operator. In general, dents of depth more than the plate thickness may be repaired by insert plates. Parts of web frames and side shell longitudinals in way of indentation more than the plate thickness may be renewed. Sharp dents will usually not be permitted, and the affected structure needs to be renewed. All visible cracks will be repaired. In addition, critical areas will be inspected and all crack-like defects therein should be repaired regardless of their size.

All pits and grooves found are either to be weld-repaired or repaired by insert plates. In the case of pitting affecting welds, any weld overfill (crown) lost by pitting will be replaced. Pitting or grooving of a depth 15–33.3 percent of the thickness may be repaired by infill welding provided at least, say, 6mm of the original plating thickness remains at the bottom of the pit and the nominal diameter of any pit does not exceed certain limits (say, 300mm), and individual pits are spaced a minimum of 75mm apart, as an example. Pitting or grooving damage outside these stated criteria will be repaired with insert plates.

12.2.2.4 Residual Strength Assessment

The residual strength assessment of the various structural components and the hull girder in terms of ultimate limit states needs to be carried out considering the effects of fatigue cracking, corrosion wastage, and any local denting damage. Residual fatigue life would also be analyzed considering historical voyage data noted previously. For the condition assessment scheme for ship structures, see ISO Final Draft International Standard (FDIS) 18072-1 (2006) and ISO Draft International Standard (DIS) 18072-2 (2006).

12.2.3 Reusability of Existing Machinery and Equipment

Existing equipment and machineries of tankers may be reusable with and without refurbishment to varying extents for the expected service as ship-shaped offshore units. Such reusability is important in terms of reducing the capital cost, but it also involves certain detailed considerations. Components and systems that may potentially be considered for refurbishment and upgrading (Parker 1999) include the following:

- Main and auxiliary machineries
- Electrical generators

- Boilers and economizers
- Starting air and instrument air systems
- Piping systems (cargo, ballast)
- Deck hydraulic systems
- Bilge, seawater, and fireman systems
- Steam, inert gas, and crude oil washing systems
- Lubricating oil systems
- Cargo or ballast pumps and related control systems
- Communication systems
- Electric cables and switchgear
- Heaters, motors, lighting fittings
- Fire and gas detection systems
- Firefighting systems and lifesaving appliances
- Corrosion protection systems (cathodic protection)
- Accommodation spaces

It is important to ensure that the components involved will be adequately available during the required service life after conversion, and this requires detailed examinations of condition and the necessary follow-up actions that include refurbishment and testing. Although this may depend on some aspects, such as previous levels of maintenance, the age of the tanker, the operational requirements, and constraints of the new service, several of the systems and/or components may be suitable for reuse with or without modifications, perhaps providing a unique cost advantage to a conversion in contrast to a new build.

Often, one might be successful in such refurbishment, modifications, or reuse of components and systems for power generation, cargo handling, inert gas, ballast, crude oil washing, steam, utility, firefighting, and accommodations to varying degrees.

12.2.4 Addition of New Components

The following components typically must be added considering the service needs of the installation:

- Process plant
- Turret and riser porch
- Flare system
- Mooring system; for example, spread mooring, or turret mooring
- Control and instrumentation systems
- Offloading system
- Helideck
- Cranes and their coverage
- Green-water protection; that is, bulwarks and breakwaters

Design requirements for topsides, process facilities, and operational systems are presented in Chapter 9. Although the decision may depend on the vessel size, environment, water depth, number and type of riser paths, and any disconnectability requirements, conversions typically have external turrets, fixed spread moorings, or perhaps smaller forward internal turrets. The following options may be possibilities for the mooring system of a conversion (see Parker 1999 and Chapter 9):

- Fixed spread moorings, forward and aft
- Internal turret forward – fixed or disconnectable
- Submerged turret production buoy
- External bow or stern turret – cantilevered at deck or keel
- External stern turret – yoke to CALM (catenary anchor leg mooring) buoy
- Rigid and articulated yokes to buoys
- Articulated buoyant column and yoke
- Mooring tower and yoke

With most mooring systems except for the fixed spread mooring type, the riser paths typically terminate via a fluid swivel and weathervaning into the environment is applied. In some cases, specifically when active heading control is required for offloading operation to shuttle tanker and/or due to environmental conditions, a tunnel thruster may be fitted aft. However, this requires complex conversion work and also requires space at the aft end for the tunnel.

The process plant is usually installed as skids, packages, or modules, and similar preassembled units that are ready for onboard hook-up and precommissioning. The size of preassembled units may depend on the shipment, crane, or load-out facilities available at the conversion yard. As described in Chapter 9, the location of process equipment on the upper deck will be supported by evaluations of various aspects such as longitudinal hull girder strength, stability, deck structure deflection, and green water. Topsides weight and center of gravity are important parameters to be controlled; as such, they must be determined at an early stage of conversion and monitored throughout the entire conversion so that changes at later stages can be avoided to the extent possible.

The maximum extreme wave-induced bending moment, which is related to longitudinal strength and shear force, may be determined for the design environment-based onsite-specific data, for example, with waves of 100-year return period. The still-water bending moment and shear force must be considered for various loading/offloading conditions including full load, ballast, and intermediate conditions.

To allow any in-situ inspection of tanks with minimal decrease of production, the condition that certain sets of tanks are empty in turn needs to be considered. Even if the additional weight of process equipment may not significantly increase the still-water loads, it is possible that they may affect the stability aspects because the center of gravity may be increased, and also it may become harder to meet the requirements for damage stability. The free-surface effect in slack cargo storage tanks is another issue related to stability that must be considered. Design for sloshing is also important.

In some cases where green-water loading is serious, such as in harsh environments and in low-freeboard situations, it may be required to add appropriate structure for the safety of process equipment. Also, deck structures must invariably be strengthened in way of the process plant, that can be heavy. In particular, the supports of process equipment may be vulnerable to deformation and overstress, and related detailed structural analysis needs to be undertaken. Such structures must be designed considering hull girder bending and the interaction between vessel hull and topsides structure.

12.2.5 Appraisals of Conversion Yard

While the vessel is under conversion, the process plant facilities may be fabricated in the same yard or elsewhere. In the case of the latter, the preassembled units will be transported to the conversion yard to place on board or the vessel may transit to the general location of the process plant fabrication.

Some important factors that must be evaluated during conversion yard appraisals are as follows (Parker 1999):

- Health, safety, and environmental issues
- Related past conversion experience
- Physical facilities and trade resources
- Staffing, discipline, and labor issues
- Corporate considerations, including management experience and fiscal stability
- Ability to manage technical issues and changes during design and conversion
- Fast-track project execution planning experience
- Ability to manage complex projects

12.3 Decommissioning

The decommissioning of ship-shaped offshore units occurs when oil or gas production from a field is exhausted and/or when an installation reaches the end of its available life. Decommissioning issues are becoming as important as those related to construction or operation; the major aim is to determine and design for the best way to shut down the installations at the end of their field life and for removing and perhaps reusing or disposing of some or all of the facilities and components as appropriate.

Decommissioning is usually accompanied by work to contain and remove harmful materials that may cause environmental pollution and health and safety risks. These include materials such as asbestos, and it is immensely helpful in this regard to avoid those materials in the design originally. Technological and economical issues may also occur and need to be solved during decommissioning activities.

International or national regulations pertaining to the decommissioning process and issues pertaining to health, safety, and environmental impact will need to be addressed. Usually, the planning, approval, and undertaking of decommissioning will need to be effectively monitored in detail by all parties concerned because the process can be technically difficult, dangerous to the personnel involved, and potentially harmful to the environment.

It is important to develop the decommissioning strategies for a particular case by achieving an acceptable balance between the issues of technical feasibility, safety, health, environmental impacts, and cost effectiveness. Useful information and practices related to decommissioning may be found in the literature, although most of them are for fixed platforms offshore or relatively small size facilities on land. For regulatory frameworks for decommissioning offshore installations, see Hoyle and

Griffin (1989), Griffin (1998a), HSE (2001), and Garland (2002, 2005). For general aspects and practices of decommissioning, see Hustoft and Gamblin (1995), Passard (1997), Gorman and Neilson (1998), Prasthofer (1998), Griffin (1998b, 1998c), Anthony et al. (2000), and Dempsey et al. (2000).

Issues associated with hazards and safety measures during decommissioning activities are studied by Bamidele (1997), Griffin (1998b, 1998c), and HSE (2001). Heavy-lift operations are often required in particular during decommissioning of fixed offshore platforms and HSE (1999) reviews the safety of lifting procedures that may be involved. UKOOA (2006) provides guidelines for decommissioning activities of offshore structures.

12.3.1 Regulatory Framework

International authorities have recognized that the decommissioning of offshore installations have an impact on health, safety, and the environment. Subsequently, a number of regulations have been established by various international authorities. Also, various regional authorities have dedicated themselves to making similar regulations, either now or in the immediate future. Noteworthy conventions and considerations may include the following, in the yearly order:

- Geneva convention, 1958
- London dumping convention, 1972
- Oslo convention, 1972
- Convention for the protection of the North-East Atlantic, 1972–1998
- Convention for the protection of the marine environment and the coastal region of the Mediterranean Sea, 1976
- United Nations convention on the law of the sea, 1982
- Bonn agreement: agreement for cooperation in dealing with pollution of the North Sea by oil and other harmful substances, 1983
- Vienna convention for the protection of the ozone layer, 1985
- International Maritime Organization's guidelines and standards, 1988
- Basel convention on the control of transboundary movements of hazardous wastes and their disposal, 1989
- United Nations framework convention on climate change, 1992
- Convention on the protection of the marine environment of the Baltic Sea area, 1992
- Convention on the protection of the Black Sea against pollution, 1992
- Protocol to the 1972 London convention, 1996

International standards are more likely to set minimum rules for decommissioning of offshore installations; however, regional regulations usually provide more stringent guidelines. The coastal states review the proposal of decommissioning plans on a case-by-case basis where various issues associated with health, safety, and environmental impacts are evaluated. The international oil companies involved invariably have their own extensive considerations as well.

The concerns for health, safety, environmental impact, and economical expenditure will be very important factors in the development and execution of the decommissioning plans in specific cases and important drivers, as always, for legislating new laws or amending/strengthening existing laws regarding decommissioning. An example may be the decommissioning case of the *Brent* spar, which had a significant effect on related regulations for offshore installations specifically in the North Sea and North-East Atlantic (Griffin 1998a, 1998b).

In the beginning, it was decided that the *Brent* spar, which weighed approximately 14,500 tons, was to be decommissioned by disposal in the deep Atlantic 240km west of Scotland because considerable scientific and technological studies had pointed to the same as perhaps the best practical environmental option. However, concerns of environmental organizations and the ensuing public demands probably resulted in the initial decommissioning option being abandoned. The spar was then towed to Norway and eventually the operator's proposal to reuse the spar as a Norwegian Ro/Ro ferry quay was accepted (*Oil and Gas* 1998).

12.3.2 Technical Feasibility Issues

The technical feasibility issues related to decommissioning usually arise when no past experience for decommissioning the same types of offshore installations exists. Even with experience, however, decommissioning of some types of offshore platforms, such as deep-water fixed structures (e.g., large steel jackets, concrete gravity structures), will cause the greatest technical challenge (Bamidele 1997). One key issue with removal is about how to safely and efficiently cut heavy steel-walled sections by explosives or mechanical means (e.g., diamond wire cutting, abrasive water jets). Other issues arise due to the need for lifting and transportation of heavy sections, the behavior of the installation during toppling operation, and the removal of large integrated or hybrid topsides (Prasthofer 1998).

Although relatively greater experience with removal and disposal of shallow-water steel jackets is now available, given that many have been performed, it is recognized that the same cutting and lifting techniques can, in principle, be applied to decommissioning of larger platforms, where the platform is cut and removed in several lifts by a semisubmersible crane barge; however, the details need to be considered and fully worked out in every case.

In some remote areas, however, infrastructure (e.g., large crane vessels, barges) for deconstruction may be lacking, and suitable recycling facilities are generally far from the site. Also, barges with equipment and process facilities in the process of being decommissioned may not usually be allowed entry into ports. When onshore disposal is planned, considerable road transport will also be required. Besides, onshore recycling or disposal may also be considered undesirable in terms of economical and environmental viewpoints and, in any event, brings with it its own unique set of concerns and challenges.

Studies have indicated that deep-water disposal may be practical to consider at some point (Prasthofer 1998). Some high-value components (e.g., pressure vessels, turbines) and individual modules (e.g., topsides), rather than the entire facility, may be considered for reuse. The use of decommissioned offshore platforms as artificial reefs is also being considered.

In deep-water disposal, offshore installations, after necessary cleaning, will be transported to deeper water areas and left there instead of bringing them to recycling facilities onshore. However, the public is very concerned with deep-water disposal and several problems of deep-sea disposal remain yet to be completely resolved. For example, even with lifting vessels and equipment available in recent years to facilitate removal and transport of heavy items (Anthony et al. 2000), many technical issues to be resolved still remain for decommissioning of large fixed types of offshore platforms (*E and P Forum* 1995), as follows:

- Reliable and cost-effective cutting methods
- Remove operating vehicle (ROV) capability to reduce the need of divers
- Alternatives to large-capacity crane barges (e.g., controlled auxiliary buoyancy systems)

Ship-shaped offshore installations that are typically operated in deep-water areas may afford more flexibility in the selection of their decommissioning methods than fixed offshore platforms. The deep-water disposal may be one of the easiest options to apply because ship-shaped offshore installations can be simply sunk after relevant treatment (e.g., cleaning and removal of oily materials inside cargo tanks). Besides, ship-shaped offshore installations can be towed to conversion yards and then most equipment and process facilities may be reused, or they may be treated similar to ships for their breakup.

12.3.3 Safety and Health Issues

Safety and health issues during decommissioning activities arise for many reasons including hazardous materials, use of explosives, diver exposure, and multiple heavy lifts (Bamidele 1997; HSE 1999, 2001). Risks to safety and health may be exacerbated by uncertainties about the structural integrity, precise weights, and centers of gravity of components or equipment that might have been changed during the life of the installation. It has been said that there is considerably more risk to personnel during total removal of a structure than with a partial removal because of higher exposure to hazards (Prasthofer 1998).

12.3.4 Environmental Issues

Although land and air environments must also be considered, the most critical environmental issues during decommissioning activities of offshore installations are related to the sea environment and the presence of oily materials. A thorough assessment on the environmental impact that can arise from the decommissioning of offshore installations must then be undertaken for each project (HSE 2001).

It may be that deep-sea disposal is the best practical environmental option as long as relevant treatments such as cleaning and removal of harmful and hazardous materials are strictly performed prior to disposal, as previously described. Ideally, it is desirable to fully recycle and reuse the equipment and process facilities (e.g., topsides); this must be the general future direction. We expect that such a goal will be easier to achieve sooner with ship-shaped offshore installations than perhaps with installations of other kinds.

12.3.5 Cost Issues

The decommissioning costs depend on the complexities related to preparing the structure for removal or disposal such as strengthening to ensure structural integrity during the tow and cleaning of the structure to remove harmful and hazardous materials, and also the cost to lease large lift vessels or supporting equipment (*E and P Forum* 1996; Anthony et al. 2000). When structures are left fully or partially in situ, long-term monitoring costs will also be incurred; these may be significant depending on the amount of material remaining offshore, distance from shore, and related regulatory requirements and similar considerations.

The decommissioning costs will be different for different structural characteristics and disposal options in terms of size, weight, and amount of cleaning required, among others. These costs are also closely related to the potential availability of particular decommissioning methods. It has been seen that a large amount of decommissioning cost savings may be realized in the cases of fixed offshore platforms by adopting the options, such as partial removal and toppling in situ compared to total removal in situ considering the reduced time of offshore deconstruction and the reduced cost of large lift vessels or supporting equipment (*E and P Forum* 1995). For disposal of fixed offshore platforms, the costs increase rapidly with increase in water depth because the operation becomes more complex; therefore, in this regard, ship-shaped offshore installations perhaps can offer greater flexibility and cost advantages.

12.3.6 Decommissioning Practices for Ship-Shaped Offshore Installations

The decommissioning of ship-shaped offshore installations is perhaps easier and less expensive than fixed offshore platforms because the structures are floating and mobile in nature. Some decommissioning experiences related to ship-shaped offshore installations are presented in Meenan (1998) and Anthony et al. (2000).

During the decommissioning of ship-shaped offshore installations, all moorings, flexible flowlines, and risers will be disconnected. Wells at seabed can be capped and abandoned. The installation can be towed to a conversion yard for reuse of its parts or its entirety, or it may be disassembled and used for parts. Equipment or process facilities may be upgraded or refitted at a dry-dock of the conversion yard before they are recommissioned in another field on the same vessel or on another vessel.

It has been determined that more than two thirds of ship-shaped offshore installations that have been decommissioned worldwide have subsequently been reused in other fields. The reuse of decommissioned installations is, of course, the best desirable option for mitigating environmental impact as well as for saving the available resources and energy. It is very important to always bear in mind that in the end, it is the duty and obligation of every engineering professional to be sensitive to and to appropriately serve the needs of society in general.

REFERENCES

Anthony, N. R., Ronalds, B. F., and Fakas, E. (2000). *Platform decommissioning trends*. SPE Asia Pacific Oil and Gas Conference and Exhibition, SPE 64446, Society of Petroleum Engineers Inc., Brisbane, Australia, October 16–18.

Assayag, S., Prallon, E., and Sartori, F. (1997). *Improvements in design of converted FPSOs regarding 20 years of operation without docking.* Offshore Technology Conference, OTC 8389, Houston, May.

Bamidele, B. (1997). *Review of the hazards and management control issues in abandonment safety cases.* (Offshore Technology Report, OTH 1997/547), Health and Safety Executive, UK.

Biasotto, P., Bonniol, V., and Cambos, P. (2005). *Selection of trading tankers for FPSO conversion projects.* Offshore Technology Conference, OTC 17506, Houston, May.

da Costa Filho, F. H. (1997). *The world's biggest conversion.* Offshore Technology Conference, OTC 8407, Houston, May.

Dempsey, M. J., Mathieson, W. E., and Winters, T. A. (2000). *Learning from offshore decommissioning practices in Europe and the USA.* SPE Asia Pacific Oil and Gas Conference and Exhibition, SPE 64444, Society of Petroleum Engineers Inc., Brisbane, Australia, October 16–18.

E and P Forum (1995). *Removal/disposal of large North Sea steel structures.* (Report No. 10.14/243), The Oil Industry International Exploration and Production Forum, July.

E and P Forum (1996). *Removal and disposal of offshore platform topside facilities.* (Report No. 10.15/248), The Oil Industry International Exploration and Production Forum, August.

Garland, E. (2002). *Environmental regulatory framework in the North Sea: An update of the existing and foreseeable constraints.* SPE Annual Technical Conference and Exhibition, SPE 77390, Society of Petroleum Engineers, San Antonio, September 30–October 2.

Garland, E. (2005). *Environmental regulatory framework in Europe: An update.* SPE/EPA/DOE Exploration and Production Environmental Conference, SPE 93796, Society of Petroleum Engineers, Galveston, Texas, March 7–9.

Gorman, D. G., and Neilson, J. (1998). *Decommissioning offshore structures.* Germany: Springer-Verlag.

Griffin, W. S. (1998a). *The global and international regulatory regime for decommissioning disused platforms.* SPE International Conference on Health, Safety, and Environment in Oil and Gas Exploration and Production, SPE 46591, Society of Petroleum Engineers, Caracas, Venezuela, June 7–10.

Griffin, W. S. (1998b). *Managing the platform decommissioning process.* SPE International Conference and Exhibition, SPE 48892, Society of Petroleum Engineers, Beijing, China, November 2–6.

Griffin, W. S. (1998c). *Evolution of the global decommissioning regulatory regime.* Offshore Technology Conference, OTC 8784, Houston, May.

Hoyle, B. J., and Griffin Jr., W. S. (1989). *International standards for removal of abandoned and disused offshore oil production platforms: Negotiation and agreement.* Offshore Technology Conference, OTC 5932, Houston, May.

HSE (1999). *Decommissioning – Heavy lift operations: A review of safe lifting procedures.* (Offshore Technology Report, OTO 1998/170), Health and Safety Executive, UK.

HSE (2001). *Decommissioning topic strategy.* (Offshore Technology Report, OTO 2001/032), Health and Safety Executive, UK.

Hustoft, R., and Gamblin, R. (1995). *Preparing for decommissioning of the Heather field.* SPE Offshore European Conference, SPE 30372, Society of Petroleum Engineers, Aberdeen, Scotland, September 5–8.

IMO (2003). *Guidelines for application of MARPOL Annex I requirements to FPSOs and FSUs.* (MEPC/Circ. 406), International Maritime Organization, London, November.

ISO DIS 18072-2 (2006). *Ships and marine technology – Ship structures: Part 2. Requirements for ultimate limit state assessment.* (Draft International Standards), International Organization for Standardization, Geneva.

ISO FDIS 18072-1 (2006). *Ships and marine technology – Ship structures: Part 1. General requirements for limit state assessment.* (Final Draft International Standards), International Organization for Standardization, Geneva.

Johnson, M. (1996). *Application of the ABS/SafeHull technology to FPSO conversions.* Presented at the 23rd February 1996 Annual Meeting of the Texas Section, The Society of Naval Architects and Marine Engineers, NJ.

Lane, J. A., Bryans, R., and Preston, R. (2004). *Conversion of the TT Sahara to FPSO Fluminense: A low cost solution for the Bijupira and Salema field development, offshore Brazil.* Offshore Technology Conference, OTC 16707, Houston, May.

Meenan, P. A. (1998). "Technical aspects of decommissioning offshore structures." In *Decommissioning offshore structures*, D. G. Gorman and J. Neilson, eds. London: Springer-Verlag.

Mones, M. (2004). Hull structural assessment and refurbishment for conversion FPSOs. Proceedings of ASMR-PTI FPSO Integrity Workshop, American Society Mechanical Engineers/Petroleum Technology Institute, Houston, August 30–September 2.

Neto, T. G., and de Souza Lima, H. A. (2001). *Conversion of tankers into FPSOs: Practical design experiences.* Offshore Technology Conference, OTC 13209, Houston, May.

Newport, A., Basu, R., and Peden, A. (2004). "Structural modifications to the FPSO Kuito cargo tanks." *Proceedings of OMAE–FPSO 2004 – OMAE Specialty Symposium on FPSO Integrity, OMAE–FPSO'04-0085,* Houston, August 30–September 2.

O'Neil, W. A. (2001). *Reaching the breaking point.* (DNV Forum, No. 3), Det Norske Veritas, Oslo, Norway.

Oil and Gas Journal (1998). "Shell picks re-use option for Brent Spar." *Oil and Gas Journal,* 96, February 6.

Park, I. K., Jang, Y. S., Shin, H. S., and Yang, Y. T. (1998). *Conceptual design and analyses of deep-sea FPSO converted from VLCC.* Offshore Technology Conference, OTC 8809, Houston, May.

Parker, G. (1999). *The FPSO design and construction guidance manual – A reference guide to successful projects designing and constructing FPSOs in all water depths.* Houston: Reserve Technology Institute.

Passard, J. P. (1997). *Environmental impact from removal of installations: The North East Frigg field installations.* SPE/UKOOA European Environmental Conference, SPE 37857, Society of Petroleum Engineers, Aberdeen, Scotland, April 15–16.

Pearce, F. (1996). "101 things to do with an old oil rig." *New Scientist Magazine,* Issue 2058, November 30 (http://www.newscientist.com).

Prasthofer, P. H. (1998). *Decommissioning technology challenges.* Offshore Technology Conference, OTC 8785, Houston, May.

Terpstra, T., d'Hautefeuille, B. B., and MacMillan A. A. (2001). *FPSO design and conversion: a designer's approach.* Offshore Technology Conference, OTC 13210, Houston, May.

Terpstra, T., Schouten, G., and Ursini, L. (2004). *Design and conversion of FPSO Mystras.* Offshore Technology Conference, OTC 16198, Houston, May.

UKOOA (2006). *Guidelines on stakeholder engagement for decommissioning activities.* Offshore Operators Association, UK.

van Voorst, O., de Baan, J., van Loenhout, A., and Krekel, M. (1995). *Conversion of existing tanker to North Sea FPSO use.* Offshore Technology Conference, OTC 7724, Houston, May.

CHAPTER 13

Risk Assessment and Management

13.1 Introduction

Typically, the term "risk" is defined as either the product or a composite of (a) the probability or likelihood that any accident or limit state leading to severe consequences, such as human injuries, environmental damage, and loss of property or financial expenditure, occurs; and (b) the resulting consequences. In the design and operation of ship-shaped offshore units, as in many other types of structures, there are a number of hazards that must be dealt with in the process of risk assessment. Wherever there are potential hazards, a risk always exists.

To minimize the risk, one may either attempt to reduce the likelihood of occurrence of the undesirable events or hazards concerned, or contain, reduce, or mitigate the consequences, or both. In the lifecycle of a ship-shaped offshore installation, assessing managing, and controlling the risk is required so that it remains under a tolerable level. The risk management and control should, in fact, be an ongoing process throughout the lifecycle of an installation – that is, involving feasibility study, concept, or front-end design, detailed design, operation, and decommissioning. The different stages of the lifecycle will offer different opportunities for risk management and control, as may be expected.

Substantial efforts, such as the SAFEDOR project (http://www.safedor.org), are being directed by the maritime industry toward the application of the risk-assessment techniques together with risk-evaluation criteria to offshore design, operation, and human and environmental safety (e.g., Skjong et al. 2005).

IMO (2002) released guidelines for a formal safety assessment (FSA), which we use in this chapter for illustration and explanation of methodology, although numerous other guidelines and variations also exist. This FSA from IMO is a rational and systematic process for the proactive management and control of risks, consisting of hazard identification, risk analysis, risk-management option, cost–benefit assessment, and related decision-making recommendations. Although the IMO FSA process has been developed for possible application to the shipping industry in the future development of *goal-based standards*, it can be applied to ship-shaped offshore installations with appropriate adjustments and modifications.

This chapter describes the FSA process for risk assessment and management of ship-shaped offshore installations mostly in conceptual terms. Some selected measures for risk correction or reduction during design and operation are also presented; for specific examples related to inspection and maintenance, see Chapter 11 as well.

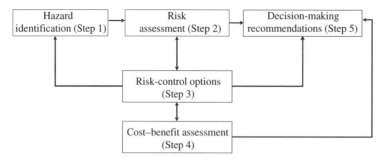

Figure 13.1. Process for formal safety assessment (FSA).

The treatment here considers "accidental scenarios" for convenience; however, note that similar methodology and procedures should also be applied to routine operational considerations because human enterprise is replete with risks for various kinds.

A vast number of publications dealing with risk management and control of offshore installations as well as other types of structures are available in the literature; the following publications can be obtained for students' personal use:

Formal safety assessment, prepared by International Maritime Organization (IMO): http://www.imo.org/safety

Formal safety training modules, prepared by International Association of Classification Societies (IACS): http://www.iacs.org.uk

Risk-based decision-making guidelines, prepared by the U.S. Coast Guard (USCG): http://www.uscg.mil/hq/g-m/risk/e-guidelines/rbdm.htm

Risk management, prepared by UK Health and Safety Executive: http://www.hse.gov.uk/risk/

Risk management, prepared by Chester Simmons: http://sparc.airtime.co.uk/users/wysywig/risk_1.htm

Health, safety, and environmental management system, prepared by Baker Hughes: http://www.bakerhughes.com/hse/default.htm

Health and safety in the UK oil and gas industry, prepared by UK Offshore Operators Association (UKOOA): http://www.ukooa.org.uk/issues/health/faq.htm

13.2 Process for Formal Safety Assessment

The FSA process consists of the following five major steps (Figure 13.1):

(1) Hazard identification (HAZID)
(2) Risk assessment (RA)
(3) Risk-management options (RMO)
(4) Cost–benefit assessment (CBA)
(5) Decision-making recommendations (DMR)

The characterization of hazards and risks can be both qualitative and quantitative consistent with the available data, and one should always consider a broad enough range of risk-control options. As examples, these may include the design changes, changes in the operational procedures, implementation of appropriate training, and provision of alarms. In Sections 13.2.1–13.2.6, each step of the FSA process is described briefly.

13.2.1 System Definition

The FSA will start with the system definition that reviews all functional and operational characteristics of the offshore installation under consideration. The following parameters need to be clearly defined:

- Installation category: type, size, new-build, or converted
- System designs: layout, mooring system, subdivision, structural features, and hazardous materials
- Operational plans: storage and offloading
- Accident categories: unintended flooding, collision, fire, explosion, and accelerated degradation
- Risk categories: injuries, damage to property/financial expenditure, and environmental impacts

For detailed risk assessment, the installation may be divided further in terms of subsystems, activities, materials, and the like that are associated with the hazards concerned. For example, Table 13.1 indicates a possible classification of some of the significant hazards that may be involved in ship-shaped offshore installations, listed by the subsystem(s) associated with it (HSE 2000a).

13.2.2 Hazard Identification

A hazard is a situation with potential for causing harm to human safety, the environment, or property, or a potential result of significant financial expenditure. A hazard may be a physical situation (e.g., a shuttle tanker is a hazard because it may collide with the installation), or an activity (e.g., crane operations are a hazard because the object may drop), or a material (e.g., fuel oil is a hazard because it may catch fire).

The purpose of hazard identification (HAZID) is to identify such hazards with some reference to how likely or unlikely each is to occur and also to give some consideration to the degree of consequences involved, usually in a qualitative manner. Through the HAZID, hazards are prioritized and accident scenarios ranked. Accident scenarios are typically the sequence of basic events from the initiating event up to the consequence.

Although this section describes only some selected methods, there are a number of methods available for HAZID and/or risk assessment of offshore installations (http://www.uscg.mil/hq/g-m/risk/e-guidelines/RBDM.htm), as follows:

- Checklist analysis
- What-if analysis
- Preliminary risk analysis
- Preliminary hazard analysis
- Hazard and operability analysis (HAZOP)

Table 13.1. *Examples of subsystems and associated hazards in ship-shaped offshore installations*

Subsystem	No.	Hazard
	1.	Hull failure due to extreme wave load
	2.	Hull failure or marine accident due to ballast failure or failure during loading/offloading operations
	3.	Leak from cargo tank caused by fatigue
	4.	Accident (e.g., explosion, during tank intervention)
Hull and marine	5.	Collision with passing vessel
	6.	Collision with shuttle tanker or supply vessel
	7.	Collision with other vessels
	8.	Collision during offloading
	9.	Rapid change of wind direction
	10.	Multiple anchor failure
	1.	Leak leading to fire or explosion in process facility
	2.	Leak from turret systems causing fire or explosion in turret
	3.	Leak or rupture of rise
	4.	Impacting loads due to crane operations (swing loads)
Hydrocarbon systems	5.	Dropped objects from retrieval of cargo pumps
	6.	Severe rolling during critical operations (e.g., crane operations)
	7.	Topside fire threatening cargo tank
	8.	Emergency flaring with approaching shuttle tanker or during offloading
	9.	Unintended release of riser
	10.	Work in open-air spaces during windy conditions
	1.	Failure of cargo tank explosion prevention function during normal operation
	2.	Fire or explosion in pump room
Auxiliary systems	3.	Oil spill from offloading system (e.g., hoses, valves)
	4.	Fire or explosion in engine room
	5.	Helicopter impact or crash

- Change analysis
- Relative ranking/risk index analysis
- Event and casual factor chart analysis
- Fault tree analysis (FTA)
- Event tree analysis (ETA)
- Failure modes and effect analysis (FMEA)

Each of these methods has its strengths and weaknesses. AICE (1992) presents various HAZID methods used in the process industry. CMPT (1999) and HSE (2001) summarize the HAZID methods available for general offshore installations. The HAZID methods available for offshore safety cases are described by HSE (1997a).

As an appropriate starting point for a HAZID process, an intuitive review of hazards may be useful. The following are data that must be reviewed in this type of HAZID:

- Existing HAZID data
- Historical accident data

Table 13.2. *Example of a generic hazard checklist, following CMPT (1999)*

Event	Cause
Blowouts	Blowout during drilling, completion, production; blowout during work over, abandonment; underground blowout; blowouts due to well-control incidents (usually less severe than blowouts), fire in the drilling system (e.g., mud pits, shale shaker)
Riser/pipeline leaks; leaks of gas and/or oil from:	Import flowlines; export risers; subsea pipelines, well head manifolds
Process leaks; leaks of gas and/or oil from:	Well-head equipment; separators and other process equipment; compressors and other gas treatment equipment; process piping, flanges, valves, pumps, etc.; topsides flowlines; flare/vent system; storage tanks; loading/unloading system; turret swivel system
Nonprocess fires	Fuel gas fires, electrical fires; accommodation fires; methanol/diesel/aviation fuel fires; generator/turbine fires; heating system fires; machinery fires; workshop fires
Nonprocess spills	Chemical spills; methanol/diesel/aviation fuel spills; bottled-gas leaks; radioactive material releases; accidental explosive detonation
Marine collisions; impacts from:	Supply vessels; stand-by vessels; other support vessels (diving vessels, barges, etc.); passing merchant vessels and fishing vessels; naval vessels (including submarines); flotel; drilling rig, drilling support vessel (jack-up or barge); offshore shuttle tankers; drifting offshore vessels (e.g., semisubmersibles, barges, storage vessels); icebergs; for each vessel category, different categories of events, such as powered and drifting may be separated
Structural events	Structural failure due to fatigue or design error; extreme weather, earthquakes; foundation failure; derrick, crane, and mast collapse; disintegration of rotating equipment
Marine events	Anchor loss/dragging (including winch failure); capsize (due to ballast error or extreme weather); incorrect weight distribution (due to ballast or cargo shift); collision, grounding or loss of tow during transit; icing
Dropped objects; objects dropped during:	Construction; crane operations; cargo transfer; rigging up derricks and drilling
Transport accidents involving a crew change or in-field transfers	Helicopter crash into sea/platform/ashore; fire during helicopter refueling; aircraft crash on platform (including military); personal accidents during transfer to boat; road traffic accident during mobilization
Construction accidents; occurring during:	Construction onshore or offshore; marine installation; hook-up and commissioning; pipe laying

- Past experience in terms of risk management and control obtained during design and operation
- Hazardous material data
- Recognized guidelines and standards

The method using a generic checklist may be useful because it may promote standard hazard categories. Table 13.2 indicates an example of the *generic hazard checklist* (CMPT 1999). To these we must add a new category: terrorism.

Table 13.3. *Example of a what-if checklist*

What if . . .?	Causes	Immediate consequences	Ultimate consequences	Recommendations
What if {a specific accident} occurs?				
What if {a specific subsystem} fails?				
What if {a specific human error} occurs?				

The process of using a *what-if checklist* usually involves a brainstorming session making use of broad, loosely structured questioning to postulate initiating events resulting in accidents, system performance issues, and related consequences to ensure that relevant safeguards or risk corrective/preventive measures against those issues can be identified and implemented. Table 13.3 shows an example of a what-if checklist.

Another method, a hazard and operability (HAZOP) study, may also be useful to identify hazards in a systematic and comprehensive manner considering all hazardous process deviations. It is also more effective in taking account of human errors as well as technical faults. A team of experts with different expertise in various subsystems or activities as those indicated in Table 13.1 makes a systematic check to identify deviations from design intent, referring to process and instrumentation diagrams (PIDs) in the case of process systems. Based on a standard list of guidewords, possible hazard causes, and their consequences, recommendations for corrective/preventive measures are recorded in a standard format. Table 13.4 is an example of a HAZOP checklist. The guidewords may include the following:

- Failed
- Impaired/damaged
- Fails during
- Not done
- Inadequate/insufficient
- Incorrect/inappropriate
- Too late/soon
- Congested/overloaded

Other methods, such as failure modes and effects analysis (FMEA), may also help provide a systematic identification of hazards in a HAZOP exercise, considering the failure modes of a system, their effects, and criticality or severity. Possible

Table 13.4. *Example of a HAZOP checklist.*

Property words and guidewords	Causes	Consequences	Safeguards	Recommendations

Table 13.5. *Example of a FMEA checklist*

Failure modes/causes	Effect	Detection	Safeguards	Overall assessment	Overall criticality

failures for all components in the system are evaluated together with recommended corrective/preventive measures and recorded in a systematic list as indicated in Table 13.5. Failure modes will be rated in terms of their frequency and severity.

Based on HAZID, all potential accident scenarios can be listed in a relevant form, as indicated in Table 13.6. Important information including the consequence and its criticality, which will be used for risk assessment, must also be included, although this may be qualitative. Extensive HAZID for ship-shaped offshore installations is presented by HSE (1997c).

13.2.3 Risk Assessment

The primary tasks of risk assessment are to determine frequency, consequence, and risk for the various accidental scenarios identified in a HAZID and indicated in Table 13.6. The levels of frequency and consequence for any accident scenario will differ depending on the site-specific functional and operational requirements and other factors. Although the frequency categories may be defined by the probabilities of occurrence, the consequence categories may be defined by various means including the degree of impairment of safety functions, injuries or fatalities, and, of course, cost consequences.

For risk assessment, two broad classes of techniques are usually relevant in more detail: qualitative and quantitative techniques, which are described in the following sections. A qualitative technique involves an intuitive representation of an accident scenario together with its consequence intensities. A quantitative technique provides quantitative outcomes of an event that are typically found as a function of physical circumstances relating to a given accident scenario as identified from sophisticated analyses such as limit-state assessments (e.g., impact energy due to side collisions, impact-pressure actions arising from explosion) or other possible consequence analyses.

The early design stages will typically employ the qualitative approach. The later design stages or the phases of operation may apply the quantitative approach for more refined characterization of risks, particularly where local regulations require the same. Depending on the types of accident scenarios, different techniques and models for quantitative assessment of frequencies and consequences may, of course, be used.

Table 13.6. *Example of a HAZID checklist*

Accident scenario	Initiating event	Frequency	Consequence	Criticality assessment	Recommendations

Once the frequency and consequence of the ith event are determined, the risk may be calculated as a product of frequency and consequence,

$$R_i = F_i \times C_i, \tag{13.1}$$

where R_i, F_i, C_i = risk, frequency, and consequence for the ith event, respectively. The total risk for any accidental scenario, R, which may be composed of a number of basic events, can then be calculated by appropriate risk summation.

13.2.4 Risk-Management Options

A measure to control a single element risk is called a *risk-control measure*. For a system installation, a number of risk-control measures must be relevant because there might be a number of different risk aspects due to the various consequential effects of hazards and accident scenarios. Therefore, risk-control options will consist of appropriate combinations of risk-control measures. The task of deriving risk control options may comprise three stages, as follows:

Stage 1: Identify the necessary areas for risk control by constructing a risk-root (basis) map, usually with the focus on high risk, high frequency, high consequence, and high uncertainty hazardous events.

Stage 2: Identify risk-control measures. As necessary, new risk control measures are required unless existing risk-control measures are sufficient. The use of casual chains and risk-control attributes in the sequence of casual factors $\rightarrow$ failure $\rightarrow$ circumstance $\rightarrow$ accident $\rightarrow$ consequence is useful in developing appropriate measures.

Stage 3: Derive risk-control options. Decide first whether risk-corrective or preventive measures are necessary, and then appropriately select from and group the risk-control measures developed at stage 2.

13.2.5 Cost–Benefit Analysis

The cost–benefit analysis represents the risks in terms of costs and provides a basis for decision making about risk-control options. In practice, two terminologies are typically used for this purpose: cost–benefit analysis and cost-effectiveness analysis. In the cost–benefit analysis, risk-control options cease when their benefits simply outweigh their costs. In contrast, the cost-effectiveness analysis uses certain ratios of costs to benefits, which may be useful, for example, to consider evaluating risk-control options associated with essential safety functions (e.g., lifesaving) in purely financial terms for comparison purposes.

The risk indices associated with costs of averting a fatality are often useful for evaluating related risk-control measures in terms of their cost effectiveness. Two typical indices used for this purpose are as follows:

$$\text{Gross cost of averting a fatality: GCAF} = \frac{\Delta C}{\Delta R}, \tag{13.2a}$$

$$\text{Net cost of averting a fatality: NCAF} = \frac{\Delta C - \Delta B}{\Delta R}, \tag{13.2b}$$

where ΔR = risk reduction in terms of the number of fatalities averted, implied by the risk-control option (i.e., the risk corrective/preventive measures); ΔC = cost of the risk-control option; and $\Delta B =$ economic benefit resulting from the implementation of the risk-control option.

13.2.6 Decision-Making Recommendations

With the risk level known, risk acceptance is judged by whether or not a risk is tolerable or any measure for risk correction or reduction is required. Risk acceptance may be affected by many factors, such as:

- Familiarity: People are more comfortable and accepting of risk when they are personally familiar with the operation. For example, is a traveler more fearful of a bus accident or a plane crash? Which has the greater risk?
- Frequency: The belief in the frequency of an accident influences the risk acceptance. If it is not believed that any accident will happen, the risk may be more likely accepted.
- Control: When any accident is controllable, more risk may be accepted.
- Public relations: Our awareness of risk impacts can be increased by media coverage of accidents.
- Severity of consequence: The acceptable risk level is stringently affected by severity of consequences.
- Suddenness of consequence: If the consequence of the risk is sooner, the acceptable risk level must be lower.
- Personal versus societal: The acceptable risk level for the public must be lower than that for any individual person.
- Benefit: With more benefits obtained from the operation, the acceptable risk level may be higher.
- Dread: When the related dread or fear is stronger, the acceptable risk level would be lower.

Considering so many factors affecting the risk acceptance noted here, it is not straightforward to define acceptable risk levels for a particular case. Tolerable risk levels are generally determined from those of similar offshore installations with successful past experiences, but they must provide a balance between absolute safety requirements and costs and benefits of proposed risk-reduction measures for the specific installation.

The as low as reasonably practicable (ALARP) concept is a very common way to judge risk acceptance. This concept was first applied to the nuclear industry (HSE 1988) and was later adopted for offshore installations (HSE 1992). In this concept, risks above one particular acceptable level are considered intolerable; that is, they are so high as to call for immediate action to reduce them regardless of cost (e.g., banning that operation). Risks below another particular acceptable level are considered negligible or broadly acceptable, meaning that they do not require any further action for their correction or reduction.

In the intermediate region between these two levels, risk may be considered tolerable but not negligible. In this case, it is required to reduce the risk further as low as reasonably practicable, that is, to the so-called ALARP region. The remaining risks, therefore, may be either in the broadly acceptable or the ALARP regions after implementation of risk-control options.

13.3 Qualitative Risk Assessment

Qualitative risk management involves an intuitive assessment of risks arising from the accidental scenarios, when considered together with their consequence intensities.

Using HAZID results, the accident scenarios can be ranked. For this purpose, frequency and consequence indices are often defined on a logarithmic scale so that a risk index can be expressed by adding the frequency and consequence indices (IMO 2002):

$$RI = FI + CI, \tag{13.3}$$

where $RI = Log(R_i) = $ risk index; $FI = Log(F_i) = $ frequency index; $CI = Log(C_i) = $ consequence index. For example, when $FI = 3$ and $CI = 2$, then $RI = 5$ for ranking purposes of the accident scenarios.

Historical databases and expert opinions may be employed for the determination of frequency and consequence for any accidental scenario. Useful historical accident databases pertinent to offshore installations and operations may be found in DNV (1980–2006, 1997) or HSE (1997b). CMPT (1999) reviews other data sources available for offshore installations. HSE (1999) provides a comprehensive evaluation of ship impacts with offshore installations in terms of likelihood and consequences. Nesje et al. (1999) present useful information regarding the process leaks from ship-shaped offshore installations in terms of leak frequency and consequences together with recommendations for risk corrective/preventive measures. They also provide similar information regarding explosion risk for ship-shaped offshore installations.

However, we caution that when data from historical databases are used, the same data may need to be adjusted for the present circumstances; for example, because the site-specific environment and specific facility features are not always adequately represented by historical data.

The level of frequency of occurrence for an individual scenario is often divided into several categories (IMO 2002):

- Frequent
- Reasonably probable
- Remote
- Extremely remote

Table 13.7 shows an example of the frequency rating together with a logarithmic frequency index provided by IMO (2002). In each case, a risk-assessment team would appropriately define for themselves what these frequency ratings mean or how they would relate to actual data. ISO 17776 (1999) uses five rating categories for the frequency of occurrence based on more factual likelihood terminology:

(1) Happens several times per year in location.
(2) Happens several times per year in operating company.
(3) Has occurred in operating company.
(4) Happens several times per year in industry.
(5) Rarely occurs in the industry.

Table 13.7. *Example related to the frequency rating (IMO 2002)*

Rating	Definition	Frequency F_i (per year)	Frequency index (FI)
Frequent	Likely to occur once per month on an installation	10	7
Reasonably probable	Likely to occur once per year in a world fleet of 10 installations, that is, likely to occur a few times during the entire life cycle	0.1	5
Remote	Likely to occur once per year in a world fleet of 1,000 installations, that is, likely to occur in the total life of several similar installations	10^{-3}	3
Extremely remote	Likely to occur once in the lifetime of a world fleet of 5,000 installations	10^{-5}	1

Also, the consequence is divided into several categories depending on the severity of the accident impacts (IMO 2002):

- Catastrophic
- Severe
- Significant
- Minor

In each case, the risk-assessment team would appropriately define what these consequence ratings mean under their particular circumstances. These characterizations are by no means universal. For example, the consequence severity of human injury has been categorized for a particular use, as follows (HSE 2000b):

- Fatality
- Permanent total disability
- Permanent partial disability
- Lost workday
- Restricted work case
- Medical treatment
- First aid case

Table 13.8 shows an example of the consequence rating in terms of safety issues, provided by IMO (2002). Compared to this, ISO 17776 (1999) provides six ratings of

Table 13.8. *Example related to the consequence rating for safety issues (IMO 2002)*

Rating	Consequence on human safety	Consequence on the system	Consequence C_i (equivalent fatality)	Consequence index (CI)
Catastrophic	Multiple fatalities	Total loss	4	10
Severe	Single fatality or multiple severe injuries	Severe system damage	3	1
Significant	Multiple or severe injuries	Nonsevere system damage	2	0.1
Minor	Single or minor injuries	Local equipment damage	1	0.01

Table 13.9. *Example related to the consequence rating for four significant aspects (ISO 17776 1999)*

Rating	People (injury)	Assets (damage)	Environment (effect)	Reputation (impact)
0	Zero injury	Zero damage	Zero effect	Zero impact
1	Slight injury	Slight damage	Slight effect	Slight impact
2	Minor injury	Minor damage	Minor effect	Limited impact
3	Major injury	Local damage	Local effect	Considerable impact
4	Single fatality	Major damage	Major effect	Major impact
5	Multiple fatalities	Extensive damage	Massive effect	Major international impact

consequence severity covering various issues affecting people, assets, environment, and reputation, as indicated in Table 13.9.

Table 13.10 shows an example of results from a qualitative risk assessment following the definitions suggested by IMO (2002) for some 10 accident scenarios that are not described here. Table 13.11 shows an example of the risk matrix based on the risk levels or the logarithmic risk indices indicated in Table 13.10. Three regions of risk acceptance – intolerable, ALARP, and negligible regions – are also identified in this example.

Table 13.12 shows another example of qualitative risk-assessment results where the consequence part of the risk is considered in terms of associated costs. Although details are not given here, in this case the frequency of occurrence included the likelihood of possible hazard escalation into a chain of basic events, which was calculated as follows:

$$F_i = f_i \times p_{i1} \times p_{i2} \times \ldots, \tag{13.4}$$

where F_i = frequency of the ith accident; f_i = frequency of incident (cause) for the ith accident scenario; and p_{i1}, p_{i2} = probabilities of occurrence of basic initiating events No. 1 or No. 2 causing the ith accident scenario. The total *expected risk* is, therefore, 15,300 US$ per year.

Table 13.10. *Example of qualitative risk assessment with 10 accident scenarios*

Scenario	F_i	FI	C_i	CI	R_i	RI
S1	10^{-3}	3	10.0	4	10^{-2}	7
S2	10	7	10.0	4	10^2	11
S3	10^{-4}	2	0.01	1	10^{-6}	3
S4	0.1	5	10.0	4	1.0	9
S5	10^{-3}	3	10.0	4	10^{-2}	7
S6	10^{-3}	3	1.0	3	10^{-3}	6
S7	10^{-4}	2	10.0	4	10^{-3}	6
S8	10^{-5}	1	0.1	2	10^{-6}	3
S9	0.1	5	10.0	4	1.0	9
S10	10^{-3}	3	0.1	2	10^{-4}	5

Table 13.11. *Example of the risk matrix for qualitative risk assessment*

13.4 Quantitative Risk Assessment

The quantitative approach is, in theory, more sophisticated than the qualitative approach because the quantitative approach deals with all aspects of frequency, consequence, and risk in a more refined and numerical manner. The use of the quantitative approach to risk has been encouraged for risk management and control of offshore installations in some sectors, for instance UK and Norway in particular (e.g., UKOOA 1999, 2000).

The quantitative risk assessment (QRA) also involves the selection of accidental events that represent specific hazards, and it is important to select a comprehensive set in the range of accidents that may occur in reality. A hazard may be represented by single- or multiple-accidental events identified in advance. The QRA then attempts to determine the frequencies and consequences of these prescribed accidental events by various means.

Table 13.12. *Example of qualitative risk assessment in terms of associated costs*

Accident scenario	f_i (per year)	p_{i1}	p_{i2}	F_i (per year)	C_i (US$)	R_i (US$/year)
Scenario 1	1.0 (e.g., valve leaks)	0.1 (e.g., flow not stopped)	0.01 (e.g., oil enters water)	0.001	100,000	100
Scenario 2	0.5 (e.g., hose leaks)	0.2 (e.g., flow not stopped)	0.1 (e.g., oil enters water)	0.001	200,000	200
Scenario 3	0.1 (e.g., hose rupture)	0.5 (e.g., flow not stopped)	1.0 (e.g., oil enters water)	0.05	300,000	15,000
		Total				15,300

13.4.1 Frequency Analysis

The purpose of the frequency analysis is to estimate the likelihood of occurrence of each accidental event. The following are useful techniques for frequency analysis:

- Use of existing generic accident frequency data
- Use of historical accident frequency data
- Use of artificially developed accident frequency data
- Application of fault tree analysis (FTA)
- Application of event tree analysis (ETA)

Many QRAs use existing generic accident frequency data instead of developing new data. CMPT (1999) provides extensive compilations of such data. When generic accident data is not available, however, new data must be developed. The use of the historical accident frequency data is a simple approach that is easy to understand, although strictly speaking, it may be applicable only to existing technology with sufficient experience of accidents with appropriate records. In this case, frequencies could, for instance, be simply calculated by combining accident experience and population exposure, typically measured in terms of installation years, as follows:

$$\text{Frequency of an event per offshore unit-year} = \frac{N_e}{N_i \times Y_e}, \tag{13.5}$$

where N_e = number of event; N_i = number of installation; and Y_e = years of exposure. The historical data provided by the DNV worldwide offshore accident databank (DNV 1997 and 1980–2006), HSE (1997b), or CMPT (1999) may be useful for this purpose. AICE (1989) presents useful guidelines related to the collecting and processing of frequency data for QRAs in the chemical industry. CMPT (1999) provides some simpler guidelines for the frequency analysis of accidental events involved in offshore installations.

A major challenge in the use of historical data is that available data may not include all accidents that have occurred in the world or can conceivably occur, unless of course the accidents are very severe. Thus, the historical data may not always be sufficient to determine the frequencies of occurrence of various possible "accidents" or the entire range of accidents that may be of interest. When the data for only major accidents (e.g., fatalities) is available, the frequency of less-severe accidents may be determined by "guesstimation," using interpolation or extrapolation techniques cast in terms of the ratios of the frequencies of such accidents and/or using artificial accident data generated by numerical simulation. Paik et al. (2003) review some insights on the development of artificial accident data relating to collisions and grounding of trading tankers.

FTA is a logical combination of many "basic events" that cause one critical "top event," for example, a system failure. It uses logic gates such as AND or OR gates to represent how basic events combine to cause the critical top event, as illustrated in Figure 13.2. Usually, it starts with the critical top event and then works down toward the basic events. If multiple basic events must occur simultaneously to cause a higher

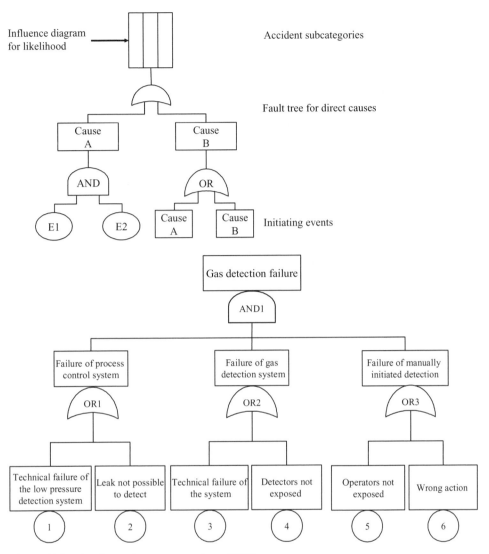

Figure 13.2. A schematic representation of FTA.

event, then they will be combined with an AND gate. If any one of several basic events may cause the higher event, then the events will be joined with an OR gate. For a more detailed description on FTA, see Lees (1996).

ETA is a logical representation of various events that may follow from an initiating event or basic event. Various possibilities of higher events are represented by branches. This technique is also useful to relate an event to various consequence models. The questions defining the branches are positioned across the tree, with one branch indicating "yes" or the other branch for "no." Figure 13.3 shows an example of ETA.

Various parameters may be involved in the likelihood of the accidental scenarios. For example, the following parameters should be considered for the frequency

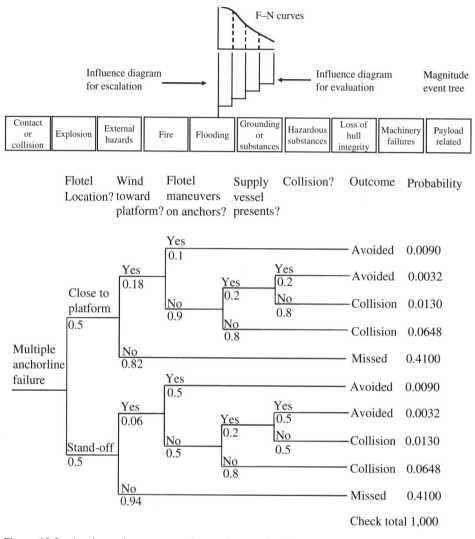

Figure 13.3. A schematic representation and example of ETA.

analysis when developing risk-control options against accelerated structural degradation in the context of operations and inspection of offshore installations:

- Design standards
- Age
- Historical performance
- Degradation
- Time interval of inspection

The likelihood scoring of all events per installation may be obtained using the
frequency categories indicated in Section 13.3. Table 3.13 indicates an example of
the link between frequency category and score. Figure 13.4 is an example of the
outcome of the frequency analysis that represents the accumulated likelihood score

Table 13.13. *Example of the link between likelihood (frequency) and consequence scorings*

Frequency		Consequence	
Category	Score	Category	Score
1	$\leq L_1$	A	$\leq C_1$
2	$> L_1$ and $\leq L_2$	B	$> C_1$ and $\leq C_2$
3	$> L_2$ and $\leq L_3$	C	$> C_2$ and $\leq C_3$
4	$> L_3$ and $\leq L_4$	D	$> C_3$ and $\leq C_4$
5	$> L_5$	E	$> C_5$

of various damage events as per installation arranged in the order of the larger likelihood scoring.

13.4.2 Consequence Analysis

The method for the consequence analysis of an accidental event may differ for the different hazard types. Although ETA technique can also be useful for this purpose, CMPT (1999) provides guidance to determine the consequence of some selected accidental events in offshore installations. For example, a range of consequences associated with loss of stability in offshore installations may be determined using ETA where branch probabilities are defined using, in part, damage stability calculations.

The parameters considered for the consequence analysis must usually include:

- Health (human injury)
- Safety (damage to property or financial expenditure)
- Environmental impacts (marine pollution)
- Damage to reputation

The consequence scoring should also be made using techniques similar to those used for the frequency analysis. Table 13.13 also indicates an example of the link between consequence category and score in this regard. Figure 13.5 is an example of the outcome of the consequence analysis that represents the accumulated consequence (cost) score of various damage events as per installation.

Figure 13.4. Example of likelihood scoring of events for an installation.

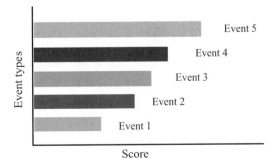

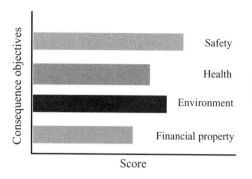

Figure 13.5. Example of consequence scoring of events for an installation.

13.4.3 Risk Representation

In QRAs, risks may be represented at various levels, for example:

- Individual risk
- Group risk
- Impairment frequency
- Risk of damage to installation
- Risk of damage to environment

The individual risk indicates the risk experienced by individuals on the installation; for example, injuries or death. Usually, this is expressed as an individual risk per annum (IRPA) or a fatal accident rate (FAR) per, say, 100 million exposed hours. This risk measure will be important for those individuals who almost permanently stay on the installation. Figure 13.6 shows an example of individual risk.

The group risk is the risk experienced by the whole group of personnel working on the installation. It is usually expressed as an average number of fatalities per installation-year (or annual fatality rate) or potential loss of life (PLL). Alternatively, the F–N curve showing the cumulative frequencies (F) against number of fatalities (N) may be used to represent this risk. Figure 13.7 shows an example of the F–N curve.

The impairment frequency indicates the frequency at which essential safety functions (e.g., temporary refuge and escape routes) fail when the accidents occur.

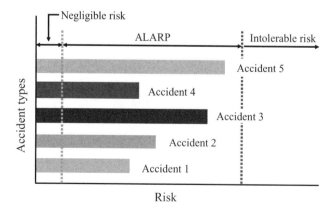

Figure 13.6. Example of individual risk for accidents.

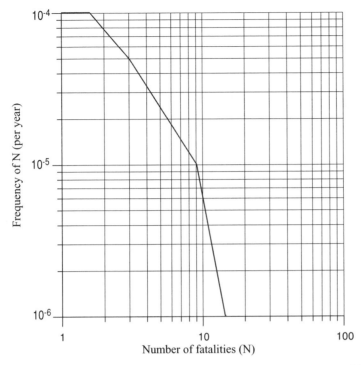

Figure 13.7. An F–N cumulative frequency against number of fatalities plot.

Therefore, the frequency of any safety function impairment from any accidental event must be such that the risk falls in the ALARP region.

The risk of damage to the installation may be expressed as the frequency per year of previously defined levels of damage; for example, total loss or severe damage. Alternatively, the damage levels may be converted to financial losses or the costs for averting a fatality in the form of, say, average damage cost per year. Such a risk index may be useful for cost–benefit analysis to develop risk-control options.

The risk of damage to the environment (e.g., an oil spill) can be represented in forms similar to the group risk case previously considered for people. For example, it may be expressed as an average amount of environmental damage (e.g., oil spill amount) per year or the cumulative frequency of different sizes of environmental damage, as the case may be.

A comprehensive risk assessment for oil spills pertinent to ship-shaped offshore installations was carried out by Gilbert et al. (2001). Figure 13.8 shows some selected results of that risk assessment in terms of transportation oil spills for various deep-water production installations. There are two notable differences in oil transportation. Spars, tension leg platforms (TLPs), and jackets use the pipelines and floating, production, storage, and offloading systems (FPSOs) apply in-field storage and offloading to shuttle tankers.

Figure 13.8 shows that the frequency of spills for an FPSO could, in relative terms, be greater than from pipelines when the spill size is small. This is because there is always the potential of small spills occurring during offloading from hoses and valves. For medium spill sizes, however, the annual frequency of spills for an FPSO may be smaller than that for pipelines. One reason for this is that the potential of spills from

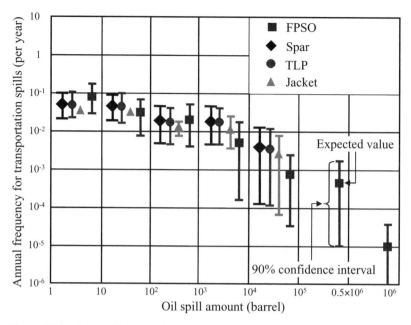

Figure 13.8. Annual frequency of transportation oil spills for various deep-water production systems (Gilbert et al. 2001).

pipelines may remain constant as long as there is oil in the pipeline regardless of the production rate. However, the potential of spills from the shuttle tanker may decrease as the production rate decreases because, presumably, fewer offloading events are required.

13.5 Risk Management during Design

Although two categories of risks are typically considered – remaining risks and risks that can be reduced or eliminated – risk should be managed and controlled in many different ways throughout the life cycle of the installation, that is, from initial planning, detailed design, and operation to decommissioning. Implementation of risk-control options at an earlier stage (e.g., at the design stage rather than operational stage) is much more beneficial particularly when the risk level is relatively high, because risk corrective/preventive measures can then be adopted at lower cost.

Two terminologies for risk management during design are relevant: *robustness* or *redundancy*. The robustness in a design is necessary so that the system will be able to withstand loads or demands up to a particular limit without damage. The redundancy in a design better allows for the loads beyond a limit to be absorbed in some manner with some subsequent damage although it must be ensured that loss of the global system will not happen while repairs are being undertaken within a specified time period to achieve the degree of robustness needed.

For example, although structural damage can be caused by ship collisions, it may eventually lead to the total loss of the system if it is left unrepaired, but the system could keep robust as long as relevant repairs are undertaken within a given time period. Also, accidental actions arising from fire or explosion can exceed the limit loads causing severe damage, but it should be ensured that the system must maintain its integrity for a long enough period until all personnel can at least evacuate safely. The risk-management and risk-control options during design are then to be decided on depending on the particular types of hazards and the severity of their consequences that can be tolerated.

A method for environmental quantitative risk assessment (EQRA) together with risk-acceptance criteria of an offshore installation is presented by Wiig et al. (1996). Ballard et al. (2004) review the application of risk assessment with a particular emphasis on the structural design issues of ship-shaped offshore installations in terms of water-impact loading, a collision with a shuttle tanker, and structural inspection. Haughie (2004) discusses many design challenges of risk management application to health, safety, and the environment of offshore installations. Lassagne et al. (2001) emphasize the necessity of risk-assessment approaches for design of ship-shaped offshore installations. Application of risk assessment to design of topsides, structural, and marine systems of ship-shaped offshore installations is presented by Wolford et al. (2001). It is also recognized that risk-based decision making is now becoming more permanent among classification societies in their certification role (Verzbolovskis 2004). Smith and Schmidt (2005) describe the marine accident-investigation process, including the skills and special knowledge that is often required for a complete investigation.

In Sections 13.5.1–13.5.5, some considerations to manage the risks associated with hazards during design are addressed. In most cases, design considerations either implicitly assume or need to be supplemented by operational measures such as implementation of relevant procedures and training, as will also be evident.

13.5.1 Selection of Materials

Less hazardous or flammable materials must be selected to build the system. Also, less hazardous processes must be developed and used so as to enable the materials to be used in a less hazardous environment.

13.5.2 Layout for Hazard Impact Minimization

It is important to minimize escalation of any hazard impact. In particular, emergency systems for rescue and evacuation must always be protected. For example, improved layout such as reorientation of equipment or removal of walls and cladding can reduce congestion in explosion and thus blast-pressure actions can be limited. Also, fire or blast walls can be arranged so as to prevent escalation of the consequence from the process area into the accommodation or temporary refuge.

13.5.3 Limit-State Design

The limit-state design approach can presumably more effectively manage the risk than the allowable working stress design approach. In particular, accidental limit states must be a basis for design of the installation to reduce the severity of hazard consequences or the risk itself.

13.5.4 Passive Safeguards for Fire and Explosion

Within design, the use of some passive safeguards for fire and explosion can be considered specifically on nonredundant parts of the installation. The use of a passive fire protection system can limit the impacts of a fire, prevent escalation of a fire, and mitigate the impact on personnel, for example, by protecting the temporary refuge. However, selection and use of this kind of risk-control option may need to be based on the risk assessment in terms of cost effectiveness as previously described in Eq. (13.2). Boyle and Smith (2000) studied emergency planning using the evacuation, escape, and rescue (EER) HAZOP technique.

13.5.5 Accelerated Degradation Protection

Accelerated degradation caused by corrosion and others must be prevented. Within the design process, the addition of a corrosion margin is to be considered necessary. Cathodic protection and anodes are also applied where anodes provide reverse electrolysis near a joint or structural component so that the anode corrodes instead of the structure.

13.6 Risk Management during Operation

More active measures are more relevant to manage the risks during operation. The use of procedural measures is typically applied. For example, in bad weather conditions, helicopter landings, unloading from supply vessels, crane movements, riser repairs, or drilling operations must be prohibited to reduce the risks.

A summary of risk-control options available for various hazards involved in offshore installations is presented by HSE (1998). Nesje et al. (1999) present qualitative recommendations for risk corrective/preventive measures for process leaks and explosions in ship-shaped offshore installations. Williams et al. (1999) present risk-management considerations for offloading of ship-shaped offshore installations. More refined inspection and maintenance strategies should also be developed by risk-management approaches (e.g., Xu et al. 2001; Conachey 2004).

We will discuss considerations to manage the risks during operation of the installation.

13.6.1 Collisions

A comprehensive study for developing risk-control options against ship collisions with offshore installations was carried out by HSE (1999), among others. Ship-shaped offshore units may collide with shuttle tankers, supply vessels, or passing vessels. In the process of offloading to shuttle tankers, dynamic positioning equipment can help ensure that such vessels maintain correct separation distance from the installation.

The supply vessels may need to be small enough or stay far enough so as not to cause significant damage in the event of an impact, and vessels may be permitted to unload at a location away from risers or in a position away from the installation depending on the weather conditions. To avoid collisions with passing vessels, an exclusion zone may need to be provided so that unauthorized vessels must not enter the zone. The use of a radar detection system may also be arranged to enable early detection of unauthorized vessels approaching the exclusion zone and to provide additional time to advise the vessel or to warn the offshore installation personnel.

Daughdrill and Clark (2002) present a summary of key operational procedures to reduce risks of accidents and oil spills in FPSO and shuttle tanker operations. The risk of collision between FPSO and a shuttle tanker during offloading strongly depends on human and organizational factors. Vinnem et al. (2003) present the likelihood of a collision between an FPSO and a shuttle tanker in terms of risk-influencing factors, which are based on data from incidents and near-misses as well as expert opinions.

13.6.2 Dropped Objects

Falling objects or even aircraft impacts can cause structural damage and will need to be considered as relevant. To prevent damage from dropped objects, the moving loads carried by cranes should not be over a certain limit. Also, the object should not be moved over vulnerable areas of the deck. A relevant warning system needs to be employed to assist crane operators. In addition, navigation lights and fog horns may help warn and, hence, prevent damage in many cases.

13.6.3 Active Safeguards for Fire and Explosion

Active measures such as fire and gas detection and fighting systems will be helpful to mitigate the risks involved. Shut-down may also be necessary when fire, heat, or gas is detected. Shut-down can be achieved automatically, triggered by a threshold value of gas concentration, for instance, or by manual intervention using remote valves. To control the spread of smoke, fire, or gas to spaces where personnel can be at risk, appropriate measures such as emergency shut-down of heating, ventilation, or air-conditioning should also be considered.

13.6.4 Inspection and Maintenance

Relevant inspection and management strategies must be established with the focus on appropriate areas at appropriate times to keep the structural integrity at required levels. Suspicious areas prone to known or past significant corrosion and fatigue cracking or damage most likely need inspection more frequently. The condition of primary structural components that have the greatest impact on the structural integrity if damaged should also be carefully monitored. After an incident or bad weather possibly accompanied by structural damage due to collisions or dropped objects, additional inspections for damaged areas are required. Periodic inspection and maintenance (including repair and replacement) of the various corrosion-protection system components including coatings, impressed cathodic corrosion protection (ICCP), and anodes is needed to avoid accelerated degradation of the structure due to corrosion. For a detailed description of risk-based inspection and maintenance, see Chapter 11. For a risk-based underwater inspection of offshore structures, see DeFranco et al. (1999).

REFERENCES

AICE (1989). *Chemical process quantitative risk analysis*. Center for Chemical Process Safety, American Institute of Chemical Engineers, New York.

AICE (1992). *Guidelines for hazard evaluation procedure*. Center for Chemical Process Safety, American Institute of Chemical Engineers, New York.

API RP 2FPS (2001). *Recommended practice for planning, designing, and constructing floating production systems*. American Petroleum Institute, USA.

Ballard, E. J., Frieze, P. A., Fyfe, A. J., and Smedley, P. A. (2004). "Update of risk and reliability issues for FPSO structural design." *Proceedings of OMAE–FPSO 2004, OMAE FPSO'04–0053, OMAE Specialty Symposium on FPSO Integrity*, Houston, August 30–September 2.

Boyle, P., and Smith, E. J. (2000). *Emergency planning using HSE's evacuation, escape, and Rescue (EER) HAZOP technique*. (Hazards XV, Symposium Series No. 147), Institution of Chemical Engineers, Rugby, England.

CMPT (1999). *A guide to quantitative risk assessment for offshore installations*. The Center for Maritime and Petroleum Technology (CMPT), UK.

Conachey, R. M. (2004). *Development of RCM requirements for the marine industry*. American Bureau of Shipping, Houston.

Daughdrill, W. H., and Clark, T. A. (2002). *Considerations in reducing risks in FPSO and shuttle vessel lightening operations*. Offshore Technology Conference, OTC 14000, Houston, May.

DeFranco, S., O'Connor, P., Tallin, A., Roy, R., and Puskar, F. (1999). *Development of a risk based underwater inspection (RBUI) process for prioritizing inspections of large numbers of platforms*. Offshore Technology Conference, OTC 10846, Houston, May.

DNV (1980–2006). *Worldwide offshore accident databank statistical report*. Det Norske Veritas, Norway.

DNV (1997). *Offshore reliability data bank*. Det Norske Veritas, Norway.

Haughie, I. J. (2004). *Application of risk-based design relative to health, environment, and safety*. Offshore Technology Conference, OTC 16990, Houston, May.

HSE (1988). *The tolerability of risk from nuclear power stations*. Health and Safety Executive, UK.

HSE (1992). *A guide to the offshore (safety case) regulations*. Health and Safety Executive, UK.

HSE (1997a). *Approaches to hazard identification*. (Offshore Technology Report, OTO 1997/068), Health and Safety Executive, UK.

HSE (1997b). Offshore hydrocarbon release statistics. (Offshore Technology Report, OTO 1997/050), Health and Safety Executive, UK.

HSE (1997c). *Close proximity study*. (Offshore Technology Report, OTO 1997/055), Health and Safety Executive, UK.

HSE (1998). *Hazard management in structural integrity: Vol. 3 Hazard management measures*. (Offshore Technology Report, OTO 1998/150), Health and Safety Executive, UK.

HSE (1999). *Effective collision risk management for offshore installations*. (Offshore Technology Report, OTO 1999/052), Health and Safety Executive, UK.

HSE (2000a). *Operational safety of FPSOs: Initial summary report*. (Offshore Technology Report, 2000/086), Health and Safety Executive, UK.

HSE (2000b). *Injuries on offshore oil and gas installations: An analysis of temporal and occupational factors*. (Offshore Technology Report, OTO 1999/097), Health and Safety Executive, UK.

HSE (2001). *Marine risk assessment*. (Offshore Technology Report, OTO 2001/063), Health and Safety Executive, UK.

IMO (2002). *Guidelines for the application of formal safety assessment (FSA) for use in the IMO rule-making process*. Marine Safety Committee MSC/Circ. 1023, MEPC/Circ. 392, International Maritime Organization, London, April.

Gilbert, R. B., Ward, E. G., and Wolford, A. J. (2001). *A comparative risk analysis of FPSOs with other deepwater production systems in the Gulf of Mexico*. Offshore Technology Conference, OTC 13173, Houston, May.

ISO 17776 (1999). *Petroleum and natural gas industries – Offshore production installations – Guidelines on tools and techniques for the identification and assessment of hazardous events.* International Organization for Standardization, Geneva.

Lassagne, M. G., Pang, D. X., and Vieira, R. (2001). *Prescriptive and risk-based approaches to regulation: The case of FPSOs in deepwater Gulf of Mexico.* Offshore Technology Conference, OTC 12950, Houston, May.

Lees, F. P. (1996). *Loss prevention in the process industries.* Oxford, UK: Butterworth Heinemann.

MacDonald, A., Cain, M., Aggarwal, R. K., Vivalda, C., and Lie, O. E. (1999). *Collision risks associated with FPSOs in deep water Gulf of Mexico.* Offshore Technology Conference, OTC 10999, Houston, May.

Nesje, J. D., Aggarwal, R. K., Petrauskas, C., Vinnem, J. E., Keolanui, G. L., Hoffman, J., and McDonnell, R. (1999). *Risk assessment technology and its application to tanker based floating production storage and offloading (FPSO) systems.* Offshore Technology Conference, OTC 10998, Houston, May.

Paik, J. K., Amdahl, J., Barltrop, N., Donner, E. R., Gu, Y., Ito, H., Ludolphy, H., Pedersen, P. T., Rohr, U., and Wang, G. (2003). "Collision and grounding." *International Ships and Offshore Structures Congress*, Specialist Committee V.3, 2: 71–107.

Skjong, R., Vanem, E., and Endresen, Ø. (2005). *SAFEDOR project: Design, operation, and regulation for safety – Risk evaluation criteria.* Det Norske Veritas, Oslo (http://www.safedor.org).

Smith, K. M., and Schmidt, F. E. (2005). "Marine accident investigation techniques: Finding the root causes." *Marine Technology*, 42(4): 210–219.

UKOOA (1999). *A framework for risk related decision support.* Offshore Operators Association, UK.

UKOOA (2000). *Guidelines for quantitative risk assessment uncertainty.* Offshore Operators Association, UK.

Verzbolovskis, M. (2004). *ABS approach to classification using risk analysis.* American Bureau of Shipping, Houston.

Vinnem, J. E., Hokstad, P., Dammen, T., Saele, H., Chen, H., Haver, S., Kieran, O., Kleppestoe, H., Thomas, J. J., and Toennessen, L. I. (2003). *Operational safety analysis of FPSO – shuttle tanker collision risk reveals areas of improvement.* Offshore Technology Conference, OTC 15317, Houston, May.

Wiig, E., Nesse, E., and Kittelsen, A. (1996). *Environmental quantitative risk assessment (EQRA).* SPE International Conference on Health, Safety, and Environment, Society of Petroleum Engineers, SPE 35945, New Orleans, LA, June.

Williams, J. S., Brown, D. A., Shaw, M. L., and Howard, A. R. (1999). *Tanker loading export systems for harsh environments: A risk-management challenge.* Offshore Technology Conference, OTC 10905, Houston, May.

Wolford, A. J., Lin, J. C., Liming, J. K., and Lidstone, A. (2001). *Integrated risk based design of FPSO topsides, structural, and marine systems.* Offshore Technology Conference, OTC 12948, Houston, May.

Xu, T., Bai, Y., Wang, M., and Bea, R. G. (2001). *Risk based optimum inspection for FPSO hulls.* Offshore Technology Conference, OTC 12949, Houston, May.

APPENDIX 1

Terms and Definitions

For the purpose of this book, the following terms and definitions are relevant. The terminologies described here generally follow international standards, regulations, recognized guidelines, and handbooks, for example:

API (American Petroleum Institute: http://www.api.org)
HSE (Health and Safety Executive, UK: http://www.hse.gov.uk)
IACS (International Association of Classification Societies: http://www.iacs.org)
IMO (International Maritime Organization: http://www.imo.org)
ISO (International Organization for Standardization: http://www.iso.org)
USCG (U.S. Coast Guard: http://www.uscg.mil)
Handbook of fire and explosion protection engineering principles for oil, gas, chemical, and related facilities, by D. P. Nolan, Noyes Publications, Westwood, NJ, 1996

Abnormal actions: Actions larger than expected or normal actions, for example, rogue or freak waves.

Accident: A circumstance that gives rise to injury, fatality, environmental damage, property damage, production losses, or loss of facility. According to IMO, an accident is an unintended event involving fatality, injury, vessel loss or damage, other property loss or damage, or environmental damage. An accident scenario consists of a specific sequence of events from an initiating event to an undesired consequence. An accident category is a designation of accidents according to their nature, for example, fire, explosion, collisions, and grounding. Examples include the following:

- Contact: striking any fixed or floating object, other than those included under "collision"
- Collision: striking or being struck by another vessel, regardless of whether under way, anchored, or moored
- Fire or explosion: accidents where fire or explosion is the initial event
- Loss of structural integrity: structural failure that can result in the ingress of water and/or loss of strength and/or stability
- Flooding: the ingress of water that can result in foundering or sinking of the vessel

Accidental actions: Actions applied in the event of accidents such as capsizing, collisions, grounding, fire, and explosions.

Action: External load applied to the structure (direct action) or an imposed deformation or acceleration (indirect action); for example, an imposed deformation can be caused by fabrication tolerances, settlement, or temperature change. Also called "load."

Action effect: Effect of actions on a global structure or structural component, for example, internal force, moment, deformation, stress, or strain. Also called "load effect."

Aged structure: An existing structure that may suffer age-related degradation such as corrosion and fatigue cracks.

ALARP: As low as reasonably practicable; refers to a level of risk that is neither negligibly low nor intolerably high, for which further investment of resources for risk reduction is not justifiable. The ALARP principle states that risk should be reduced to an ALARP level considering the cost effectiveness of the risk-management options.

ALP: Articulated loading platform. Also termed "articulated loading column" (ALC).

Availability: Availability of a system or equipment is the probability that it is not in a failed state at a point in time.

Barrel: The standard liquid quantity of measurement in the petroleum industry. One barrel of oil equals 159 liters.

Basic variable: One of a specified set of variables representing physical quantities that characterize actions, material properties, or geometrical parameters.

Beaufort scale: A numerical scale first adopted by Admiral Beaufort in 1808 relating wind speed to the state of the sea.

Blast: The transient change in gas density, pressure, and velocity of the air surrounding a gas explosion point.

CALM: Catenary anchor leg mooring.

Catenary mooring: Mooring system where the restoring action is provided by the distributed weight of mooring lines.

Certainty: Confidence regarding a set of assumptions or results, for example, in frequency assessments carried out as part of risk analysis.

Certification: This is the process of obtaining written evidence that a structure, item of equipment, or other arrangement has been designed, constructed, installed, or maintained in a prescribed manner. The "prescribed manner" is normally documented in the regulations that are defined as orders or rules issued by a regulatory regime, such as a governmental authority. The "written evidence" may be in the form of any certificate or other document issued by the certifier who is authorized to do so by the appropriate cognizant body or agency.

Characteristic value: Value assigned to a basic variable associated with a prescribed probability of not being violated by unfavorable values during some reference period. The characteristic value is the main representative value. In some design situations, a variable can have more than one characteristic value, for example, an upper and a lower value.

Charterer: A buyer of vessel transportation services who is a vessel owner's customer. "Charter" is a contract between vessel owner and charterer by which vessel owner agrees to provide cargo transportation services. For trading tankers, charters can be for a specified number of voyages called "voyage charter" or for a period of time called "term charter." A voyage charter for a single voyage commencing within a few weeks of the charter agreement is often termed a "spot charter."

Classification: This is a form of design and construction oversight carried out by a classification society in accordance with the published rules and guidelines of the selected classification society. The classification process may consist of the review and approval of technical submissions in accordance with relevant class rules; physical confirmation of manufacture; fabrication; assembly or installation of components or finished item in accordance with approved drawings and details; and subsequent testing as required by the rules and associated issuance of documents attesting to the degree of compliance with the requirements of the classification society.

Combustion: A rapid chemical process that involves reaction of an oxidizer (usually oxygen or air) with an oxidizable material, sufficient to produce radiation effects, that is, heat and light.

Condition monitoring: Scheduled diagnostic technologies used to monitor the system condition to anticipate and detect a potential failure.

Consequence: The effects of an accidental event, such as injuries, fatalities, environmental damage, and property damage.

Corrective measures: Engineering or administrative and operational procedures activated to reduce the likelihood of a failure.

Crude oil: Hydrocarbon mixtures that have not been processed in a refinery. It is a liquid petroleum coming out of the ground, ranging from very light (high in gasoline) to very heavy (high in residual oils). Sour crude has a high sulfur content, but sweet crude has a low sulfur and is more valuable.

Current: A flow of water past a fixed location – more precisely described as an Eulerian current. A Lagrangian current is measured by following the movement of a water particle. Currents are usually measured by speed and direction; measurements are usually analyzed in terms of the tidal current and residual currents.

Current speed: The horizontal speed of the current for any direction. The speed varies throughout the water column. Depth-averaged current speed is the speed of the current averaged throughout the water column or over a specified depth of it.

Damage stability: Stability of a vessel in damaged condition due to accidental flooding, and also termed "damaged vessel stability"; see *stability*.

Decommissioning: Process of shutting down a structure and removing hazardous materials at the end of its service life.

Design condition: Set of physical situations occurring over a certain time interval for which the design needs to demonstrate that relevant limit states are not exceeded.

Design criteria: Quantitative formulations with acceptability threshold values that result in acceptable or unacceptable designations for a set of limit states governing a design, and used in support of decision making.

Design service life: Assumed period for which a structure is to be used for its intended purpose with anticipated maintenance, but without substantial repair being necessary.

Design situation: Set of physical conditions representing real conditions during a certain time interval for which the design will demonstrate that relevant limit states are not exceeded.

Design value: Value derived from the representative value for use in the design checking procedure, which accounts for the uncertainties associated with the representative value.

Design wave: "Extreme amplitude design wave" is the periodic wave having the same height as the extreme wave with the return period required by the specification. Used as the initial design condition for an offshore structure. It may have a range of wave periods associated with it, and its expected direction(s) may be specified. Extreme response design wave(s) may be different from the extreme amplitude design wave and produce a different loading on the structure. Extreme amplitude design waves are derived to represent certain levels and types of response or, more commonly, ship motions and sea loads.

DICAS: **D**ifferential **c**ompliance **a**nchoring **s**ystem is a system that achieves partial weathervaning through uneven mooring line tensioning.

Dynamic action: Action that induces acceleration of a structure or a structural component of a magnitude sufficient to require specific consideration.

Dynamic positioning: Station-keeping technique consisting primarily of a system of automatically controlled onboard thrusters, which generate appropriate thrust vectors to counter the mean and slowly varying actions induced by wind, waves, and current.

Dynamic pressure: Pressure due to water impact or explosive events, which give rise to a pressure-time history acting on the exposed area of a structure; for example, pressure loads arising from slamming or sloshing.

Environmental severity factor: A severity measure of historical or intended environment relative to the unrestricted service environment in terms of extreme or fatigue actions.

Event tree: A graphical representation of the relation between a primary cause (initiating) event and the final undesired events. Event trees are generally time-dependent and rely on a sequence of events.

Expected value: First-order statistical moment of the probability density function for the considered variable; in the case of a time-dependent parameter, the expected value may be obtained to represent a specific reference period.

Explosion: A release of energy that causes a blast.

Extreme action: Maximum action applied to a structure during its design service life.

Extreme value: An estimate of the value of a metocean variable with a stated return period. Other definitions are also possible.

Failure criterion: Threshold value considered in a limit state that separates success and failure.

Fault tree: A graphical method of describing the combination of events leading to a defined failure.

Fire: A combustible vapor or gas combining with an oxidizer (usually oxygen or air) in a combustion process manifested by the evolution of light, heat, and flame.

Flag of convenience: A country that admits the vessel registration regardless of the owner's nationality (perhaps for reasons of tonnage fees).

Flag state: The country where the vessel is registered.

Flash point: The minimum temperature of a liquid at which it gives off sufficient vapors to form an ignitable mixture with air immediately above the surface of the liquid or within the vessel used on the application of an ignition source under specified conditions.

Floating structure: Structure where the full weight is supported by buoyancy. The full weight includes lightship weight, mooring system weight, any riser pretension, and operating weight.

Fluid transfer system: A system to transport the produced oil of FPSO (via the subsea pipeline and through the single point mooring's own piping arrangement) to the shuttle tanker.

FPSO: Floating, production, storage, and offloading vessel. A marine vessel usually moored to the seabed or using an internal or external turret system, allowing direct production, storage, and offloading of process fluids from subsea installations.

Frequency: A score indicating the expected number of occurrences per unit time (e.g., year) of an event such as an accident. Also termed "likelihood."

FSA: Formal safety assessment. A formal, structured, and systematic methodology, developed to assist risk assessment and to facilitate proactive risk management.

FSO: Floating, storage, and offloading vessel. Similar to FPSO, but no production capability is provided. It stores oil and gas products that can be offloaded to pipelines or shuttle tankers.

FSU: Floating storage unit. It is generic term for floating installations, including FPSOs and FSOs.

Gale ballast tank: A cargo tank that is allowed to put ballast water for reasons of safety.

General corrosion: A corrosion wastage in which the thickness of structural members is idealized to be uniformly reduced. Also called "uniform corrosion."

Global structure: An entire structure or an assembly of structural components.

Hawsers: Marine ropes. The most common materials used for hawsers are polyester, polypropylene, and nylon (polyamide). Polyester offers the best fatigue properties of the three materials, and it does not lose strength when submerged in water. But it has the lowest elastic properties, and it is not buoyant as with nylon. Polypropylene is the only material that is self-floating and can be inspected without having floaters removed, but it is the weakest and least abrasion resistant of the three fibers. Polypropylene is less elastic than nylon and degrades in strength when exposed to sunlight, but it makes winching-in of the hawser easy so that it is often used for mooring pick-up ropes. Nylon offers high strength and good abrasion resistance together with good elastic properties, but it is less fatigue resistant than the other two fibers, and its breaking strength can be reduced when it is wet. Nylon is not buoyant and needs buoyancy tanks or floaters if it is required to float.

Hazard: An event with the potential to cause injuries, fatalities, environmental damage, and/or property damage. According to IMO, examples of an event with potential to threaten human life, health, property, or the environment are as follows:

- Hazards external to the vessel: storms, lightning, poor visibility, uncharted submerged objects, other ships, war, and sabotage
- Hazards on board a vessel:
  - In accommodation areas: combustible furnishings, cleaning material in stores, and oil/fat in galley equipment
  - In deck areas: cargo, slippery deck due to paint/oils/grease/water, hatch covers, and electrical connections
  - In machinery spaces: cabling, fuel and diesel oil for engines, boilers, fuel oil piping and valves, oily bilge, and refrigerants
  - Sources of ignition: naked flame, electrical appliances, hot surface, sparks from hot work or funnel exhaust, and deck and engine room machinery
- Operational hazards to personnel: long working hours, working on deck at sea, cargo operation, tank surveys, and onboard repairs

Heat flux: The rate of heat transfer per unit area normal to the direction of heatflow. It is the total heat transmitted by radiation, conduction, and convection.

Heeling: Inclination of the vessel from the upright in any direction. A heeling moment exists at any inclination of the vessel where the forces of weight and buoyancy are not aligned and act to move the vessel away from the upright position.

Heeling moment: See *heeling*.

Hotwork: Welding or steel flame cutting.

Hydrocarbon: An organic compound containing only hydrogen and carbon. The simplest examples of hydrocarbons are gases at ordinary temperatures; but, with increasing molecular weight, they change to the liquid form and finally to the solid state. Hydrocarbons form the principal constituents of petroleum and natural gas.

Ignition: The process of starting a combustion process through the input of energy. Ignition occurs when the temperature of a substance is raised to the point at which its molecules react simultaneously with an oxidizer and combustion occurs.

Impact: Action(s) due to accidental or abnormal events usually during a very short period of time, for example, dropped objects, collisions.

Impulsive action: Dynamic action or impact action with a very short duration (t) of action persistence, for example, an explosion compared to the natural period (T) of the exposed structure. The impulsive-action profile can usually be characterized by two parameters: an impact load and its duration time. A dynamic action problem needs to be dealt with in the impulsive domain when $t \leq 0.3T$.

Inerting: The process of removing an oxidizer (usually oxygen or air) to prevent a combustion process or corrosion progress from occurring normally accomplished by purging. Inert gas contains less than 5 percent oxygen, usually obtained from the boiler exhaust.

Initiating event: Hardware failure, control system failure, human error, extreme weather, or geophysical event, which could lead to a hazard being realized.

In situ: A specific offshore field or location.

Inspection: Activity to detect and evaluate deterioration in components, structure, equipment, or plants by visual, electronic, or other means.

Intact stability: Stability of a vessel in intact condition and also termed "intact vessel stability"; see *stability*.

Jet fire: A turbulent diffusion flame resulting from the combustion of a liquid or gas continuously released under pressure in a particular direction.

Joint probability: When two or more variables interact to producing a response on a structure, it may be necessary to determine the probabilities with which various combinations of variables occur, that is, their joint probability of occurrence.

Likelihood: See *frequency*.

Limit state: A condition for which a structure or structural component fails to perform its intended function.

Localized corrosion: A type of corrosion wastage that occurs in local regions, for example, pitting or grooving.

Maintenance: Total set of activities performed during the design service life of a structure to enable it to remain fit for purpose, which includes replacement, repair, or adjustment.

MARPOL tanker: A single-skin tanker meeting the requirements of IMO MARPOL/78 for segregated ballast. A pre-MARPOL tanker is a single-skin tanker built before the MARPOL/78 segregated ballast requirements.

Mean sea level: The average level of the sea over a period of time long enough to remove variations in level due to waves, tides, and storm surges.

Mean water depth: See *water depth*.

Mean zero up-crossing period: See *wave period*.

Merchant cargo ship: A ship carrying merchant cargo.

Metacenter: The center of buoyancy of a heeled vessel is moved off of the vessel's centerline as the result of the inclination, and the lines along which the resultants of weight and buoyancy act are separated by a distance corresponding to righting arm. A vertical line through the center of buoyancy will intersect the original vertical through the center of buoyancy at a point in the vessel's centerline plane, called the "metacenter." The distance from the original center of buoyancy to the metacenter is called the "metacentric height." At relatively small angles of heeling, the metacentric height is used as an important index of stability. The metacentric height is positive when the metacenter is above the original center of buoyancy, and negative when the metacenter is below the original center of buoyancy.

Metacentric height: See *metacenter*.

Metocean: Abbreviation of *meteorological* and *oceanographic*.

Minimum breaking strength: Breaking strength of chain and for wire and other materials. It is the certified breaking strength.

Mobile mooring system: Mooring systems, generally retrievable, intended for deployment at a specific location for a short-term operation, such as those for mobile offshore units.

Mobile offshore drilling units: Floating structure capable of engaging in drilling operations for exploration or exploitation of subsea petroleum resources.

Mobile offshore unit: Floating structure intended to be frequently relocated to perform a particular function.

Mooring components: General class of components such as chain, steel wire rope, synthetic fiber rope, clump weight, buoy, winch/windlass, fairlead, and anchor.

Natural gas: A mixture of hydrocarbon compounds and small amounts of various non-hydrocarbons (e.g., carbon dioxide, helium, hydrogen sulfide, and nitrogen) existing in gaseous phase or in solution with crude oil in natural underground reservoirs.

Neutral equilibrium: See *stability*.

Nominal value: Value assigned to a basic variable determined on a nonstatistical basis, typically from acquired experience or physical conditions.

Overpressure: A pressure relative to ambient pressure caused by a blast, both positive and negative.

Owner: Representative of the owner or owners of a development, who may be the operator on behalf of co-licensees.

Partial safety factor: A factor normally greater than unity applied to a representative value of a strength or action to determine its corresponding design value.

Peak frequency: See *wave energy spectrum*.

Permanent mooring system: A system normally used to moor floating structures deployed for long-term operations, such as those for a floating production system.

PLEM: Pipeline end manifold.

Pool fire: A release of a flammable liquid and/or condensed gas that accumulates on a surface forming a pool, where flammable vapors burn above the liquid surface of the accumulated liquid.

Port city: The city where the vessel loads or offloads.

Pre-MARPOL tanker: See *MARPOL tanker*.

Probability: The likelihood of occurrence of a specific event.

Probability of joint occurrence: See *joint probability*.

Purging: The process of replacing the inert gas or hydrocarbon mixture in a tank with normal air for tank inspection or repair.

Quasistatic action: Action that will not cause significant acceleration of the structure or structural component. Dynamic structural action problems may be approximately dealt with in the quasistatic domain, when the action duration time (t) is long enough compared to the natural period (T) of the exposed structure, say, $t \geq 3T$.

Random failure: Failure caused by sudden stresses, extreme conditions, and random errors.

Recognized classification societies: Societies that are members of the International Association of Classification Societies (IACS), with recognized and relevant competence and experience in floating structures and with established rules and procedures for classification/certification of installations used in petroleum-related activities.

Reference period: Period of time used as basis for determining the values of basic variables.

Reliability: Probability that an item or system will adequately perform a required function under stated conditions of use and maintenance for a stated period of time. Reliability is a probability of desired performance over time in a specified condition, for example, machinery or system reliability, structural reliability, or human reliability. Reliability = 1− Failure probability.

Repetitive action: Repeatedly applied action.

Representative value: Value assigned to a basic variable for verification of a limit state.

Residual current: The component of a current other than tidal current. The residual currents may be, predominantly, storm surge currents.

Return period: Reciprocal of the probability of exceeding an event during a particular period of time, that is, the average period of time between exceedance of a stated value of a variable.

Righting moment: A righting moment exists at any angle of inclination of the vessel where the forces of weight and buoyancy act to move the vessel toward the upright position.

Riser: Piping connecting the process facilities or drilling equipment on the floating structure with the subsea facilities, pipelines, or reservoirs.

Risk: The product of the probability of occurrence (the "frequency" or "likelihood") of a hazardous event and its consequence(s); that is, Risk = Frequency × Consequence. Risk is a measure of the likelihood that an undesirable event will occur together with a measure of the resulting consequence(s) within a specified time; that is, the combination of the frequency and the severity of the consequence(s). This can be either a quantitative or a qualitative measure.

Risk assessment: An array of techniques that can integrate successfully diverse aspects of design and operation to assess risk; it may address three key questions: (1) Scenario – what can go wrong; (2) Likelihood – how likely is it; and (3) Impact – what are the impacts. Risk-assessment tools comprise a vast array of techniques, including those for reliability, availability, and maintainability of engineering, statistics, decision theory, systems engineering, human behavior, performance, and error. Often called "safety case study."

Risk-based decision making: A process that organizes information about the possibility for one or more unwanted outcomes into a broad, orderly structure that helps decision makers make more informed management choices on the basis of risk.

Risk communication: Interaction process of exchanging information and opinion among individuals, groups, and institutions involving multiple messages about the nature of risk.

Risk-index number: A relative measure of the overall risk associated with a deviation.

Risk matrix: A matrix depicting the risk profile. Each cell in the matrix indicates the number of accidents having that frequency and consequence.

Running fire: A fire resulting from a burning liquid fuel that flows by gravity to lower elevation. The characteristics of running fire are similar to those of pool fire, although running fire is moving or draining to a lower level.

Safeguard: An engineered system (hardware) or administrative control for reducing the frequency of occurrence of accidents or mitigating the severity of accidents.

Safety: Relates to the degree of risk. Because no activity is free of risk, an activity is considered safe when the level of risk is within acceptable limits.

Safety factor: A simple measure of safety margin normally defined as the ratio of design strength to design action effect.

Safety margin: A difference between structural strength and maximum value of an action effect, by which a structure or component reserve with respect to failure can be quantified.

SALM: Single-anchor leg mooring.

Scantlings: Size and thickness of structural components.

Sea state: A general term for the wave conditions at a particular time and place. Parameters such as significant wave height and mean zero up-crossing wave period are often referred to as *sea state parameters*. A sea state is usually assumed to stay statistically stationary for a period of a few hours depending on the underlying data analyses; for example, 3 hours.

Sea surface variance: The mean-square elevation of the sea surface (with respect to still water level) due to waves. Proportional to the energy density per unit area of sea surface.

Segregated ballast tank system: Sea water ballast should not be carried in cargo oil tanks.

Semisubmersible: Floating structure normally consisting of a deck structure with a number of widely spaced, large cross-section, and supporting columns connected to submerged pontoons.

Serviceability: Ability of a structure or structural component to perform adequately for normal functional use for a period of time. Also called "operability."

Ship-shaped offshore structure: Floating offshore structure that has a similar shape to a tanker-type ship in terms of geometry and functions.

Shuttle tanker: Trading tanker that transports oil from an FSU or FPSO to shore. The shuttle tanker will be moored to a single buoy or in tandem to an FSU or FPSO. Although ordinary trading tankers are only equipped for oil transfer at midships, the shuttle tanker accommodates a long floating hose from a single buoy to the loading position. It is required to have the capacity to provide a reverse thrust in order to prevent collision with the single-point mooring or FSU and to keep the hawser tight.

Significant wave height: See *wave height*.

Single-buoy mooring: Mooring by a single buoy.

Single-point mooring (SPM): Mooring system at a single location (point) that allows the floating structure to which it is connected to vary its heading (weathervaning).

Site-specific environments: Metocean phenomena and their variables observed at the location or proposed location of an offshore installation.

Slamming: A phenomenon in which the local area of a structure is exposed to dynamic water pressure. Bottom slamming occurs when the trading ship's bottom emerges from the water because of pitching, possibly combined with the occurrence of a wave trough. Bow flare slamming occurs due to the plunging of the upper flared portion of the bow deeper into the water.

Slew: A complex irregular yaw motion. A slewing ring is required in most single-point mooring applications to control slew motion during weathervaning.

Sloshing: Actions arising from the motion of liquids in partially filled tanks.

Spar platform: Deep draught, vertically floating structure.

Spectrum: See *wave energy spectrum*.

Sponson: Enclosed attachments for anti-collisions to the side structure of a single-skin tanker to be converted to an FPSO to be equivalent to double-sided arrangements.

Spread mooring: Mooring system consisting of multiple mooring lines terminating at different locations on a floating structure and extending outwards and anchored to the seabed, providing an almost constant structure heading.

Stability: Ability of the vessel to maintain proper floating attitude, usually upright, when subject to weight and buoyancy actions. In dealing with static floating body stability, equilibrium is a condition in which the resultant of all gravity forces (weights) acting downward and the resultant of the buoyancy forces acting upward on the body are of equal magnitude and are applied in the same vertical line. Stable equilibrium is a situation in which a floating body with an angular attitude change due to an external moment can return to its original position upon removal of the external moment. On the other hand, if the inclined body remains in that displaced position even after the external moment is removed, it is called "neutral equilibrium." Unstable equilibrium is a situation in which a floating body, displaced from its original position by an external moment, continues to move after the force is removed.

Stable equilibrium: See *stability*.

Station-keeping system: System capable of limiting the excursions of a floating structure within prescribed limits.

Still-water level: The level of the surface of the sea in the absence of surface waves generated by the wind. Variations in still-water levels are principally due to tides and storm surges.

Still-water load: Action arising from the difference between weight and buoyancy for a stationary structure.

Storm surge: Irregular movement of the sea brought about by wind and atmospheric pressure variations. Storm surge elevation is the change from the predicted tidal level as a result of a storm surge. It can be positive or negative, and for design purposes it is usually defined by an extreme value. Storm surge current is the current resulting from a storm surge. An extreme value is usually required for design purposes.

Structural component: Structural element or physically distinguishable part of a structure, for example, column, beam, stiffeners, plating, or stiffened panel. In terms of terminology for structural mechanics, a plate panel between stiffeners is called a "plating" or a "plate," and the plating with stiffeners is called a "stiffened panel." A one-dimensional member is called a "column" under axial compressive loads, and it is called a "beam" under bending or lateral loads. When a one-dimensional member is used for resisting combined axial compression and bending (or lateral loads), it is called a "beam-column." A large stiffened panel supported by heavy support members is sometimes idealized by a grillage, which, in concept, is essentially a set of intersecting beam members. Strong main support members are normally called "girders" when they are located in the primary loading direction (i.e., longitudinal direction in a box girder or a vessel hull), but they are sometimes called "frames" when they are located in the direction normal to the primary load direction (i.e., in the transverse direction in a box girder or a ship hull).

Structural failure: Insufficient strength or inadequate serviceability of a structure or structural component.

Structural resistance: Equivalent to *structural strength*.

Structural strength: Capacity of a structure or structural component to withstand an action effect.

Structural system: Load-bearing components of a structure including the way in which these components function together.

Structure: Organized combination of connected parts designed to withstand actions and provide adequate rigidity.

Surface wind drift: The current, in the top few meters of the water column, generated in direct response to the local wind blowing over the surface of the sea.

Survey: The inspection of classification society. "Surveyor" is the classification society inspector.

Swash bulkhead: Nonwatertight partition in a tank designed for avoiding sloshing.

Swivels: A swivel assembly that allows a vessel to weathervane during loading or offloading, typically using electrical and hydraulic systems to power and control subsea equipment through the riser system.

Taut-line mooring: Mooring system where the restoring action in tension is provided by elastic deformation of mooring lines.

Thruster-assisted mooring: Station-keeping system consisting of mooring lines and thrusters. The thrusters contribute to the control of the structure's heading and/or to adjust the mooring forces and the structure offset.

Tides: Regular and predictable movements of the sea generated by astronomical forces. They can be represented as the sum of a number of harmonic constituents, each with different but known periods. Tidal current is the current resulting from the tides. During a characteristic tidal current period, the current vector may describe an ellipse with a maximum current speed and associated direction and a minimum speed and direction.

Turret: A cylindrical single-point mooring system geo-stationary with the seabed allowing rotation of FPSO or FSO vessel in response to wave and wind conditions. See also *weathervaning*.

Uncertainty: A general description of the randomness of a basic variable, action, or strength parameter that is quantified, for example, by bias and standard deviation, provides a basis by which characteristic values, partial action, and strength factors may be determined.

Underwriters: A vessel's insurers.

Unstable equilibrium: See *stability*.

Verification: This is the process of performance of certification-related activities by a third-party agent before the said agent issues a certificate attesting to the degree of compliance of a structure, item of equipment, or other arrangement with any regulation or standard.

Vetting: Tanker inspections by oil company personnel.

Water depth: The vertical distance between the seabed and a defined datum near the sea surface, for example, mean sea level.

Waves: Movements on the sea surface generated by wind and with wave periods of less than 25 seconds.

Wave crest elevation: The vertical distance between the crest of a wave and still-water level.

Wave direction: The mean direction from which wave energy is traveling.

Wave-energy spectrum: A frequency domain description related to the energy in a whole wave system (or sea state). The wave system is assumed to consist of a large number of long-crested sinusoidal wave trains traveling independently but superimposed on each other. The omnidirectional spectral density function $S(f)$ is defined such that $S(f)\delta(f) =$ the sum of the sea surface variances (proportional to energy per unit of area) of the wave trains with frequencies between f and $f + \delta f$, where δf is a small frequency interval. Peak frequency of a spectrum is the wave frequency corresponding to the maximum value of the omnidirectional spectral density function.

Wave-exceedance diagram: A plot of the proportion of time for which the wave height is less than the value specified on the abscissa. This plot can be presented on a seasonal or all-year basis. Also called the cumulative frequency distribution of wave height.

Wave frequency: The number of waves passing a fixed point in unit time.

Wave load: Action on a structure or component arising from waves.

Wave height: The vertical distance between the crest of one wave and the preceding trough. In idealized circumstances, it is exactly twice the wave crest elevation. Height of zero up-crossing wave is the vertical distance between the highest and lowest points on the water surface of a particular zero up-crossing wave. Significant wave height (H_s) is $4\sqrt{m_o}$, where m_o is the sea surface variance. In sea states with only a narrow band of wave frequencies, H_s is approximately equal to $H_{1/3}$, which is the mean height of the largest third of the zero up-crossing waves. Extreme significant wave height (H_{sN})

is the significant wave height with a return period of N years (e.g., 100 years for H_{s100}). Extreme wave height (H_N) is the individual wave height (generally the zero up-crossing wave height) with a return period of N years (e.g., 100 years for H_{100}).

Wave hindcasting: Estimating the wave characteristics using meteorological data and a numerical model for wave generation and energy transport.

Wave period: The time interval between successive waves. The period of a zero up-crossing wave is the time interval between the two zero up-crossings that bound it. See also *wave frequency*. Mean zero up-crossing period (T_z) is calculated for a random sea by dividing the wave sampling period by the number of zero up-crossing waves in the sampling period.

Wave-sampling period: The relatively short period of time (usually 1,000 seconds) for which wave elevation and/or other wave variables are measured in order to define the sea state.

Wave spectrum: See *wave energy spectrum*.

Wave steepness: The ratio of the wave height to the wave length. Significant wave steepness in deep water is the ratio of the significant wave height to the wave length of a periodic wave whose period is the mean zero up-crossing wave period.

Wear-in failure: Failure of weak components associated with problems, such as manufacturing defects and installation/maintenance errors, also defined as "burn in" or "infant mortality."

Wear-out failure: End of useful life failure.

Weathervaning: Rotation of the vessel about the turret in response to environmental actions such as winds and waves. Partial weathervaning can be a characteristic of some non-turret-moored systems as well.

Yoke: A rigid-articulated connection system in a form of a lattice or box structure between a mooring column and a vessel.

Zero up-crossing wave: The portion of a wave record (the time history of wave elevation) between adjacent zero up-crossings. A zero up-crossing occurs when the sea surface rises (rather than falls) through the still-water level. Wave records are conventionally analyzed on the basis of the zero up-crossing waves they contain. For height of a zero up-crossing wave, see *wave height*. For a zero up-crossing period, see *wave period*.

APPENDIX 2

Scale Definitions of Winds, Waves, and Swells

A2.1. Beaufort Wind Scale

Beaufort number	Definition	Mean wind speed (knots)	Mean wind speed (m/s)	Probable mean wave height (m)
0	Calm	<1	<0.5	0.0
1	Light air	1–3	0.5–1.7	0.1
2	Light breeze	4–6	1.8–3.3	0.2
3	Gentle breeze	7–10	3.4–5.4	0.6
4	Moderate breeze	11–16	5.5–8.4	1.0
5	Fresh breeze	17–21	8.5–11.0	2.0
6	Strong breeze	22–27	11.1–14.1	3.0
7	Near gale	28–33	14.2–17.2	4.0
8	Gale	34–40	17.3–20.8	5.5
9	Strong gale	41–47	20.9–24.4	7.0
10	Storm	48–55	24.5–28.5	9.0
11	Violent storm	56–63	28.6–32.6	11.5
12	Hurricane	>64	>32.7	14.0

A2.2. Wave Scale

Definition	Wave height (m)
Calm–glassy	0.0
Calm–rippled	0.0–0.1
Smooth wavelets	0.1–0.5
Slight	0.5–1.25
Moderate	1.25–2.5
Rough	2.5–4.0
Very rough	4.0–6.0
High	6.0–9.0
Very high	9.0–14.0
Phenomenal	>14.0

A2.3. Swell Scale

Definition (length)	Length (m)	Definition (height)	Height (m)
Short	0–100	Low	0–2
Average	100–200	Moderate	2–4
Long	>200	Heavy	>4

Probability of Sea States at Various Ocean Regions

A3.1. Identification of Ocean Areas Using Marsden Squares

(Source: http://www.bmt.org and http://www.geomar.de and also several prior publications)

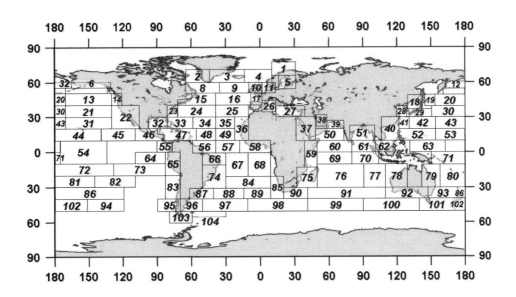

A3.2. Probability of Sea States in the North Atlantic

(Derived from BMT's Global Wave Statistics: http://www.bmt.org)

H_s/T_z	1.5	2.5	3.5	4.5	5.5	6.5	7.5	8.5	9.5	10.5	11.5	12.5	13.5	14.5	15.5	16.5	17.5	18.5	Sum
0.5	0.0	0.0	1.3	133.7	865.6	1,186.0	634.2	186.3	36.9	5.6	0.7	0.1	.0	0.0	0.0	0.0	0.0	0.0	3,050
1.5	0.0	0.0	0.0	29.3	986.0	4,976.0	7,738.0	5,569.7	2,375.7	703.5	160.7	30.5	5.1	0.8	0.1	0.0	0.0	0.0	22,575
2.5	0.0	0.0	0.0	2.2	197.5	2,158.8	6,230.0	7,449.5	4,860.4	2,066.0	644.5	160.2	33.7	6.3	1.1	0.2	0.0	0.0	23,810
3.5	0.0	0.0	0.0	0.2	34.9	695.5	3,226.5	5,675.0	5,099.1	2,838.0	1,114.1	337.7	84.3	18.2	3.5	0.6	0.1	0.0	19,128
4.5	0.0	0.0	0.0	0.0	6.0	196.1	1,354.3	3,288.5	3,857.5	2,685.5	1,275.2	455.1	130.9	31.9	6.9	1.3	0.2	0.0	13,289
5.5	0.0	0.0	0.0	0.0	1.0	51.0	498.4	1,602.9	2,372.7	2,008.3	1,126.0	463.6	150.9	41.0	9.7	2.1	0.4	0.1	8,328
6.5	0.0	0.0	0.0	0.0	0.2	12.6	167.0	690.3	1,257.9	1,268.6	825.9	386.8	140.8	42.2	10.9	2.5	0.5	0.1	4,806
7.5	0.0	0.0	0.0	0.0	0.0	3.0	52.1	270.1	594.4	703.2	524.9	276.7	111.7	36.7	10.2	2.5	0.6	0.1	2,586
8.5	0.0	0.0	0.0	0.0	0.0	0.7	15.4	97.9	255.9	350.6	296.9	174.6	77.6	27.7	8.4	2.2	0.5	0.1	1,309
9.5	0.0	0.0	0.0	0.0	0.0	0.2	4.3	33.2	101.9	159.9	152.2	99.2	48.3	18.7	6.1	1.7	0.4	0.1	626
10.5	0.0	0.0	0.0	0.0	0.0	0.0	1.2	10.7	37.9	67.5	71.7	51.5	27.3	11.4	4.0	1.2	0.3	0.1	285
11.5	0.0	0.0	0.0	0.0	0.0	0.0	0.3	3.3	13.3	26.6	31.4	24.7	14.2	6.4	2.4	0.7	0.2	0.1	124
12.5	0.0	0.0	0.0	0.0	0.0	0.0	0.1	1.0	4.4	9.9	12.8	11.0	6.8	3.3	1.3	0.4	0.1	0.0	51
13.5	0.0	0.0	0.0	0.0	0.0	0.0	0.0	0.3	1.4	3.5	5.0	4.6	3.1	1.6	0.7	0.2	0.1	0.0	21
14.5	0.0	0.0	0.0	0.0	0.0	0.0	0.0	0.1	0.4	1.2	1.8	1.8	1.3	0.7	0.3	0.1	0.0	0.0	8
15.5	0.0	0.0	0.0	0.0	0.0	0.0	0.0	0.0	0.1	0.4	0.6	0.7	0.5	0.3	0.1	0.1	0.0	0.0	3
16.5	0.0	0.0	0.0	0.0	0.0	0.0	0.0	0.0	0.0	0.1	0.2	0.2	0.2	0.1	0.1	0.0	0.0	0.0	1
Sum	0	0	1	165	2,091	9,280	19,922	24,879	20,870	12,898	6,245	2,479	837	247	66	16	3	1	100,000

Notes: H_s = significant wave height (m); T_z = spectral peak period (s).
Source: IACS (2001). *Standard wave data*. URS 11, No. 34, International Association of Classification Societies, November.

A3.3. Annual Sea States in the North Atlantic

Sea state number	Significant wave height (m)		Sustained wind speed[1] (knots)		Probability of sea states (%)	Modal wave period (sec)	
	Range	Mean	Range	Mean		Range[2]	Most probable[3]
0.1	0.0–0.1	0.05	0–6	3.0	0.70	–	–
2	0.1–0.5	0.3	7–10	8.5	6.80	3.3–12.8	7.5
3	0.5–1.25	0.88	11–16	13.5	23.70	5.0–14.8	7.5
4	1.25–2.5	1.88	17–21	19.0	27.80	6.1–15.2	8.8
5	2.5–4.0	3.25	22–27	24.5	20.64	8.3–15.5	9.7
6	4.0–6.0	5.0	28–47	37.5	13.15	9.8–16.2	12.4
7	6.0–9.0	7.5	48–55	51.5	6.05	11.8–18.5	15.0
8	9.0–14.0	11.5	56–63	59.5	1.11	14.2–18.6	16.4
>8	>14.0	>14.0	>63	>63	0.05	18.0–23.7	20.0

Notes: (1) Ambient wind sustained at 19.5m above surface to generate fully developed seas. To convert to another altitude (H_2), apply $V_2 = V_1(H_2/19.5)^{1/7}$. (2) Minimum is 5 percent and maximum is 95 percent for periods given wave height range. (3) Based on periods associated with central frequencies included in hindcast climatology. *Source:* Lee, W., Bales, W. L., and Sowby, S. E. (1985). *Standardized wind and wave environments for North Pacific Ocean areas.* R/SPD-0919–02, DTNSRDC, Washington, DC.

507

A3.4. Annual Sea States in the North Pacific

Sea state number	Significant wave height (m) Range	Mean	Sustained wind speed[1] (knots) Range	Mean	Probability of sea states (%)	Modal wave period (sec) Range[2]	Most probable[3]
0.1	0.0–0.1	0.05	0–6	3.0	1.30	–	–
2	0.1–0.5	0.3	7–10	8.5	6.40	5.1–14.9	6.3
3	0.5–1.25	0.88	11–16	13.5	15.50	5.3–16.1	7.5
4	1.25–2.5	1.88	17–21	19.0	31.60	6.1–17.2	8.8
5	2.5–4.0	3.25	22–27	24.5	20.94	7.7–17.8	9.7
6	4.0–6.0	5.0	28–47	37.5	15.03	10.0–18.7	12.4
7	6.0–9.0	7.5	48–55	51.5	7.60	11.7–19.8	15.0
8	9.0–14.0	11.5	56–63	59.5	1.56	14.5–21.5	16.4
>8	>14.0	>14.0	>63	>63	0.07	16.4–22.5	20.0

Notes: [1] Ambient wind sustained at 19.5m above surface to generate fully developed seas. To convert to another altitude (H_2), apply $V_2 = V_1 (H_2/19.5)^{1/7}$. [2]Minimum is 5 percent and maximum is 95 percent for periods given wave height range. [3]Based on periods associated with central frequencies included in hindcast climatology. *Source:* Lee, W., Bales, W. L., and Sowby, S. E. (1985). *Standardized wind and wave environments for North Pacific Ocean areas.* R/SPD-0919–02, DTNSRDC, Washington, DC.

A3.5. Characteristics of 100-Year Return Period Storms at Various Ocean Regions

Property		Gulf of Mexico	North Sea	West Africa	Brazil	Timor Sea
Wind speed (knots)		80	83	42	60	80
Current speed (knots)		2.1	2.8	1.7	3.2	4.2
Wave	H_s (m)	12.2	16.5	2.7	7.6	9.8
	H_{max} (m)	22.8	30.8	5.1	14.2	18.2
	T_z (s)	14.0	17.5	7.6	14.3	12.4
	H_s (m)	–	–	3.7	–	–
Swell	H_{max} (m)	–	–	6.8	–	–
	T_z (s)	–	–	13.9	–	–

Notes: H_s = significant wave height; H_{max} = maximum wave height; T_z = period. This data is indicative of metocean conditions for usual areas of operation of ship-shaped offshore installations.

A3.6. Extremes of Environmental Phenomena at Various Ocean Regions

Item	Gulf of Mexico	Southern North Sea	Northern North Sea	Canadian Georges Bank	N.W. Australia	Beaufort Sea	Newfoundland	Bass Strait
Wave height (m)	22	20	32	25	22	8	30	23
Wind (m/s)	70	40	45	50	70	40	60	50
Current (cm/s)	100	50	50	120	180	75	150	130
Tide (m)	1.5	2	2	2	4	0.5	2	1
Icing	No	No	No	Yes	No	Yes	Yes	No
Fog (%)	5	5–15	2–5	30–40	1	20	30–40	1

Source: Sharples, B. P. M., Stiff, J. J., Kalinowski, D. W., and Tidmarsh, W. G. (1989). *Statistical risk methodology: Application for pollution risks for Canadian Georges Bank drilling program.* Offshore Technology Conference, OTC 6082, Houston, May.

APPENDIX 4

Scaling Laws for Physical Model Testing

Ship-shaped offshore units exhibit complex dynamic responses to the metocean environment. Physical model testing is usually required for some aspects of their design, although theoretical and numerical simulations are becoming increasingly adopted with reasonable confidence. It is, of course, essential to consider and keep the correct scaling laws for both hydrodynamics and structural mechanics model tests using small-scale models.

A4.1 Hydrodynamics Model Tests

A4.1.1 Froude Scaling Law

The Froude scaling law is to ensure the correct relationship between inertial and gravitational forces (except for viscous roll damping forces) when the full-scale vessel is scaled down to model dimensions. It is recognized that the Froude law is ensured if the following Froude number F_n is the same at both a small-scale model and a full-scale prototype:

$$F_n = \frac{V}{\sqrt{gL}}, \tag{A4.1}$$

where F_n = Froude number; V = velocity; L = length; and g = acceleration of gravity.

To get the correct Froude mumber scaling, all lengths in a particular model test must be scaled by the same factor, as indicated in Table A4.1. For example, if the water depth is considered at a scale of 1:κ, then the same scale should be considered for the vessel's length, breadth, draught, wave height, and wave length. Although the model test is usually undertaken in fresh water, the full-scale unit will be used in salt water. The density ratio of salt water to fresh water is considered to be r = 1.025.

A4.1.2 Reynolds Scaling Law

The scaling effect associated with viscous forces, for example, viscous roll damping moments on a vessel, risers, and mooring lines, are not consistent with the Froude scaling law, but follows the Reynolds scaling law, which must be the same at both model and full scales:

$$R_e = \frac{VL}{\mu}, \tag{A4.2}$$

Table A4.1. *The Froude scaling laws for various physical quantities*

Quantity	Typical units	Scaling parameter
Length	m	κ
Time	s	$\kappa^{1/2}$
Frequency	1/s	$\alpha^{-1/2}$
Velocity	m/s	$\kappa^{1/2}$
Acceleration	m/s^2	1
Volume	m^3	κ^3
Water density	ton/m^3	r
Mass	ton	$r\kappa^3$
Force	kN	$r\kappa^3$
Moment	kNm	$r\kappa^4$
Extension stiffness	kN/m	$r\kappa^2$

Note: κ = scale factor and r = density ratio of fresh water to sea water that may usually be taken as r = 1.025, when the model testing is performed in fresh water, while the full-scale unit will be used in salt water.

where R_e = Reynolds number; μ = kinematic viscosity; and L, V = as defined in Eq. (A4.1).

In reality, it is not straightforward to achieve both the Froude and Reynolds scaling laws simultaneously in a particular model test. This is because the Froude scaling law requires the model velocity to vary with the square root of length, but the Reynolds scaling law requires an inverse relationship.

In practice, the model testing may need to be performed with a high Reynolds number so that a larger model with a faster flow speed must be applied specifically when the free surface condition is not relevant due to currents and wind. However, this is, again, not easy to achieve because of physical limitations on the model flow speed and also on the model Reynolds number. Note that the differences between the model and the full-scale Reynolds numbers may not be significant if the Reynolds number values at the model and full scales are both high enough.

A4.1.3 Vortex-Shedding Effects

Vortex-shedding effects are important in flow around bluff bodies, where instabilities in the wake flow result in the periodic creation and shedding of eddies and vortices. Due to vortex shedding, the body can be subjected to the forces that have the largest components in the direction transverse to the flow and the smallest components in the line of the flow. Flexible structures such as risers and mooring lines with low damping can be subject to the phenomenon of vortex-induced vibration, when the excitation frequency corresponds to one of the natural modes of the flexible structure.

Vortex-shedding effects may be considered in model testing by keeping the following quantity the same at both model and full scales:

$$S = \frac{\lambda D}{V} \quad \text{or} \quad V_r = \frac{1}{S}, \tag{A4.3}$$

where S = Strouhal number; V = flow speed; D = diameter of the body; λ = frequency of the eddy shedding; and V_r = reduced flow speed.

A4.1.4 Surface Tension Effects

The effects of surface tension can be important in model testing with a very small-scale compared to the full-scale prototype. The primary source of the surface tension effects arises from the properties of small waves. This is because a significant straightening effect on the surface of the water can be caused when the waves become small

Table A4.2. *Similarity considerations for a small-scale model and a full-scale prototype*

Quantity	Full-scale prototype	Small-scale model	Relationship
Length	L	ℓ	$\ell = \alpha L$
Displacement	Δ	δ	—
Strain	$E = \Delta/L$	ε	$\varepsilon = E$
Stress	$\Sigma = EE$	$\sigma = E\varepsilon$	$\sigma = \Sigma$
Pressure[1]	P	p	$p = P$
Dynamic force	$F = M\Delta/T^2$	$f = m\delta/t^2$	$f = \alpha^2 F$
Mass	M	m	$m = \alpha^3 M$
Time	T	t	$t = \alpha T$
Velocity	$V = \Delta/T$	v	$v = V$
Acceleration	$A = \Delta/T^2$	$a = \delta/t^2$	$a = A/\alpha$
Stress wave speed	C	c	$c = C$

Note: [1] Force on the boundary of the body under hydrostatic pressure is related to PL^2 in the full-scale prototype, while it is related to $p\ell^2 = p\alpha^2 L^2$.

enough. This can change the relationship between wave length and phase velocity, so that the surface tension behaves as if an additional effect of gravity.

The surface tension effects are considered important when the wave length of the model is less than 0.1m and the waves are referred to as "ripples." In offshore engineering, the waves with a period shorter than 4 seconds equivalent to a wave length less than 25m may, in general, not be of interest. Therefore, the surface tension effects may not be important as long as the model scale is larger than 1:250.

A4.1.5 Compressibility Effects

The effects of water and air compressibility are usually not considered for design of ship-shaped offshore units, although these effects may be important for propellers of trading ships and thrusters of dynamic positioning systems.

A4.2 Structural Mechanics Model Tests

In physical model testing for the purpose of structural mechanics, a small-scale test model and a full-scale prototype must have certain similarities including the geometrical similarity and the values of Young modulus (modulus of elasticity) (E), mass density (ρ), and Poisson ratio (v).

The relationship between the characteristics of a small-scale model and a full-scale prototype is given by the geometrical scale factor that is defined as:

$$\frac{\ell}{L} = \alpha, \tag{A4.4}$$

where α = geometrical scaling factor, ℓ = length dimension for the small-scale model; L = length dimension for the full-scale prototype. Typically, $\alpha \leq 1$ will be considered although the test model can, of course, be larger in size than the full-scale prototype. Table A4.2 indicates the relationships of various quantities for a small-scale model and a full-scale prototype.

APPENDIX 5

Wind-Tunnel Test Requirements

The determination of wind effects, such as forces and heeling moments for the hull, topsides, accommodation areas, and helideck of a ship-shaped offshore unit, can be an essential task for the analysis of intact and damage stabilities and other strength aspects. Wind forces and wind moments should also be predicted for the analyses of mooring and thruster systems. Although theoretical and numerical simulations including computational fluid dynamics (CFD) methods may be employed, wind-tunnel tests are highly desirable to get more reliable estimates in this regard.

Wind-tunnel tests are also usually used to analyze smoke ingress and ventilation problems on board a ship-shaped offshore unit, aspects that are always involved in various environmental and safety risk assessments. Examples include assessment and optimization of the areas over the helideck, which are affected by disturbed flow and by temperature rises due to turbine exhaust emissions. To model emergency gas releases and fire scenarios and to identify the regions of poor ventilation, wind-tunnel tests may be required. The natural ventilation within the process areas of an FPSO can also be assessed by wind-tunnel testing.

For a detailed description of the wind-tunnel testing of ship-shaped offshore units involving test procedures, measurement techniques, and assessment criteria, refer to the UK HSE report titled *Review of model-testing requirements for FPSOs*, Offshore Technology Report, 2000/123, Health and Safety Executive, UK, 2000.

APPENDIX 6

List of Selected Industry Standards

The following list is a source of ready reference, particularly for students and beginning practitioners. As with all such lists, this list is by no means either complete or comprehensive. We caution that some of the references may be undergoing revision. One is always cautioned to obtain and use the latest information in such cases. In this regard, website addresses are given wherever possible.

ABS (American Bureau of Shipping: http://www.eagle.org)

ABS Rules for building and classing mobile offshore drilling units (column-stabilized units)
ABS Rules for building and classing single-point moorings
ABS Rules for building and classing steel vessels
ABS Guide for certification of drilling systems
ABS Guide for building and classing facilities on offshore installations
ABS Guide for building and classing floating production, storage, and offloading systems
ABS Guide for underwater inspection in lieu of a dry-docking survey
ABS Guide for construction of shipboard elevators
ABS Guide for cranes
ABS Guide for undersea pipeline systems and risers
ABS Guide for surveys using risk-based inspection for the offshore industry
ABS Guidance notes on reliability-centered maintenance
ABS Guidance notes on risk and reliability
ABS Guidance notes for synthetic rope
ABS Guidance notes on nonlinear finite-element analysis of side structures subject to ice loads
ABS Guidance notes on ice class

AISC (American Institute of Steel Construction: http://www.aisc.org)

AISC Standard specification for structural steel for buildings, ASD

ANSI (American National Standards Institute: http://www.ansi.org)

ANSI Z41. Personal protection – protective footwear

ANSI Z87.1. Practice for occupational and educational eye and face protection

ANSI Z88.2. Respiratory protection

ANSI Z359.1. Safety requirements for personal fall – arrest systems, subsystems, and components

ANSI/SME Boiler and pressure code, Section I (power boilers)

ANSI/SME Boiler and pressure code, Section IV (heating boilers)

ANSI/SME Boiler and pressure code, Section V (nondestructive testing)

ANSI/SME Boiler and pressure code, Section VIII (pressure vessels)

ANSI/SME Boiler and pressure code, Section IX (welding and brazing qualifications)

ANSI/SME SPPE quality assurance and certification of safety and pollution prevention equipment used in offshore oil and gas operations

ANSI/IEEE (Institute of Electrical and Electronics Engineers) C37.4 Rating structure for AC high-voltage circuit breakers rated on a symmetrical current basis

ANSI/UL (Underwriters Laboratories) 1581 Reference standard for electrical wires, cables, and flexible cords

API (American Petroleum Institute: http://www.api.org)

API RP 2A-LRFD. Recommended practice for planning, designing, and constructing fixed offshore platforms – load and resistance-factor design

API RP 2A-WSD. Recommended practice for planning, designing, and constructing fixed offshore platforms – working stress design

API RP 2D. Operation and maintenance of offshore cranes

API RP 2FP1. Recommended practice for design, analysis, and maintenance for mooring for floating production systems

API RP 2FPS. Recommended practice for planning, designing, and constructing floating production systems

API RP 2G. Recommended practice for production facilities on offshore structures

API RP 2I. In-service inspection of mooring hardware for floating drilling units

API RP 2L. Production facilities on offshore structures

API RP 2R. Design rating and testing of marine drilling riser couplings

API RP 2RD. Design of risers for floating production systems (FPSs) and tension leg platforms (TLPs)

API RP 2SK. Design and analysis of station-keeping systems for floating structures

API RP 2SM. Design and analysis of synthetic moorings

API RP 2T. Planning, designing, and constructing tension leg platforms

API RP 2X. Ultrasonic and magnetic examination of offshore structural fabrication and guidelines for qualification of technicians

API RP 5L. Marine transportation of line pipe

API RP 7G. Drill stem design and operating limits

API RP 8B. Inspection, maintenance, repair, and remanufacture of hoisting equipment

API RP 9B. Application, care, and use of wire rope for oil field service

API RP 14B. Design, installation, repair, and operation of subsurface safety valve system

API RP 14C. Analysis, design, installation, and testing of basic surface safety systems on offshore production platforms

API RP 14E. Design and installation of offshore production piping systems

API RP 14F. Design and installation of electrical systems for offshore production systems

API RP 14G. Fire prevention and control on open-type offshore production platforms

API RP 14H. Installation, maintenance, and repair of surface safety valves and underwater safety valves offshore

API RP 14J. Design and hazards analysis for offshore production facilities

API RP 16C. Specification for choke and kill systems

API RP 16E. Design of control systems for drilling well control equipment

API RP 16Q. Design, selection, operation, and maintenance of marine drilling riser systems

API RP 17A. Recommended practice for design and operation of subsea production systems

API RP 17B. Flexible pipe

API RP 17D. Specification for subsea wellhead and Christmas tree equipment

API RP 17G. Design and operation of completion/workover riser systems

API RP 17I. Installation of subsea umbilical

API RP 53. Blowout prevention equipment systems for drilling wells

API RP 57. Offshore well completion, servicing, workover, and abandonment operations

API RP 75. Development of a safety and environmental management program for outer continental shelf operations and facilities

API RP 500. Classification of locations for electrical installations classified as class I, Div. 1 and Div. 2

API RP 500B. Classification of areas for electrical installations at drilling rigs and production facilities on land and on marine fixed and mobile platforms

API RP 505. Classification of locations for electrical installations classified as class I, zone 0, 1, 2

API RP 520. Design and installation of pressure-relieving systems in refineries, Parts I and II

API RP 521. Guide for pressure-relief and pressuring systems

API RP 530. Calculation of heater tube thickness in petroleum refineries

API RP 550. Manual on installation of refinery instruments and control systems

API RP 1110. Pressure testing of liquid petroleum pipelines

API RP 1111. (Reballot copy) design, construction, operation, and maintenance of offshore hydrocarbon pipelines

API RP T1. Orientation programs for personnel going offshore for the first time

API RP T4. Training of offshore personnel in nonoperating emergencies

API RP T7. Training of personnel in rescue of persons in water

API 510. Pressure vessel inspection code

API 615. Sound control of mechanical equipment for refinery service

API Bulletin 2J. Comparison of marine drilling riser analysis

API Bulletin 2N. Bulletin for planning, designing, and constructing fixed-offshore structures in ice environments

API Bulletin 2U. Stability design of cylindrical shells

API Bulletin 2V. Design of flat-plate structures

API Bulletin D16. Development of a spill prevention control and countermeasure

API Spec 2B. Fabrication of steel pipe

API Spec 2C. Specification for offshore cranes

API Spec 2F. Mooring chain

API Spec 4F. Drilling and well-servicing structures

API Spec 5L. Line pipe

API Spec 6A. Well head and Christmas tree equipment

API Spec 6AV1. Verification test of well head surface safety valves and underwater safety valves for offshore service

API Spec 6FA. Fire test for valves

API Spec 6D. Pipeline valves end closures, connectors, and swivels

API Spec 6H. End closures, connectors, and swivels

API Spec 7. Specification for rotary drilling equipment

API Spec 8A and 8C. Drilling and production hoisting equipment

API Spec 9A. Wire rope

API Spec 14A. Subsurface safety valve equipment

API Spec 14D. Well-head surface safety valves and underwater safety valve systems

API Spec 16A. Specification for drill-through equipment

API Spec 17E. Subsea production control umbilicals

API Spec 17J. Specification for unbonded flexible pipe

API Spec Q1. Specification for quality programs for the petroleum and natural gas industry

API Std 610. Centrifugal pumps for general refinery service

API Std 616. Gas turbines for the petroleum, chemical, and gas industry services

API Std 617. Centrifugal compressors for general refinery service

API Std 618. Reciprocating compressors for general refinery service

API Std 650. Welded steel tanks for oil storage

API Std 661. Air-cooled heat exchangers for general refinery services

API Std 1104. Welding of pipelines and related facilities

ASTM (American Society for Testing and Materials: http://www.astm.org)

ASTM A20. Steel plates for pressure vessels

ASTM A36. Structural steel

ASTM A47. Malleable iron castings

ASTM A53. Pipe: steel, black, hot-dipped, zinc-coated, welded, and seamless

ASTM A106. Seamless carbon steel pipe for high-temperature service

ASTM A126. Gray iron casings for valves, flanges, and pipe fittings

ASTM A134. Pipe, steel, electric-fusion arc-welded

ASTM A135. Electric-resistance-welded steel pipe

ASTM A139. Electric-fusion arc-welded steel pipe

ASTM A178. Electric-resistance-welded carbon steel boiler tubes

ASTM A179. Seamless cold-drawn low-carbon steel heat-exchangers and condenser tubes

ASTM A182. Forged or rolled alloy-steel pipe flanges, forged fittings, and valves, and parts for high-temperature service

ASTM A192. Seamless carbon steel boiler tubes for high-pressure service

ASTM A194. Carbon and alloy-steel nuts for bolts for high-pressure and high-temperature service

ASTM A199. Seamless cold-drawn intermediate alloy-steel heat-exchangers and condenser tubes

ASTM A203. Pressure vessel plates, alloy steel, and nickel

ASTM A210. Seamless medium carbon steel boiler and super-heater tubes

ASTM A213. Seamless ferritic and austenitic alloy-steel boiler, super-heater, and heat-exchanger tubes

ASTM A214. Electric-resistance-welded carbon steel heat-exchangers and condenser tubes

ASTM A226. Electric-resistance-welded carbon steel boiler and super-heater tubes for high-pressure service

ASTM A234. Piping fittings of wrought carbon steel and alloy steel for moderate and elevated temperatures

ASTM A249. Welded austenitic steel boiler, super-heater, heat-exchanger, and condenser tubes

ASTM A268. Seamless and welded ferritic stainless steel tubing for general service

ASTM A276. Stainless and heat-resisting steel bars and shapes

ASTM A307. Carbon steel externally threaded standard fasteners

ASTM A312. Seamless and welded austenitic stainless steel pipe

ASTM A316. Specification for low-alloy steel covered filler metal arc-welding electrodes

ASTM A320. Alloy-steel bolting materials for low-temperature service

ASTM A333. Seamless and welded steel pipe for low-temperature service

ASTM A334. Seamless and welded carbon and alloy-steel tubes for low-temperature service

ASTM A335. Seamless ferritic alloy-steel pipe for high-temperature service

ASTM A350. Forgings, carbon and low-alloy steel, requiring notch toughness testing for piping components

ASTM A351. Steel castings, austenitic, for high-temperature service

ASTM A352. Steel castings, ferritic and martensitic, for pressure-containing parts suitable for low-temperature service

ASTM A358. Electric-fusion-welded austenitic chromium nickel alloy steel pipe for high-temperature service

ASTM A369. Carbon and ferric alloy steel-forged and bored pipe for high-temperature service

ASTM A370. Mechanical testing of steel products

ASTM A376. Seamless austenitic steel pipe for high-temperature central station service

ASTM A395. Ferritic ductile iron pressure retaining castings for use at elevated temperatures

ASTM A403. Wrought austenitic stainless steel and alloy steel for low-temperature service

ASTM A420. Piping fittings of wrought carbon steel and alloy steel for low-temperature service

ASTM A430. Austenitic steel, forged and bored pipe, for high-temperature service

ASTM A520. Supplementary requirements for seamless and electrical-resistance welded carbon-steel tubular products for high-temperature service conforming to ISO recommendations for boiler construction

ASTM A522. Forged or rolled 8 percent and 9 percent nickel alloy steel flanges, fittings, valves, and parts for low-temperature service

ASTM B858M. Standard test method for determination of susceptibility to stress corrosion cracking in copper alloys using an ammonia vapor test

ASTM D4541. Pull-off adhesion testing of paint, varnish, and other coatings, and films with the PAT adhesion tester and the DFD method

ASTM D9394. Flash point by Pennsky–Martens closed-cup tester

ASTM E23. Notched bar impact testing of metallic materials

ASTM E208. Conducting drop-weight test to determine nil-ductility transition temperature of ferritic steels

ASTM F682. Wrought carbon-steel sleeve-type pipe couplings

ASTM F1003. Standard specification for searchlights on motor lifeboats

ASTM F1006. Entrainment separators for use in marine piping application

ASTM F1007. Pipeline expansion joints for use in marine piping applications

ASTM F1014. Standard specifications for flashlights on vessels

ASTM F1020. Line blind valves for marine applications

ASTM F1120. Circular metallic bellows-type expansion joints for use in marine piping

ASTM F1121. International shore connections for marine fire applications

ASTM F1123. Nonmetallic expansion joints for use in marine piping applications

ASTM F1139. Steam traps and drains

ASTM F1196. Sliding watertight door assemblies

ASTM F1197. Sliding watertight door-control systems

ASTM F1273. Vent flame arresters

ASTM F1321. Standard guide for conducting a stability test (lighweight survey and inclining experiment) to determine the light ship displacement and centers of gravity of a vessel

ASTM F1387. Standard specifications for performance of mechanically attached fittings, including supplementary requirements and annex

ASTM F1476. Standard specification for performance of gasketed mechanical couplings for use in piping applications, including annex

ASTM F1548. Standard specification for performance of fittings for use with gasketed mechanical couplings for use in piping applications

ASTM F1626. Symbols for use in accordance with regulation II-2/20 of 1974 SOLAS

AWS (American Welding Society: http://www.aws.org)

AWS D1.1. Structural welding code-steel

AWS D3.5. Guide for steel hull welding

AWS D1.4. Structural welding code-reinforcing steel

BSI (British Standards Institute: http://www.bsi-global.com)

BS 449. The use of structural steel in buildings

BS 1113. Water-tube steam-generating plant

BS 2790. Shell boiler of welded construction

BS 2853. The design and testing of steel overhead runway beams

BS 3351. Specification for piping systems for petroleum refineries and petrochemical plants

BS 5512 19991/ISO 281. Dynamic load ratings and rating life of roller bearings

BS 5500. Unfired fusion-welded pressure vessels

BS 5950. Structural use of steelwork in buildings

BS 6755. Testing of valves: Part 2. Specification for fire type testing requirements

BS 7910. Guide on methods for assessing the acceptability of flaws in fusion-welded structures

BS 8010. Code of practice for pipelines: Part 3. Pipelines subsea – design, construction, and installation

BS 8100. Lattice towers and masts

BV (Bureau Veritas: http://www.bureauveritas.com)

BV Offshore unit rules

BV Steel ships rules

BV Rule Note No. 497. Hull structure of production storage and offloading surface units

BV NI 184. Lifting appliances

BV NI 187. Classification of dynamic positioning systems

BV NI 205. Certification of well control equipment

BV NI 364. Nonbonded flexible steel pipe flowlines

BV NI 432. Guidance note for synthetic fiber rope

BV OU NR 428. Electrical systems

BV OU NR 458. Piping systems

BV OU NR 459. Process systems

BV OU NR 460. Safety features

DNV (Det Norske Veritas: http://www.dnv.com)

DNV Rules for classification of ships

DNV Rules for planning and execution of marine operations

DNV CN (Classification Notes) 4.3. Repair of surface/dimensions by means of metal coating

DNV CN 7. Nondestructive testing

DNV CN 8. Conversion of ships

DNV CN 10.2. Guidance for condition monitoring

DNV CN 30.1. Buckling strength analysis of bars, frames, and spherical shells

DNV CN 30.4. Foundations

DNV CN 30.5. Environmental conditions and environmental loads

DNV CN 30.6. Structural reliability analysis of marine structures

DNV CN 30.7. Fatigue assessments of ship structures

DNV CN 33.1. Corrosion prevention of tanks and holds

DNV CN 72.1. Allowable thickness diminution for hull structure

DNV OS (Offshore Standards) A101. Safety principles and arrangement

DNV OS B101. Metallic materials

DNV OS C101. Design of offshore steel structures, general (LRFD method)

DNV OS C102. Structural design of offshore ships

DNV OS C103. Structural design of column-stabilized units (LRFD method)

DNV OS C104. Structural design of self-elevating units (LRFD method)

DNV OS C105. Structural design of TLP (LRFD method)

DNV OS C106. Structural design of deep draught floating units

DNV OS C301. Stability and watertight integrity

DNV OS C401. Fabrication and testing of offshore structures

DNV OS D101. Marine and machinery systems and equipment

DNV OS D201. Electrical installations

DNV OS D301. Fire protection

DNV OS E101. Drilling plant

DNV OS E201. Oil and gas processing systems

DNV OS E301. Position mooring

DNV OS E401. Helicopter decks

DNV OS F201. Dynamic risers

DNV OSS (Offshore Service Specifications) 101. Rules for classification of drilling and support units

DNV OSS 102. Rules for classification of floating production and storage units

DNV OSS 103. Rules for classification of LNG/LPG floating production and storage units or installations

DNV OSS 121. Classification based on performance criteria determined from risk-assessment methodology

DNV OSS 300. Risk-based verification

DNV RP (Recommended Practices) A202. Documentation of offshore projects

DNV RP B401. Cathodic protection design

DNV RP C102. Structural design of offshore ships

DNV RP C103. Column-stabilized units

DNV RP C201. Buckling strength of plated structure

DNV RP C202. Buckling strength of shells

DNV RP C203. Fatigue strength analysis of offshore steel structures

DNV RP C204. Design against accidental loads

DNV RP C206. Fatigue methodology for offshore ships

DNV RP E304. Damage assessment of fiber ropes for offshore mooring

DNV RP F201. Design of titanium risers

DNV RP F204. Riser fatigue

DNV RP F205. Global performance analysis of deep-water floating structures

DNV RP G101. Risk-based inspection of offshore topside static mechanical equipment

DNV RP H101. Risk management in marine and subsea operations

DNV RP H102. Marine operations during removal of offshore installations

DNV RP O401. Safety and reliability of subsea systems

EN (European Committee for Standardization: http://www.cenorm.be)

EN 13852. Offshore cranes: Part 1. General purpose offshore cranes

HSE (Health and Safety Executive, UK: http://www.hse.gov.uk)

HSE Offshore installations – guidance on design and certification

HSE The Health and Safety at Work Act (1974)

HSE A guide to the integrity, workplace environment, and miscellaneous aspects of the offshore installations and wells (design and construction, etc.) regulations (1996)

HSE SI 1992/2885. The offshore installations (safety case) regulations

HSE SI 1994/3246. The control of substances hazardous to health regulations (COSHH)

HSE SI 1995/743. The offshore installations: prevention of fire and explosion, and emergency response (PFEER)

HSE SI 1996/825. Pipeline safety regulations

HSE SI 1996/913. The offshore installations and wells: design and construction (DCR)

HSE SI 1996/2154. Merchant shipping (prevention of oil pollution) regulations

HSE SI 1999/3242. Management of health and safety at work regulations

HSE SI 2001/1091. Offshore combustion installations (prevention and control of pollution) regulations

HSE OTI 95–634. Jet fire resistance test of passive fire materials

IACS (International Association of Classification Societies: http://www.iacs.org)

IACS Shipbuilding and repair quality standard No. 47 (1996)

IACS Common structural rules (CSR) for double-hull oil tankers

ICS (International Chamber of Shipping: http://www.marisec.org/ics)

ICS/OCIMF International safety guide for oil tankers and terminals (ISGOTT) – Revision 4 (1997), published by Witherby & Co., Ltd.

IEC (International Electrotechnical Commission: http://www.iec.org)

IEC 56. High-voltage alternating-current circuit breakers

IEC 68–2–52. Basic environmental testing procedures

IEC 79–0. Electrical apparatus for explosive gas atmospheres: Part 0

IEC 79–1. Electrical apparatus for explosive gas atmospheres: Part 1

IEC 79–2. Electrical apparatus for explosive gas atmospheres: Part 2

IEC 79–5. Electrical apparatus for explosive gas atmospheres: Part 5

IEC 79–6. Electrical apparatus for explosive gas atmospheres: Part 6

IEC 79–7. Electrical apparatus for explosive gas atmospheres: Part 7

IEC 79–11. Electrical apparatus for explosive gas atmospheres: Part 11

IEC 79–15. Electrical apparatus for explosive gas atmospheres: Part 15

IEC 79–18. Electrical apparatus for explosive gas atmospheres: Part 18

IEC 92–3. Electrical installation in ships: Part 3. Cables

IEC 92–101. Electrical installation in ships: Part 101. Definitions and general requirements

IEC 92–201. Electrical installation in ships: Part 201. System design – general

IEC 92–202. Electrical installation in ships: Part 202. System design – protection

IEC 92–301. Electrical installation in ships: Part 301. Equipment – generators and motors

IEC 92–302. Electrical installation in ships: Part 302. Equipment – switchgear and control gear assemblies

IEC 92–303. Electrical installation in ships: Part 303. Equipment – transformers for power and lighting

IEC 92–304. Electrical installation in ships: Part 304. Equipment – semiconductor converters

IEC 92–306. Electrical installation in ships: Part 306. Equipment – luminaries and accessories

IEC 92–352. Electrical installation in ships: Part 352. Choice and installation of cables for low-voltage power systems

IEC 92–401. Electrical installation in ships: Part 401. Installation and test of completed installation

IEC 92–501. Electrical installation in ships: Part 501. Special features – electric propulsion plant

IEC 92–502. Electrical installation in ships: Part 502. Tankers – special features

IEC 92–503. Electrical installation in ships: Part 503. Special features – AC supply systems with voltages in the range above 1KV up to and including 11 KV

IEC 92–504. Electrical installation in ships: Part 504. Special features – control and instrumentation

IEC 331. Fire-resisting characteristics of electric cables

IEC 331–1. Tests on electric cables under fire conditions: Part 1

IEC 331–3. Tests on electric cables under fire conditions: Part 3

IEC 363. Short-circuit current evaluation with special regard to rated short-circuit capacity of circuit-breakers in installations in ships

IEC 529. Degrees of protection provided by enclosures

IEC 533. Electromagnetic compatibility of electrical and electronic installations in ships

IEC 947–2. Low-voltage switch gear and control gear: Part 2. Circuit breakers

IEC 61508. Functional safety of electrical, electronic, and programmable electronic safety-related systems (2000)

IEEE (Institute of Electrical and Electronics Engineers: http://www.ieee.org)

IEEE Std C37.13. Low-voltage AC power circuit breakers used in enclosures

IEEE Std C37.14. Low-voltage DC power circuit breakers used in enclosures

IEEE Std 45. Recommended practice for electric installations on shipboard

IEEE Std 100. The new IEEE standard dictionary of electrical and electronics terms

IEEE Std 320. Application guide for AC high-voltage circuit breakers rated on a symmetrical current basis

IEEE Std 331. Application guide for low-voltage AC nonintegrally fused power circuit breakers

IEEE Std 1202. Flame testing of cables for use in cable tray in industrial and commercial occupancies

IMO (International Maritime Organization: http://www.imo.org)

IMO Convention on the international regulations for preventing collisions at sea (COLREG 1972)

IMO International convention for the Safety of Life at Sea (SOLAS 1974)

IMO International convention for the prevention of pollution from ships (MARPOL 73/78)

IMO International convention on oil pollution preparedness, response, and cooperation (OPRC)

IMO International convention on load lines (ICLL 1966)

IMO Load line convention for ship type A: tankers (1996)

IMO MSC Circular 699 Revised guidelines for passenger safety instructions

IMO MSC.4(48). International code for the construction and equipment of ships carrying dangerous chemicals in bulk

IMO MSC.5(48). International code for the construction and equipment of ships carrying liquefied gases in bulk

IMO MSC.35(63). Adoption of guidelines for emergency towing arrangement on tankers

IMO Resolution A.265(VIII). Stability

IMO Resolution A.414(XI). Code for construction and equipment of MODUS, 1979

IMO Resolution A.468(XII). Code on noise levels onboard ships, 1981

IMO Resolution A.520(13). Code of practice for the evaluation, testing, and acceptance of prototype novel life-saving appliances and arrangements

IMO Resolution A.535(13). Recommendations on emergency towing requirements for tankers

IMO Resolution A.649(16). Code for construction and equipment on MODUS, 1989

IMO Resolution A.654(16). Graphical symbols for fire control plans

IMO Resolution A.657(16). Instructions for action in survival craft

IMO Resolution A.658(16). Use and fitting of retroreflective materials on lifesaving appliances

IMO Resolution A.749. Code on intact stability criteria

IMO Resolution A.760(18). Symbols related to life-saving appliances and arrangements

IMO Resolution A.753(18). Guidelines for the application of plastic pipes on ships

INTERTANKO (Association of Independent Tanker Owners: http://www.intertanko.com)

INTERTANKO Risk-minimization guidelines for shuttle tanker operations worldwide at offshore locations (2000)

IP (Institute of Petroleum)

IP Code Part 15. Area classification code for petroleum installations

ISA (The Instrumentation, Systems and Automation Society: http://www.isa.org)

ISA RP 12.6. Wiring practices for hazardous (classified) locations instrumentation: Part I. Intrinsic safety

ISO (International Organization for Standardization: http://www.iso.org)

ISO 2314. Gas turbine acceptance tests

ISO 3730. Mooring winches

ISO 4624. Same as ASTM D4541

ISO 6067. Winches for lifeboats

ISO 7365. Towing winches for deep-sea use

ISO 7825. Deck machinery general requirements

ISO 8501–1. Preparation of steel substrates before application of paints and related products – visual assessment of surface cleanliness: Part 1. Rust grades and preparation grades of uncoated steel substrates and of steel substrates after overall removal of previous coatings

ISO 9089. Anchor winches

ISO 13702. Petroleum and natural gas industries – control and mitigation of fires and explosions on offshore production installations – requirements and guidelines

ISO 13819–1. Petroleum and natural gas industries – offshore structures: Part 1. General requirements

ISO TR 14564. Marking of escape routes

ISO 17776. Petroleum and natural gas industries – offshore production installations – Guidelines on tools and techniques for the identification and assessment of hazardous events

ISO FDIS 18072–1. Ships and marine technology – ship structures: Part 1. General requirements for their limit state assessment

ISO DIS 18072–2. Ships and marine technology – ship structures: Part 2. Requirements for their ultimate limit state assessment

ISO 19900. Petroleum and natural gas industries – general requirements for offshore structures

ISO 19901–1. Petroleum and natural gas industries – specific requirements for offshore structures: Part 1. Metocean design and operating considerations

ISO 19901–2. Petroleum and natural gas industries – specific requirements for offshore structures: Part 2. Seismic design procedures and criteria

ISO 19901–3. Petroleum and natural gas industries – specific requirements for offshore structures: Part 3. Topsides structure

ISO 19901–4. Petroleum and natural gas industries – specific requirements for offshore structures: Part 4. Geotechnical and foundation design considerations

ISO 19901–5. Petroleum and natural gas industries – specific requirements for offshore structures: Part 5. Weight control during engineering and construction

ISO 19901–6. Petroleum and natural gas industries – specific requirements for offshore structures: Part 6. Marine operations

ISO 19901–7. Petroleum and natural gas industries – specific requirements for offshore structures: Part 7. Station-keeping systems

ISO 19902. Petroleum and natural gas industries – fixed steel offshore structures

ISO 19903. Petroleum and natural gas industries – fixed concrete offshore structures

ISO 19904–1. Petroleum and natural gas industries – floating offshore structures: Part 1. Monohulls, semisubmersibles, and spars

ISO 19905–1. Petroleum and natural gas industries – site-specific assessment of mobile offshore units: Part 1. Jack-ups

ISO 19906. Petroleum and natural gas industries – arctic offshore structures

LR (Lloyd's Register: http://www.lr.org)

LR Rules and regulations for the classification of mobile offshore units

LR Rules and regulations for the classification of a floating installation at a fixed location, and ship-type FPSO hull structural appraisal

LR Rules and regulations for the classification of ships

LR OS/GN/99002. Ship-type FPSO hull structural appraisal (1999)

LR Provisional rules for the construction and classification of submarine pipelines

LR Code for lifting appliances in a marine environment

LR Rules and regulations for the classification of fixed offshore installations: Part 9. Provisional rules for floating offshore production, storage, and offloading installations

NACE (National Association of Corrosion Engineers, USA: http://www.nace.org)

NACE MR 0175. Sulfide stress cracking resistant materials for oilfield equipment

NACE RP 0175. Control of internal corrosion in steel pipelines and piping systems

NACE RP 0176. Corrosion control of steel fixed offshore platforms associated with petroleum production

NACE RP 0188. Discontinuity (holiday) testing of new protective coatings on conductive substrates

NACE RP 0287. Field measurement of surface profile of abrasive blast-cleaned steel surfaces using a replica tape

NACE RP 0675. Control of corrosion on offshore steel pipelines

NEMA (National Electrical Manufacturers Association, USA: http://www.nema.org)

NEMA Std ICS 2. Industrial control and systems controllers, contractors, and overload relays rated not more than 2,000 volts AC or 750 volts DC

NEMA Std 2.3. Instructions for the handling, installation, operation, and maintenance of motor control centers

NEMA Std 2.4. NEMA and IEC devices for motor service – a guide for understanding the differences

NEMA Std 250. Enclosures for electrical equipment

NEMA Std WC-3. Rubber-insulated wire and cable for the transmission and distribution of electrical energy

NEMA Std WC-8. Ethylene-propylene-rubber-insulated wire and cable for the transmission and distribution of electrical energy

NFPA *(National Fire Protection Association, USA: http://www.nfpa.org)*

NFPA 1. Fire prevention code

NFPA 10. Standard for portable fire extinguishers

NFPA 11. Standard for low-expansion foam and combined agent systems

NFPA 11A. Medium- and high-expansion foam systems

NFPA 11C. Mobile foam apparatus

NFPA 12. Carbon dioxide systems

NFPA 12A. Halon 1301 systems

NFPA 13. Standard for installation of sprinkler systems

NFPA 14. Standpipe hose systems

NFPA 15. Standard for water spray fixed systems

NFPA 16. Deluge foam-water system

NFPA 16A. Closed head foam-water sprinkler systems

NFPA 17. Dry chemical extinguishing systems

NFPA 17A. Wet chemical extinguishing systems

NFPA 20. Standard for the installation of centrifugal fire pumps

NFPA 25. Water-based fire protection systems

NFPA 37. 1975 Stationary combustion engines and gas turbines

NFPA 68. Venting of deflagrations

NFPA 69. Explosion prevention system

NFPA 70. National electrical code

NFPA 72. National fire alarm code

NFPA 77. Recommended practice on static electricity

NFPA 80. Fire doors and fire windows

NFPA 99. Standard for health care facilities

NFPA 170. Fire safety symbols

NFPA 251. Standard methods of tests of fire endurance of building construction and materials

NFPA 252. Standard method of fire tests of door assemblies

NFPA 306. Standard for the control of gas hazards on vessels (1997)

NFPA 496. Standard for purged and pressurized enclosures for electrical equipment

NFPA 704. Fire hazards of materials

NFPA 750. Standard for installation of water-mist fire suppression systems

NFPA 2001. Standard on clean-agent fire extinguishing systems

NIOSH *(National Institute for Occupational Safety and Health, USA: http://www.cdc.gov/niosh)*

NIOSH 87–116. Guide to industrial respiratory protection

NORSOK *(Standardization Organizations in Norway: http://www.nts.no/norsok)*

NORSOK M-101. Structural steel fabrication

NORSOK N-001. Structural design

NORSOK N-003. Actions and action effects

NORSOK N-004. Design of steel structures

NPD (Norwegian Petroleum Directorate: http://www.npd.no)

NPD Regulations for compliant FPSOs

OCIMF (Oil Companies International Marine Forum: http://www.ocimf.com)

OCIMF Offshore loading safety guidelines (1999)
OCIMF Guidance manual for the inspection and condition assessment of tanker
    structures – Revision 1 (1986)
OCIMF Effective mooring (2005)
OCIMF Guide for prediction of wind and current loads on VLCC
OCIMF Hose standards
OCIMF Handling, storing, inspecting hoses in the field
OCIMF Standards for oil tanker manifolds and associated equipment
OCIMF Ship-to-ship transfer guide (petroleum)

SSPC (Society for Protective Coatings: http://www.sspc.org)

SSPC PA 2. Measurement of dry coating thickness with magnetic gages (steel struc-
    tures painting manual: Chapter 5. Paint application specs)
SSPC SP 5. White metal blast cleaning NACE No.1–2000 (steel structures painting
    manual: Chapter 2. Surface preparation specs)
SSPC VIS 1. Visual standard for abrasive blast cleaned steel (standard reference
    photographs) editorial changes September 1, 2000 (steel structures painting man-
    ual: Chapter 2. Surface preparation specs)

TSCF (Tanker Structures Co-operative Forum)

TSCF Condition evaluation and maintenance of tanker structures, published by
    Witherby & Co. Ltd.
TSCF Guidance manual for the inspection and condition assessment of tanker struc-
    tures, published by Witherby & Co. Ltd.

UKOOA (UK Offshore Operators Association: http://www.ukooa.co.uk)

UKOOA FPSO design guidance notes for UKCS service (2002)
UKOOA The management of offshore helideck operations (1997)
UKOOA Guidelines for the management of safety critical elements – a joint industry
    guide (1996)
UKOOA FPSO committee – tandem loading guidelines: Volume 1. FPSO/tanker
    risk control during offtake
UKOOA Guidelines for fire and explosion hazard management (1995)
UKOOA Guidelines for instrument-based protective systems (1999)

UKOOA Specification and recommended practice for the use of GRP piping offshore

UKOOA Safety-related telecommunications systems on normally attended installations

UKOOA Guidelines on stakeholder engagement for decommissioning activities (2006)

Index

accidental flooding, 259, 308
accidental limit-state design, 257
  active safeguards for fire and explosion, 485
  collisions, 261
  damaged vessel stability, 259
  design principles and criteria, 257
  considerations for ALS applications to
    ship-shaped offshore structures, 313
  dropped objects, 277
  fire and heat, 296
  gas explosion and blast, 299
  passive risk control for fire and explosion, 484
  progressive collapse of heeled hulls, 308
  risk control for collisions, 484
  risk control for dropped objects, 485
accommodation, 16
  design issues, 48
actions and action-effects analysis, 113, 150
  action-effect analysis issues, 43
addition of new components, 454
age-related deterioration, 401
ALARP (as low as reasonably practicable),
  471
allowable stress, 56
ALPS/HULL, 201, 309
ALPS/SPINE, 166
ALPS/ULSAP, 133, 168, 180
analytical methods, 159, 173
annualized corrosion rates, 374
Arco, 10
Association for Structural Improvement of the
  Shipbuilding Industry of Japan (ASIS), 274

ballast water deoxygenation, 393
beam-column type collapse, 171
Beaufort wind scale, 503
benign environmental areas, 10
biaxial compressive collapse, 171
Biggs method, 303
bottom-supported platforms, 2
bow slamming, 99
building material issues, 39
  yield stress, 39
  fracture toughness, 41
  risk control, 483

capacity, 65
  characteristics value, 68
  design capacity, 65
cargo handling systems, 324
cathodic protection, 391
Castellon, 10
chemical inhibitors, 395
classification societies, 78
classification society rules, 78
classing issues, 52
coating, 386
  coating-life prediction, 390
  selection criteria of coating material, 389
  types of coating, 387
Cognac platform, 2
collisions, 261
  energy absorption characteristics, 265
  nonlinear finite-element modeling techniques,
    265
  practices for collision assessment, 263
  prescriptive procedure, 263
  risk control, 484
commissioning issues, 50
compressibility effects, 513
computational fluid dynamics (CFD) methods, 297
concrete gravity platform, 2
condition assessment of aged tanker structures,
  450
condition assessment scheme, 408
  emergency response services, 411
  enhanced survey programme, 409
  ship inspection report programme, 411
consequence analysis, 479
consequence rating, 473, 474
consequence severity, 473
construction issues, 49
contracting strategies, 32
conversion yard, 456
corrosion assessment and management, 356
  ballast water deoxygenation, 393
  cathodic prediction, 391
  chemical inhibitors, 395
  coating, 386
  corrosion issues, 48
  corrosion margin addition, 383

corrosion assessment and management (*cont.*)
  design issues, 48
  risk control, 484
  serviceability limit-state design, 145
corrosion margin addition, 383
corrosion models, 364
  mechanical models, 366
  phenomenological models, 379
corrosion rates, 374
corrosion wastage, 48
  corrosion wastage examination, 402
  corrosion wastage prediction, 364
  effect of corrosion wastage on plate ultimate
    strength, 428
cost–benefit analysis, 470
Cowper–Symonds equation, 271
crack growth rate, 252
crack initiation, 251
crack propagation, 251
critical buckling strength, 57
critical buckling strength design (CBSD),
  57
critical fracture strain, 267, 270
critical joints and details, 231
cross-stiffened plate structures, 132
currents, 91
  Det Norske Veritas classification notes, 92
cyclic stress ranges, 221

damaged vessel stability, 259
decision-making recommendations,
  471
decommissioning, 447, 456
  cost issues, 460
  decommissioning practices, 460
  environmental issues, 459
  regulatory framework, 457
  safety and health issues, 459
  technical feasibility issues, 458
demand, 65
  characteristic value, 68
  design demand, 65, 67
dented plates, 281
design capacity, 65
design criterion, 60
design demand, 65, 67
design principles, 55
  for accidental limit-state design, 257
  for environment, 77
  for fatigue limit-state design, 218
  for health, 77
  for safety, 75
  for serviceability limit-state design, 112
  for stability, 74
  for station-keeping, 74
  for towing, 74
  for ultimate limit-state design, 148
  for vessel motions, 75
development drilling, 3
DNV PULS, 168

double-hull arrangements, 22
  double sides, 22
  double bottoms, 22
dropped objects, 277
  risk control, 485
  ultimate strength of dented plates, 281, 293
DYNA3D nonlinear dynamic finite-element
  simulations, 133
dynamic fracture strain, 272, 273
dynamic/impact-pressure actions, 9, 37, 45
dynamic load analysis, 43
dynamic material properties, 271
dynamic positioning systems, 5
dynamic yield stress (strength), 271, 272

effective width, 125
elastic buckling limits, 118
elastic deflection limits, 114
elastic plate buckling, 120
elastic stiffener flange buckling, 128
elastic stiffener web buckling, 123
elastic tripping of stiffener, 125
elevated deck, 327
environmental data, 83
  metocean design parameters, 348
environmental phenomena, 82
  design basis environmental conditions, 107
environmental severity factor (ESF), 151
equipment testing issues, 49
equivalent yield stress (strength), 116
event tree analysis (ETA), 478
exploration, 3
exploratory drilling, 3
export systems, 318, 350
  design considerations, 352
  design issues, 47
  offloading system, 47
  shuttle tankers, 351

fatal accident rate (FAR), 480
fatigue cracks, 404
  effect of fatigue crack on plate ultimate
    strength, 431
  fatigue crack examination, 404
  fatigue crack propagation models, 250
fatigue damage accumulation, 225
  fatigue damage calculations, 245
fatigue limit-state design, 217
  design principles and criteria, 218
  fatigue design issues, 44
  fatigue safety factors, 44
fault tree analysis (FTA), 477
field development concepts, 4
field installation, 50
fire, 296
  active risk control, 485
  passive risk control, 484
  practices for fire assessment, 297
first-order reliability methods (FORM), 66
fixed-type offshore structures, 2, 5

floating-type offshore structures, 5
F–N curve, 480
fold length, 267
formal safety assessment, 464
FPSOs (floating, production, storage, and
    offloading systems), 3
    FPSO project cost, 34
    installation requirements, 28
    major parts, 28
    parameters driving the cost, 34
    post-bid schedule, 39, 40
    progress curve, 41
    project management, 38
    project management organization, 38
FPSO hull ultimate strength reliability, 433
frequency analysis, 476
frequency index, 472
frequency of occurrence or likelihood, 472
frequency rating, 473
front-end engineering, 31
    front-end engineering and design (FEED), 33
Froude's scaling law, 511
Galerkin method, 161
gas compression facilities, 323
gas explosion, 299
    active risk control, 485
    passive risk control, 484
    practices for gas explosion action analysis, 301
    practices for gas explosion consequence
        analysis, 303
    prescriptive methods, 301
    probabilistic methods, 302

general arrangement, 16
    general arrangement drawing, 26
general corrosion, 359
geometrical scaling factor, 513
global structural analysis, 230, 237
green water, 100
    fundamentals, 100
    measures for green-water risk mitigation, 102
    practices for green-water assessment, 101
grooving, 361
gross cost, 470
gross yielding, 170

hazard identification, 465
heading control, 45
    serviceability limit-state design, 139
heeled hulls, 308
high-cycle fatigue, 250
high-tensile-strength steel factor, 56
hot spot stress, 222, 238
    finite-element analysis modeling, 241
hull structural scantling issues, 42
hydrodynamics model tests, 511

ice loads, 94
idealized structural unit method (ISUM), 185
    ISUM beam-column element, 195

ISUM plate element, 188
ISUM structural modeling, 187
    test hull models under vertical bending, 201
Ifrikia, 10
impact-pressure actions, 9, 37, 45
    green water, 100
    slamming, 99
    sloshing, 96
individual risk per annum (IRPA), 480
initial planning, 32
inspection, maintenance, and repair, 400
    inspection and maintenance issues, 51
    inspection practices, 423
    maintenance and repair practices, 425
    considerations for repair strategies, 439
    repair strategies, 439
    risk-based inspection, 411
    risk-based maintenance, 416
    risk control, 485
    selected experience for repairs, 426
intact vessel stability, 134
international organizations, 78
international standards, 78

jacket-type offshore structures, 2
Johnson–Ostenfeld equation, 57

layout, 16
    field layout, 17
    risk control, 483
    topsides layout, 18
limit-state criteria, 65
limit-state design (LSD), 56
limit states, 56
    accidental limit states, 257
    fatigue limit states, 217
    limit-state design requirements, 36
    risk control, 483
    serviceability limit states, 111
    ultimate limit states, 148
linear (knock-down factor) approach, 183
load-carrying capacity, 56
local buckling of stiffener web, 171
local structural analysis, 238
longitudinal strength, 23
low-cycle fatigue, 250

MAESTRO modeler, 25
marginal fields, 5
marine corrosion mechanisms, 357
    types of corrosion, 358
marine growth, 95
Marsden squares, 505
MaxWave project, 70
mechanical damage examination, 405
metocean design parameters, 83
midship section configuration, 25
    three-dimensional midship configuration, 25
    midship section drawing, 27
Miner sum, 220

mooring systems, 318, 336
  design considerations, 349
  design issues, 47
  DICAS mooring system, 349
  mooring line vortex-induced resonance
    oscillation, 143
  single-point moorings, 338
  spread moorings, 337
  mooring system selection, 348
  turret moorings, 342

natural period of a rectangular plate, 131
net cost, 470
new build, 15
  building cost, 15
nominal stress, 222
non-collinear environmental actions, 228
nonlinear finite-element methods, 166, 177
nonlinear governing differential equations of
    plates, 159
  incremental forms, 164
nonlinear governing differential equations of
    stiffened panels, 174
  incremental forms, 176
notch stress, 222

oil and water separation facilities, 319
orthotropic plates, 174
overall collapse, 170
owner requirements, 33

Paris-Erdogan law, 252
partial safety factors, 68
partial safety factor format, 67
performance requirements, 5
permanent set deflection limits, 128
phenomenological corrosion models, 379
pipeline infrastructure, 5, 350
pitting corrosion, 360
plate-induced failure, 178
plate-stiffener combination (PSC) model, 179
polar trading ship designs, 95
post-bid schedule, 39
  sample schedule, 40
potential loss of life (PLL), 480
preassembled unit (PAU), 332
principal dimensions, 19
  average principal dimensions of FPSOs, 19, 20
  relationship between breadth and depth, 21
  relationship between freeboard and depth, 21
  relationship between length and depth, 20, 21
  sample data, 24
probabilistic format, 65
probability of detection and sizing, 405
probability of sea states, 506
process facilities, 36
  cargo handling systems, 324
  design requirements, 36
  gas compression facilities, 324
  offshore/onshore processing options, 36

oil and water separation facilities, 319
  optimum processing options, 4, 36
  safeguard systems, 326
  utility and support systems, 325
  water injection facilities, 324
processing options, 4, 36
project management, 38
  project management organization, 38
progressive hull collapse analysis, 185
  heeled hulls, 308
pseudo-LSD approach, 59
Ramberg-Osgood model, 268
qualitative risk assessment, 472
quantitative risk assessment, 475

recommended practices, 78
regular wave theory selection diagram, 87
regulations, 78
  regulations issues, 52
reliability index, 66
renewal thickness, 452
residual strength assessment, 453
response amplitude operators (RAOs), 221
return period, 103
reusability of existing machinery and equipment,
    453
Reynold's scaling law, 511
risers, 6
  flexible risers, 6
  rigid risers, 6
risk, 463, 470
risk assessment, 463
  qualitative risk assessment, 472
  quantitative risk assessment, 475
  risk-assessment requirements, 37
risk-based inspection (RBI), 411
risk-based maintenance (RBM), 416
risk-control options, 470
  risk control during design, 482
  risk control during operation, 484
risk corrective/preventive measures, 482
risk index, 472

safety factors, 68
safeguard systems, 326
safety, health, and environment, 75
scaling laws, 511
seakeeping analysis, 232
second order reliability methods (SORM), 66
SAFEDOR project, 463
self-contained systems, 2
semianalytical methods, 164, 173–175
semisubmersibles, 5
serviceability limit-state design, 111
  actions and action-effects analysis, 113
  corrosion wastage, 145
  design principles and criteria, 112
  elastic buckling limits, 118
  elastic deflection limits, 114
  intact vessel stability, 134

mooring line vortex-induced resonance oscillation, 143
permanent set deflection limits, 128
vessel motion exceedance, 140
vessel station-keeping, 137
vessel weathervaning and heading control, 139
vibration and noise, 141
ship-shaped offshore units, 7, 13
shuttle tanker export, 351
  tandem export, 352
  side-by-side export, 353
  CALM buoy, 353
simple beam theory, 150
Single Buoy Moorings, Inc., 9
site-specific metocean data, 35
slamming, 99
  fundamentals, 99
  measures for slamming risk mitigation, 100
  practices for slamming assessment, 99
sloshing, 96
  acceptance criterion, 98
  fundamentals, 96
  measures for sloshing risk mitigation, 99
  practices for sloshing assessment, 96
Smith method, 186
S–N curves, 223
  selection of S-N curves, 243
snow and icing, 93
  ice loads, 94
  mean ice thickness, 94
spar, 6
spectral analysis, 226, 227
stability, 74
  intact stability criteria, 136
  intact vessel stability, 134
station-keeping, 45
  design principles, 74
  design issues, 45
  serviceability limit-state design, 137
stiffener-induced failure, 180
stiffened panel (SP) model, 179
storage capacity, 34
  factors affecting the storage capacity, 34
stress concentration area, 241
stress concentration factor, 223
stress intensity factor, 252
stress ranges, 221
stress range transfer functions, 236
structural adequacy, 67
structural details, 231
structural mechanics model tests, 513
surface preparation, 386
surface tension effects, 512
swell scale, 504

tank design, 22
tanker conversion, 447, 448
  tanker conversion option, 15
temperatures, 93
template, 2

tension leg platform (TLP), 7–8
Terra Nova FPSO, 95
test hull models under vertical bending, 504
tether, 6
  tensioning effect, 7
  tether-mooring system, 6
  vertical tethers, 7
tidal levels, 91
time-variant fatigue crack propagation models, 250
time-variant ultimate hull strength reliability, 438
topsides fabrication, 331
  built-in grillage deck, 332
  pre-assembled units (PAU), 332
topsides facilities, 3, 17, 318, 319
  cargo handling systems, 324
  design issues, 46
  computer graphics, 320
  gas compression facilities, 323
  oil and water separation facilities, 319
  safeguard systems, 326
  topsides design issues, 46
  utility and support systems, 325
  water injection facilities, 324
topsides flooring, 330
topsides modules, 18
topsides supports, 327
  multipoint support columns, 328
  sliding/flexible support stools, 329
  transverse girder supports, 330
topsides and their interfaces with hull, 327, 332
towing condition, 43
  design principles, 74
  towing issues, 50
trading tankers, 13
tripping of stiffener, 172
turret-mooring system, 342
  buoyant turret, 346
  clamped-riser turret, 345
  disconnectable turret, 345
  external turret, 344
  internal turret, 344

ultimate hull girder strength interaction relationship, 183
ultimate hull strength reliability, 438
ultimate limit-state design, 148
  actions and action-effects analysis, 150
  design principles and criteria, 148
  ultimate strength of plates, 153
  ultimate strength of stiffened plate structures, 168
  ultimate strength of vessel hulls, 182
ultimate strength of dented plates, 293
ultimate strength of plates, 153
ultimate strength of stiffened plate structures, 168
  primary modes of overall failure, 170
ultimate strength of vessel hulls, 182
ultimate test hull models under vertical bending, 201

unified design requirements, 71
    common structural rules (CSR), 71
    minimum still-water bending moments, 73
    minimum wave-induced bending moments, 73
utility and support systems, 325

vessel motions, 45
    design issues, 45
    design principles, 75
    serviceability limit-state design, 140
vibration and noise, 141
vortex-shedding effects, 512

water depths, 91
water injection facilities, 324
waves, 84
    American Petroleum Institute recommended
      practices, 86
    Det Norske Veritas classification notes, 86
    regular wave theory selection diagram, 87
    UKOOA FPSO design guidance notes, 85

wave energy spectra, 105
    directional wave spectra, 106
    generalized Pierson-Moskowitz spectrum,
      105
    JONSWAP spectrum, 106
wave scale, 503
wave scatter diagrams, 14
weather routing, 14
weathervaning, 139
Weibull stress range scale distribution parameter,
    247
welding connection types, 244
weld metal corrosion, 361
winds, 88
    American Petroleum Institute recommended
      practices, 89
    Det Norske Veritas classification notes, 89
    UKOOA FPSO design guidance notes, 89
    wind force, 89
wind tunnel test requirements, 514
working stress design (WSD), 56